CASEBOOK ON INSURGENCY AND REVOLUTIONARY WARFARE: 23 SUMMARY ACCOUNTS

Primary Research Responsibility:

Paul A. Jureidini
Norman A. La Charité
Bert H. Cooper
William A. Lybrand

SPECIAL OPERATIONS RESEARCH OFFICE
The American University
Washington 16, D.C.

December 1962

Published by Conflict Research Group

First published by Special Operations Research Office in 1962

ISBN: 978-1-925907-21-6

CONFLICT
RESEARCH
GROUP

FOREWORD

The United States of America, born of revolution, finds itself deeply interested to this day in revolutionary movements around the world. Vitally concerned with maintaining international tranquility, we seek to minimize internal unrest within the boundaries of other nations, while fostering orderly growth and change. At the same time that we seek to extend education and economic opportunities for many peoples of the world, we must accept the risk of causing aspirations and expectations to rise more rapidly than the means to meet them. We must accept the associated risk that the frustrations thus engendered could develop into forces of explosive proportions.

Thus it behooves us to extend our understanding of those processes of violent social change usually called revolutions, whether we see them as Communist inspired or emerging from the clash of popular hopes with a political system that resists change.

The US Army's concern with revolutions derives from two sources: First, the Army is a participant, through the many mutual security pacts between the United States and other countries, in the total United States effort to bring about change. Wisdom in the application of military means in the growth and development of nations is certainly to be desired. Second, the Army may, despite our best national efforts, have to cope directly, or through assistance and advice, indirectly, with revolutionary actions. It need not be emphasized that here again a better understanding of the processes of revolution might well be of critical significance.

Thus we perceive a requirement to extend our knowledge of how revolutions are born, grow, succeed, or fail. This book, prepared by the staff of the Special Operations Research Office, is the first part of an effort to so extend our knowledge. The book briefly describes 23 recent revolutions, using a standard outline to facilitate comparisons. Companion volumes will treat four of them in much greater detail. These volumes will comprise a good portion of the data for a further study of revolutions which will aim to specify the terms of a general model of violent political change. Hopefully this step would aid in the invention of a set of criteria for more accurately assessing unrest and revolutionary potential within a country.

Readers are urged to at least skim the Introduction as the first step toward understanding the rationale of the standard outline and toward becoming familiar with the definition of revolution used in this study. Because of the Special Operation Research Office's continuing research interest in revolutionary processes, correspondence with readers will be welcomed.

[signature: Theodore R Vallance]

T. R. Vallance,
Director.

PREFACE

This casebook has been developed as a "reader" in insurgency and revolutionary warfare. Its major functions are to provide a general introduction to revolutionary warfare and to serve as a consolidated source of background information on a number of relatively recent revolutions. A standard outline was used for each summary account in order to facilitate comparisons among the revolutions. Readers are urged to at least skim the Introduction as the first step in order to become familiar with the definition of terms used in the casebook and to gain insight into the rationale for the standard outline.

All of the sources used in preparation of this casebook are unclassified, and for the most part secondary sources were used. Certain advantages and disadvantages accrue. As an unclassified document, the study will be more widely distributed. Reliance on unclassified secondary sources, however, may have led to the exclusion of certain significant considerations or to the use of unreliable information and thus to factual and interpretative errors. It is believed, however, that the advantages outweigh the disadvantages. If, because of its sources, the casebook adds no new information about the revolutions covered, it does claim that systematic ordering, for comparative purposes, of already available open information is a meaningful contribution to the study of insurgency and revolutionary warfare.

It should be noted that the intent of this casebook is not to present any particular "slant" on the revolutions covered, or the actors and parties in them. Rather, the intent is to present as objective an account as possible of what happened in each revolution, in terms of the standard outline used. To this end, the aim was to prepare the summary accounts from the viewpoint of an impartial, objective observer. Perhaps such an aspiration is beyond grasp—the events may be too recent, the sources too unreliable, the "observer" too biased in implicit ways.

For these reasons, no infallibility is claimed and it is readily conceded that this casebook cannot be the final word on the revolutions covered. Subsequent events always have a way of leading to reinterpretation of prior events. But any efforts of omission or commission are not deliberate, but truly errors—and not a result of intent to foster any particular "slant." Little is to be gained in terms of increased understanding of revolutions if justification of any preconceived notions about the revolutions were the intent under the guise of objective analysis.

Beyond the resolve of objective analysis in the preparation of the summary accounts, sources were selected on the basis of their judged reliability. A balance was sought among sources of known persuasion in order not to unwittingly bias the casebook in one direction or another. As a final check, the study draft was submitted to area specialists, who are identified in the Introduction. The experts reviewed the summaries for accuracy of fact and reasonableness of interpretations and their comments and criticisms provided the basis for final revisions. Although their contributions were substantial, final responsibility for the manuscript, both with respect to substantive content and methodology, rests solely with the Special Operations Research Office.

TABLE OF CONTENTS

LIST OF MAPS

INTRODUCTION

PURPOSE OF CASEBOOK

This casebook provides summary descriptive accounts of 23 revolutions[a] that have occurred in seven geographic areas of the world, mostly since World War II. Each revolution is described in terms of the environment in which it occurred, the form of the revolutionary movement itself, and the results which were accomplished.

The casebook is designed:

(1) to present a comprehensive introduction to the subject of revolutions;

(2) to illustrate the types of political, military, cultural, social, and economic conditions under which revolutions have occurred;

(3) to examine the general characteristics of prior revolutionary movements and some operational problems experienced in waging, or countering, revolutionary warfare.

ORGANIZATION OF CASEBOOK

The 23 summary accounts are grouped under the following seven sections according to the geographic area in which the revolution occurred:

Section I: *Southeast Asia*

      1. Vietnam (1946–1954)

      2. Indonesia (1945–1949)

      3. Malaya (1948–1957)

Section II: *Latin America*

      1. Guatemala (1944)

      2. Venezuela (1945)

      3. Argentina (1943)

      4. Bolivia (1952)

      5. Cuba (1953–1959)

Section III: *North Africa*

      1. Tunisia (1950–1954)

---

[a] For purposes of this Casebook, the terms insurgency and revolution are considered synonymous.

1

Each section is preceded by a separate Table of Contents covering all the revolutions summarized in that section. In addition, the individual summary accounts are preceded by an overview of revolutionary developments in that geographic area as a whole.

STANDARD FORMAT OF SUMMARY ACCOUNTS

Each summary account of a revolution in the casebook follows a standard format. The purpose of using a standardized system was twofold: (1) to insure that information considered important was included; (2) to facilitate comparisons among revolutions.

The major parts of each summary account are preceded by a *SYNOPSIS*—a very brief overview of the major events in the revolution. The synopsis is designed to provide the reader unacquainted with the revolution with an introductory contextual frame of reference to which to relate the more specific information in the separate parts of the summary account.

The first part of each summary is a short description of the *Major Historical Events* leading up to, and culminating in, the revolution. This statement acquaints the reader with the basic facts that form the background to the discussion that follows.

The second part describes the *Environment of the Revolution,* examining thegeographic, political, economic, and social setting in which the revolution took place. Major conflicts and weaknesses of the societal system are pointed out. In many cases these conflicts and weaknesses can be presumed to have been "causes" of the revolution; in others it is only possible to state that the conflicts and weaknesses existed and that a revolution occurred.

The third part is concerned with the *Form and Characteristics* of the revolutionary struggle itself. It discusses the actors in the revolution, the organization of the revolutionary forces, the goals of the revolution—both the long range social and economic goals and the immediate practical goals—the techniques adopted by the revolutionaries, and the countertechniques of the government to defeat the revolutionary effort. Finally, it deals with the manner in which the revolutionaries, if successful, assumed the responsibility of government.

The fourth part examines and summarizes the *Effects of the Revolution* as they have become apparent. A distinction is made between the immediate changes caused by the transfer of power and the long range political, social, and economic changes that can be attributed to the revolution.

By definition, each summary account necessarily treats only major factors and general categories of information. As a service to readers who may be interested in examining particular aspects of the revolution in more detail, a *Recommended Reading List* is included at the conclusion of each summary.

ON THE DEFINITION OF REVOLUTION AND RELATED TERMS

Revolution is a word that is often used but seldom explicitly defined. It is frequently used interchangeably with such other terms as rebellion, coup d'etat, insurgency, or insurrection. Thus, *Webster's Dictionary,* the *Encyclopedia of the Social Sciences,* and various writers on the subject, do not agree on a common meaning of the word. Because of this lack of agreement, the manner in which the term revolution is used in any systematic study must be delineated.

Outside of the natural sciences, revolution usually is used to refer to any sudden change with far-reaching consequences. Occasionally, revolution is used to indicate gradual change that has suddenly been recognized as having had far-reaching consequences. The type of change involved is usually indicated by such adjectives as cultural, scientific, economic, industrial, or technological. Without such qualifying adjectives, the word revolution is most frequently applied to the concept of political revolution and it will be so used in this study.

More precisely, revolution will be used in this study to mean the modification, or attempted modification, of an existing political order at least partially by the unconstitutional or illegal use, or threat of use, of force. Thus, in this study, the term encompasses both a delineation of the type of effort involved and the achieved or strived-for effects. In this definition, two key words are "illegal" and "force."

The Term "Illegal." The use of the word "illegal" needs further clarification. The fact that the executive power of a government against which the revolution is directed may have been used unconstitutionally and that the revolution may be an attempt to restore constitutional government are considered irrelevant for purposes of this study. The laws and decrees as promulgated or interpreted by the existing *de facto* executive power of government are the sole criteria of "legality" in this context. Whether these laws or decrees, or their interpretation, are contrary to international law, moral law, or the law of the land before the executive power of government decided to change or set aside the constitution will not concern us.[b]

In addition, illegality, as used, has the important attribute that it applies only to persons subject to the laws of the state in which the revolution takes place. These persons may hold official positions within the government or they may have no official connection at all. Thus the term would cover foreign subjects operating within the territorial jurisdiction of the country, but would not apply to foreign armies during a declared or *de facto* state of war.

If a government is overthrown, a region made independent, or a country subjugated through the application of force by an outside power with no, or only nominal, participation of indigenous persons, it would not be considered a revolution. This qualification does not, however, exclude the situation in which the subjects of a political entity engaged in a revolutionary effort may be supplied or otherwise strongly supported by an outside power.

The Term "Force." An effort to effect political modification is considered a revolution only if the means which it employs to alter the political *status quo* include the use of force or the threat, overt or implied, to use force to achieve the objectives of the revolution (if the execu-

---

[b] The Bolivian revolution of 1952 is a case in point. Bolivia, at that time, was ruled by a military junta which had taken over the government illegally about a year earlier. This takeover had occurred after an election, the results of which displeased conservative elements, for the specific purpose of nullifying the results of the election. This was clearly an illegal act. Yet the revolution of 1952, carried out by the party which was denied its rightful place in 1951, must nevertheless also be considered illegal.

tive power of the *de facto* government does not grant such objectives on peaceful demand). The revolutionary effort may include any kind of legal or illegal nonviolent action along with its use of force: overt and covert propaganda effort, passive resistance or illegally aggressive agitation; corruption, or conspiracy. The use of force may include all or any of many forms: unorganized violence through underground organizations; overt armed action, whether at the level of guerrilla warfare, or that of civil war between regular armed forces.

While revolutions may differ in the kind, amount, and manner in which force is used, its use in one of these forms, or the readiness to use it, is considered a general characteristic of all revolutions. If the threat of force does not suffice to achieve the objectives of the revolution and the effort reaches the stage of the actual use of force, such force may be applied by revolutionary forces within the territorial jurisdiction of the government to be overthrown or by revolutionary forces who are externally based and invade such territory.

Subcategories of Revolutionary Efforts. Some writers do not classify a "coup d'etat" or a "rebellion" as a revolution. In this study both coup d'etat and rebellion will be considered special subcategories of revolutions that can be distinguished from revolutions per se. Most writers who exclude these events as types of revolutions do so on the basis of the *effects* of the political action that modified the government. These writers will call a revolution only that effort which results in far-reaching changes in the political and/or social institutions of the country, or that effort which is preceded by important changes in the social structure of the country. The disadvantage of imposing this reaction lies in the difficulty of determining what constitutes "far-reaching changes."

Those who use this additional qualification to define revolution will consider a change in the personnel of government without corresponding changes in the political or social institutions as a coup d'etat. For this study a *coup d'etat* will be defined as a revolution in which a change of government or governmental personnel is effected suddenly *by holders of governmental power* in defiance of the state's legal constitution or against the will of the chief executive. Usually, one of the characteristics of a coup d'etat is that public support for the coup itself is not sought until after it has succeeded, or at least until after the coup has been initiated.[c] In cases where holders of government power are not the sole executors of the revolutionary attempt, but seek the support of a political party or other mass group, it is difficult

---

[c] A coup d'etat, in this sense, occurred in Iraq in 1936 and 1958, and in Egypt in 1952. In all three cases, the "holders of governmental power" were members of the Armed Forces.

to define such a revolution as a coup d'etat. In Bolivia, for instance, a major organ of the government, the police department, conspired with a major political party. The revolution was a result of this conspiracy. In this study, therefore, it would not be considered a coup d'etat, since the revolution did not rely on governmental power alone. Also, in the case of Iraq in 1936 there was a conspiracy between a political group and the army; however, the coup itself was carried out entirely by the army and there was no mass action to support the army in the application of force. For that reason, the Iraqi coup of 1936 would be classified as a coup d'etat.

A *rebellion* is considered as a revolution in which the revolutionary effort aims at territorial autonomy, or independence, for a part or parts of a political entity, but in which no attempt is made to alter or overthrow the central government itself. In this sense, a colonial uprising for independence is a rebellion, as were the actions of the Confederate States in the American Civil War.

A *postaccessional revolution* is not usually discussed as a distinct type of revolution. However, it is a concept that is very useful to denote a revolution which is carried out by a revolutionary part or group *after* it has succeeded in gaining governmental power by legal means and after it has some or all of the law-enforcing agencies under its control. The revolution is thus carried out, in effect, by the government but against the established constitutional order. It differs from a coup d'etat in that the "holders of government power" executing the revolution are really the dominant power of the government and the revolution has the outward appearance of legality. This type of revolution was successfully completed in Germany in 1933. After Hitler had become chancellor, and after some members of the Nazi Party had been placed into official positions, he systematically, through the illegal use of force, eliminated opposition within the government and transformed a parliamentary democracy into a totalitarian state. A similar revolution took place in Czechoslovakia in 1948.

A *counterrevolution* may take any of the forms discussed above. It is essentially an attempt, through revolution, to restore a previously existing condition. However, a pure counterrevolution exists in theory only. The realities of the historical processes, invariably, make it impossible for a counterrevolution to restore the past completely. Just as there is no revolution that has signified a complete break with the past—not even the French or the Russian Revolutions—there is no counterrevolution which has not incorporated some of the changes brought about by the revolution or, at least, recognized the need to change some of the conditions which were responsible for the first revolution.

An *insurrection* is not considered a type of revolution but is a term frequently used in connection with revolution. It will be used in this study to denote the initial stage of a revolution. "An insurrection may be thought of as an incipient revolution still localized and limited to securing modifications of governmental policy or personnel and not as yet a serious threat to the state or government in power."[d] Using this definition, as insurrection is an overt revolutionary effort that may become a revolution if it gathers momentum and receives wide public support. The action of the French Army in Algeria in April 1961 was an insurrection, defying the central authority of the government in Paris. It might have developed into a revolution if it had succeeded in uniting the Armed Forces against President de Gaulle and in forcing his resignation as head of state.

In this study, the expression *revolutionary warfare* will be used to encompass all aspects of the revolutionary efforts to displace an existing government by force as well as the efforts of the *de facto* government to defend itself against such efforts. "Revolutionary warfare" will be examined in a general sense and will not be restricted to Communist methods, although Communist techniques will be, of course, closely examined. However, there is no *a priori* assumption that "revolutionary warfare" is peculiarly and exclusively a Communist phenomenon.

Thus, the term will not be used in the more specific sense evolved by French writers. The term "revolutionary warfare," as it has been popularized by the French, is applied specifically to Communist techniques and methods of organization. According to the French theory, there is a specific way in which such destructive techniques as strikes, riots, mass demonstrations, sabotage, guerrilla warfare, and, finally, open warfare are used in accordance with a preconceived plan. These techniques are, again according to the French, supplemented by "constructive" techniques to win over the masses and include propaganda, training, agitation, and the organization of "parallel hierarchies." Differences in the application of "revolutionary warfare" from country to country are explained as conscious adaptation to the special conditions which prevail in the area in which "revolutionary warfare" is being fought.

METHODOLOGICAL NOTES AND ACKNOWLEDGMENTS

From a research viewpoint, the casebook is one end-product of an initial effort in a longer range research task aimed at improving knowledge of the nature of revolutions and revolutionary warfare. A

---

[d] *Encyclopedia of the Social Sciences.* Ed. Edwin R. Seligman (New York: Macmillan Company, 1957), VIII, p. 117.

major research purpose of using a standardized format in preparing the summary accounts was to facilitate the development of hypotheses regarding general relationships among factors, and patterns of factors, for later more intensive analytic study and evaluation. Shortly to be published—separately—are detailed case studies in depth of four of the revolutions summarized in this casebook.[c]

The standardized format was developed on the basis of a review of prior general analyses of revolutions, consideration of the types of operational questions to which revolutionary knowledge can be applied, and the anticipated use of the casebook as a text or reference. Although a straight chronological account may have been more readable, it would not have allowed ready systematic comparisons among revolutions, a major research consideration, as well as a practical consideration in terms of using the casebook as a text or reference as a study tool.

The selection of revolutions for the summary accounts was done on a country and area basis, as well as on the basis of coverage of a wide range of different social, political, and environmental conditions under which revolutions have taken place. For countries which had more than one revolution during the last 40 years, selection was guided by considerations of: (1) time of occurrence, (2) importance, (3) availability of information. Thus, revolutions were selected which occurred as recently as possible; which were judged important in terms of the political and social changes which resulted; and on which there was sufficient information in published sources to permit an adequate description of the revolutionary processes.

The research approach used for the preparation of a summary account was essentially the same for each revolution. Area experts were consulted to identify the major secondary documentary sources (unclassified) which covered aspects of the subject revolution. Using the standardized format as a guide, information was collected, synthesized, and summarized from the secondary sources. Gaps in coverage were filled in by further search of secondary sources, use of primary sources when readily available, or through interviews with knowledgeable persons. After intensive internal review and revision, draft copies of the summaries were submitted to area experts for final technical review for accuracy of fact and soundness of description and interpretation. When differences in interpretation could not be resolved on the basis of available evidence, both interpretations were included and clearly identified as conflicting views, unresolvable within the time and resources available.

---

[c] Guatemala (1944–54); Cuba (1953–59); Vietnam (1946–54); Algeria (1954–62).

The research on the casebook was completed between 1 April 1961 and 30 July 1962, with varying cutoffs within those bounds for different summaries. Thus, caution must be exercised in terms of the implications of events after cutoff dates on earlier conclusions.

SORO wishes to acknowledge its indebtedness to the following consultants who technically reviewed the sections of the casebook identified. While much of what may be useful can be attributed to them, they are in no way to be held responsible for any errors of commission or omission which may appear in the casebook.

Southeast Asia
 Dr. Bernard B. Fall
 Professor of Government
 Howard University

Latin America
 Dr. Harold E. Davis
 Professor of Latin American
  Studies
 The American University

Middle East and North Africa
 Dr. Abdul A. Said
 Associate Professor of Inter-
  national Law
 The American University

 Dr. Hisham B. Sharabi
 Associate Professor of His-
  tory and Government
 Georgetown University

Europe
 Dr. Karl H. Cerny
 Associate Professor of
  Government
 Georgetown University

 Mr. Martin Blumenson
 Senior Historian
 Office of The Chief of Mili-
  tary History

Far East
 Dr. Peter S. H. Tang
 Lecturer, Department of
  Government
 Georgetown University

 Michael Lindsay
 (Lord Lindsay of Birker)
 Professor of Far Eastern
  Studies
 The American University

In addition, SORO wishes to express its appreciation to Mr. Slavko N. Bjelajac, Office of the Director of Special Warfare, Department of the Army, for his patient guidance and counsel as Project Monitor.

Finally, it should be noted that Mr. Ralph V. Mavrogordato was associated with the project for some time as Acting Chairman of the research team. He made a significant contribution to the organization of the casebook and the development of the standardized format for the summary accounts, as well as supervising preparation of many of the summaries.

SECTION I

....

SOUTHEAST ASIA

General Discussion of Area and Revolutionary Developments
The Revolution in Vietnam: 1946–1954
The Indonesian Rebellion: 1945–1949
The Revolution in Malaya: 1948–1957

TABLE OF CONTENTS

GENERAL DISCUSSION OF AREA AND REVOLUTIONARY DEVELOPMENTS

THE REVOLUTION IN VIETNAM: 1946–1954

14

THE REVOLUTION IN MALAYA: 1948–1957

TABLE OF CONTENTS

17

SOUTHEAST ASIA

GENERAL DISCUSSION OF AREA AND REVOLUTIONARY DEVELOPMENTS

GEOGRAPHICAL DEFINITION

Southeast Asia is a term applied loosely to the area extending south and east of Communist China, including the adjacent islands. Prior to World War II, it contained one independent state, Siam or Thailand; and five colonial territories: Burma, Malaya, Indochina (Laos, Cambodia, and Vietnam), the Philippine Islands, and the Netherlands East Indies or Indonesia. Since the end of the war, all the major colonial territories have acquired their independence. With a population of nearly 180,000,000 people[1] forming loosely organized and relatively underpopulated societies, the area is divided into three cultural areas according to religious differences: Burma, Thailand, Laos, and Cambodia are predominantly Buddhist; Malaya and Indonesia Muslim; and the Philippines principally Christian. Vietnam, the exception, is Taoist-Confucianist. The Indonesian island of Java, Lower Thailand, the Red River Delta in Vietnam, Lower Burma, Central Luzon in the Philippines, and the Lower Mekong valley in Cambodia and Vietnam are densely populated, while the rest of the areas are relatively underpopulated.

The richness of the resources in the area was discovered soon after the era of exploration began. Southeast Asia became a target for economic expansion of the Western nations. Rice, rubber, petroleum, tin, and copra production, developed by modern capitalism, "was grafted onto an indigenous, precapitalist economic social system"[2] characterized by primitive forms of agriculture, mining, and other regional enterprises. The impact of the West eventually created antagonism that turned into a strong desire to evict the foreigners and establish independent indigenous institutions. The area became the site of revolutions, uprisings, demonstrations, and general unrest which reached the peak after World War II and have not yet subsided.

BACKGROUND OF COLONIAL RULE

POLITICAL

In all areas of Southeast Asia, with the exception of Thailand (formerly known as the Kingdom of Siam) the Western colonial powers had established their rule by the end of the 19th century. Portuguese and Spaniards first explored the South Asian seas, but English, French, Dutch, and Spanish traders became the influential elements in the area. As the competition for prime areas grew more intense, the governments intervened in order to offer better protection for their economic interests. France had established itself in Indochina; Great Britain in Burma and Malaya; the Netherlands in Indonesia; and Spain in the Philippine Islands, which were taken over by the United States following the Spanish-American War.

Reactions to Western expansion varied widely from one area to another. There was some immediate resistance to French administration in Indochina. Some of the causes of the later violent military conflict between the French and the Vietnamese go back to this time. The British introduced features of modernization in both Burma and Malaya. The governmental institutions established by the British in these two countries were, according to one authority, "far more rationalized and modernized than any other institutions within their society."[3] Burma was a quiet colonial holding and a slow growth toward modernization took place. It achieved independence much sooner than expected and without violence. Malaya, although modernized in other ways, was the least advanced on the road to political independence. This was partly due to its large transient Indian and Chinese population and the establishment of protective policies toward the Malays. Indonesia developed very unevenly. The social development of Java was in sharp contrast to the retention of tribal culture and economics in New Guinea. U.S. policy was to attempt to prepare the Philippine Islands for political freedom in the shortest possible time. However, the political maturity of the Philippines appears to be far ahead of its economic and social conditions. Thailand never experienced colonial status, but it, too, was strongly influenced by the West.

Some colonial powers, such as the United States, dominated the political rather than the commercial aspects of the colonies. The administration was humane and liberal, and it introduced democratic methods intended to lead eventually to self-government. As a result, many authors and commentators find much to commend in colonial rule. Thus, it is claimed that British efficiency and justice in Malaya resulted in prosperity which pushed the idea of Malayan independence into the background until the disruption of World War II. The Dutch in Indonesia had an "Ethical Policy," the French in Indochina a policy of "Assimilation," and the United States in the Philippines also had a policy designed to benefit indigenous society.

The colonial administrations at the central and intermediate levels were Western type political institutions. The actual rulers in the colonies were governors general or high commissioners, responsible to the Western governments. The nominal heads of institutions representing local interests acted in advisory capacities only. But at the village levels, which included the mass of the agricultural population, the traditional methods of governing often remained. The higher administrative positions, where the power and responsibility of ultimate decision rested, were filled by Westerners and a small number of indigenous people trained or experienced in the operations of their own government institutions. Through this practice only a small number of indigenous administrative experts and political leaders emerged. The old colonial political and social institutions are still the basic structures around which most of the Southeast Asian countries are attempting to build modern states.

ECONOMIC

Southeast Asia produced some raw materials and foodstuffs for home consumption, and some raw materials and agricultural products for export. The production of subsistence foodstuffs was carried out by traditional and backward techniques. Export production, in contrast, was on a large-scale production basis. The capital and managerial skills were contributed by Europeans, Indians, and Chinese and the labor force by the local population. The Western countries made large capital investments. This system created a dual economy in which most profits went to the West and a "subsistence modified by the amount of welfare necessary and useful to the ongoing entrepreneurial system remained behind."[4]

The colonial powers depended on the increased production of natural resources and raw materials. Prewar Burma, Thailand, and Indochina fed themselves and exported approximately 6 million tons of rice annually. Malaya and Indonesia accounted for 800,000 tons of exported rubber. Malaya, Indonesia, and Thailand mined 90,000 metric tons of tin in concentrates. Malaya alone produced 65,000 metric tons of tin. Indonesia, the Philippine Islands, and Malaya produced 1.6 million tons of copra for coconut oil. Seventy-eight percent of the world's rubber exports and 73 percent of copra exports came from Malaya in 1938. The colonies were rich in raw materials and products, but had little industrial power and capital goods.[5] Southeast Asia maintained its subsistence economy side by side with the newer export economy.

The colonial areas of Southeast Asia were greatly affected by the Western economic depression of the late 1920's and 1930's. The effects of the crisis were felt in Southeast Asia as early as 1928. The crisis revealed economic shortcomings of the colonial system and tended to weaken the image of the colonial powers.

EDUCATION

Before the First World War little education was offered the indigenous population of Southeast Asia. Universities were established in Manila and Hanoi; but there was no system of national education that reached down to the people before World War II. Higher education was usually offered in a Western language and literary in character. Even where there were two systems of education, one local and the other foreign, the masses were left illiterate. Thus, in Southeast Asia, there was a large illiterate base, a small number of indigenous people with an education in their own language, and an even smaller minority educated in a foreign language. Although some education was needed to keep pace with industrialization, the literary character of education did not favor the development of technical skills and retarded appreciation for scientific development.

ETHNIC MINORITIES

Chinese and Indians are the largest and most important minority groups in Southeast Asia; there are numerous other minorities, such as the Arakanese, Karens, Shaus, Chins and Kachins, in Burma; the Malays in southern Thailand; the Ambonese, Dutch, Eurasians, Arabs, and others, in Indonesia; and aboriginal tribes throughout the area. The Chinese pose by far the greatest problem, particularly in Malaya and Thailand.

SUMMARY

For political, economic, and racial reasons European rule proved unacceptable to Southeast Asia after World War II. Its alien elements had not been well digested by the Asian countries. Democracy was slowly, and at times quite reluctantly, introduced into the area. For the population, Western rule often appeared as a slow erosion of everything that was traditional while offering little to replace traditional values. The intelligentsia developed an inferiority feeling as the inevitable result of being ruled by a foreign power and this was "intensified by the tactless behavior of some Europeans who were persuaded that their position was due not to an accident of history but to innate racial superiority."[6] Viewed by today's standards, much of colonial rule seems highly objectionable; however, for the time in which colonialism flourished, a different standard must be applied. It is also true that the colonial "exploitation" of the people often differed little from previous "exploitation" by indigenous elites.

THE DEVELOPMENT OF INDEPENDENCE

THE INTERWAR PERIOD

The intensity of dissatisfaction with Western rule increased greatly after the First World War. In the interwar period "four forces which had grown up in Southeast Asia interacted on each other and developed in the direction of the inevitable explosion." These forces were: a western-oriented intelligentsia becoming more and more antagonistic to the Western world; the masses made aware of their material poverty and stubbornly resisting Western control and guidance as the traditional structure of their social environment faded; the Western powers competing with each other in their economic expansion; and the anti-Western force of a Western ideology, communism, attempting to provide its own solution to these problems. Finally, a fifth force, Japanese imperialism, irrevocably destroyed the entire Western structure, leaving a clear road for the other three forces (the Western-oriented intelligentsia, the masses, and communism) to interact upon each other in different ways in each country of the area.[7]

WORLD WAR II

Japanese rule established after the invasion in 1940 and 1941 gave the nationalist movements in the colonial areas an immense stimulus and "accelerated the development of indigenous administrative organs."[8] Prior to the war, the colonial administrations had suppressed nationalism by imprisoning or exiling those leaders who threatened European rule. The Japanese freed these leaders, who were soon able to find "status and power within their own societies."[9] The impetus toward independence given by the Japanese occupation was no less effective for the fact that the Japanese themselves were utilizing the Southeast Asian countries for their own military and economic purposes much as the European powers had done.

With the possible exception of Malaya, the Southeast Asian countries did not want the return and restoration of European order. The war had given them a taste for self-administration and greatly strengthened their desire for self-rule, especially in Indonesia, Burma, and Vietnam. The political passions of the indigenous population entirely overshadowed the existing economic needs. The Southeast Asian countries did not want to reestablish trade with the Europeans on the prewar basis, although they were desperately in need of economic recovery. Most indigenous politicians felt that economic objectives could be achieved only through political independence.[10]

When the Allied occupation forces arrived in Southeast Asia after the Japanese capitulation in 1945, they found the area in better order than expected. It was apparent that forceful methods would be required if the Western powers were to reestablish their authority. The new republican governments in Vietnam and Indonesia organized military units, declared their independence, and challenged the authority of the French and Dutch. Eventually, they freed themselves of their prewar colonial rulers by combining international pressure, subversion, and guerrilla and open warfare.

GENERAL DISCUSSION OF THE REVOLUTIONS IN SOUTHEAST ASIA

One fundamental fact about Southeast Asia is that, following World War II, it became the scene of revolutions which have not yet run their course. The ideology, form, and intensity of the revolutions varied from one country to another, but the basic motivation was the same: to dissolve the old political order and establish a new government more desirable to the revolutionary leaders and the people. In the postwar period Western influence over the Asian countries sharply declined.

There were four factors involved in these revolutionary movements. The first and most important was nationalism. It played a dual role, as an essential element for the attainment of independence and as a means of unifying a country. The second was the apparent concern of the revolutionary leaders for individual liberty and constitutional governments. The third was the desire to establish a centralized economic planning organization for state control of foreign trade and industrial development. The fourth was the trend toward unitary

24

control to prevent political disunity and to provide a central direction for economic planning.[11]

The desire for independence was universal in Southeast Asia, but the methods of achieving it differed considerably from country to country. In Burma emancipation was accomplished by peaceful means. In May 1945 Britain reaffirmed her intention to grant Burma full independence within the Commonwealth after a short period of continued direct rule. The Burmese demanded and received complete independence through negotiations conducted by Burmese Nationalists and pro-Burmese British officials. Burma was not at rest, however; the country faced insurrections by the Karens and the Communists, both in 1949. Conditions had improved by 1951, but some unrest continues to the present time.

In Indonesia nationalist troops met stronger Dutch columns in their fight for independence, but independence was achieved in 1949 through U.N. and U.S. intervention. In Vietnam, a coalition of Communists and nationalists engaged the French Union Forces in 1946 in a jungle war which brought defeat to the French and a Korean-type division of Vietnam in 1954. The Malayans gained their independence in cooperation with the British, but not without considerable difficulty. Shortly after the war the British advocated the establishment of a Malayan Union. This proposal was rejected by the Malays. Later an agreement was reached which resulted in the formation of a Malayan Federation in 1957. The Communist rebellion in Malaya was not responsible in any way for Malayan independence, but was a serious challenge for several years. The United States fulfilled her promise by granting independence to the Philippines in 1946. In Thailand a "bloodless" military coup d'etat replaced Marshal Pibul's rule in 1957, changing the form of government but little. A 1947 military coup apparently had eliminated the possibility of a more leftist political development.

In all these countries the threat of communism existed to some degree. The revolution in Vietnam established a Communist People's Republic north of the 17th Parallel. In Malaya British Security Forces had a long, stubborn and costly struggle to stamp out armed Communist activity. In Indonesia and the Philippine Islands, the republican governments were also able to defeat a Communist attempt to gain control. Peking-directed propaganda found the Thais, with the possible exception of the Chinese minority, largely unresponsive to communistic appeals. However, the Communist threat in all of these areas has not been eliminated. South Vietnam and Laos, in particular, are currently the scenes of open revolutionary warfare.

A recent problem, anti-Communist rather than Communist in nature, has plagued the Indonesian Government. This was the 1958 uprising staged by the conservative "Government of the Revolutionary Indonesian Republic (PRRI)," a revolutionary group that tried to overthrow the government from bases in Celebes (Sulawesi) and central Sumatra. In July of the same year the national Indonesian Government forces were reportedly entering the final stages of the struggle against the scattered PRRI forces in northern Celebes. However, guerrillas are still active in that area, although the effort of the PRRI forces was weakened by dissension in the leadership and by religious discord. Another problem facing Indonesia today is the Darul Islam movement, which originated in west Java during the Indonesian struggle against the Dutch in the late 1940's, and became antirepublican in 1948. The movement is organized along theocratic lines, advancing the "ideal Islamic state." Darul Islam soon came under control of fortune hunting politicians and guerrilla bands. The organization continues a terrorist campaign against the republican government and there are indications that it recently increased in strength.

In Laos a Communist challenge has become increasingly serious through subversion and guerrilla warfare and appears to be close to accomplishing its goals. The struggle against the Pathet Lao (Communist) has been waged with U.S. military aid while Pathet Lao troops benefit by support from the North Vietnamese (Communist) Government. Pro-Western forces in Laos are in a precarious position and the country, at best, may become "neutralist." The situation is still very much unsettled at the time of this writing (October, 1961).

A revolutionary struggle is also taking place in South Vietnam. During the past 7 years the Viet Cong (Communist) guerrilla force is reported to have increased from 3,000 to 12,000. Unlike the Laotians, South Vietnamese troops, with considerable American military aid, are fighting a determined battle against the Communists. However, the government's procrastination in instituting urgently needed reforms is reported to be exasperating senior army officers and playing into the hands of the Communists.[12] Overt military operations have been on the increase in 1961.

RESULTS AND OUTLOOK

The Southeast Asian countries have all achieved independence since World War II. A few areas, such as the Dutch West Irian and the Portuguese part of Timor, remain disputed and are still under European control. Although comparatively unimportant as such, these areas tend to keep alive the issue of "colonialism," which has been and still is a rallying point for nationalists.

The real problems facing Southeast Asian countries are related to the "colonialism" of the past rather than to the present. Through the exposure of Southeast Asia to European rule and European influence, European institutions and standards have been inherited. The education, economy, political institutions, and even the philosophy of Southeast Asian nationalism are products of Western thought and practice. Ideas of Western democracies are competing with Marxism and religious mysticism, while the great majority of the people still live in the traditional village manner. As yet, no political force appears to have emerged which has been able to amalgamate these diverse influences and create a movement truly representative of the needs of the new nations. Moreover, "political parties, except where they have been all-embracing social movements, have not generally been the key units in the political processes of the Southeast Asian countries. Except for the Communists, none of the parties in the region has a strong organizational structure."[13] This is true even for Indonesia. In Thailand, South Vietnam, Laos, and Cambodia the political parties are almost like "public relations organizations" of the "real political actors who control them."[14] In Thailand and Indonesia the army plays an important role in political life; in South Vietnam old established elite groups are the dominant political force. As a result of these factors and trends, governments tend to be unstable.

The dominant political philosophy to which everyone in Southeast Asia pays at least lip service is socialism. Western-oriented or neutral nationalists are competing with the Communists in their attempts to impose their particular version of "socialism." In both cases the trend is toward authoritarianism. The choice appears to be between an authoritarianism of the right, as in Thailand and South Vietnam, or of the left, as in North Vietnam. "Only in the Philippines can it be said that the authoritative institutional groups do not dominate the political process."[15]

The following four factors appear to strengthen the authoritarian structure of the non-Communist countries:

(1) The government tends to be an interest group representing not the broad mass of the people but a special urban minority.

(2) Politics, manipulated by an urban elite, provides protection and security to those who recognize the leaders and denies the masses a share in political life.

(3) Special interest groups, such as trade unions, peasant associations, and student unions, either lack real power or have become tools of the governments.

(4) The government controls the radio and news releases on which the press has to depend.

In addition to the problem of developing a sense of national unity, and political processes which would lend stability to such unity, there is a considerable problem of minorities in several of the Southeast Asian countries. By far the greatest and most troublesome minority group is the Chinese, who are widely regarded "as an extension of China itself and as a potential fifth column in the event of a Chinese advance into the region."[16] However, their loyalties toward their countries of residence as well as their political orientations differ widely.

The Chinese minority in Thailand has been used as a scapegoat and has been the butt of Thai discriminatory practices. Similar conditions prevail in Malaya and South Vietnam. British policy in Malaya had always favored the Malays while treating the Chinese as aliens, even though the Chinese had been there for generations. The antagonisms between the Malay, Chinese, and Indian communities had forced the British to delay self-government in Malaya. A *modus vivendi* has now been reached between the three communities, and some sort of constructive alliance does exist. Possible new political alignments between Malaya and Indonesia would probably weaken the Malayan entente since the position of the Chinese element would be weakened.

Chinese commercial interests in Southeast Asia are still disproportionately large whatever nationality the Chinese may claim. Tensions between the Chinese and other races are probably aggravated because local governments do not offer them equality of citizenship and security of land tenure, or give them a stake in the country in which they reside.

The Indian minority in Southeast Asia is not feared as is the Chinese minority, but neither is it respected for the economic power it holds. Indian "collaboration" with the Japanese during World War II and the remittance of wealth accumulated by the Indian moneylenders to India have been used to justify nationalization of Indian-owned land in Burma.

Former colonial powers tend to sympathize with other ethnic minorities of Southeast Asia. They see the Karen separatist movement in Burma and the Ambonese separatist movement in Indonesia as a result of premature withdrawal of the Western powers. Often the minority groups feel antagonistic toward the ethnic majority. Thus, the Mois, Thais, and other minority groups in Vietnam are known to dislike the Vietnamese strongly. The importance of the minority groups is enhanced by their control over vast land areas. In Vietnam, for example, the minorities control a much larger area than do the Vietnamese.

Three colonial rebellions are briefly described in the following section. The rebellion in *Vietnam* was included to illustrate the success of an inferior but well-adapted indigenous force against a superior but alien force in a prolonged jungle war which culminated in the establishment of a Communist state. The *Indonesian* rebellion was characterized by a combination of guerrilla skill and foreign intervention. The *Malayan* Communist rebellion was chosen as an example of one of many attempts to establish a Communist regime in Southeast Asia. The Communist effort in Malaya failed and the events illustrate effective countermeasures taken by the British Security Forces in combatting Communist guerrillas and the creation of an environment which was not conducive to a successful revolutionary effort.

<u>NOTES</u>

1. Lucien W. Pye, "The Politics of Southeast Asia," *The Politics of the Developing Areas*, eds. Gabriel A. Almond and James S. Coleman (Princeton: Princeton University Press, 1960), p. 66.
2. Frank N. Trager, *Marxism in Southeast Asia* (Stanford: Stanford University Press, 1959), p. 279.
3. Pye, "The Politics of Southeast Asia," p. 99.
4. Trager, *Marxism*, p. 280.
5. Ibid.; see also Charles Robequin, *Malaya, Indonesia, Boreno, and the Philippines* (New York: Longmans, Green and Co., 1955), pp. 298–299.
6. J. H. Brimmell, *Communism in Southeast Asia* (New York: Oxford University Press, 1959), p. 75.
7. Ibid.
8. Trager, *Marxism*, p. 253.
9. Ibid.
10. D. G. E. Hall, *A History of Southeast Asia* (New York: St. Martin's Press, Inc., 1955), pp. 698–699.
11. John F. Cady, "Evolving Political Institutions in Southeast Asia," *Nationalism and Progress in Free Asia*, ed. Philip W. Thayer (Baltimore: The Johns Hopkins Press, 1956), p. 116.
12. "Review of the Situation in the Far East (Revised up to 16th May, 1961)," *The Army Quarterly*, LXXXII, 1 (April 1961), 148–149.
13. Pye, "The Politics of Southeast Asia," pp. 114–115.
14. Ibid., p. 115.

15. Ibid.

16. Victor Purcell, "The Influence of Racial Minorities," *Nationalism and Progress in Free Asia*, ed. Philip W. Thayer (Baltimore: The Johns Hopkins Press, 1956), p. 237.

RECOMMENDED READING

BOOKS:

Brimmell, J. H. *Communism in Southeast Asia.* New York: Oxford University Press, 1959. The book attempts to present an account of the impact of communism in Southeast Asia and to assess the significance of the movement.

Emerson, Rupert. *Representative Government in Southeast Asia.* Cambridge: Harvard University Press, 1955. This book is an examination of the functions of representative governments and the difficulties involved for countries emerging from colonial status.

Hall, D. G. E. *A History of Southeast Asia.* New York: St. Martin's Press, Inc., 1955. Although this book may have too much background for summary studies of contemporary revolutions, it is nevertheless a good scholarly account of the history of the area.

Pye, Lucian W. "The Politics of Southeast Asia," *The Politics of Developing Areas*, eds. Gabriel A. Almond and James S. Coleman. Princeton: Princeton University Press, 1960, pp. 65–152. The article presents a highly technical aspect of an area in transition. It is well worth reading.

Thayer, Philip W. (ed.). *Nationalism and Progress in Free Asia.* Baltimore: The Johns Hopkins Press, 1956. The book is a selection of articles dealing with postrevolutionary problems.

Trager, Frank N. *Marxism in Southeast Asia.* Stanford: Stanford University Press, 1959. This also is a selection of articles on the development of Marxism in four specific countries of Southeast Asia, with an "historical overview" presented by Mr. Trager as a summary.

PERIODICALS:

"Review of the Situation in the Far East (Revised up to 16th May, 1961)," *The Army Quarterly*, LXXXII, 1 (April 1961), 148–149.

THE REVOLUTION IN VIETNAM: 1946–1954

SYNOPSIS

The revolution in Vietnam began in December 1946 after the Vietminh Government and the French authorities failed to arrive at a mutually agreeable compromise concerning Vietnam's future political status. The rebellion began with guerrilla warfare, but later developed into a combination of guerrilla warfare and regular warfare. After 7 years of military, political, and psychological warfare the French found themselves in an untenable position and withdrew their forces from Vietnam.

BRIEF HISTORY OF EVENTS LEADING UP TO AND CULMINATING IN REVOLUTION

By the end of the 19th century the French had established complete control over Vietnam. This control was periodically challenged by sporadic demonstrations and uprisings which reached a peak in the early 1930's. The French were able to suppress these outbreaks, and during the period of the "Popular Front" before World War II, they initiated some reforms designed to allow greater autonomy to the Vietnamese. French rule in Vietnam wavered following the German attack on France and the Japanese attack on Indochina in 1940. The Japanese Government forced the representative of Vichy France to sign a "common defense" accord in 1941, later supplemented by economic agreements.

A strong Vietnamese nationalist movement developed prior to and during World War II, with some encouragement from the Chinese Nationalists. Spurred by the defeat of the French, a coalition of Vietnamese nationalist groups, dominated by Communist leaders, declared Vietnam independent in 1945. This Vietnamese organization, known as the Vietminh, was the only anti-Japanese movement of consequence, and was therefore able to fill the power vacuum existing between the time of the Japanese capitulation in August 1945 and the Allied landings in September. While smaller nationalist groups were wrangling over idealistic details, the Communist-controlled Vietminh took over the country. Subsequent events represented the efforts of the Vietminh to solidify a Communist regime and defend its position against the French and some of their Vietnamese supporters.

The French returned soon after the war and attempted to reestablish their sovereignty. As a result, they became involved in a pro-

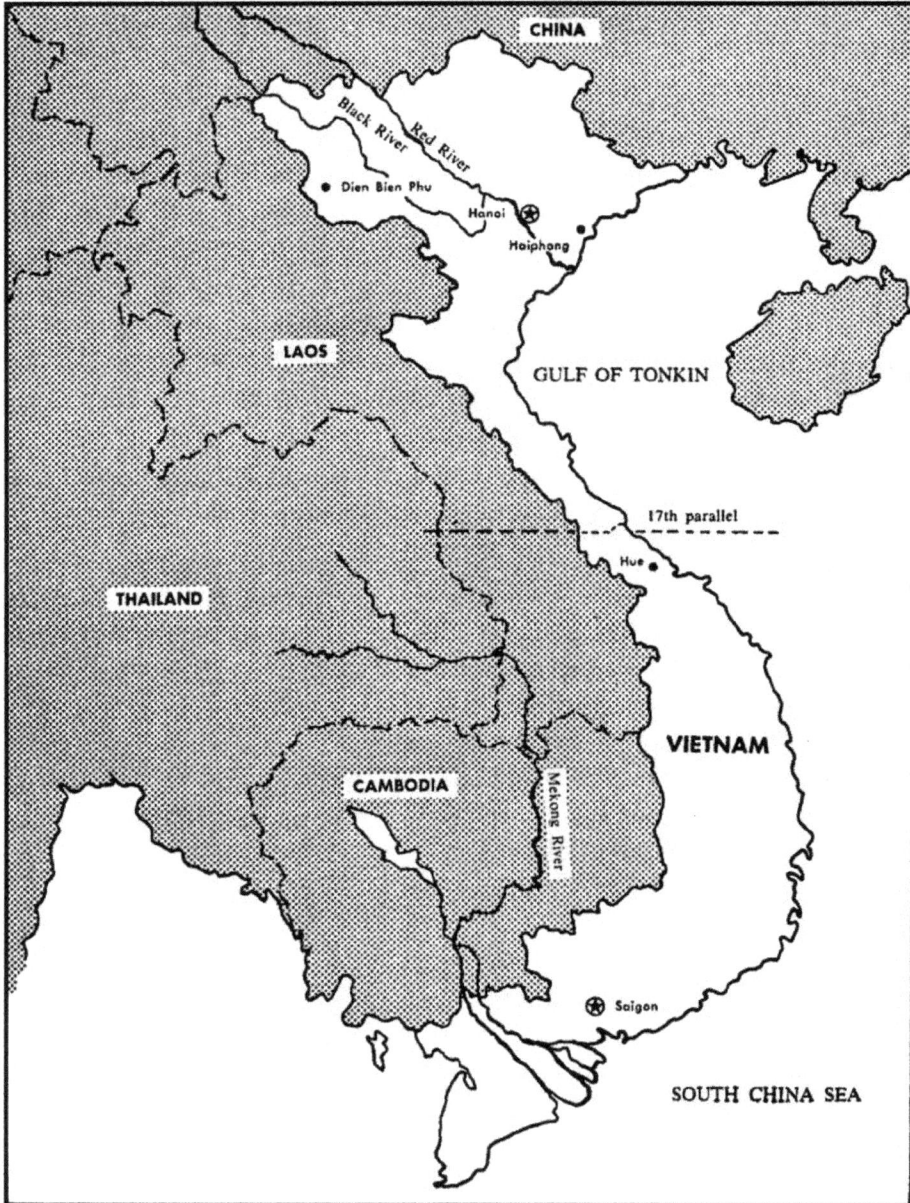

longed struggle with the Vietnamese, who were trying to defend their newly proclaimed independence. The French forces were defeated in 1954 by the use of propaganda, planned uprisings, guerrilla warfare, and other unconventional warfare techniques. The French defeat at Dien Bien Phu, in May 1954, symbolized France's untenable position in Vietnam. At an international conference in Geneva, convoked to settle the status of all three former French possessions in Indochina, an agreement was reached between the Mendes-France government in France and the Ho Chi Minh government in Vietnam. The independence of Vietnam from the French Union was recognized, but the country was divided into a "people's democracy," allied with the Communist bloc, and a pro-Western autocratic republic. This division followed the pattern set in Korea in July 1953 and was symbolic of the struggle between East and West.

THE ENVIRONMENT OF THE REVOLUTION[a]

DESCRIPTION OF COUNTRY

Physical characteristics

Vietnam extends from the Chinese border to the Gulf of Siam on the eastern half of the Indochinese peninsula, covering an area slightly larger than New Mexico. About 1,000 miles long, with a maximum width of 250 miles, Vietnam is predominantly mountainous and heavily forested. The climate is mostly tropical, particularly in the Mekong River Delta in the south and the Red River Delta in the north. The rainy season usually lasts from April to September.

The people

According to official prewar figures, approximately 18 million people live in Vietnam, which is not a densely populated country. With the exception of Khmers, Thais, and other racial minorities, they all speak Vietnamese. The racial minorities, composed largely of tribes occupying the large, sparsely populated inland mountain areas, have been very antagonistic toward the Vietnamese.

The overwhelming majority of people lived and worked in rural areas along the coast, where they cultivated rice fields. At the time of the French rule, there were fairly good-sized segments of the French and Chinese populace who were part of the merchant class and resided in the large merchant centers. In addition, there was a sizable French administration whose personnel lived apart from the indigenous population in the exclusive districts of the large cities. In the colonial period Vietnam was divided into three provinces: Tonkin in the north, Annam in the center, and Cochin China in the south. Tonkin and Annam were French protectorates, whereas Cochin China was a French colony. The city of Saigon in Cochin China, one of the major cities in Vietnam, was the administrative center of French rule. Other cities of importance included Hué, in Annam, the seat of

---

[a] Unless otherwise indicated the environment described refers to the French colonial period.

old Vietnamese dynasties; Hanoi, in Tonkin, the resident city of the French Governor; and Haiphong, also in Tonkin.

Communications

Most of Vietnam was accessible by some means of transportation during the rebellion—particularly during periods of good weather. There were a few asphalt-covered roads, most of them in the south radiating out from Saigon. However, there was an important hard-surfaced all-weather road, the Colonial Highway No. 1, extending from north to south all along the coast, which connected major coastal cities. Some metaled all-weather roads ran in an east-west direction and connected the inland cities and villages with the coastal cities; they also connected Laos and Cambodia with Vietnam. Most other roads were unsurfaced and became treacherous during the monsoon season. A modern rail system, the Trans-Indochinese Railway, covered the north-south distance from Saigon to Hanoi and extended across the border into China. Another railway from Haiphong to Kunming followed the Red River Valley and connected north Vietnam with southwest China.[4]

Air travel was the most effective means of transport. Most of the airports were near the major cities along the coast; but there were several airports inland—for example, at Dalat and at Ban Me Thuat in the southern Annam high plateaus. There were port facilities all along the coast of Vietnam. The major ports were Saigon and Haiphong.

Natural resources

Rice, rubber, and fish were the most important resources. The mineral deposits were not of primary importance to the country's economy, although the Hongay coal mines in the north have produced anthracite of the highest quality. Iron, bauxite, manganese, lead, zinc, tin, and phosphates also are mined in the north. The southern areas of Vietnam are relatively poor in mineral resources.

SOCIO-ECONOMIC STRUCTURE

Economic system

Rice-growing and fishing were the two major industries of Vietnam during French rule. Rice and fish were also the mainstays of the Vietnamese diet. Because of the primitive methods used, large concentrations of labor were needed in rice-growing areas. In the primary rice-growing paddies of the Tonkinese (Red River) Delta, some six or seven million people were concentrated in an area slightly under

6,000 square miles.[5] Rubber plantations were developed—by the French before, and by the Japanese during, World War II.

The French introduced Western economic methods to a very primitive country, and had developed a prosperous enterprise by 1939. Vietnam's economy became an extension of French mercantilistic policies: the French bought Vietnamese raw materials at low prices, and sold the Vietnamese French industrial commodities at high prices. Rice and rubber became Vietnam's most profitable exports, enriching the wealthy landlords, most of whom were French or Chinese.[6] During the interwar period the Vietnamese economy had five major characteristics. (1) Most of the capital came from abroad. Prior to 1920 funds came largely from the state budget, but following World War I the funds came from French private investors, whose goals were immediate high returns. Only a small fraction of the high profits were reinvested into the Vietnamese economy. (2) The economic policy was geared to the exploitation of rice, rubber, and some rare minerals for export. Vietnamese industry produced goods only for immediate consumption. Vietnamese economy became restrictive, tending toward a market for overpriced, tariff-protected products of the metropolitan industries. (3) The fiscal policies of the French did not work to the advantage of the Vietnamese. All public works programs were paid for by taxing the small incomes of the peasants while foreign concerns were taxed little or not at all. Some French industries were subsidized by public funds. (4) Public works programs benefited the investor. (5) The Chinese middleman, the small Vietnamese landed class, and the French exporters benefited from rice production. Although there was an increase in the production of rice, the disproportionate increase in population resulted in a decline in individual rice consumption.[7] In the Northern Province almost all the land was owned and tilled by farmers, whereas in the South, most land, prior to World War II, belonged to absentee landlords.

Class structure

The largest sector of the population was made up of poor peasants, most of whom worked in the rice fields. Sometimes they were conscripted by French authorities to work in mines or on public works.[8] A middle class emerged from the predominantly peasant class shortly after the French had established their authority and continued to grow under French rule. It was made up of merchants, officials, and intellectuals, and had virtually no political or economic power.[9] Along with the military, this middle class held an intermediate position on the Vietnamese class scale. The privileged class comprised approximately 10 percent of the population and consisted of Europeans (mostly French), Chinese, and very few Vietnamese. They were the administrators and businessmen.[10]

Literacy and education

The mass of the indigenous population was illiterate. Students who completed secondary school had little opportunity to go on to higher education. Instruction in the technical fields was almost entirely nonexistent, and, for the most part, an education at college or university level was either a legal, medical, or liberal arts education. Some Vietnamese intellectuals obtained their education in foreign schools. French was the official language. In most cases the educated Vietnamese found themselves unable to take maximum advantage of their training.[11] Higher administrative positions were denied them and other opportunities were limited.

Major religions and religious institutions

Most of the Vietnamese people had retained a certain amount of Confucianism and Taoism. In mountain areas, animism was still prevalent. Converted Catholics, although they held important positions under French rule, often adhered to ancestor worship and certain magical practices retained from the older religions.

In 1926 a group of Vietnamese officials established the Cao Dai religious sect, which amalgamated a number of faiths. Its hierarchy of priesthood was headed by a pope, and its adherents numbered into the hundreds of thousands. The Cao Dai sect and the less important Hoa Hao sect were both strongly nationalistic.

GOVERNMENT AND THE RULING ELITE

Description of form of government

The French administration was highly centralized and headed by a governor general or high commissioner. Chief residents were responsible to him. Vietnamese policy was determined by the colonial officials in Paris.[12] The Vietnamese played a very minor role in the government and the administration of their own country. Vietnam, along with Laos and Cambodia, was represented in the Grand Council of Economics and Financial Interests, of which half the members were French. Cochin China had a Colonial Council which was partly French and partly Vietnamese. Tonkin and Annam, on the other hand, had separate councils for Frenchmen and for the indigenous population. All these bodies were concerned generally with local economic affairs, and had advisory powers only. Vietnamese members were either appointed by the government or elected under a system of very restricted suffrage.[13]

The status of Cochin China differed from Tonkin and Annam. Cochin China was a colony while Tonkin and Annam were protectorates. The highly centralized French administration in Vietnam was dependent upon the Paris government, and its policy reflected the political fluctuations and the changing patterns of that government. During the Japanese occupation, the French administration became dependent on Japanese authorities.

Description of political process

Political groups and political parties were organized shortly after the French had gained complete control over Vietnam. Most of them represented opposition to French rule and were declared illegal, except during the time of the "Popular Front" governments in France between 1933 and 1939, when all political parties were allowed to function. Parties such as the Constitutionalist Party, which was established in 1923 and was the first legal political organization, were sterile in their limited power and never gained wide popular support. The nationalist groups in Vietnam operated clandestinely most of the time.

Recognizing French rule as an imposition and as a sign of their own weakness, the Vietnamese began to organize terror movements within Vietnam and to issue propaganda from foreign bases. The terroristic activities were easily suppressed by the French. The early groups were Robin Hood type bands which had no political goal except to oppose the existing regime. It was not until the late 1920's that the nationalist groups began concentrating on organization and developing political motives. The Cao Dai religious sect, for instance, was organized in Saigon by several disgruntled Constitutionalist Party members in 1926, and in the 1930's it took on the characteristics of the protest movement. The group maintained secret liaison with the Japanese in the late 1930's.

The Vietnam Nationalist Party (VNQDD) was founded in 1927 by the young nationalist teacher, Nguyen Thai Hoc, and became the most important non-Communist nationalist organization in Vietnam. In its formative years it was responsible for some of the most terroristic acts ever committed against the French authorities. At a 1930 uprising in Yen Bay, the VNQDD exhausted its initial drive and was suppressed into near inactivity by the French. The structure of the organization had been modeled after the Kuomintang, and the party often looked to China for support. Most of the VNQDD leaders remained in China after the Yen Bay incident to recuperate and rally their forces.[14]

Nguyen Ai Quoc, better known as Ho Chi Minh (and thus referred to throughout this section) was instrumental in organizing the Indochinese Communist Party. He had "little more than his mandate from Moscow to guide him in his choice of means for building a communist movement in Vietnam" wrote Milton Sacks.[15] Yet he succeeded so well that from the early 1930's on, communism dominated Vietnamese nationalism through its internal disputes. Much of this time other nationalist groups vegetated in exile. In the later 1930's the Communist Party joined the "Popular Front" movement composed of all French and Vietnamese democratic elements, to combat Japanese imperialism.[16] After the German-Soviet Pact was signed in 1939, the party was declared illegal and its leaders went into exile.

The outbreak of war limited French authority, as Japanese forces marched in to occupy certain strategic areas of Vietnam in 1940. During the war most leaders of the nationalist groups were in exile reorganizing their forces. The Communist element of the nationalists, however, remained active at home and engaged in such activities as establishing cells in youth organizations. As a result of the war and the Japanese occupation, many Vietnamese intellectuals were able to assume new roles of administrative levels never before available to them. Other Vietnamese, primarily under Communist leadership, were offering some token resistance to Japanese forces.

In 1943, the Chinese Government forcibly persuaded Ho Chi Minh, after jailing him for 18 months, to reorganize the Vietnamese nationalist groups on Chinese soil. A coalition of nationalists called the Vietminh was created under the auspices of the Kuomintang, and a provisional government of the Democratic Republic of Vietnam was established, made up of leaders of the VNQDD. The revolutionary organization that fought the French for 7 years had its origin in these developments.[b]

Under a wartime agreement Chinese forces were to occupy that portion north of the 16th parallel on the defeat of the Japanese, and British force were to move into the south. With British aid, the French authorities returned *en force* to reestablish French rule, and out of political and military weakness the new republican government consented to negotiations with representatives of the French Government. In March of 1946 a temporary agreement was reached between the two parties pending further discussion in Paris during the summer. Throughout the summer of 1946, Ho Chi Minh represented his government at the Fontainebleau Conference, during which a *modus vivendi* was reached in September. However, relations between the

---

[b] See below p. 42 and 43.

French and the Vietminh soon deteriorated, as both charged breach of faith. On December 19, 1946, Vietminh guerrilla units attacked French posts, initiating open warfare.

Legal procedure for changing government institutions

Prior to World War II Vietnam was under the complete authority of the French Government. Any legal changes within the institutions had to be initiated by the government in Paris. During World War II, the French administrator, Admiral Jean Decoux, had to be given full authority to act independently.

Relationship to foreign powers

Relationships with foreign powers and foreign policy were prerogatives of the French during the colonial period. During the war, Japan became the dominant force in Indochina and greatly contributed to Vietnamese nationalism and desire for independence.

The role of military and police powers

Military and police power had been under French control until 1940, when it was shared with the Japanese occupation troops. During the interwar period, force and coercion had been used successfully by the French to suppress any form of demonstration or illegal activity. Vietnamese served in the French Army and police forces, but a few of them were not dependable and sided with anti-French demonstrators.

WEAKNESS OF THE SOCIO-ECONOMIC-POLITICAL STRUCTURE OF THE PREREVOLUTIONARY REGIME

History of revolutions or governmental instabilities

On the whole, the French were able to govern the country effectively. However, during periods of unrest, military and police action were needed to ensure French control. As early as March and April 1908, mass demonstrations were staged in Bienh Dinh to persuade the French authorities to reduce high taxes. The demonstrators were fired on and many were arrested and sent to prison at Poulo Condore. In June of the same year an attempt was made to poison the French garrison in Hanoi. This started a wave of repression against the nationalists.[17]

The only uprising of any consequence during French rule took place at Yen Bay in Tonkin during February of 1930. The rebellion was led primarily by the VNQDD and its success was short-lived. French planes fired upon villages and the uprising was successfully put down. The leaders were either executed or imprisoned.

Sporadic and scattered mass demonstrations occurred between 1930 and 1932, particularly in northern Annam, where famines and Communist agitation led the peasants to protest against local conditions.

Economic weaknesses

The major economic weakness of the French administration in Vietnam during the interwar period was its monocultural dependence on rice. If rice prices dropped for any reason the Vietnamese were seriously affected. Rice requisitions as taxes in kind placed a burden on the Vietnamese peasantry, especially at times of crop failures. The worldwide economic crisis of the late 1920's led to a serious depression in Vietnam as early as 1928.

Other economic weaknesses of the French regime in Vietnam were noticeable in industry. Some French industries were subsidized entirely by Vietnam taxes in order to create high returns for the investors, and a small elite group, mostly non-Vietnamese, benefited from the Vietnamese economic growth.

Social tensions

Social tensions were largely the result of the disparity in the standards of living between the masses and a small upper class. Inequalities existed everywhere—for example, in the availability of medical facilities. Although the French had made progress in medicine and in hospital construction, the indigenous poor were left with little medical aid. The same conditions existed in education. Few Vietnamese students were able to complete their education, and the vast majority remained illiterate. Lack of opportunity to participate in shaping Vietnamese affairs caused resentment, particularly among the small educated elite, and increased social tensions in Vietnam.

Government recognition of and reaction to weaknesses

The French were quite aware of the existing antagonisms. The demonstrations and uprisings, at Yen Bay in particular, boldly announced that the revolutionary movement in Vietnam had a head and a body. French reaction to the regime's weaknesses came first in the form of suppression. In 1930 there were 699 executions without trial, and 3,000 arrests were made resulting in 83 death sentences and 546 life sentences. From January to April of 1931, 1,500 more Vietnamese were arrested.[18]

In Paris during the middle 1930's the French recognized the need for reforms to pacify the country. These reforms were limited and confined to the economic and social spheres. A rise in the export of rice greatly eased the economic crisis. The "Popular Front" movement ushered in a great wave of hope and confidence. Bao Dai, of the Nguyen Dynasty, returned to claim his throne. Although not significant in itself, Bao Dai's return symbolized the change which was taking place. There seemed to be greater unity among the nationalist groups, who were now more inclined toward negotiating directly with the French rather than demanding complete independence. The authorities relaxed their controls when they allowed associations to form in Vietnam and political parties to function in Cochin China. However, the events just before the outbreak of World War II disrupted the trend toward liberalism, and once again political freedoms were curtailed.

FORM AND CHARACTERISTICS OF REVOLUTION

ACTORS IN THE REVOLUTION

The revolutionary leadership

The nationalist groups had the common goal of opposing the French regime but varied in ideology, intensity of opposition, and size of revolutionary following. Most of the revolutionary leaders, however, were Marxist-Leninist in orientation. The most active and most deliberately revolutionary leader in Vietnam was Ho Chi Minh (Nguyen Ai Quoc). The son of a Vietnamese mandarin, Ho received his high school education in Indochina. After World War I he went to France, where he made contact with French Socialists. Later he studied in Moscow and in 1923 served his apprenticeship in China with Borodin, whose sole mission was to reshape the Chinese Kuomintang with Communist support. Ho was a very well-disciplined Communist, time and again proving his loyalty to the party. He helped to organize the Indochinese Communist Party during its formative years in the late twenties. He organized soviets in Vietnam, and instituted a revolutionary youth group in 1927. During the war he played a major role in shaping the nationalist and Communist groups into a "front" known as the League for the Independence of Vietnam (Vietminh). His Marxist convictions have never prevented him from soberly evaluating his political alternatives.[19]

Gen. Vo Nguyen Giap, the military leader of the revolution, was also a professed Communist. He received a doctoral degree in history

in Vietnam, continued his studies at the Chinese Communist stronghold of Yenan, and returned home well versed in Mao Tse-tung's theories on guerrilla warfare. Most of Giap's work consisted of organizing
and establishing Communist cells in Vietnam. He spent much time
in French prisons, where both his wife and sister-in-law died. In 1945
Giap was raised to the rank of Commander-in-Chief of the Vietminh
Armed Forces, a position in which he proved himself a master tactician in paramilitary activities.[20]

The revolutionary following

The nationalist groups in the twenties consisted primarily of intellectuals from the middle class with virtually no mass support. By the
late twenties, however, the VNQDD began to have a following among
the peasants. It was not until the 1930's that the nationalist movement
gained wide peasant support. By the time open hostilities began in
1946, the country was united against French rule. Disunity continued,
however, within the nationalist movement. Especially in the south,
more conservative nationalist leaders remained active, while the north
became a stronghold for the Communist-dominated Vietminh.

ORGANIZATION OF REVOLUTIONARY EFFORT

Internal organization

The revolutionary organization had its beginning in May 1941
on Chinese soil. Initially, under Chinese Nationalist auspices, a coalition between various anti-French and anti-Japanese Communist and
non-Communist nationalist groups was formed that came to be called
the "League for the Independence of Vietnam," better known as the
"Vietminh." By 1944, Ho Chi Minh had assumed leadership over the
Vietminh; he was assisted by Gen. Vo Nguyen Giap, who had organized anti-Japanese guerrilla units in northern Vietnam. Following
the Japanese collapse, the Vietminh fought a three-cornered "diplomatic" battle against the Chinese occupying northern Vietnam, the
French, and against other nationalist leaders. By August 1945 Ho Chi
Minh had become the leader of a "provisional government" which
proclaimed its independence on August 25, 1945, with the blessing of
the former emperor Bao Dai, who renounced his claim to the throne
to promote unity and avoid civil war. The Vietminh had succeeded in
gaining control over the government apparatus with the appearance
of legality. Ho Chi Minh and the Communist Party were in control.

The National Assembly of the newly proclaimed Democratic
Republic of Vietnam approved Ho Chi Minh's position as Premier,

President, and Minister of Foreign Affairs. The Vietminh was controlled by Communists and only three minor cabinet posts were allotted to non-Vietminh nationalists. The Indochinese Communist Party, which had necessarily established a highly knit organization, voluntarily dissolved itself in November 1945 and the political and military work of the party was turned over to the Vietminh. Later the Vietminh was gradually absorbed in a "United National Front." However, at no time did Ho Chi Minh and the Communist elements lose control. Just the opposite, the organizational changes appear to have been tactical moves to strengthen Ho Chi Minh's control, while, at the same time, avoiding the appearance that the new government was anything but a nationalist group trying to establish and protect an independent democratic regime. "Interpretations may vary as to whether the Viet Minh was 'really Communist' at that time. Nobody, however, can fairly contest the fact that it was on the road to one-party rule," observed a leading expert on Vietnam.[21]

Elections in January 1946 resulted in a new government which included a number of non-Communist nationalists in important positions. Gradually these were eased or forced out and before long the Communist leaders were again in complete control. By 1949 the Vietminh had broken with the West, and by February 1951 the Communist Party reappeared officially under the name of the "Vietnam Workers' Party."

When the French reoccupied the country the Democratic Republic at first proclaimed its administration of Vietnam from the capital at Hanoi. Open hostilities broke out between the Democratic Republic and French troops in December 1946, and the entire government apparatus was transferred from Hanoi into deep caves in the Tuyen Quang-Bac Kan-Thai Nguyen redoubt. The need to coordinate the fight against the French resulted in a strong concentration of power in the hands of the executive. On various administrative levels from small villages to larger units, the Vietminh administered the areas not directly under French rule through "committees for resistance and administration." These committees were composed of appointed and trusted political officials and tightly controlled all political, economic, and social activities of the Vietnamese population. The cellular structure of government that was established had the advantage of being decentralized and flexible, while at the same time allowing decisions taken by the Central Committee to be rigidly enforced on all levels down to the smallest village.

In addition to the regular governmental machinery, other organizations helped to maintain complete control over individuals. These organizations, called "parallel inventories" by the French, ranged from male and female youth groups, farmers' and trade unions to groups as specialized as a flute players' association. These organizations were all effectively used for purposes of indoctrination and propaganda. In some areas which were administered by pro-French officials and under nominal French control, the Vietminh was able to set up a parallel administration ready to take over the functions of government any time and exercising considerable control over the inhabitants. On the surface, many a village may have appeared safe, but as soon as the military situation permitted it, the secret Vietminh administration would emerge and assume open control.

Closely paralleling the political machinery was the military, which also operated on various levels. It is estimated that the revolutionary army consisted of 400,000 troops. The army was organized on a local, regional, and national basis. There were (1) irregular local guerrilla units, composed of peasants and other "civilians," (2) regional military units, and (3) a hard core of well-disciplined and trained "regular" troops not organized on a regional basis. All three types coordinated their activities, and often the regional units and "regular" units combined forces for large-scale military action, assisted by local guerrilla units. The system was flexible and efficient. Replacements for the regular "elite" forces came from the regional units, and the regional units recruited their replacements from the local guerrilla groups. Close cooperation with civilian administrative committees resulted in first rate intelligence information and provided labor by thousands of civilians when needed as supply carriers, or in other capacities. The final victory over the French forces in 1954 was made possible only through the efficient mobilization of every available body under the control of Ho Chi Minh and his organization.

External organization

After the success of the Communist Chinese revolution, the Vietminh Government received military aid from Communist China. The extent of this aid is not precisely known, but it is safe to assume that it was a decisive factor in the Communist victory. Molotov trucks and larger caliber weapons were sent from Moscow via Peking after 1950. Some Vietnamese units found sanctuary behind the Chinese border, while artillery received via China greatly increased the fighting power of Giap's army.

GOALS OF THE REVOLUTION

Throughout the pre-World War II era, the main nationalistic objective was to evict the French from Vietnam in order to establish a Vietnamese national government and a national economy. There were certain periods when the Vietnamese nationalists were willing to settle for political reforms, but for the most part, the main objective was never lost. There was a rapproachement with the French Government during the anti-Japanese "Popular Front" movement, but anti-imperialism (anti-French as well as anti-Japanese) was again intensified by the Communists after the Nazi-Soviet Pact of 1939 and particularly after the Vichy Government came into being in 1940. The destruction of French rule, the obstruction of the Japanese invasion, the realization of Vietnamese independence, and later the alliance of Vietnam with the Sino-Soviet bloc were the main aims of the Communist Party.

The specific goals at various times were adapted to the realities of the political and military situation. Immediately after World War II, for instance, Vietminh appeared willing to accept independence within a French union. When negotiations over this issue broke down, the Vietminh aimed for a complete victory over the French and the establishment of a United Communist Vietnamese State.

REVOLUTIONARY TECHNIQUES AND GOVERNMENT COUNTERMEASURES

Methods for weakening existing authority and countermeasures by government

Historical

The Vietminh's independence declaration of 1945 was formulated in a fashion obviously intended to gain international recognition: it contained all the terminology of Western liberal philosophy. But it left the Western nations unimpressed. The new Vietnamese Government was then faced with three alternatives: a long-term resistance campaign until final victory, a short war to obtain a better settlement, or full negotiations with the French. Because of its weak position, both politically and militarily, the Vietnamese Government chose at first to negotiate.

An agreement was reached during the summer of 1946, but it fell far short of the desired independence. In December 1946 Vietnamese guerrillas attacked several French garrisons to spark the revolution. The Vietminh held that whatever agreements had been concluded were abrogated unilaterally by the French in November when the

French interfered with some commercial activity under Vietnamese jurisdiction. The French, on the other hand, maintained that the Vietminh voided the agreements by its unprovoked attack on the French garrisons in December. During this period and up to 1950, the Vietminh continued sporadic guerrilla attacks on French garrisons, while training units of a regular army in the mountain areas of southwest China. Ho continued to denounce French claims that he was collaborating with the Communists and was taking part in an international plot to overthrow democracy in Southeast Asia, all the while making efforts to negotiate peace with France.

As the guerrillas attacked French garrisons, the Vietminh planned a three-phase war strategy—a period of defense, a period of equilibrium of forces, and a general counteroffensive. The Vietminh established bases in the countryside of Vietnam which were entirely independent of each other, and which engaged in agricultural pursuits whose surpluses bought arms for the guerrilla fighters. In the small villages, some guns, bazookas, and other military items were being manufactured by the peasants. At the same time the Vietminh attempted to maintain friendly diplomatic relations with the Chinese Communist and Thai Governments.

Consisting of approximately 400,000 well-disciplined troops, and described by their Commander in Chief, General Giap, as "the military arm of the Vietminh government," the Vietminh Army units were responsible for breaking the French line on the Chinese border in the 1950 offensive. They captured the Thai country in 1953, and finally crushed the French forces at Dien Bien Phu in 1954. Unlike the French forces, some of the Vietminh units were organized on a regional rather than an operational basis. The supply problem was thus reduced. The chief weaknesses of the Vietminh were the lack of heavy artillery, transport facilities, and air support; however, their intelligence system was quite superior.

The French had a total force of about 500,000 troops, the security of operating from firm bases, and complete mastery of air operations. Misled in their intelligence information and incapable of forming effective defensive positions, the French, however, were at a fatal disadvantage.

In 1953 General Navarre took over the French forces in Vietnam and found the situation stagnant. He reorganized the units so as to increase their mobility and began a series of offensives which were designed to break up the Communist forces by 1955. General Navarre involved his troops in too many encounters in many different areas too often and was never able to prepare himself for a large-scale operation at the proper time. Failing to defeat Giap's troops on French territory, Navarre decided to take his troops into Giap's territory. At the first site chosen, Dien Bien Phu, French paratroopers were dropped into the valley. This maneuver was unsuccessful. The Vietminh went into the surrounding hills and waited for reinforcements. Heavy artillery was positioned beyond the perimeter of the French stronghold. The barrage of artillery fire combined with wave after wave of Vietminh units decreased the perimeter of the French garrison. Most supplies that were airdropped to the French units were captured by the Vietminh, and when the monsoon weather curtailed French air activity, Dien Bien Phu was lost.

Operations continued in the Tonkin Delta, which had the appearance of being a French stronghold. It was soon evident, however, that the Vietminh infiltrated the area and neutralized the French forces. The last French operation took place on June 30, 1954. Hostilities ended on July 21, 1954, when a cease-fire order arrived from Geneva.[c]

Functional

Within each phase of the Vietminh's total war strategy, tactics which have been developed and employed successfully by Communist organizations elsewhere were applied. The final objective of the Communist strategy was to overthrow French authority and place Vietnam under the complete control of the Vietminh. However, in the initial phase only small target areas were firmly established as Vietminh bases from which the military units operated against French forces. Taking possession of areas to convert into bases was a tactical step. The Communists made use of propaganda techniques to win over the population—secretly if possible—and this greatly facilitated taking possession of areas. Appeals were made to non-Communists and ethnic minorities, using the rallying symbol of "popular democratic nationalism," which combined ideology with technique. Propaganda was most effective on organized masses—such as the politico-military organizations, or organizations of the family, of religion, of trade unions—and organizations facilitated the use of " 'psychological techniques' which are applied more easily to homogeneous categories of human beings."[22]

The object of the "psychological techniques" was to control the minds of the masses by means of indoctrination. These techniques took many forms: for example, whispered propaganda, conferences, assemblies and meetings, directed discussions led by specialists, rumors, pamphlets, radio, and plays.

---

[c] For a detailed study of the revolutionary organization see Bernard Fall, *The Vietminh Regime* (New York: Institute of Pacific Affairs, 1956).

The "morale technique" of "self-criticism" was most effective. "Self-criticism" is a form of confession in which an individual verbally "expels heretical ideas" before a group. This technique greatly aided in maintaining strict party discipline. Vietminh troops particularly employed this technique; and the increased sessions of "self-criticism" during the hard campaigns in northwest Vietnam in 1953 restored the morale of soldiers and officers who might have been thinking of deserting. The application of the above techniques varied according to the composition of organizations or types of individuals.

A definite pattern was followed by the Communists in taking possession of areas and establishing bases. Propagandists were the first to enter an area, preferably deep in the jungle. These Communist agents then propagandized the inhabitants, showed them much respect, and developed "sympathizers" among groups which showed no outward sign of hatred for the French. "Propagandized" villages at times led the French to misinterpret the outward "calmness" of the Vietnamese.

Coolies from large plantations were induced to desert their work, and were regrouped in the jungle villages where firearms appeared. Trials were held for convicted murderers and traitors, which greatly impressed the people, as they were subjected to strict physical and moral control. Thus, through terror, the Vietminh convinced the population that it had a much stronger organization than did the French.

The Communist movement then had freedom of action within the bases, and within these bases enemy agents were easily detected. All resources were at the disposal of the rebel groups, and their troops and depots were safely concealed by the population. Then Vietminh troops went into action, attacking and ambushing French troops.

The French were exasperated by the slight results of their own attacks. When they entered enemy villages, they often found them deserted. The peasants had evacuated and taken refuge in the jungle, and barred their trails with traps. Vietnamese guides became worthless, and it became impossible for the French to use the element of surprise.

Not all of the Communist guerrilla units were based on Vietminh-held soil. Many of the units operating from French-held bases were involved in minor skirmishes which ranged from urban terrorism and rice field warfare to hill, mountain, and jungle warfare. Although not as well-equipped as the main force, these "second line" troops "screened" for the main force offensives and infiltrated other important French-held areas. Their ability to disperse and "blend into the landscape" made these units very elusive and difficult to capture.

Attacks and ambushes were carefully prepared weeks, sometimes months, in advance. The peasants volunteered or were coerced to act as agents for the Vietminh, keeping their superiors informed on all French military activities. Agents were thus innumerable, and the information was transmitted to the Vietminh military authorities.

Recruiting and training techniques employed by the Vietminh produced light mobile units. The regular forces had an 8-month training program, which included all forms of combat, handling of arms, mines and explosives, use of heavy weapons, assaults against fortified positions, light combat, and intelligence handling. The regional and local forces were trained in more "irregular" types of warfare. Many classes were held and the troops were encouraged in their learning by being promoted from local to regional forces, or from regional to regular troops.

Discipline was rigid. The troops were taught strict obedience and the worship of materiel and arms. "Self-criticism," as described above, was employed to enable them to understand better the reasons for defeat.

Methods for gaining support and countermeasures taken by government

The propaganda techniques used to weaken existing authority contributed also, of course, to enlisting the active support of the populace for the Vietminh. Still more effective was the conduct of the troops themselves, mingling with the population and helping peasants with the harvests and other chores.

In soliciting foreign support and aid, the Vietminh initially looked to the West. However, the Communist success in China in 1949 prompted Ho Chi Minh to break with the West and to openly declare his allegiance to international communism. The new government in China supported Ho's regime, and this support was followed by a recognition from the Soviet Union. The Vietminh forces of General Giap received military aid from Moscow via Peking, mostly in the form of Molotov trucks and large caliber guns. This aid was a significant factor in the Vietminh victory.

The French also attempted to gain support from both the Vietnamese population and foreign powers. On March 9, 1949, an agreement was reached between Bao Dai, the son of the last Nguyen emperor, and the French Government in which Paris conceded limited independent status to a Vietnamese National Government headed by Bao Dai himself. The French were trying to swing Ho Chi Minh's popular support toward a pro-French Vietnamese government. The United States recognized the Bao Dai government on February 7, 1950. The French by this time had persuaded the United States Government that the Vietminh represented a threat to Western interests in South-east Asia. The United States aided the French cause in Vietnam, but by June 1950, United States aid was being given directly to the Bao Dai government. The United States effort in Vietnam had totaled almost one billion dollars by 1954.

MANNER IN WHICH CONTROL OF GOVERNMENT WAS TRANSFERRED TO REVOLUTIONARIES

During the latter part of 1953, Chou En-lai, the Chinese Communist Foreign Minister, acting in behalf of Ho Chi Minh, and Prime Minister Nehru of India arranged for a conference to settle the French-Vietnamese dispute. Ho Chi Minh had made public pronouncements to the effect that he wished to negotiate. As a result a conference was called at Geneva to settle the whole Indochina problem.

It was decided at the Geneva Conference in 1954 that Vietnam was to be divided at the 17th parallel, the northern half under the regime of Ho Chi Minh's Communist-oriented government, and the southern half under a Western-oriented regime. Ngo Dinh Diem became its first President. The representatives to the Conference suggested that a general election take place in 1956, to unify Vietnam under one government. The agreements that were reached were satisfactory only to the members of the Communist bloc. The French retained only certain protective rights in South Vietnam. The agreements changed the Vietnamese political boundaries, creating two independent legal entities—each approaching the task of revising the political and social structure created by French rule.

THE EFFECTS OF THE REVOLUTION

CHANGES IN THE PERSONNEL AND INSTITUTIONS OF GOVERNMENT

The Vietminh Government in the north has established a Communist-controlled state. This result fell somewhat short of Ho Chi Minh's goal in 1945, when he took over control of the nationalist movement and temporarily united all of Vietnam under the Vietminh. South Vietnam has established an autocratic pro-Western government which pays some lip service to Western concepts of democracy.

MAJOR POLICY CHANGES

The Ngo Dinh Diem regime in South Vietnam, protected under the "Umbrella Clause" of the Southeast Asia Treaty Organization, has sought and received massive U.S. assistance against continued North Vietnamese Communist efforts to unify the country under Vietminh control. Much of this aid went into the creation of a well-equipped South Vietnamese Army, which has not been notably effective thus far in combating Communist guerrillas.

The People's Republic of Vietnam has become a permanent member of the Soviet bloc. Recently, in the disputes between China and the Soviet Union, Ho Chi Minh seems to have sided with the Soviet Union. The historic threat of Chinese domination over Vietnam may well have entered into the Vietminh's decision to rely more closely on the Soviet Union in her "ideological" dispute with Communist China.

LONG RANGE SOCIAL AND ECONOMIC EFFECTS

It is difficult to measure the long range social and economic effects, since the country is still very much in its developing stages. In the north a Communist state is following Communist economic and political practices; but the regime is considerably more tolerant toward middle-class elements than are Western Communist countries. Recent developments seem to indicate that the Vietminh believes itself capable of progressing directly to a socialist state without going through the capitalist stages. The Five-Year Plan for the period 1961–65 places heavy emphasis on the development of light—and some heavy—industry. The attempt of the Vietminh to initiate collectivization of agriculture in 1956 met with failure.[23]

In South Vietnam the government has been following a more conservative policy in its social and economic legislation. Most of the large landholdings are in the South and land reform remains one of the most pressing problems. South Vietnam's failure to institute effective land reforms is being exploited by Communist propaganda and agitation.

OTHER EFFECTS

The problem of "reunification" is as far from a solution in Vietnam as it is in Germany and Korea. However, the possibility that communism may succeed in taking over South Vietnam from "within" is strong. The Ngo Dinh Diem regime is under attack from Communist guerrillas and is subjected to Communist subversion from within. South Vietnam has resisted a general election in 1956 to unify the country, as the Geneva agreements suggested. Ho Chi Minh, on the other hand, has been demanding that Vietnam prepare itself for the election. Ho Chi Minh seems prepared to unify Vietnam by force. The United States has declared that it will actively support the maintenance of a "free" and independent South Vietnam. The crisis continues.

NOTES

1. Bernard Fall, *Street Without Joy* (Harrisburg, Pa.: The Stackpole Company, 1961), pp. 24–25.
2. Joseph Buttinger, *The Smaller Dragon: A Political History of Vietnam* (New York: Fredrick A. Praeger, 1958), p. 40.
3. Ellen J. Hammer, *The Struggle for Indochina* (Stanford, Calif.: Stanford University Press, 1954), p. 11.
4. Ibid., p. 303.
5. Ibid., p. 69.
6. Philippe Devillers, *Histoire du Viet-Nam de 1940 à 1952* (Paris: Editions de Seuil, 1952), p. 47.
7. Buttinger, *The Smaller Dragon*, pp. 429–433.
8. Hammer, *The Struggle for Indochina*, p. 68.
9. Ibid., p. 71.
10. Devillers, *Histoire du Viet-Nam*, p. 47.
11. Ibid., p. 53.
12. Allan B. Cole, *Conflict in Indo-China and International Repercussions: A Documentary History, 1945–1955* (Ithaca, N.Y.: Cornell University Press, 1956), p. xx.
13. Hammer, *The Struggle for Indochina*, p. 72.
14. Ibid., pp. 83–84.
15. I. Milton Sacks, "Marxism in Vietnam," *Marxism in Southeast Asia*, ed. Frank N. Trager (Stanford, Calif.: Stanford University Press, 1959), p. 116.
16. Devillers, *Histoire du Viet-Nam*, p. 44.
17. Hammer, *The Struggle for Indochina*, p. 60.
18. Buttinger, *The Smaller Dragon*, pp. 436–437.
19. Hammer, *The Struggle for Indochina*, pp. 74–78.
20. Ibid, pp. 97–98.
21. Bernard Fall, *The Vietminh Regime* (New York: Institute of Pacific Affairs, 1956), p. 16.
22. Anonymous Group of Officers, "Concrete Cases of Revolutionary War," in *Revue Militaire D'Information No. 281*, tr. Department of the

Army, Office of the Assistant Chief of Staff, Intelligence, Washington 25, D.C., intelligence translation No. H–2060, p. 26.

23. P. J. Honey, "North Viet Nam's Party Congress," *The China Quarterly,* No. 4 (Oct.–Dec. 1960), 66–75.

RECOMMENDED READING

Buttinger, Joseph. *The Smaller Dragon—A Political History of Vietnam.* New York: Frederick A. Praeger, 1958. There is a highly detailed historical chronology presented in this book, but it is of limited value for the story of the revolution.

Cole, Allan B. *Conflict in Indo-China and International Repercussions—A Documentary History, 1945–55.* New York: American Institute of Pacific Relations, 1955. The selection of documents provides the student with a documentary substance for the study of the Vietnamese revolution.

Devillers, Philippe. *Histoire du Viet-Nam du 1940 à 1952.* Paris: Editions de Seuil, 1952. This work is widely circulated and considered a well-documented source.

Fall, Bernard B. *Street Without Joy—Indochina at War, 1946–54.* Harrisburg, Pa.: The Stackpole Company, 1961. The revolutionary techniques of the Vietnam forces and the countermeasures employed by the French are treated in this source.

Fall, Bernard B. *The Vietminh Regime.* New York: Institute of Pacific Affairs, 1956. This is a detailed examination of the government and administration of the Democratic Republic of Vietnam.

Farley, Miriam S. *United States Relations with Southeast Asia . . . With Special Reference to Indo-China 1950–1955.* New York: American Institute of Pacific Affairs, 1955. Covering the period from the Chinese intervention in Korea in late 1950 to the end of the French rule in Indo-China, Farley examines the extent of U.S. military aid to the French and to the Vietnamese forces in Vietnam.

Hammer, Ellen J. *The Struggle for Indochina.* Stanford, Calif.: Stanford University Press, 1954. Hammer offers one of the better sources for a general account of the revolution in Vietnam, particularly for detailed background material.

Sacks, I. Milton. "Marxism in Vietnam," *Marxism in Southeast Asia.* Ed. Frank N. Trager. Stanford, Calif.: Stanford University Press, 1959. The development and growth of Marxism and how it affected the ideology of the Vietnamese revolution is well presented by Sacks.

THE INDONESIAN REBELLION: 1945–1949

SYNOPSIS

In August 1945 the Republic of Indonesia declared itself an independent entity, unilaterally severing its colonial relationship with the Netherlands Government. However, the Dutch returned with armed units to reclaim their prewar colony and launched two "police actions" against the republican army, which were countered with resistance by Indonesia. Independence was finally achieved, after many fruitless negotiations between the Indonesian nationalists and the Dutch, through the intervention of the United States and the United Nations. The Netherlands Government transferred its authority over the Indonesian islands to the republican government in December 1949.

BRIEF HISTORY OF EVENTS LEADING UP TO AND CULMINATING IN REVOLUTION

The Dutch merchants arrived in Indonesia in the 17th century to establish what was to be only an economic enterprise. Before long, the Dutch Government assumed control over a number of the Indonesian islands. Beginning in 1908, several indigenous socio-religious groups dedicated themselves to bettering the lives of the Indonesian people and exposing the social and economic weaknesses of the Dutch regime. A multi-party system developed after 1912 and political activity led to peasant uprisings and demonstrations. A major Communist-led uprising occurred in 1927. The Japanese invasion in 1942 united the nationalist movement into an anti-imperialist and anticapitalist struggle against both the Japanese and the Dutch.

During the war an anti-Japanese resistance movement developed which turned against the Dutch forces when they returned to claim their lost colony. However, in the weeks between the Japanese capitulation and the landing of British occupation forces, the Indonesians had established a free and autonomous government and, in August 1945, had declared their independence. The well-trained and well-armed Dutch forces that occupied the country a few weeks later proved superior to the Indonesian insurgents.

Armed conflict and negotiations under United Nations auspices lasted approximately 5 years. Several agreements were signed between the Indonesian Republican Government and the Netherlands which reflected a willingness to compromise on the part of the Indonesians but caused some disunity among the revolutionary groups. The Dutch had signed the agreements under international pressure and did not live up to all their stipulations.

Toward the end of 1949 the United States finally persuaded the Netherlands to accept the U.N. recommendations of January 1949. Sovereignty was transferred to the Republic of the United States of Indonesia.

THE ENVIRONMENT OF THE REVOLUTION

DESCRIPTION OF COUNTRY

Physical characteristics

The Indonesian archipelago is in the area of Southeast Asia, northwest of Australia, and south of Indochina and the Philippines. It is composed of approximately 3,000 islands, and extends 5,110 miles in an east-west direction, and 1,999 miles from north to south. The total land area is 1,482,395 square miles—more than twice the size of Alaska. Sumatra, Java, Borneo, and the Celebes (Sulawesi) are the largest islands.

The islands are masses of volcanic mountains and wide open plains. Since they lie near the equator, the climate is tropical, but ranges from warm to hot and from dry to humid. The rainy season generally lasts from November to January.

The people

Indonesia experienced waves of immigration from Siam, Burma, and Malaya which brought many types of peoples, speaking many languages. A new language called the *bahasa Indonesia*[1] was developed in 1933 and taught in schools in order to bridge the difficulty of communication between ethnic groups who spoke many tongues.

In 1940 there were 70 million people in Indonesia. By 1954 the number had increased to over 80 million. Java, about the size of Illinois, was the most densely populated of the islands, containing 75 percent of the total population, and averaging 1,400 persons per square mile. Although occupying only 7 percent of the total land area, Java contained 76 percent of the cultivated land. Sumatra, the largest island, had in 1954 a population of only 12 million. Djakarta, the capital city, located on the northwest tip of Java, had a population of three million persons in 1954. Prior to World War II, it had only one-half million people. As a result, many of these people live in huts built of

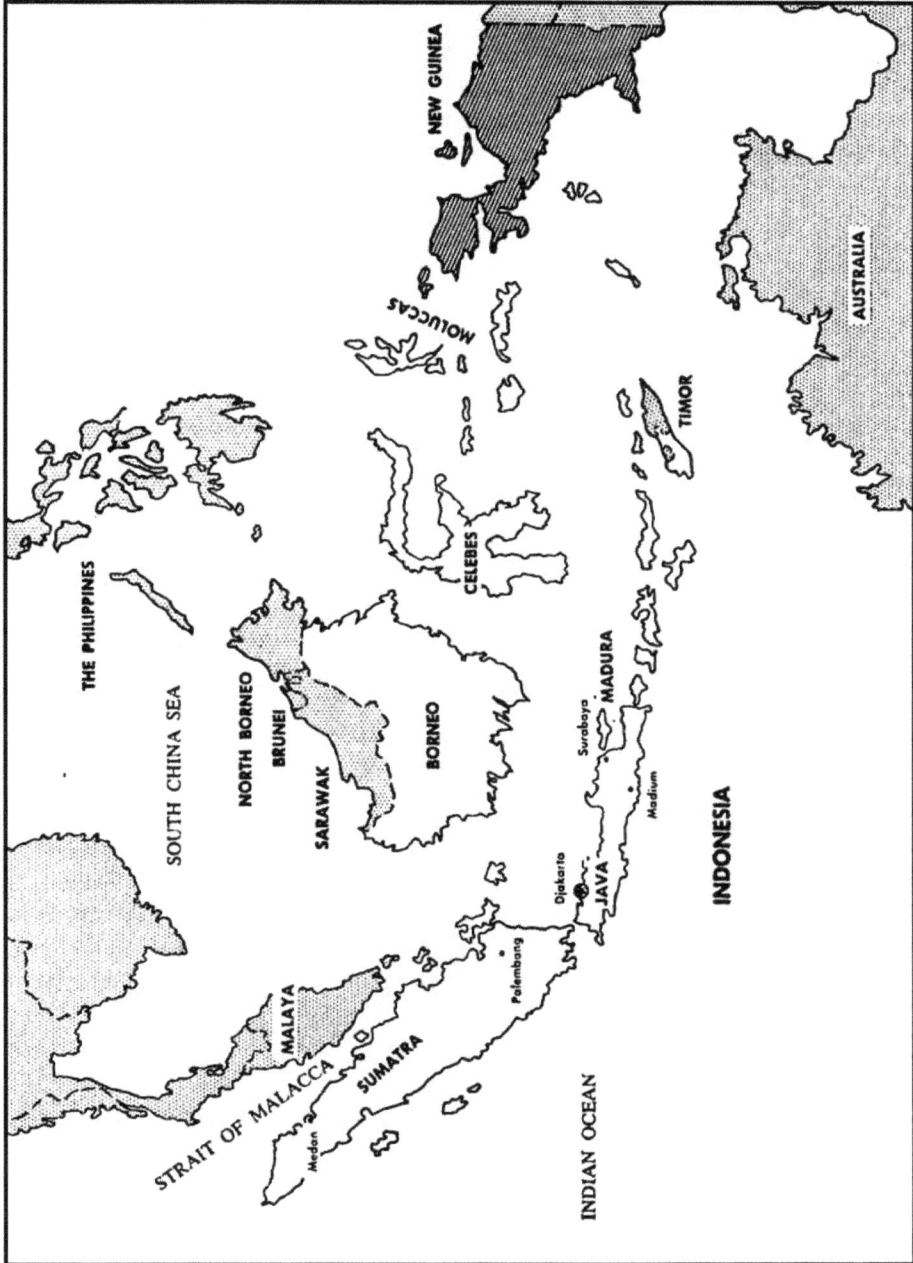

grass and mats. In 1954 there were three million Chinese, Arabs, and Indians living in Indonesia.[2]

Communications

The islanders are dependent upon sea transport when traveling from one island to another. Most of the islands have fair port facilities, and major ports are located on the larger islands.

Prior to World War II, there were 43,500 miles of road good enough for motorized traffic, 16,000 miles of it on the island of Java. Approximately 30 percent of these roads were asphalt. The Great Post Road, which extended the length of Java, was excellent, but most of the back roads offered poor driving conditions, particularly during the rainy season. Inland travel on all the islands in the postwar period was difficult. There were 4,000 miles of railroad track on Java after the war, but delivery of goods was still unreliable. Postal and telegraph services were available.

Natural resources

There are rich deposits of minerals in Indonesia, of which tin, bauxite, nickel, and coal are the most important. Large oil deposits also exist.

An important raw material is volcanic ash, valuable in the fertilization of the land.

SOCIO-ECONOMIC STRUCTURE

Economic system

Indonesia was part of the Netherlands colonial system, in which Europe furnished the capital and Indonesia furnished the labor for developing export crops. A hierarchical order developed which placed Western elements on top, and some of the native aristocracy and Chinese in the intermediary position of middlemen. The middlemen in turn administered, or delegated to village headmen and their supervisors the authority to administer the large estates or plantations formed by the villages. Following the pattern of communes or cooperatives, practically all of the industries were on large estates and were European-owned. The Dutch developed a system by which a certain fixed percentage of locally produced commodities were turned over to them at fixed prices.

Most of the peasants had to work a specified number of days during the year for an indigenous or Chinese landlord, who gave them little or no compensation. The introduction of agrarian laws in 1879 did help the peasants somewhat. Some of the small holders, however, became very proficient in their profession and increased their production to such an extent that by 1948 their exports equaled those of the estates, whereas in 1935, they were only one-third those of the estates.

Rice was one of the major products; others included cinchona, cocoa, rubber, tobacco, sugar, palm oil, coffee, and tea. In dollar trade, rubber was the chief export followed by copra and, lately, tobacco and tea. Major Indonesian imports are textiles, raw cotton, and iron.

Class structure

The indigenous population had no middle class of its own to speak of. Seventy-five percent of the working population were peasants in 1939. The Chinese, who had settled early in Indonesia, became the middlemen between the Dutch regime and the local peasants. They managed the production of entire villages and were able to extract large sums from the peasants in heavy bazaar fees, road tolls and customs, and the sale of salt.[3] Members of the indigenous aristocracy were at times able to secure profitable positions, similar to those held by the Chinese, from the Dutch Government.[4] Agrarian laws had existed since 1875, but it was nevertheless difficult for the peasant to better his status. The increased production of sugar cane after World War I reemphasized the need for communal land ownership, and this, in turn, tended to retard the growth of a strong and prosperous peasantry. However, on Java the system of communes had never worked too well and by 1932, 83 percent of the land was owned by the peasants. Nevertheless, the peasants continued to work the land in a cooperative manner.

The communal system has played an important role in shaping Indonesian society. The village community tolerated no economic difference. It acted as a "leveler," regarding the individual as an integral part of the whole. This factor worked largely against the development of agrarian-based Indonesian capitalism.

Literacy and education

At the outbreak of World War II, only 7 percent of all Indonesians were literate.[5] In 1941, the last year of Dutch domination, one out of 10 children attended school, but their attendance was seldom long enough to allow them to become literate. Most of the prewar schools were run by Christian missionaries.

The reasons for such a high percentage of illiteracy vary. Next in importance to the lack of teachers was the fact that Indonesia had no national language under Dutch rule until *bahasa Indonesia* was developed in the 1930's; therefore, the complexity of the various local dialects made it necessary to teach in the Dutch language. Another reason for illiteracy was the lack of schools for Indonesians. In 1940 secondary schools were still largely reserved for Europeans. In the same year, only 637 Indonesians attended college, but jobs were difficult to find even for this small group. Some Western-educated Indonesians found employment in the civil service, but most educated Indonesians had difficulty attaining positions commensurate with their educational levels. This situation, too, inhibited the development of an Indonesian middle class.[6]

Major religions and religious institutions

Ninety percent of the Indonesians are nominally Muslim. The average Indonesian held very strong Islamic convictions, and most official acts by Indonesian authorities reflected the Islamic faith. Even though the population is overwhelmingly Muslim, Hindu, and Buddhist ideas have modified traditional Islamic values. Approximately 4 percent of the population were Christian.[7]

GOVERNMENT AND THE RULING ELITE

Description of form of government

The Dutch Governor General had the authority to administer the Netherlands East Indies in the name of the Crown with the aid of a General Secretariat. A general advisory body, chosen by the Crown largely from former civil servants, formed the Council of the Indies, but the concurrence of the Council was rarely required.

The People's Council (*Volksraad*) was established in 1918 by the Dutch Government. It was a representative body of 60 members, some elected and some appointed, half of them Indonesians. The *Volksraad* was a legislative body, with very limited functions initially. In 1925 its powers were expanded, but the decisions made by the Council were always subject to the veto of the Governor General. The Indonesians maintained their own ancient democratic form of government in the village, though it was subject to Dutch control. Until the transfer of sovereignty at the end of 1949, the Dutch-controlled central government at Batavia remained subject to directives from The Hague.[8]

Locally, village headmen, acting as agents of the government, served as rent collectors and performed certain police functions. Village councils dealt only with the local problems of administration. Over 90 percent of the high administrative posts were held by Europeans, while over 90 percent of the lower administrative levels were filled with Indonesians. During the Japanese occupation the country was under military rule; however, considerably more administrative autonomy was granted the Indonesians than had been allowed under the Dutch.

Description of political process

Indonesians were permitted only token participation in the administration of their country through the People's Council. The Dutch discouraged political organizations and, in some instances, suppressed them if they represented a threat to the security of the Dutch colonial administration. Nevertheless, the people of Indonesia developed a pronounced political consciousness and political organizations prospered. This process was aided by the emergence of a small intellectual class, by the development of a vernacular press and radio, and by an increase in geographic mobility. Nationalism developed despite language and regional differences and the fact that the Indonesian aristocracy tended to side with the Dutch for personal gain. All political groups that developed in Indonesia combined ideas of religion, nationalism, and Marxism.[9] The main differences in the ideologies of the groups lay in the relative emphasis given any of these three categories.

The first mass political party, the Association of Muslim Traders (*Sarekat Islam*), was established in 1912. It combined religious and political aspects by devoting itself to the promotion of religious and national unity and the elimination of foreign control. Branches of the association led to the formation of other parties.

In 1927, Sukarno organized the Indonesian Nationalist Party (PNI), which became a purely nationalist mass party. It promoted cooperatives, labor unions, and education. A Communist Party of the Indies (PKI) was formed in 1920 and was the first group to advocate complete independence for Indonesia. The PKI staged an uprising in 1927 following a series of strikes; but it was successfully put down. Because of its antireligious character, the party found it difficult to enlist mass support. In 1935, Marxist study groups joined together to form the Greater Indonesian Party (PIR). Not being doctrinaire Marxists, the leaders of this party were nationalists first and socialists second. Also important to the nationalist movement in Indonesia was the Indonesian Union (PI). This organization was formed in Holland in 1922 by Indonesian students and it became a training ground for many of the top nationalist leaders.

During the Japanese occupation all political parties disbanded. The political life of the Indonesians did not, however, come to an abrupt end. On the contrary, the Japanese saw nationalism as a "real and powerful force,"[10] and realized that a *modus vivendi* had to be reached with the leaders. Indonesian political activity continued at two levels. At the first level it was legal and there was collaboration with the Japanese. At the second level, however, political activity was

61

underground and promoted anti-Japanese resistance. Both groups worked for the advancement of nationalism and independence.[11]

Legal political activity in Indonesia during the Japanese occupation was directed by the Center of People's Power (*Poetera*), formed in 1943 with the encouragement of Premier Tojo. The *Poetera*, whose authority was limited by the Japanese to the islands of Java and Madura, was primarily a means of rallying Indonesian support behind the Japanese war effort. The *Poetera* promised the Indonesians eventual self-government. The military arm of the *Poetera* was the "Volunteer Army of Defenders of the Fatherland" (*Peta*), described later under the section "Organization of Revolutionary Effort."

There were four outstanding political developments during the Japanese occupation. The first was an increase in national consciousness and in the desire for political independence. The second was the development and the spreading of an Indonesian language, which became a national symbol and, to some extent, offset the parochial tendencies existing prior to the war. The third was an increase in the general self-confidence of the Indonesians concerning their ability to govern. Many of the intellectuals were able to hold administrative positions under the Japanese that had not been available to them under the Dutch. The fourth was the advancement of those holding positions under the Dutch to better positions under the Japanese.[12]

Legal procedure for amending constitution or changing government institutions

The Dutch administered Indonesia through the Governor General, who was responsible to the Minister of Colonies of the Netherlands. Legal changes in government institutions had to be initiated by Dutch authorities.

Relationship to foreign powers

Under Dutch rule all foreign affairs relating to Indonesia were conducted by Dutch officials. During World War II, *Poetera* had contacts with the occupation authorities, but its policy-making function was limited. The British forces, stationed in Indonesia following the Japanese capitulation, prepared for the return of the Netherlands authority particularly in the Bandung and Surabaya areas.

The role of military and police powers

The Dutch Government or its agents possessed the sole power of coercion in Indonesia before World War II. The army and the police units were largely officered by Dutch and Eurasians, while the rank and file were drawn primarily from Christian Indonesians. Both the army and police units were used to suppress nationalism and as a whole they were loyal to the Dutch. However, in February 1946 the Indonesian soldiers of the KNIL (Royal Netherlands Indies Army) rebelled against the Dutch. In most demonstrations and uprisings, before World War II, the police were strong enough to restore order, but at times army units had to be employed as well.

The Governor General was vested with broad police powers—to arrest people without court authorization, to keep them under arrest, and to exile all Indonesians engaged in activities which were not "in the interests of peace and order."[13]

WEAKNESSES OF THE SOCIO-ECONOMIC-POLITICAL STRUCTURE OF THE PREREVOLUTIONARY REGIME

History of revolutions or governmental instabilities

The earlier forms of peasant unrest were probably spontaneous protests against prevailing conditions and were not nationalistic in nature. Nationalist leaders later channeled peasant dissatisfactions in a more nationalistic direction. As early as 1870 the peasants in Java expressed a strong desire to be left unmolested by government authorities. Later, passive resistance led to serious disturbances, forcing the government to use armed force. The leaders of such disturbances were generally sent into exile.[14]

In 1925 Communist organizations planned a revolution that was to be precipitated by a strike of the railway workers and was later to be developed into a general strike. Disorder broke out prematurely in Batavia rather than in Sumatra as planned. The government reacted by outlawing the right of assembly and by combining army and police units in severe repressive measures. Most of the Communist and labor leaders were arrested; as a result, the revolution deteriorated into an uncoordinated series of actions. The failure of the rebellion of 1926–27 was followed by intraparty disputes and disrupted the Communist movement.[15]

Economic weaknesses

The Dutch Government controlled the whole of Indonesian economic life and possessed the greater part of the capital. But more than this, it offered little opportunity for educated Indonesians to take an active role in directing their own economy. Moreover, even if they were given an opportunity to hold jobs previously reserved for Europeans, the Indonesians ordinarily received a much lower rate of pay.

The peasant was handicapped by harsh fiscal policies, and his obligation to perform a certain amount of work for the large plantations or estates earned him little or no compensation. Economic inequalities, high taxes, and the duty to support and pay for civic functions and public works all encouraged growth of the nationalist movement among the peasants.[16]

Social tensions

The social tensions increased with the failure of the system established by the regime to absorb the indigenous elite created by Western education. The social discriminatory practices employed by the Europeans placed the Indonesians in an inferior position in their own society. This attitude of superiority on the part of the Dutch continued to strengthen nationalism until the end of their rule. The judicial administration and the penal legislation also discriminated against the Indonesians. Two sets of courts were provided, one for Europeans and one for Indonesians. The latter were not set up according to Western standards. The Indonesian courts relied on their personnel to perform both administrative and legal functions. They had the arbitrary power to hold Indonesians under preventive detention.

Discriminatory practices also existed in the field of education. Indonesian students were charged high tuition fees, and higher education was made difficult because the government feared that too much education was politically dangerous. Thus, the Dutch inhibited the growth of an educated indigenous elite.[17] European-dominated and -oriented professional associations resisted the establishment of professional schools for the indigenous population. Despite the many difficulties that Indonesians encountered in receiving a Western education, a few leaders still benefited by their limited opportunities. They became acquainted with Western political concepts and turned this knowledge to good use in directing the nationalist movement.[18]

Government recognition of and reaction to weaknesses

One of the early voices which translated parochial Indonesian grievances into integrated nationalism was that of Islam. The government became aware of the great threat that Pan-Islamism represented, and introduced the Western system of education even at the elementary level to counter this threat.

Two forces helped the development of nationalism in the 1920's: (1) the modern political and social ideas then known to a few Indonesian intellectuals; and (2) the fact that in the period following World War I the Dutch Government was faced with difficult internal problems. As a result it sought to pacify its Indonesian colony by granting some concessions to Indonesian demands. The People's Council was created in 1918 and its powers were expanded in 1925. However, these reforms only increased the appetite of the nationalists, who now embarked on an even more extreme nationalistic program. To protect their position, the Dutch felt it necessary to ban nationalist literature, exile many of the nationalist leaders, and revert to the use of armed force in repressing demonstrations and uprisings. The Communist rebellion of 1926–27 resulted in the arrest of 13,000, the imprisonment of 4,500, and the internment of an additional 1,300.[19] The Communists were never again able to make a strong comeback against Dutch rule.

FORM AND CHARACTERISTICS OF REVOLUTION

ACTORS IN THE REVOLUTION

The revolutionary leadership

Most of the leaders of the various revolutionary groups were Islamic leftwingers who, during the struggle for independence, expressed their opposition to any form of autocratic rule. They revised Marxist doctrines to adapt socialism to Indonesian conditions. They believed that in time Indonesia could become a socialist state, but that it first needed to develop a capitalist class. They believed that Indonesia was capable of becoming a democracy.

Achmed Sukarno was probably the most important leader of the radical nationalist movement. He was born in 1902, studied engineering at the University at Bandung, and received his doctorate from that institution. In 1927 he founded the Indonesian Nationalist Party (PNI). He was arrested in 1929 as a result of his political activities and remained imprisoned until 1932. On his release he immediately resumed his work for independence. Arrested a second time in 1933, he was sent into exile to some remote region of Indonesia. In 1942 he was released by the Japanese occupation forces. Sukarno cooperated with the Japanese occupation forces while continuing his work for independence. In 1945 he became the first President of the Republic, and directed the fight for independence until the creation of the United States of Indonesia.

Sutan Sjahrir was born in 1909 and received his secondary education in Bandung. He then studied law in Leiden, Holland, where he was active in student groups. He returned in 1932 and assumed coleadership of a group working for independence. He was arrested in 1934 and also exiled within Indonesia. In 1942, he contacted underground leaders and helped to organize a resistance movement. At first he did not support Sukarno's declaration of independence but later he became aware of the mass support behind this declaration and joined the leadership of the government.

Mohammed Hatta was born in Sumatra in 1902, the son of a prominent Muslim teacher. He attended school in Padang, where he was active in youth groups. In 1921 he entered the Commercial College of Rotterdam, and traveled in Europe. There he was arrested by Dutch officials in 1928, charged with inciting unrest, and released. He returned to Indonesia in 1932. He was arrested and exiled along with Sjahrir in 1934 for organizing opposition to Dutch rule. Hatta returned in 1942 to help Sukarno head the legal arm of the fight for independence which collaborated with the Japanese.

The revolutionary following

The popular response to the Indonesian declaration of independence in 1945 "was tremendous."[20] It was particularly strong among the members of the youth organizations that had been established by the Japanese in 1943. In all areas of Indonesia where fighting broke out, large sections of the masses joined with the army. The older generation of professional people and a majority of the civil servants, plus a sprinkling of the Indonesian commercial and industrial middle class, helped to organize the new Indonesian National Party, which also amassed a large peasant following. Some of the older prewar party leaders organized the *Masjumi*, and within one year it grew into the largest political party supported by the large nonpolitical Muslim social organizations. The irregular armed organizations, the old party leaders and civil servants, and the larger sector of the peasant class all supported Indonesian independence.

ORGANIZATION OF REVOLUTIONARY EFFORT

Internal organization

The organization that later directed the revolutionary effort against the Netherlands Government was first developed under the auspices of Japanese occupation forces and was composed of two branches: a political organization and a military organization.

An "Independence Preparatory Committee" was established in August 1945 to declare the independence of Indonesia and establish its first government. The Committee elected Sukarno President of Indonesia, and Hatta, Vice President; a Cabinet was formed. Within two weeks of its first meeting, the Preparatory Committee was dissolved by Sukarno and in its place a "Central Indonesian National Committee" (KNIP) was organized. Initially the KNIP was composed of 135 outstanding nationalist leaders; later its membership fluctuated and at one point reached several hundred. The functions of the KNIP were controlled by a "Working Committee" headed by Sjahrir. At first the KNIP was primarily an advisory body, but by November it enjoyed full legislative powers and the cabinet became responsible to this body. Differences between the President, the Cabinet, and the Working Committee were supposed to be resolved by the KNIP as a whole.[21] Sukarno remained President throughout and controlled the reins of government even during periods of dissension among the nationalist leaders.

The Preparatory Committee had divided Indonesia into eight provinces, with an appointed governor and an administrative body for each. Shortly after the KNIP was created, it appointed one member from each province to establish provincial KNI's (Indonesian National Committees), which in turn were to assist the governors in their administration. In addition to these, many local and revolutionary committees had sprung up spontaneously at the instigation of local leaders. By the end of 1945 these committees had been brought under loose control of the government.[22] Thus the pattern was established: the country and the revolution were supervised by a network of committees from the local KNI's up to the national KNIP, in its turn controlled by the Working Committee.

The backbone of Indonesia's republican armed forces was a volunteer force, called the "Volunteer Army of Defenders of the Fatherland" (*Peta*). It had been organized under the Center of People's Power (*Poetera*) in September 1943 and consisted of approximately 125,000 Japanese-trained and -armed troops.[23] The regular army (TNI),[a] assisted by independent guerrilla bands under the command of local leaders, resisted the Allied Forces in 1945.

The basic revolutionary force, however, was the People's Auxiliary Organization (TNI *Masjarakat*), established in 1947 to supplement the efforts of the regular army on a local basis. As autonomous units of the army, these small bands were controlled and supplied by the local KNI's. They helped the local administrators to organize and train the communities to which they were assigned in setting up defenses against Dutch attacks. At night they executed small raids

---

[a] The republican army had its official title changed at various times and finally became known as the Indonesian National Army (TNI).

against Dutch communication lines, and were also involved in clashes with small Dutch patrols.

The Indonesian Air Force (AURI) was first organized as an arm of the army. Air equipment consisted of outdated Japanese fighter, bomber, reconnaissance, and trainer planes captured after the Japanese capitulated. In 1946 the air detachment was separated from the army. Shortages of pilots, mechanics, equipment, and maintenance facilities forced AURI to play a minor role in the revolution.

The Indonesian Navy was confined in its sea activities. A few launches and torpedo boats were used to blockade Dutch shipping and to supply guerrilla units with guns and supplies. The navy was greatly reduced during the revolution, and most of its members were transferred to the army. As a land force, the navy established numerous arms and munitions caches in mountainous areas in anticipation of greater Dutch military action.

In 1947 the coalition of the three major parties forming the republican government broke down, greatly weakening the central government. The rejuvenated Communist Party attempted to take advantage of this disunity and overthrow the republican government. In September 1948 the Communists gained control of the city of Madiun and its surrounding area in Java. A struggle for power within the party and its inability to make a well-coordinated bid for power resulted in the failure of the insurrection. The plan for a popular uprising in all sections of the Republic was not carried out. The republican army reoccupied Madiun after a few days, forcing the Communists to take refuge in the hills, where they formed guerrilla groups which were later rounded up by the republican army.[24]

External organization

Organization of revolutionary groups outside of country

Indonesian student groups in Holland were very active prior to World War II in agitating for nationalist goals. Aside from the connection with student groups, there does not appear to have been any organizational connection with other countries.

Support of revolutionary cause by foreign powers

The German invasion of the Netherlands in 1940 severed the connection with the Dutch colony in the East Indies. The Japanese attacked Sumatra on February 14, 1942, and after giving battle for a few days the Dutch Commander in Chief of the Allied Forces surrendered. The Japanese gave some support to the Indonesian nationalist cause and Indonesians were permitted to take over higher administrative posts vacated by the Dutch. By the time of the Japanese surrender in August 1945, Indonesia had a provisional government, a national army, district administrations, a national flag, and a national anthem.

However, it was not until 1949 that the Indonesians received strong support from a foreign power, this time the United States. The United States had been helping the Dutch cause financially, and had remained out of the Dutch-Indonesian dispute except in its official capacity in the United Nations and in offering its good offices. In 1949 the United States applied diplomatic pressure on the Netherlands Government to relinquish its hold on Indonesia and even froze Marshall aid funds. In 1947 Dutch military actions had provoked India, Australia, and some other countries to censure the Netherlands Government. In the summer of that year both India and Australia brought the matter before the United Nations.[b] It was under the auspices of the United Nations that Indonesia finally received her independence.

GOALS OF THE REVOLUTION

Concrete political aims of revolutionary leaders

The concrete political aims of the revolutionary leaders varied somewhat; but an independent and unified Republic of Indonesia based on Islamic fundamentals and socio-democratic principles was generally accepted as the ultimate goal.

Social and economic goals of leadership and following

The nationalist leaders also wanted to eliminate foreign control over the social and economic aspects of Indonesian life. The nationalists felt that the indigenous population had been given only a small share in its own resources. A number of Muslim nationalist leaders wanted religious as well as national unity, and aimed at establishing an Islamic state. The Indonesians wanted to better their own conditions through an indigenous-controlled economy and improved educational opportunities.

---

[b] Nehru stated that "no European country has any business to set its army in Asia against the people of Asia."[25]

REVOLUTIONARY TECHNIQUES AND GOVERNMENT COUNTERMEASURES

Methods for weakening existing authority and countermeasures by government

When Allied Forces—first the British, later the Dutch—reoccupied Indonesia in the fall of 1945, the new Republic of Indonesia found itself in a weak position. The Dutch were not prepared to relinquish control of their wealthiest colony, and they had powerful and well-equipped military forces at their disposal. Unable to challenge the Dutch in frontal combat, the Indonesians were compelled to rely chiefly on diplomatic skill, resorting to hit-and-run guerrilla tactics when the Dutch undertook what they called "police actions."

The Indonesians, however, had a powerful ally in world opinion. The 4 years of conflict that ensued before they won recognition of their independence witnessed a succession of negotiations, each entered into reluctantly by the Dutch under pressure from outside; a succession of agreements, each of which represented further concessions to Indonesian self-determination; a succession of breaches of these agreements leading to renewed violence—the "police actions" mentioned above; and a renewal of international pressure for fresh negotiations.

The British learned the strength of the independence movement as soon as they landed. Their troops became involved in house-to-house combat with nationalist troops in the port city of Surabaya, and both sides incurred losses before the British gained the upper hand. Realizing that an attempt to reimpose Dutch authority would precipitate war, the British prevailed upon Netherlands officials and nationalist leaders to enter into negotiations concerning the future status of Indonesia. As a result of these negotiations, the republican government was given limited authority over the islands of Java, Sumatra, and Madura. The Dutch also made a number of important commitments. They agreed to establish a representative Indonesian Parliament composed largely of Indonesians, to abolish racial discrimination, to recognize the new Indonesian language as coofficial with the Dutch, and to expand the school system.

Though these concessions were considerably short of self-government, the Indonesians were constrained to accept them. They remained unsatisfied, however, and sporadic disturbances continued. Members of Indonesian youth organizations engaged in terroristic activities which interfered with the restoration of the Dutch economic system. The disorders became serious enough to be placed before the United Nations, which designated a Good Offices Commission.

It instituted negotiations which led to the Linggajati Agreement of 1947. This established the United States of Indonesia, consisting of the Republic of Indonesia, whose authority over Java and Sumatra was recognized by the Dutch, and a number of other Indonesian states under Dutch authority.

The agreement also provided that differences between the Dutch and the Indonesian state should be submitted to arbitration. Ignoring this provision, the Dutch Commissioner General sent the republican government an ultimatum, with which it reluctantly complied. Further demands followed, and the Indonesians rejected them. The Netherlands Government then launched its first "police action," an all-out attack against the Republic, on July 20, 1947.

The Dutch forces consisted of 109,000 well-seasoned troops, supported by air and sea units. The attack was launched from seven main bridgeheads.[26] Dutch propaganda organs justified the action as one designed to establish peaceful conditions and liberalize the Indonesian Government. The republican army entered the battle with the 200,000 troops, 150,000 rifles, small arms, machine guns, homemade land mines, and hand grenades, and 40 ex-Japanese planes with half as many pilots to man them.[27]

Again the United Nations intervened, and in October negotiations between the Indonesians and the Dutch were opened. These resulted in the Renville Agreement, signed in January 1948, which provided for the transfer of sovereignty over Indonesia from the Netherlands to the United States of Indonesia after a stated interval.

There was a period of relative peace. But the Dutch continued to plague the republican government with political and economic pressures. In December 1948 the Dutch issued another ultimatum, and followed it up with an air attack on the Djogga Airport, in direct violation of the agreement. Many of the nationalist leaders were arrested.

The fighting went on for several months, while the United Nations tried in vain to intercede. Finally the United States was able to persuade the Netherlands Government to reopen discussions with the Nationalist leaders. The Roem-Van Royen Agreement of May 7, 1949—the third in the series—ordered the restoration of peace and provided for a conference to take place in The Hague. This conference, held under the auspices of the United Nations, granted the United States of Indonesia complete independence.

During the various phases of the struggle, the Dutch employed various combinations of military and civil authorities in the areas which they controlled. "Reviewing the structure and the actions of the [Dutch] central Government in the field of administration throughout the archipelago in the years from 1945 to 1949 makes it evident that no long-term plan was worked out. The period was characterized rather by a series of fluctuating experiments. Evidence seems to indicate that a process of trial and error was continuously employed."[28]

During the period of British occupation in 1945–46, the Dutch administration consisted of a military government of commanders who were recruited from the civil service of the central government. These officers for civil affairs were officially part of the Allied military forces. The administrative hierarchy was the same as the one that existed during the interwar period, but its source of power now was derived from the authority of the Allied Supreme Command. A state of siege declared by the Dutch authorities in 1940 was still in effect in Indonesia during this period.

During the Allied withdrawal the state of siege was lifted in some areas. The administration was transferred gradually to civil administrators. Decrees issued by the military administration during the state of siege, however, remained in force. In some areas where the state of siege was lifted and a civil administration was established, a military government was reimposed when the threat of Nationalist interference was felt. In this way the Dutch were able to end or at least limit the activity of Republic supporters and their agents in those areas. When the danger passed, the military government was again replaced by the civilians. For example, a state of siege was declared in the southwest area of Celebes in December 1946 when serious disturbances instigated by nationalists could not be put down by regular police action. A military administration took command but did not disturb the normal acts of the civil administration unless the military command judged it necessary. When the situation returned to what the military command interpreted as normal in January 1948, the state of siege was lifted.

During the second police action, a number of strategic areas were placed under Dutch military control, with a stipulation that the civil administration be left undisturbed unless the military command felt it necessary to intervene and impose greater restrictions upon the Indonesians. On the islands of Java and Sumatra, the administration established after the first police action differed completely from those in all the other Dutch-administered areas in Indonesia. This administration was replaced with an entirely new one after the second police action brought new areas of Java and Sumatra under Dutch control. After the Roem-Van Royen Agreement, calling for a cessation of hostilities, was signed, the type of Dutch administration that held authority over most areas was not determined in advance but was worked out in practice.

Republican military forces did not engage in frontal combat except in very rare instances. At the beginning of the first police action, the republican units withdrew from the flat country, suitable for mechanized warfare, and retreated to the mountains and hills to conduct guerrilla warfare. Some of these mountain areas held large towns and populations where the guerrilla units were able to find supplies and refuge.

The guerrilla units conducted small hit-and-run raids and ambushes against Dutch communications. During the period of relative truce between January and December 1948, most of the activity consisted of clashes between small Dutch and Indonesian patrols. The republican guerrillas preferred night fighting. Heavy roads utilized as supply routes by Dutch forces during the day were plagued continuously by Indonesian guerrillas at night.

During the Dutch attacks in 1947 and 1948, the Indonesians attempted to transport accumulated export crops, farm machinery, and other equipment from the attacked areas to safer areas in the hills and mountains. All that could not be moved was destroyed by the army. However, the Indonesians, in their scorched earth tactics, destroyed buildings and equipment rather than export crops, with the result that the Dutch were able to capture large quantities of export produce.

Methods for gaining support and countermeasures taken by government

The Indonesian Republic created the Ministry of Information, which became an official department in the first republican cabinet in 1945. News of republican government activity was passed down to the people from this department through the radio and other news media. During World War II the Japanese had built an extensive radio network that reached most of the Indonesian islands. The original purpose of the network was to gain mass support for the Japanese occupation forces. However, speeches made by nationalist leaders during the war were predominantly nationalistic and attracted Indonesians to the nationalist cause. In the second police action, the Dutch destroyed the radio station in the republican city of Djogjakarta to prevent Sukarno and Hatta from speaking to the masses.

The Indonesian revolution succeeded almost entirely because of the pressure applied by the United Nations and the United States, and not because of the actions of the revolutionary organization or its Armed Forces. When the Dutch attacked the Indonesians in the first police action in 1947, Prime Minister Nehru of India spoke strongly against Dutch efforts to reestablish control over an Asian country. His

government and the government of Australia supported the Indonesian cause, and under their auspices an Indonesian delegation headed
by Prime Minister Sjahrir was sent to the United Nations. Sjahrir
spoke before the Assembly and strengthened the case of Indonesia
before the United Nations. The United States offered its good offices,
but the offer was refused by the Republic, which preferred to use the
United Nations organization instead. In April 1949 the United States
persuaded the Netherlands Government to concede to Indonesian
demands and United Nations proposals.[c]

MANNER IN WHICH CONTROL OF GOVERNMENT WAS TRANSFERRED TO REVOLUTIONARIES

The United Nations guaranteed the transfer of sovereignty by the
Netherlands to Indonesia. In May 1949 the United Nations Commission for Indonesia proposed that the parties reconvene at The Hague
to draw up plans for the transfer. The Working Committee of the
republican Parliament reconvened in July and a majority agreed to
support the U.N. Commission proposal on the condition that Dutch
troops be immediately withdrawn from the positions they occupied.
Soon the Dutch troops began to withdraw from some areas. From July
to August the Republic-held inter-Indonesian conferences between
delegations representing the Indonesian Republic and the various
federal units established by the Dutch. From August 23 to November
2 the Republic met with the Netherlands officials at The Hague and a
December 30 deadline was established for the transfer of sovereignty
to the Republic of the United States of Indonesia.

EFFECTS OF THE REVOLUTION

CHANGES IN THE PERSONNEL AND INSTITUTIONS OF GOVERNMENT

The primary objective of the Indonesian revolution was achieved
with the transfer of sovereignty. The constitution that was drawn up
changed the structure of the United States of Indonesia that had been
formally established by the Linggajati Agreement. It provided for a federation of 19 states, of which the Republic of Indonesia was the largest.
Sukarno was elected president. A unitarian movement was immediately initiated by the Republic to unify all the states under one central

---

[c] For a more detailed account of the revolution, see George McTurnan Kahin, *Nationalism and Revolution in Indonesia* (Ithaca: Cornell University Press, 1955)

government. The movement was a direct reaction against the federal system established by the Dutch during the revolution and inherited by the Indonesians at The Hague Conference. On May 19, 1950, all units of the federation were combined in the Republic of Indonesia.

MAJOR POLICY CHANGES

The greatest task facing Indonesia since its independence has been to reorient its economy from the raw material export type to a better balanced one. The government has made great strides toward solving its illiteracy problem, but has been less successful in consolidating its control over the large area of the archipelago. The Republic is also faced with a Chinese minority problem which has been the subject of uneasy negotiations with the Peking Government.[29] Indonesia's foreign policy orientation has been one of neutrality in the cold war.

LONG RANGE SOCIAL AND ECONOMIC EFFECTS

Up to December 1957, the Indonesian leaders maintained the Dutch economic system. In 1950 the leaders had been conservative and hesitant to eliminate foreign economic influence. However, political pressure for more socialism resulted in the nationalization of Dutch enterprises between 1957 and 1959. Indigenous middle-class elements had hoped that these enterprises would be turned over to them, but in April 1959 an official statement made it clear that the government would not support the claims of private individuals to take over the Dutch firms. Since that time, economic planning has brought about a closer working relationship between private and state enterprises. A land reform in October 1960 distributed the land, formerly part of large estates, among the peasants.

A recently instituted 8-year plan concentrated on the basic needs of the Indonesian people. It was hoped that during the years 1961–62, Indonesia would become self-sufficient in food and textiles.

OTHER EFFECTS

Diplomatic relations were severed between the Indonesian and the Dutch Governments on August 17, 1960, as a result of the West Irian (West New Guinea) issue. The West Irian question was not solved at The Hague Conference, or at other conferences which followed. The Dutch have continued to occupy the area and have refused to leave, except under conditions that would prevent the area from becoming an integral part of Indonesia. A point has now been reached wherein a minor incident could set off a serious reaction. A similar situation exists with regard to the island of Timor, the western half of which has always been held by the Portuguese.

The republican government is also faced with the problem created by a schism between nationalist and Muslim political movements. Darul Islam, a theocratic organization eager to establish an Islamic state, has persistently engaged in terrorist acts against the government. The preponderance of Javanese in the legislative branch has caused some of the outer islands, the Moluccas particularly, to be dissatisfied. As a result, a separatist movement has developed in these islands.

NOTES

1. Marguerite Harmon Bro, *Indonesia: Land of Challenge* (New York: Harper and Brothers, 1954), p. 110.

2. Ibid., pp. 10–28.

3. Ibid., p. 102.

4. Philip W. Thayer (ed.), *Nationalism and Progress in Free Asia* (Baltimore: The Johns Hopkins Press, 1956), p. 175.

5. Bro, *Indonesia,* p. 47.

6. George McTurnan Kahin, *Nationalism and Revolution in Indonesia* (Ithaca: Cornell University Press, 1955), p. 102.

7. Frank N. Trager, *Marxism in Southeast Asia* (Stanford, California: Stanford University Press, 1959), p. 173. See also Bro, *Indonesia,* pp. 158–175.

8. A. Arthur Schiller, *The Formation of Federal Indonesia, 1945–1949* (The Hague: W. Van Hoeve, Ltd., 1955), pp. 30–31.

9. Thayer, *Nationalism and Progress,* p. 129.

10. Ibid.

11. Kahin, *Nationalism and Revolution,* p. 104.

12. Ibid., pp. 101–133.

13. Ibid., p. 61.

14. Ibid., p. 43.

15. Ibid., pp. 80–87.

16. Ibid., pp. 43–45.

17. Ibid., pp. 52–56.

18. Ibid.

19. Ibid., p. 86.

20. Ibid., p. 136.

21. Ibid., pp. 168–169.

22. Ibid., p. 140.
23. Ibid., p. 189.
24. Ibid., pp. 285–300.
25. Ibid., p. 215.
26. Bro, *Indonesia*, p. 65.
27. Ibid.
28. Schiller, *The Formation of Federal Indonesia*, p. 79.
29. Robert C. Bone, "Indonesia: Retrospect and Prospect," *Journal of International Affairs*, X, 1 (1956), 23–24.

RECOMMENDED READING

BOOKS:

Bro, Marguerite Harmon. *Indonesia: Land of Challenge.* New York: Harper and Brothers, 1954. Mrs. Bro is the wife of a former U.S. cultural attaché in Indonesia, and the accounts of the area and the people are well-seasoned with the personal research of the author.

Kahin, George McTurnan. *Nationalism and Revolution in Indonesia.* Ithaca: Cornell University Press, 1955. The author's personal observations in Indonesia and his research have resulted in an excellent general account of the Indonesian revolution.

Robequain, Charles. *Malaya, Indonesia, Borneo, and the Philippines.* New York: Congmans, Green and Co., 1955. A comprehensive description of the natural layout of the region, its economic and cultural development, and the political trends.

Schiller, A. Arthur. *The Formation of Federal Indonesia, 1945–1949.* The Hague: W. Van Hoeve, Ltd., 1955. This is an analysis of the enactments, memoranda, and reports of the postwar years which led to the fashioning of the federal state. The analysis is structural and is of limited value.

Soedjatmoko. "The Role of Political Parties in Indonesia," in *Nationalism and Progress in Free Asia,* ed. Philip W. Thayer. Baltimore: Johns Hopkins Press, 1956, pp. 128–140. A short account of the development of political parties.

Trager, Frank N. "Indonesia," in *Marxism in Southeast Asia.* Stanford: Stanford University Press, 1959, pp. 171–239. This article deals with the development of the Communist Party in Indonesia and the part it played in the growing nationalism which led to the revolution.

OTHER:

Bone, Robert C. "Indonesia: Retrospect and Prospect," *Journal of International Affairs*, X, 1 (1956), 19–27.

Special Operations Research Office, *Area Handbook for Indonesia*, The American University: Washington, D.C., August, 1959.

THE REVOLUTION IN MALAYA: 1948–1957

SYNOPSIS

In the summer of 1948 the Communists increased terrorism in Malaya to such an extent that it constituted a rebellion against British rule and against the form of independence that the British had proposed. The British countered the Communist move with the "Emergency." Between 1948 and 1951, both the British Security Forces and the Malayan Communist Party (MCP) committed themselves to policies of guerrilla warfare in which the MCP scored fewer successes than did the British. Eventually the combination of British military superiority and the pacification measures instituted under the "Emergency" program turned back the threat of the MCP and drove their military forces which had not surrendered farther into the jungle. However, the "Emergency" did not end until 1960.

BRIEF HISTORY OF EVENTS LEADING UP TO AND CULMINATING IN REVOLUTION

Simultaneously with the attack on Pearl Harbor in 1941, the Japanese invaded the Malayan Peninsula. A wartime alliance between the British High Command and the Malayan Communist Party organization was formed in an effort to coordinate resistance against the Japanese occupation forces. The relationship was not without mutual distrust, however, and shortly after the war the British took steps to demobilize the Communist guerrilla units, which posed a threat to the British administration. The demobilization program was not successful.

After the war the British had devised a plan to create a Union of Malaya in order to strengthen British central authority over the area. This plan was announced as a *fait accompli* in the British Commons, but was not acceptable to the Malay elite and old Malayan civil servants. A federal system was instituted instead which discriminated heavily against the Chinese and the Indian communities. Violent riots broke out between the Malay and Chinese communities and the Communist Party used them to promote its own aim of overthrowing the British rule and establishing a People's Democracy. In 1948, the British instituted the "Emergency," which became an all-out drive against the Communists. The rebellion reached a peak in 1951 with the assassination of Sir Henry Gurney, the High Commissioner. He was replaced in January 1952 by Sir Gerald Templer, the first military man to be appointed to that post. Between 1952 and 1953 there was an intensive drive to eliminate the Communist guerrilla forces and stop terroristic activities. Through an artful combination of political, psychological, and military measures, this drive was quite successful. Terroristic activities were more and more reduced, but even at the end of the "Emergency" in July 1960, some guerrillas were still holding out. In August 1957 the situation was so well under control that the British acceded to Malayan demands for independence and admitted the Federation of Malaya into the Commonwealth as an independent member.

THE ENVIRONMENT OF THE REVOLUTION

DESCRIPTION OF COUNTRY

Physical characteristics

Malaya is on the southernmost tip of a peninsula which projects from Burma and Thailand into the South China Sea. Located in the general area of Southeast Asia, it has a common boundary with Thailand in the north and is flanked by the Straits of Malacca on the west, the Straits of Johore on the south, and the China Sea on the east. Approximately the size of New York State, its area covers 50,690 square miles which is four-fifths jungle and swamp. The north-south mountain range, which reaches its highest point at 7,184 foot Mt. Gunong Tahan, is broken into small valleys which contain alluvial plains between the mountains and the lowlands.

The climate is tropical, with warm and moist air masses. The yearly rainfall varies from 50 inches in the dry localities to 259 inches in the mountains. There is a northeast monsoon blowing across the China Sea in December and January, and from May to October there is a southwest monsoon.

The people

With the exception of such aboriginal peoples as the dwarf Negritos, the Sakis, and the Jakuns, the Malayans are of many foreign strains, including Chinese, Siamese, Hindus, and Arabs. The Malays migrated from Yannan between 2500 B.C. and 1500 B.C. and their culture reflects a combination of Hindu etiquette and Muslim teaching.[a] According to the 1947 census there were 2,398,186 Malays, who, with the aborigines, made up 49.47 percent of the total population. The

---

[a] The term "Malayan" is used to designate all peoples living on the Malayan peninsula, including the Chinese and other minorities. The term "Malay" is used only to refer to the ethnic group of Malayan-speaking inhabitants.

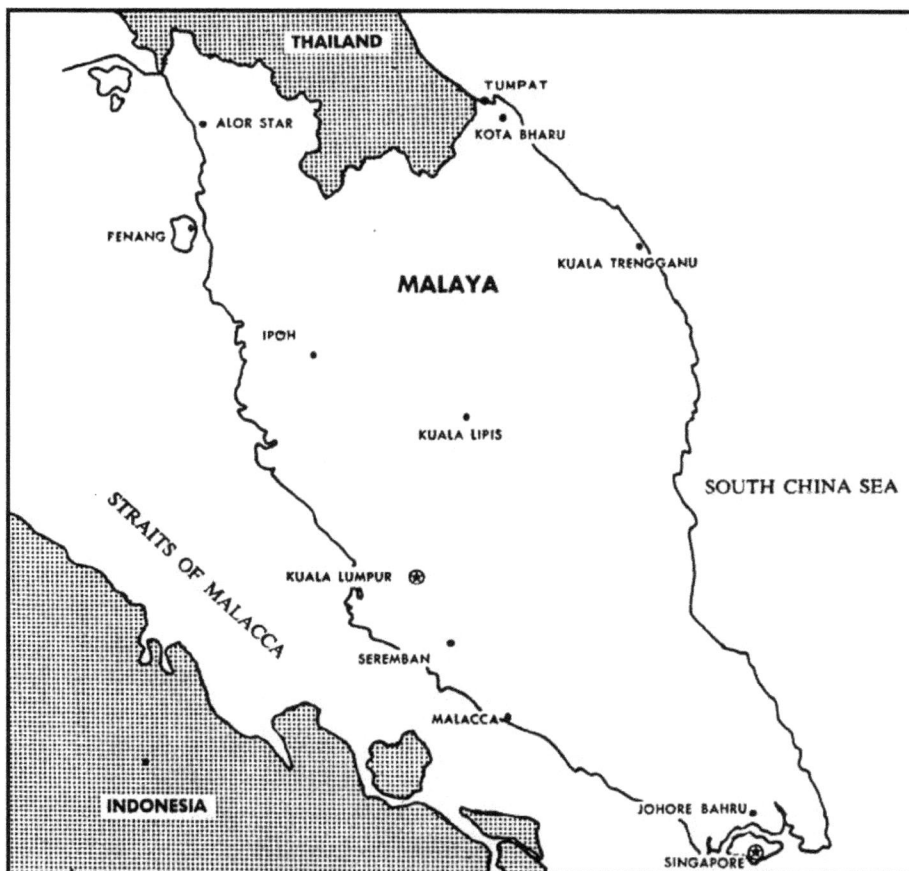

Chinese, excluding those in Singapore, numbered 1,884,534. They accounted for 38.6 percent of the population. The Indians were the third largest group, and the rest of the population was divided among the Ceylonese, Siamese, Arabs, and Jews. Indians and Ceylonese numbered 600,000.[1] In 1957 the population of Malaya was 6,278,763, an increase of 25 percent over the 1947 figure. The Malays still outnumbered the Chinese. If and when Singapore joins the Malayan Federation, the Chinese will become the most numerous ethnic group.

The figures show that there were approximately 100 inhabitants per square mile in Malaya, but the population was concentrated in highly developed economic areas. On the narrow band along the west coast, 72 percent of the population was settled in urbanized areas. In Kuala Lumpur, the capital of Malaya, and, at that time the largest city, the population increased from 176,000 in 1947 to 475,000 in 1959. Georgetown, on the Island of Penang, increased from 189,000 to 300,000 within the same period. Singapore, Malacca, and Alor Star are the other major cities on the west coast, and Tumpat and Kota

Bharu are the principal cities on the east coast. The Chinese form the bulk of the urban population.[2]

Under British rule, English was the official language in Malaya; but the Malay language, with its varied dialects, and Chinese were the most prevalent. Some Arabic and Tamil was also spoken.

Communications

The north-south direction of transportation movement is directly influenced by the north-south mountain ranges. The best-developed land transportation network is in western Malaya. There were over 6,000 miles of roads in the Federation in 1951, including 4,000 miles of paved surface roads. The most important road was Route 1, which ran from Johore Bahru to Alor Star and into Thailand on the west coast. The east coast had virtually no modern roads or railroads, but the east and west coasts were connected by two main roads. Inland transportation to the sea on the west coast was good. There were no secondary or local systems of improved roads.[3]

The postwar Malayan Railway system, one of the best in Southeast Asia,[4] covered over 1,000 miles from Singapore to Bangkok in Thailand. The major international airports were in Kuala Lumpur and Penang. There were also airdromes at Ipoh, Taiping, Alor Star, Kota Bahru, Kelantan, and Malacca, and numerous smaller landing strips throughout the country. Singapore handled 70 percent of the shipping in Malaya, having 41 lines connecting with every port of the world. Other seaports on the west coast included Penang and Port Swettenham. East coast port facilities were negligible.

Natural resources

There were four basic natural resources in postwar Malaya: the arable land; the mineral deposits, which included tin, coal, iron, and bauxite; the forests, including rubber plantations, and the fisheries. Tin, the chief nonagricultural resource, was mined in the central ranges of western Malaya. Limited visible reserves, depletion, and the hazards involved in prospecting because of the Communist guerrillas made the future of the tin industry questionable.[5] Iron was found in Trengganu, with some visible deposits in Kelantan, Johore, Perak, and Penang. Coal was a low quality type; the chief deposits were in northern Selangor. Bauxite was mined in small quantities from one active mine in Telok Ramuma on the southeast coast of Johore.

SOCIO-ECONOMIC STRUCTURE

Economic system

The postwar economy of Malaya consisted of "three clearly defined systems": 1) a subsistence economy of rice, fish, and gardening "rooted in poverty relieved only in a year of exceptional prosperity such as 1951"; 2) free trade mercantile economy in Penang and Singapore "made profitable in the past by *entrepôt* trade and, since the turn of the century, by the export of tin and rubber produced in the Federation"; and 3) a plantation and mining economy which produced one third of the world's rubber and tin and which provided the "principal source of Malaya's wealth,"[6] and also provided a livelihood for 20 percent of the population.[7]

Prior to World War II, Europeans owned and controlled most of the larger agricultural, industrial, commercial, and financial undertakings; they "supplied the technical experts and top administrators for the large enterprises." Since World War II, however, The Chinese have made "increasing inroads on the European position."[8]

At the beginning of the century the Malayan peasants had no interest in growing rice beyond their own immediate needs. Malaya produced only one-third of the rice it needed. Rice has continued to be imported from Siam and Burma.

Class structure

Although politically and economically interdependent, the Malays, Chinese, and Indians, who form the bulk of Malayan society, were separate societies. These three antagonistic groups did not mix socially with one another under British rule, and each was divided into its own rural and urban components.[9] The communal units greatly complicate the description of the class structure, mainly because class concepts, in such a case, must be determined within each communal group. For example, under British rule, a strong middle class developed among the urbanized Malays, whereas the rural Malays remained virtually classless. The Chinese and Indian communities, on the other hand, carried their pyramidal structures in both urban and rural areas. This type of plural structure has been the crux of the Malayan political difficulties, especially since World War II.

The transient character of the Chinese and the Indian populations in Malaya indicates that their main allegiance was to their country of origin. The Chinese offer the better illustration. Their objective was to accumulate some wealth as immigrants in Malaya and return to China to retire and die. Loyalty to the Chinese Government, however, was divided. Some Chinese supported the Kuomintang (Nationalist Government) and others supported the Chinese Communist Party.

The members of the Malay elite under British rule controlled the Federation. They were the rulers, aristocrats, and court officials who dominated the government and the intellectual circles, particularly on the west coast. They were the authority over the Chinese and Indian business structures and labor forces. Malays were also agriculturists and fishermen. The Chinese dominated the commercial pursuits, and also the mining and agricultural industries. The Indians concentrated more in plantation agriculture and administration, and in transport industries.[10]

Literacy and education

Sources indicate that 32 percent of all Malayans—mostly the urbanized population—were literate in 1947. Of the three main groups, the Indians had the highest literacy rate, while the Malays had the lowest. The great stride in wiping out illiteracy has since been made in the rural areas.

A number of school systems existed in Malaya, a fact which reflected the ethnic differences and the communal disagreement in terms of language instruction. There were English schools and three virtually exclusive types of vernacular schools. The English systems had both public and private schools. The Malay schools were first Koranic schools, but secularization brought restrictions in religious instruction. The Chinese were the most numerous ethnic element in English schools when their system became integrated with the government system after 1920. Indian schools were more prevalent on the estates. The University of Malaya was established in 1949.

Major religions and religious institutions

The religious preferences of the people corresponded somewhat to their ethnic origins. Islam, the official state religion, had a postwar membership of over 2.5 million, comprising over 90 percent of the Malays and some Indians. The Sultans, who headed the states and religious institutions, appointed Kadis to administer Muslim law governing the lives of the Malays. The Chinese element of the population, in general, maintained the Confucian, Taoist, or Buddhist beliefs of their ancestors.

GOVERNMENT AND THE RULING ELITE

Description of form of government

After World War II the establishment of the Union of Malay was announced in the British House of Commons. This, in effect, gave the Chinese population of Malaya greater political rights than the pre-war sultanate system. The union plan was rejected by the Malay elite and the Malayan civil servants. Pressure exerted by the two groups precipitated a December 1946 proposal to establish the Federation of Malaya; it was accepted in London and implemented in 1948. All Malayan administrative units, which included nine states and two settlements, comprised the Federation. Authority to govern was given to Malay rulers—each to devise a constitution for his state. There were no parliamentary governments established at the state levels. The High Commissioner, the representative of British interests, was the supreme authority over the whole Federation and was directly responsible to the Colonial Office at Whitehall. He had veto power over the legislature. The Federal Executive Council was an appointed body forming the cabinet of the government and helping the High Commissioner in an advisory capacity. The Federal Legislature was presided over by the High Commissioner and consisted of 3 ex-officio members, 11 ministers representing the 9 states and 2 settlements, 11 official members, and 50 unofficial members appointed to represent all the segments of Malayan society.[11] This in effect placed the British in firm control of Malaya and also reaffirmed the status of the old Malay rulers. The Chinese and the Indians remained second-class citizens or aliens and had few or no political rights.

Description of political process

Apart from the small Malayan Communist Party, Malaya never had a political organization with a history of opposition to the British rule. There were never any leaders of parties or groups who had served prison terms for illegal political activity. There was never a single leader who rose and preached nationalism to the Malayan people.[12] Malaya was a "placid country, apparently well content with British colonial rule."[13]

Prior to World War II, there was only one Malay organization which extended throughout the Malayan peninsula. This organization, known as the Malayan Union, was founded in 1926 and included both modernists and conservatives. Its main concern was to safeguard the Malays from Chinese encroachment and to improve their cultural and economic life. A Malayan Indian Association during the interwar period exerted some pressure on the government to improve Indian working conditions, but did not become involved in politics until shortly before World War II.

The demand from groups and parties for greater participation in government affairs began in 1945, when the idea for a centralized Malayan State was first proposed. The Malays fell into two groups: the Kaum Muda, which advocated a democratic government in a unified peninsula; and the Kaum Tua, which was more conservative and wanted a return of authority to Malay rulers of the states. Until this time the Malays had no real sense of unity and no desire for self-government. But when a small middle class evolved among the Malays after World War II it provided the impetus toward self-government. As a result, political groups began to flourish all over Malaya.

The Malay Union of Johore very early opposed the existing authority. Groups consisting of officials and state servants were formed in almost all the states. In March 1946, a conference of representatives of such groups established the United Malay National Organization (UMNO). The Malayan Chinese Association (MCA), founded by Tan Cheng-lock, was established, with the government's approval, to combat communism in the Chinese community. The Alliance Coalition of Communal Parties, composed of elements from the three communities, emerged in 1950 as a strong political party. The first Malayan election was held in 1955 and six major parties were represented. The Alliance won a large majority of seats.

The only organized group eager for independence immediately following World War II was the Malayan Communist Party (MCP). Established in 1930, it had four periods of growth. First, up to 1937, it operated illegally in Malaya and appealed mostly to the Chinese community. Second, during World War II, it became associated with anti-Japanese sentiment and its underground resistance movement made it the champion of the Allied cause. Third, the post-World War II period found the MCP organizing front groups and dominating the trade unions; it conducted a wave of fear and violent threats which precipitated racial riots between the Chinese and the Malays. Fourth, the "Emergency" declared in 1948 by the British officials outlawed the MCP and its associated organizations, forcing them to return to a policy of open violence and guerrilla warfare.[14]

Legal procedure for amending constitution or changing government institutions

Until independence in 1957, the High Commissioner was the highest authority in the postwar Malayan Government, and he in turn was responsible to his government in Great Britain. Government institutions could legally be changed only with his approval.

Relationship to foreign powers

The Malayan Government until 1957 was under the direction of the British Colonial Office and its foreign policies reflected British interests.

The role of military and police powers

Prior to World War II, the British were faced with some violent demonstrations and strikes. The communal riots and the Communist-inspired strikes and demonstrations after the war necessitated the employment of British army units to put the outbreaks down. The Federation Police Force, which had members from all communities, was also employed to maintain order.

WEAKNESSES OF THE SOCIO-ECONOMIC-POLITICAL STRUCTURE OF THE PREREVOLUTIONARY REGIME

History of revolutions or governmental instabilities

Although there were apparent weaknesses in the British rule in Malaya prior to World War II, this rule was never plagued with a serious rebellion. The Malay community appeared well satisfied with the *status quo;* and although there was reason for dissatisfaction in the Indian and Chinese communities, there was never an attempt to overthrow the British. The MCP, however, stirred up demonstrations and strikes both before and after the war.

Economic weaknesses

Rubber and tin, industrialization and urbanization, war and economic dislocation, brought changes to the Malayan society which the existing social institutions could not deal with adequately. Communism flourished in this atmosphere. It found strength among those who rejected the help of others in meeting the problems of social and economic changes. Because of industrialization and urbanization, a labor force developed in Malaya during the 1930's. The British failed to grasp the significance of this development and did not encourage the organization of labor. The Communists took control by default and indoctrinated Malayan labor.[15] Following the war, Malaya's near bankruptcy created difficult problems for the British. The rubber and tin industries needed some sort of rehabilitation; food was scarce; living costs soared beyond means; and the labor force became restless. These conditions furthered the development of communism.

Social tensions

Aside from the social tensions created by the plural character of Malayan society and the political predominance of the Malay community over the Chinese and the Indian communities, the Chinese produced further tensions, not by their disloyalty to Malaya, but rather by their extended loyalty to the competing political groups in China. Malayan Chinese loyalty to China was actively intensified after the Japanese invasion of the Chinese Mainland in 1931, placing the British authorities in a precarious situation vis-a-vis British-Japanese relations. Furthermore, Chinese cadres of the MCP were indoctrinating the Malayan labor force and instigating strikes and demonstrations in the 1930's. To cope with the problem, the British instituted the "Banishment Ordinance" which allowed for the deportation of Chinese activists to their homeland. Branded as Communists in Malaya, they received a bad welcome from the Nationalist Government in China.

Following the invasion of Malaya in December 1941, the Japanese occupation troops forced one-fourth of the Chinese population out of the towns and cities and reestablished them in small villages on the fringes of the jungles. The "squatter" population, as they became known, became the butt of the legal restrictions devised to discriminate against the Chinese after the war. The Chinese saw themselves as second-class citizens under the postwar federal system, which to them represented an injustice that they were committed to change.

Government recognition of and reaction to weaknesses

The British did not respond to the basic social problem that the Chinese community in Malaya represented. Their reaction was to attempt to suppress the MCP rather than to counter its appeal. In the early 1930's the Communist Party was declared illegal, and when the party set in motion a wave of strikes in the middle thirties, the "Banishment Ordinance" was instituted. The British did, however, propose a plan providing for Chinese political participation in the form of a Malayan Union after World War II, but this was rejected by the Malay community. As a consequence, a large number of Chinese succumbed to the appeals of the Communist Party.

After the war, the British authorities attempted to control Communist activities—first, by demobilizing a wartime guerrilla army composed of MCP members which had been trained and armed by the British High Command to fight the Japanese, and second, by passing certain legislation to control MCP activities. The Federation of Trade Unions, which the Communists formed, controlled the strikes and unrest, violence and intimidation. Outlawing the Federation and restricting the leadership of the trade unions, the Government tried to undermine Communist influence in labor activities.[16] The Malayan Communist Party's efforts to overthrow British rule forced the British to declare an all-out drive against the party in June 1948.

FORM AND CHARACTERISTICS OF REVOLUTION

ACTORS IN THE REVOLUTION

The revolutionary leadership

The revolutionary leaders in Malaya were not the type usually associated with nationalist movements. One leader dedicated himself to an ideology best suited to enhance his ascendancy to power, while another, a dedicated Communist, received comparatively little popular support. The outstanding, and certainly the most intriguing, character of the revolutionary leadership was Loi Tak. Supposedly of Annamite stock, Loi Tak arrived in Singapore in 1934; where he came from and what he had done previously is not recorded. He immediately made contacts with Communist Party members, and his glib use of Marxist terminology won him the office of Secretary General in 1939. During the war, he collaborated with the British authorities and helped the Allied cause. Several times during his career his party loyalty was questioned and his leadership threatened. In 1947 he absconded with party funds and has never been seen since.

Ch'en P'ing, also known as Wong Man-wa and Chin Peng, inherited the party leadership vacated by Loi Tak. Ch'en P'ing—a Malayan Chinese—had both a Chinese and English education. He was introduced to communism while in school and became a member of the party in 1940. During the war, he became Loi Tak's right-hand man. Ch'en P'ing was awarded the Order of the British Empire for his military service against the Japanese during the war and, like Loi Tak, he was a good organizer. By 1954, he was known to have had a very large reward over his head.[b]

---

[b] See Harry Miller, *Menace in Malaya* (London: George G. Harrap & Co., Ltd., 1954), or Gene Z. Hanrahan, *The Communist Struggle in Malaya* (New York: Institute of Pacific Relations, 1954) for more complete biographical sketches.

The revolutionary following

The MCP seems to have absorbed the ambitious and the restless, the undisciplined and opportunistically inclined people. From these, it created "a highly disciplined organization of professional revolutionaries."[17]

At first, the party recruits were mostly embittered and radical intellectuals, with a sprinkling of certain elements of the industrial working force. The party became widely accepted because of its anti-Japanese sentiments. After the war, the MCP found a fertile field for proselytizing among the Chinese. The Chinese community was no longer integrated and coherent; it was caught up in change and the transition was beyond the traditional bounds. Chinese interests ran counter to political developments and communism profited by this situation. The revolutionary following was approximately 95 percent Chinese. Although the party's adherents numbered many thousands, at no time was its actual membership more than 3,000 strong.

ORGANIZATION OF REVOLUTIONARY EFFORT

Internal organization

The internal organization of the revolutionary effort consisted first and foremost of the Malayan Communist Party. The Central Committee of the party, 10 to 13 members, was headed first by Loi Tak and later by Ch'en P'ing. Ch'en P'ing, Yeung Kwo, and Can Lee formed the Politburo after 1948. Below the Central Committee, the regional bureaus followed, with State Committees, District Committees, Branch Committees and finally, the cells. Relations between the various levels of the party hierarchy were maintained by having senior members of lower committees serve as members of the next highest committee.

The Malayan Races Liberation Army (MRLA) was the fighting unit of the revolutionary organization. First known as the Malayan People's Anti-Japanese Army (MPAJA), it was organized under the auspices of the British Command shortly before the Japanese invasion to conduct guerrilla attacks against the Japanese occupation forces. The British attempted to disband the guerrillas after the war, but the MRLA kept its units fairly intact by falsifying the membership lists submitted to the British, forming the "Old Comrades Associations," and hiding large quantities of arms. The party created and controlled the guerrilla units down to the platoon level through its Central Military Committee. Coordination from the MCP Executive Committee down to the smallest bandit units proved to be an impossibility. In most cases the regiments, the largest units of the MRLA, had to be controlled by the state districts, and later during the insurrection, coordination between the companies proved impossible. There were approximately 5,000 MRLA guerrillas living in jungle camps, each containing up to 600 troops. The camps were later reduced to as few as three to five men.

The *Min Yuen*, a civilian mass organization set up to assist the guerrillas in the jungle, was the auxiliary unit of the MRLA. Its main duties included furnishing the guerrillas with material supplies, functioning as an intelligence and courier network, and effecting a closer liaison with the masses. Local *Min Yuen* organizations included regional committees, peasants' unions, "liberation leagues," women's unions, and self-protecting corps which also acted as part-time guerrillas.[18] The Branch Committees of the MCP organized and controlled the *Min Yuen* within their districts.

It is difficult to assess the membership of this organization. Many of the Chinese villages established when the Japanese relocated some of the Chinese population acted as supply depots for runners carrying supplies to the guerrillas in the deeper jungles. Many of these Chinese villagers were volunteer members of the revolutionary organization. But many were also intimidated and many more refused to take part in the revolution. Some figures for the combined membership of the MRLA and the *Min Yuen* go as high as 500,000, but these estimates appear to be far too high.

External organization

The Communist uprising in the summer of 1948 has been attributed, by some sources, to an international conference held in Calcutta in February 1948. This probability is based on circumstantial evidence only and "no documentary evidence is likely to be found."[19]

GOALS OF THE REVOLUTION

Concrete political aims of revolutionary leaders

The prewar goals of the revolutionary leaders were to infiltrate and control many political and economic organizations in order to disrupt the Malayan economy. The goals of the revolution itself were to "liberate" an area of Malaya, gain control, and declare that area independent. The periphery of the controlled area was to be widened to include larger sections of Malaya. Captured documents show that the Communists had timed a declaration for the independence of a Communist Republic of Malaya for August 3, 1948.[20] As announced by the revolutionary leaders, their objectives were "national liberation," the establishment of a "democratic government" based on universal suffrage, and unification of all the "oppressed peoples of the Far East."[21] Their more immediate political aims after 1950 were the election of all the Legislative Council members, greater equality of citizenship rights in all the Malayan communities, and recognition of political parties in all discussions concerning a Malayan constitution.[22]

Social and economic goals of leadership and following

Social and economic goals of the revolutionary leadership and its following concerned the expansion of the Malayan economy and greater achievements toward social progress. This included adoption of an 8-hour workday, a social security program, an educational system allowing the three communities to use their own languages, and recognition of equal rights for women.[23]

REVOLUTIONARY TECHNIQUES AND GOVERNMENT COUNTERMEASURES

Methods for weakening existing authority and countermeasures by government

The overall Communist strategy for conducting the armed revolutionary struggle against the British in Malaya called for a protracted war to be executed in three separate phases. In the first phase the MCP planned to weaken the British forces and to preserve and expand its own forces. In the second phase the Communist guerrillas were to drive the British from the hinterland and restrict them to such strategic points as supply centers and cities. During this same phase the MCP planned to transform its activities from irregular warfare to mobile warfare. In the final phase, the MCP planned to establish strong Communist bases from which to operate, and which would provide the necessary recruits to expand these bases, and finally to join the bases together in order to control all of Malaya.[24]

A general uprising to pave the way for a rebellion was instituted by the MCP in June 1948 when communal riots, which had been occurring since the announcement of the Union Plan in 1946, were at a peak. The Communists capitalized on the chaotic situation and organized their own violence as MRLA guerrilla units began the first phase of the revolution. Small-scale raids, road ambushes, murder, robbery, sabotage, terrorist action, skirmishes with the British Security Forces, and attacks on police stations were the tactics employed by the small guerrilla bands. Rubber plantations were ruined as the Communists attempted to cripple the British economic system; and terroristic Communist activities so frightened some of the anti-Communist Chinese squatters that the MCP was able to extract large sums of money from them in the form of "protection payoffs" to help finance its war against the British.[25]

Three factors limited Communist success during the first phase and prevented the MCP from carrying out the second phase as planned. First, the MCP failed to gain mass support. Second, the MRLA was unable to mobilize a major force and therefore had no alternative but to continue its small-scale raids against government supply lines and large estates. Added to this second factor was the crude manner of providing, and the unpredictability of receiving supplies from the *Min Yuen*, which gave the MRLA units logistical problems. Third, British countermeasures successfully stymied MRLA attempts to establish "temporary bases" in the jungle. Without secure base areas from which to operate, the guerrilla units were forced to seek refuge deeper in the jungle. Thus food and supply problems were complicated and the MCP communications system was rendered ineffective.[26]

British countermeasures were not restricted to a military campaign. The entire responsibility for conducting counterguerrilla activities against the Communists rested "fairly and squarely with the civil government. The Armed Forces work [ed] in support of police, though all emergency activities [were] so interconnected that on occasions it [was] a little difficult to decide who [was] in support of whom."[27] This led one noted expert on Asian politics to comment that the British were waging "socialized warfare" against the MCP.[28] The "sociological" techniques included resettlement, detention, repatriation, and rehabilitation.

In order to control the activities of the *Min Yuen* and to reduce the maneuverability of the MRLA, the government in 1950 instituted the "Briggs Plan," which, in part, made provisions to resettle close to one-half a million Chinese squatters, who had been living close to the jungle, in new villages which could be controlled by the British.[29] This resettlement served the dual purpose of fencing in potential Communist "sympathizers," and keeping out *Min Yuen* agents who relied on these "sympathizers" for funds and supplies. As a result, the main supply line of the guerrillas was cut. At the same time, the government expended considerable sums to further economic, social, and civic development in these villages. Development was concentrated mainly on education, health, agriculture, and public works.[30]

"Emergency Regulation 17," promulgated in 1948, made provisions for the "detention" of anyone suspected of aiding and abetting those who were taking part in Communist activities. Detention was a preventive measure and was not punitive. Under the same regulation, detained persons were subject to repatriation, and these repatriated persons had to remain outside the Federation. The alternative to repatriation was rehabilitation. Under this provision of "Emergency Regulation 17," detained persons were placed in "rehabilitation" centers for 8 months to a year, after which they were to be released unconditionally.[31]

Having concentrated and isolated the squatters, both physically and psychologically, the British then isolated MCP units from the whole population. Everything, including food, clothing, supplies, and equipment, was denied them. Gen. Sir Gerald Templer, who commanded the activities of the "Emergency," described this as "securing the base."[32] The forces mainly concerned with "securing the base" were: the Regular Police Force, the Home Guard, and the Special Constable.

The Regular Police Force was used primarily for the maintenance of law and order; at the same time it attempted to win the confidence of the people. The Home Guard patrolled and guarded the perimeter of the villages and scattered huts. An Operation Section of the Home Guard was sent out from the villages for 2 or more days to patrol remote areas. The Special Constable aided the Regular Police Force in a variety of tasks such as road and food checks, guarding "labor lines," and furnishing personal bodyguards for estate owners.

Active operations to eliminate the MRLA in the jungle were conducted by the army, RAF, and the navy. Stationed according to intelligence reports of the disposition and strength of the MRLA, the army was cosmopolitan in character and contained British, Fijians, and Malays, as well as Chinese and Indian units. The RAF and the navy assisted in the war against the MRLA in a variety of ways such as bombing, strafing, transport, reconnaissance, casualty evacuation, and movement of small forces.[33] Navy helicopter operations became indispensable. These aircraft were able to land in the most inaccessible and hazardous jungle areas. Troops, shinnying down ropes from the helicopters, landed in zones deep in jungles from which operations were centered. The helicopters were also used to chase Communists and land troops to continue the chase.[34]

The government Security Force became more mobile in the jungle. With its superior military technology, the Security Force depended highly on the use of helicopters and air drops with which to maintain strong units for longer periods in the jungle where Communists thought themselves safe. Deep jungle penetration brought the British in contact with aboriginal tribes, which eventually were controlled by the Security Force, who were then able to establish jungle supply depots. The Security Force was greatly aided by an agreement made with the Thai Government in 1949 which allowed the Federation police to pursue Communist guerrillas as far as 10 miles within the Thai frontier.

By 1951, terrorism had proved insufficient as a means for the Communist forces to achieve their aims. This became the turning point for MCP tactics. Directives from the Central Committee of the MCP ordered that terrorism should become more discreet, and party members should concentrate more on the infiltration of specific legal organizations. This also largely met with failure, and the party again reversed its directives, in 1952, reverting to increased terrorism. The policy shifts of the Central Committee were not very rapidly carried out in the different areas of Malaya. This was due primarily to the inefficient and very slow system of communications which the party could operate while being harassed by the British. The MCP in the meantime repeatedly requested that the Federation recognize its organization as a legal political party. The government refused to give in to these requests, or to negotiate with the MCP as an organized group, but in 1955 it offered an amnesty to all persons who had individually taken arms against the government. All persons who had given aid to the insurgents could surrender without being prosecuted for their activities during the "Emergency."[35] The amnesty increased the number of Communist defectors, but failed to bring the MRLA activities to an end. Terrorism continued in decreasing intensity until and even beyond the end of the "Emergency" in July 1960. However, by August 1957, the situation was considered sufficiently secure that Malaya could take its place as an independent member of the Commonwealth.

Methods for gaining support and countermeasures taken by government

The main technique used by the Communists to gain a following was to exploit the vulnerable position of the Malayan Chinese. It was among this group that they were most successful.

Communist youth leagues were established by the MCP in the early years to organize the radical element among the younger Chinese. Combined with this was an all-out effort to spread propaganda among the general public by the distribution of leaflets. When the MCP turned its attention to the unorganized labor force, the British began to impose restrictive measures upon the cadres: the "Banishment Ordinance" was one such measure.

In 1945, the MCP was at its peak of popularity. Until 1948 it leaned heavily on the Chinese community and the increasing labor force for support. The party conducted its propaganda campaign with published papers and pamphlets, but the distribution of printed matter was severely restricted by the Printing and Publishing Enactment of 1945. Two Chinese newspapers were suppressed. By 1948, the beginning of the "Emergency," MCP popular support had declined and the propaganda themes had lost some of their flavor. During the "Emergency" the *Min Yuen* "runners" kept Communist propaganda in circulation among the sympathizers. Much of the Communist propaganda was countered by the legal anti-Communist Malayan Chinese Association.

Many Communists surrendered, but the surrender rate varied greatly with the successes of communism in other parts of the world, dropping when Great Britain recognized Red China, when China was scoring victories in Korea, when the Vietminh was scoring victories in Vietnam, and again when the British commander, General Templer, left Malaya. By 1956, a date for which figures are available, the Communist casualties for the "Emergency" were: 5,933 killed, 1,752 surrendered, and 1,173 captured. Ninety percent of these casualties were Chinese.[36]

THE EFFECTS OF THE REVOLUTION

The Communist retreat during 1952 and 1953 allowed political activity to assume greater importance. In 1953, the major political parties held conferences in which an Elections Committee was appointed to consider recommendations made by the parties. In the fall of 1955, Malaya held its first election, and on August 31, 1957, the Federation of Malaya became an independent member of the British Commonwealth. Singapore, a city-state with a large Chinese population, also received its independence; Singapore and the Federation of Malaya became separate political entities despite their common economic and cultural framework.

The Communists claimed that independence had resulted from their revolutionary efforts. They had lost the shooting war, however, and much of their popular support. It is not likely that they could now count on the civilian populace if they should attempt to overthrow the Federal Government. From 1955 to 1957 the MCP depended primarily upon its ability to penetrate and infiltrate a still disorganized Chinese community. Eighty-four percent of the electorate in the 1955 elections was Malay, leaving large numbers of Malayan Chinese uncertain about committing themselves to cooperation with the other racial groups.[37] By 1959, however, the electorate on the whole increased, "and the proportion of non-Malay electors also was considerably larger than in 1955." The Chinese electorate increased from 150,000 in 1955 to

750,000 in 1959 as a result of citizenship regulations stipulated in the independent Malayan constitution.[38] This, in effect, made the MCP's prime target elusive.

On the other hand, pro-Communist elements have gained much influence within the ranks of political and social organizations. Labor unions, political parties, and other organized groups in the urban areas have been affected by this influence and many have shifted to the left on the political scale. In the 1955 elections the anti-Communist Alliance Coalition of Communal Parties won 51 of the 52 seats; in 1959 it won only 74 of 104 seats.[39] Many seats were gained by the leftwing Labour Party, the Socialist Front, and the People's Progressive Party. The rural vote, still representing the bulk of the electorate, continues to support the Alliance.

Though remaining predominant, the Malay vote has continued to diminish slowly in importance. "The extent to which the Alliance can find a real basis for Chinese and Indian support may therefore well determine the outcome of Malaya's next Parliamentary election."[40]

NOTES

1. Sir Richard Winstedt, *Malaya and Its History* (London: Hutchinsons University Library, 1958), pp. 18–21.
2. *The Worldmark Encyclopedia of the Nations* (New York: Harper and Brothers, 1960), 627.
3. University of Chicago, *Area Handbook for Malaya* (HRAF Subcontractor's Monograph, 1955), pp. 72–76.
4. Ibid., p. 81.
5. Ibid., p. 64.
6. *Area Handbook for Malaya* (Washington, D.C.: Special Operations Research Office, The American University, [n.d.]), p. 796.
7. Ibid., p. 771.
8. Ibid., p. 772.
9. Ibid., p. 131.
10. Ibid.
11. Ibid., pp. 518–520.
12. Ibid., p. 572.
13. Gene Z. Hanrahan, *The Communist Struggle in Malaya* (New York: Institute of Pacific Relations, 1954), p. 4.
14. Lucien W. Pye, *Guerilla Communism in Malaya* (Princeton: Princeton University Press, 1956), pp. 47–104.
15. Ibid., p. 61.

16. Harry Miller, *Menace in Malaya* (London: George G. Harrap & Co., Ltd., 1954), p. 77.

17. Pye, *Guerrilla Communism*, p. 110.

18. Hanrahan, *The Communist Struggle*, p. 69.

19. Frank N. Trager, *Marxism in Southeast Asia* (Stanford, California: Stanford University Press, 1959), pp. 263–274.

20. D. G. E. Hall, *A History of South-East Asia* (New York: St. Martin's Press, Inc., 1955), p. 705.

21. *Malaya*, SORO, p. 683.

22. Ibid., p. 684.

23. Ibid., p. 683.

24. Pye, *Guerrilla Communism*, pp. 86–87.

25. Hall, *A History of South-East Asia*, p. 705.

26. Pye, *Guerrilla Communism*, pp. 99–100.

27. Noll, "The Emergency in Malaya," *The Army Quarterly* (April 1954), 56.

28. Maj. Paul M. A. Linebarger, "They Call 'Em Bandits in Malaya," *Combat Forces Journal* (January 1951), 29.

29. *Malaya*, SORO, pp. 702–706.

30. *Annual Report, Federation of Malaya*, 1952 (London: Her Majesty's Stationery Office, 1953), pp. 14–15.

31. Ibid., pp. 16–17.

32. Noll, "Emergency," p. 49.

33. Ibid., pp. 49–53.

34. Lt. Cmdr. Trevor Blore, "The Queen's 'Copters," *Marine Corps Gazette* (July 1954), 52–56; and Lt. Cmdr. J. E. Breese, "Rotors Over the Jungle; No. 848 Naval Air Squadron in Malaya," *Flight* (March 12, 1954), 291–292.

35. Pye, *Guerrilla Communism*, p. 107.

36. Ibid., p. 109.

37. T. E. Silcok, "Singapore in Malaya," *Far Eastern Survey*, XXIX, 3 (March 1960), 39.

38. T. E. Smith, "The Malayan Elections of 1959," *Pacific Affairs*, XXXIII, 1 (March 1960), 40.

39. Ibid., p. 39.

40. Ibid., p. 47.

RECOMMENDED READING

BOOKS:

Bartlett, Vernon. *Report from Malaya*. New York: Criterion Books, 1935. Author describes antiguerrilla warfare. The book has a good deal of material on the use of propaganda and is excellent in describing the success of the resettlement of villages.

Hall, D. G. E. *A History of South-East Asia*. New York: St. Martin's Press, Inc., 1955. Not critical of colonialism, this scholarly volume represents the historical approach to the existing general conditions in Southeast Asia.

Miers, Richard. *Shoot to Kill*. London: Faber and Faber, 1959. Author deals with the last phases of British antiguerrilla activities. The book offers a detailed description of the use of informers and helicopters.

Miller, Harry. *Menace in Malaya*. London: George G. Harrap & Co., Ltd., 1954. With little background, this is a more intimate view of the tactics employed by the MCP and its British countermeasures presented in "newsreporting style."

Pye, Lucien W. *Guerrilla Communism in Malaya*. Princeton: Princeton University Press, 1956. This book is based on interviews with members of the Malayan Communist Party, and examines the attractiveness that communism holds for the people of Malaya who are swept by the forces of a changing society.

Trager, Frank N. *Marxism in Southeast Asia*. Stanford, Calif.: Stanford University Press, 1959. This book covers the development of Marxism in four countries of Southeast Asia.

Winstedt, Sir Richard. *Malaya and its History*. London: Hutchinsons University Library, 1958. Of limited value, this work affords the student a brief background of events leading to the revolution.

PERIODICALS:

Linebarger, Major Paul M. A. "They Call 'Em Bandits in Malaya," *Combat Forces Journal* (January 1951), pp. 26–29. A review of British antiguerrilla tactics.

Noll (Pseud. of senior officer in Malaya). "The Emergency in Malaya," *The Army Quarterly* (April 1954), pp. 46–65. An excellent account of antiguerrilla measures, both military and civil, and a discussion of problems involved.

Purcell, Victor. "Colonialism in Contemporary Southeast Asia," *Journal of International Affairs*, School of International Affairs, Columbia University, X, 1 (1956), 49–59.

Silcok, T. E. "Singapore in Malaya," *Far Eastern Survey*, XXIX, 3 (March 1960), 33–39.

Smith, T. E. "The Malayan Elections of 1959," *Pacific Affairs*, XXXIII, 1 (March 1960), 38–47.

OTHER:

Hanrahan, Gene Z. *The Communist Struggle in Malaya.* New York: Institute of Pacific Relations, 1954. The monograph covers the growth and development of the MCP. It is by no means a comprehensive study of the "Emergency."

Area Handbook for Malaya. Washington, D.C.: Special Operations Research Office, [n.d.].

Area Handbook on Malaya. University of Chicago. HRAF Subcontractor's Monograph, 1955.

"Air Supply in Malaya District," in *Canadian Army Journal* (October 1950), pp. 49–50. A brief account of the role of the RAF air drops as operations against Communist guerrillas.

SECTION II

....

LATIN AMERICA

TABLE OF CONTENTS

GENERAL DISCUSSION OF AREA AND REVOLUTIONARY DEVELOPMENTS

THE GUATEMALAN REVOLUTION OF 1944

THE VENEZUELAN REVOLUTION OF 1945

THE ARGENTINE REVOLUTION OF JUNE 1943

TABLE OF CONTENTS

THE BOLIVIAN REVOLUTION OF 1952

THE CUBAN REVOLUTION: 1953–1959

TABLE OF CONTENTS

LATIN AMERICA

GENERAL DISCUSSION OF AREA AND REVOLUTIONARY DEVELOPMENTS

GEOGRAPHICAL DEFINITION

Latin America extends from the southern border of Texas southward 7,000 miles to Cape Horn. Its area is 2½ times that of the continental United States.[1] The term "Latin America" embraces 20 sovereign nations of Central and South America and the Caribbean area. The countries range in size from Brazil, larger than the United States, to Haiti, which is smaller than West Virginia. Brazil has 65.7 million inhabitants, while Panama has only 1.1 million. Racial diversity is likewise great; in Haiti the population is almost completely Negro; in Bolivia, Ecuador, Peru, Guatemala, and Paraguay the inhabitants are principally Indian or of Indian origin, and in the remaining republics they are primarily white.

BACKGROUND OF TWENTIETH CENTURY REVOLUTIONS

GEOGRAPHY

Geographical factors have had a great influence on Latin American political life. The high, rugged, and barren Andes Mountains, steaming jungles and forests, and great rivers served to intensify the clustering of people into isolated groups around some natural point of vantage. These same fractioning features serve as barriers to transportation and effective communication, further isolating these dispersed groups. This isolation inevitably affects the political life of these countries. Not only is it hard for the inhabitants of remote areas to be well-informed on national issues and events, but it is even more difficult for them to make their influence felt. A case in point is Bolivia, where the tin miners were unable to aid the dictator Villarroel when he was overthrown in 1946.[a] Because of their inability to make their voices heard, these rural inhabitants are ignored by residents of the large cities, and in turn often lose any interest in politics, thus making it

---

[a] Refer to the Bolivian Revolution, p. 189.

easier for minorities within the populous centers to manipulate or gain control over the government.

ECONOMY

There are four important economic factors that influence domestic and international politics in Latin America. However, in an area so large and varied there are always major exceptions to any possible generalization.

1. Economists describe the area as being "underdeveloped," as measured by three distinguishing characteristics. First, in an "underdeveloped" economy industrialization is either nonexistent or in a primitive state, and land cultivation remains the primary means of economic support. Second, living standards, expressed in per capita income, are low in relation to those of "industrialized" countries. Third, and most significant, the area is in a state of rapid transition toward industrialization and urbanization.

But within this general structure there is obviously a wide range of difference. Latin American countries range from the "Indian" countries with extremely underdeveloped economies, such as Paraguay, Bolivia, Peru, Guatemala, and Ecuador, through the typically underdeveloped economies such as Costa Rica, to the industrialized and fairly well-developed countries of Argentina, Venezuela, and Uruguay.[2]

2. Agriculture plays a disproportionately large role in the economy of these nations, although the productivity of the soil ranges from the highly fruitful conditions in Argentina to the bare subsistence agriculture of Ecuador. Land is regarded as the most important source of wealth, and land ownership is considered a mark of prestige. In most of these countries a very small percentage of the population owns the bulk of the arable land, and this group consequently plays a very important role in government. The great majority of the agricultural workers continue to live under conditions of sharecropping and peonage. It is, therefore, relatively easy for a gifted demagogue to gain a following merely by harping on the string of "agricultural reform"; to the poor this means taking the land away from the rich and giving it to them.

3. Latin America is one of the world's major sources of raw materials, and especially of minerals. The political implications of this fact lie mainly in the international field, but it also plays an important part in domestic politics. In Bolivia, power was for a long time in the hands of the "tin barons," and oil influences the politics of Mexico and Venezuela. Other governments are likewise influenced by the possession

of valuable minerals. The peacetime demand for these raw materials is essential to Latin American economies, and in time of war the continent is a leading supplier of strategic and critical materials.

4. Finally, foreign capital plays a large part in both the economy and the politics of Latin America and is the basis of most charges of "imperialism." British investment is concentrated in southern South America and U.S. investment in the rest of Latin America.

THE SOCIAL SYSTEM

The class system, established by the early colonizers, has become formalized and rigid with the passing of generations. In general, it takes the form of a division into three groups, based on social rather than ethnic differences. The "whites" or creoles, making up an average of 15 to 30 percent of the population,[3] are European in cultural, historic, and linguistic orientation; they constitute a species of ruling class, occupying the major offices in the government, army, and church, and owning a major portion of the country's land. In some countries a middle group, known as mestizos or Cholos, is composed of Indians or persons of mixed blood who have adopted European customs and language and attempt to emulate the upper class. These constitute roughly a third of the total population of Latin America. The Indians who have maintained a traditional semi-isolated existence in scattered villages make up the lower class. Where Negroes exist in large numbers, as in the Caribbean island republics and certain regions of Brazil, they tend to be associated with the lower class. But this is not to say that all Latin American countries contain these three groups in significant proportions. Argentina provides an example of a virtually all-"white," non-Indian republic, whereas Paraguay, Bolivia, Peru and Ecuador are largely Indian and Haiti is largely Negro.[b]

Although possible, class mobility is a slow and difficult process, for two reasons: (1) each class values its own way of life, and (2) the barriers to interclass mobility are still formidable. It is not easy for an Indian to learn to speak Spanish, let alone read or write it; it is difficult for him to move to a big city, to buy a large landed estate, or to become a priest. But once an Indian has made the transition, he is for all practical purposes considered to be a mestizo, and a mestizo able to penetrate the upper class is likewise considered to be a "white," despite his race.

---

[b] It is important to remember that the basis of this social structure usually is not races but classes. "The bases of differentiation among them are social rather than biological; an individual's learned behavior, his way of life rather than his physical characteristics, determine the group with which he is identified."[4]

115

The majority of the Latin American revolutions within the past two decades were instigated by the "white" or creole group. This group in each country is the best educated, the wealthiest, and the one in closest contact with politics. The tendency of revolutions carried out by this group has been to change the leadership, but not to overhaul any of the traditional institutions. There have been important exceptions, however, as in Mexico and, recently Cuba.

The maldistribution of land and material comforts, and the rigid class system protected by the traditional oligarchy of army, government, and church, have been in existence since the early days of the Spanish conquest. But only in fairly recent times have special pressures begun to manifest themselves in violent change—in Mexico beginning in 1910 and later in Cuba, Bolivia, and elsewhere.

COLONIAL BACKGROUND

It is commonly accepted that the seeds of present day Latin American dictatorships are found in the history of the Iberian and native American autocracies, from the Incas to the Spanish monarchy. Two institutions during the colonial era did much to cultivate authoritarian thought. One was the church, which acted as an arm of the colonial government. It introduced education and dominated it; through education it implanted the belief that authority should not be challenged. The second institution was government. Spain's absolute control over the colonies through viceroys responsible only to the Spanish crown conditioned the people to dictatorship. The results of these influences are seen in the facts that: (1) although Latin American constitutions are frequently adapted from that of the United States, they usually strengthen the power of the executive at the expense of the legislative and judicial branches,[5] and (2) when independence movements appeared in the 19th century, many of the leaders believed that monarchy should remain the basic governmental form.

REVOLUTIONS OF THE TWENTIETH CENTURY

REVOLUTIONARY TRENDS

Latin America has experienced a succession of revolutions during the 20th century. Within these a definite trend is distinguishable—a gradual broadening of the revolutionary base. Risking oversimplification, one may distinguish between three types of revolutions: (1) The traditional coup d'etat, carried on within the ruling elite, is essentially restricted to the replacement of the president and his immediate aids, leaving the basic political system intact. This type of conservative "revolution" completely dominated Latin American political history until the Mexican Revolution of 1910 broadened the revolutionary base and introduced a more progressive element. The Mexican Revolution was carried out by a combination of elite governmental factions and opposition political groups, some of whom had the support of the peasants. Dr. Getulio Vargas' "march of Rio" in 1930 utilized the working classes.[6] The Guatemalan Revolution of 1944 and the Bolivian Revolution in 1952 also attempted to consider the needs of labor and found support from this group. (2) The second type has a broader base and has resulted in some social change. (3) The Cuban Revolution of Fidel Castro may represent a third type of revolutionary movement—one with a very wide popular base and expensive social and economic changes. It was supported by a large proportion of the citizenry of the island, and resulted in significant and sudden political, economic, and social change. Although all three types occur side by side, the trend appears to be toward the mass movement, as popularized by Fidel Castro, who made it clear to all of Latin America that the goods of the rich could be expropriated to feed, clothe, and house the poor—at the expense of the personal and political liberty that frequently, under authoritarian governments, had been more nominal than actual.

INFLUENCING FACTORS

The frequency of Latin American revolutions may be explained by several factors. A prominent Peruvian lawyer once said: "When you Anglo-Saxons write a constitution, you do it as a practical guide by which you operate. We Latins write an ideal constitution to which we may aspire, but which is above and beyond us."[7] This disregard of constitutions has fostered a widespread lack of respect for law and order, from minor traffic regulations to electoral laws. Lack of respect for law has been enhanced by personal experience with fraudulent manipulation of electoral results under dictatorial regimes. Thus the individual becomes highly susceptible to the ambitious leader who argues that since it is impossible to attain civic objectives under the law, one must bypass the rules and regulations. One political analyst observed that:

> . . . constitutional fragility in the area is aggravated by
> such divisive forces among the whites as personalism and
> family rivalries, regionalism and other conflicts of intra-

class interests, sometimes translated into doctrinal and ideological terms. These conflicts are frequently suffi- ciently disruptive to give the *coup de grace* to a weak and unworkable constitution. They are normally followed by the promulgation of a new—and equally (ideal)— constitution with the political system, fundamentally unchanged, settling down to await its next revolution.[8]

The "gentlemen's agreement," whereby ousted government lead- ers are assured asylum and exile in neighboring countries, protects them from punishment for political misdeeds and provides a conve- nient place in which to conspire and organize a counterrevolt.

World War II greatly influenced the occurrence of revolutions. During the early war years the demand for goods by the United States tended to bring prosperity and, to some extent, stability to Latin America. Those revolutionary attempts that did occur up to March 1944 were usually motivated by the desire of internal pro-Axis or pro- Ally factions to better their positions. The overthrow of Castillo in Argentina in 1943 was interpreted at first as a move toward the Allies, but the new administration reaffirmed neutrality, with some conces- sions to each side.[9]

In the later war years and the early postwar period stability and prosperity faltered and unrest developed as shortages and infla- tion began to appear. The United States, embarking on a program of mutual security, supplied Latin American governments with sur- plus and other war materials; this tended to strengthen entrenched conservative and military-oriented regimes. The little man, who suf- fered most from economic fluctuation, sought more real freedom and better conditions through leftist-oriented revolutionary move- ments: Democracy, equality, and freedom were much-discussed con- cepts in the postwar world; although the rural inhabitants of the Latin American countries were largely unmoved by them, urban dwellers responded eagerly. Thus, three revolutions between May 1944 and July 1946 were strongly supported by large groups in the capital cit- ies. These were the ouster of Martínez in El Salvador and Ubico in Guatemala,[c] and the assassination of Villarroel in Bolivia.

ACTORS

Latin American revolutions have traditionally been generated by politics among the "whites," or upper class. Indians and mestizos have

---

[c] See summary of the Guatemalan Revolution, p. 125.

rarely become involved in politics; when they did participate, it was usually to join an action organized by the "whites."

The combatants were often limited to rival military chieftains and their close supporters. Sometimes interested members of the landed oligarchy or a small group of professional men would become involved in political intrigues, but the general population was not affected. Because of the control exercised by the top military officers over the manpower and weapons of the armed forces under their command it was extremely difficult for anyone to maintain himself in a dictatorial position or to stage a revolution without at least the partial support—or the calculated neutrality—of the military leaders.

In recent decades, as pointed out, the base of the revolutionary movement has broadened and the middle class as well as labor and peasant groups played an increasingly important role. Of outstanding influence are the often highly idealistic and Marxist-oriented university students and their professors. Much of the early impetus to the movements which overthrew Villarroel in Bolivia and Ubico in Guatemala was supplied by students. Even after the Latin American student leaves the university campus, he maintains the tradition of using political agitation and even revolution to achieve his desired ends. Dissatisfied urban workers under persuasive organizers have also swollen the revolutionary followings.

Within the last few years revolutionary movements have come to include also a significant proportion of the peasants and the very poor, who are easily attracted by the promise of agrarian reform and better social conditions. Fidel Castro's Marxist-oriented social revolution has become a symbol to many of the poor and underprivileged of Latin America. The peasants played a role in the Mexican Revolution of 1910, and indirectly in Bolivia in 1952.

TECHNIQUES

The internal coup d'etat, which has been the most frequent type of revolution, is characterized by a narrow internal organization; it is usually carried out purely within the government and involves only military, government, and closely associated elements. Planned and executed by a very small and closely organized group, aided by loyal supporters, it usually results merely in a change of dictators or of the leadership clique. Several significant techniques have been used with varying frequency in the broader-based Latin American revolutions. Since the 1930's the general strike has been gaining importance. It usually rests on the action of university students, often, but not always, acting in coalition with organized labor. The general strike has been especially significant in Central America, where it has been a major factor in the overthrow of governments—e.g., Martínez in El Salvador and Ubico in Guatemala.

Demonstrations also are likely to have some sort of organizational base. They are typically instruments of organized opposition groups—political parties, especially the active Communist minorities; the church, when it finds itself opposed to government measures; labor unions; and university groups. But demonstrations have also been organized by the government in power to give the impression of popular support.

Mob action and riots, more violent and less under organizational control, are infrequent in Latin America. They signify a very serious malfunctioning of the political system. An example was the great violence accompanying the downfall of the Bolivian Government in 1946, when President Vallarroel's body was hanged from a lamppost in La Paz.

Assassination of government officials is another manifestation of revolutionary violence in Latin America. Usually minor officials are the victims; the leader is given the traditional right of "asylum and exile." However, there have been a number of recent exceptions: Gen. Anastasio Somoza of Nicaragua was assassinated in 1956, Col. Carlos Castillo Armas of Guatemala was killed a year later, and General Trujillo of the Dominican Republic was murdered in 1961.

RESULTS AND OUTLOOK

As everywhere in underdeveloped areas, nationalism is a major element of today's revolutionary movements in Latin America, but their social and political impact varies widely. Some are extreme in their orientation, others are moderate in their social and political philosophy and advocate gradual change. The standard components include an appeal to national integrity and demands for economic development and social justice. Most nationalist movements also call for a broadening of the popular base in the political processes of the state.

The social welfare orientation is shared by three types of movements: (1) the Communist movements, (2) the national agrarian-populistic movements, and (3) a steadily growing movement aptly termed the "Jacobin Left" by Robert Alexander.[10] Each of these will be discussed in turn.

Communist parties—small, well-organized, and highly vocal—function openly and legally in some Latin American countries; in the others—Brazil, Costa Rica, El Salvador, Guatemala, Honduras, Nicaragua, Paraguay, and Peru—they maintain active underground organizations or ally themselves so closely with legal national parties that their aims seem to be indistinguishable. The Communist tactics are to remain in the background, manipulating those in positions of leadership and creating confusion, rather than putting forth a positive program. Their interest lies not in strengthening and bettering the lot of the people but in weakening governments, thus creating conditions deemed favorable for the eventual rise of communism.

The second type of nationalistic movement is comprised of those agrarian-populistic parties supported by the elements of the democratic left. These "Aprista parties," as they are called, seek far-reaching social and economic change, usually including agrarian reform and the integration of the poor into the country's political process. Unlike the Communists, they have international ties only within Latin America. The prototype is the APRA of Peru; other Aprista parties are the *Acción Democrática* of Venezuela,[d] the National Liberation Party of Costa Rica and, in a different way, Mexico's PRI.

The third type of nationalistic movement is made up of the "Jacobin Left."

> This group regards the checks and balances and the individual freedoms of the democratic system as obstacles to agrarian reform, mass education and "economic independence." It argues that the United States, fearful of losing a source of cheap raw materials and foodstuffs, opposes industrialization in the hemisphere. It insists that the enemies of the United States are the area's natural allies, and that only a dictatorial revolutionary regime can move with the speed and decision necessary to bring about an effective social revolution in Latin America.[11]

The Jacobin Left is not a new phenomenon. A decade ago President Juan Perón of Argentina tried unsuccessfully to rally this group around himself so that he could become a hemispheric leader. Today Fidel Castro is trying to gain the support of this steadily growing group, which, although not Communist, does not believe in political democracy and thus is willing to work with the Communists in their struggle to gain control of the nations of Latin America.

At its outset, Castro's revolution found great sympathy throughout Latin America. He overthrew the dictatorship of Fulgencio Batista, and inspired hope for the future not only of Cuba, but of all Latin

---

[d] See discussion of Venezuela, p. 147.

America. But it rapidly became apparent that the movement was controlled by Communists, temporarily masquerading as advocates of an agrarian-populist revolution of the Mexican type. Castro's statement of December 1961 that he had always been a Marxist-Leninist was an open admission that, to gain mass support, he had concealed his Communist convictions.

The turn of events in Cuba has resulted in a bitter struggle throughout Latin America between the Jacobin Left and the democratic left. Although Castro came into power with the full support of the Latin American democratic social revolutionary elements, he has since cut all ties with these groups. In fact, he has gone even further. Whereas during the early months of his regime he dispatched men, money, and arms to enable the exiles of Nicaragua, Panama, the Dominican Republic, and Paraguay to fight against their dictators, he has since concentrated his efforts on overturning regimes controlled by democratic social revolutionary elements—for instance, President Rómulo Betancourt's government in Venezuela.

Despite the "communizing" measures taken in Cuba, a large majority of the Latin American public remain pro-Castro, each part of it for different reasons: the middle class because they do not want to be considered antirevolutionary or pro-United States; the labor unions out of sympathy for Castro's social reforms; the peasants because of the propaganda-built agrarian reforms. Most governments, although knowing the facts, have been reluctant to declare themselves anti-Castro because the parties they represent are split into pro-Castro and anti-Castro factions. However, at the meeting of the Organization of American States in January 1962, 14 of 21 nations were persuaded to vote in favor of a resolution recommending the ouster of Cuba from that international body.

Instability and dissatisfaction with the existing social and economic order have been manifested with increasing frequency in Latin America. Bolivia, Ecuador, and Colombia are being racked by demonstrations of discontent, which threaten the existence of the present governments. It is a subject of conjecture whether these overt indications of displeasure have arisen as a direct result of Communist and Fidelist agitation or have been merely stimulated by the social revolution in Cuba.

All Latin American governments will have to come to grips with three important developments and their ability to do so will probably greatly affect the type of revolutions that will predominate in the Western hemisphere. (1) There is a broad trend to oust the landed aristocracy, which has occupied a position of economic, political, social, and cultural domination, from its traditional place of authority. (2) There is a general desire for greater economic development and, particularly, industrialization. (3) There is a continuing rise of nationalism, which has received impetus largely through the growth of the middle class with its nationalist loyalties as distinct from the generally international outlook of the aristocracy.[12]

NOTES

1. Hubert A. Herring, *A History of Latin America From the Beginnings to the Present* (New York: Alfred A. Knopf, 1960), p. 3.

2. George I. Blanksten, "The Aspiration for Economic Development," *The Annals of the American Academy of Political and Social Science* (March 1961), 11.

3. George I. Blanksten, "The Politics of Latin America," in *The Politics of the Developing Areas*, eds. Gabriel A. Almond and James S. Coleman (Princeton: Princeton University Press, 1960), pp. 462–463.

4. Ibid., p. 462.

5. Lee Benson Valentine, "A Comparative Study of Successful Revolutions in Latin America, 1941–1950" (Ann Arbor, Michigan: University Microfilms, Inc., 1958. PhD. Thesis, Stanford University [1958]), 25.

6. Halcro J. Ferguson, "Cuban Revolution and Latin America," *International Affairs*, XXXVII (July 1961), 288.

7. Valentine, "Study," p. 37.

8. Blanksten, "Politics," pp. 497–501.

9. Valentine, "Study," pp. 249–251.

10. Robert Alexander, "Communism in Latin America," *The New Leader*, XLIV, 5 (January 30, 1961), 19.

11. Robert Alexander, "Castroism vs. Democracy," *The New Leader*, XLIV, 6 (February 6, 1961), 3.

12. Robert Alexander, "New Directions: The United States and Latin America" Current History, XLII, 246 (February 1962), 66–67.

RECOMMENDED READING

BOOKS:

Blanksten, George I. "The Politics of Latin America," in *The Politics of the Developing Areas*, eds. Gabriel A. Almond and James S. Coleman. Princeton: Princeton University Press, 1960. Contains a general discussion of social and political trends. Not always very clear or convincing.

Valentine, Lee Benson. "A Comparative Study of Successful Revolutions in Latin America, 1941–1950." Ann Arbor, Michigan: University Microfilms, Inc., 1958. PhD. Thesis, Stanford University [1958]. Author looks at revolutions in each country.

Herring, Hubert A., *A History of Latin America from the Beginnings to the Present.* New York: Alfred A. Knopf, 1960. A standard history of Latin America brought up to date.

PERIODICALS:

Blanksten, George I. "The Aspiration for Economic Development," and K. H. Silvert, "Nationalism in Latin America," *The Annals of the American Academy of Political and Social Science* (March 1961). This issue of the *Annals* presents a series of informative articles by today's authorities on Latin America; it is especially aimed at a study of nationalism and covers all the important aspects, including economic development, revolutionary factors and forces, a study of selected Latin American revolutions.

Ferguson, Halcro J. "Cuban Revolution and Latin America," *International Affairs* (July 1961), 285–292. This article by an English observer analyzes the effect of the Cuban revolution on the rest of Latin America and criticizes the American attitude and policy toward this upheaval.

THE GUATEMALAN REVOLUTION OF 1944

SYNOPSIS

In October 1944 Guatemala experienced the second phase of a revolution that overturned the traditional conservatism that had kept the country in a semifeudal state ever since it won independence from Spain. In the previous June, the 13-year-old dictatorship of Gen. Jorge Ubico had been replaced by an equally authoritarian government under Gen. Federico Ponce Vaides. But the "forces of enlightenment"—the young middle class professionals, workers, and students—which had brought about Ubico's overthrow allied themselves with powerful military elements to remove Ponce also. On October 20, in a rapid armed attack, these elements completed the task begun in June. The government was placed in the hands of a junta, which initiated the writing of a new and more democratic constitution, and called a general election. In December, Juan José Arévalo, a middle-class intellectual with a new ideological outlook, was elected President, and started Guatemala on the path of economic and social reform. Gradually, however, it came more and more under the influence of communism.

BRIEF HISTORY OF EVENTS LEADING UP TO AND CULMINATING IN REVOLUTION

At the beginning of the 1940's Guatemala's basic power structure was similar to what it had been under Spanish colonial rule. The monarch-directed political system had given way to the military dictatorship of *caudillos*, who maintained an equally authoritarian government. The wealthy Spanish landowner had moved over to share his elitist position with the industrious German coffee grower and representatives of the United Fruit Company, a U.S. corporation. The army still served as the personal protector of the executive in power, and the basis of the political system.

The church, united by traditional bonds to conservative military men and landowners, held thousands under its sway. The Indians, who still constituted the majority of the population, maintained their traditionally isolated low-level existence on subsistence farms and on the great coffee *fincas*.

Yet there was something new in the air, generated by the 20th century ferment elsewhere, against which Guatemala could not completely isolate itself. The growing and predominantly mestizo (Ladino)

125

middle class, consisting of intellectuals, white-collar workers, professionals, students, and young army officers, were strongly impressed by the victories of democracy over totalitarianism, of the Allies over the Axis powers, of the people of neighboring El Salvador over their dictator, Hernández Martínez, in 1944.

These victories—especially that of the people of El Salvador, whose situation under Martínez was not much different from that of the Guatemalans under Ubico—served to crystallize dissatisfaction with Ubico's conservative policies and his unconstitutional and often violent methods of maintaining himself in power. Ubico's rule had been based on conservative elements in the army and the ownership class. He used systematic violence and blackmail to maintain himself in power. He distrusted the labor movement and feared Indian uprisings. Demands from students, lawyers, and professors for more autonomy in the university, political reforms, and salary increases brought on

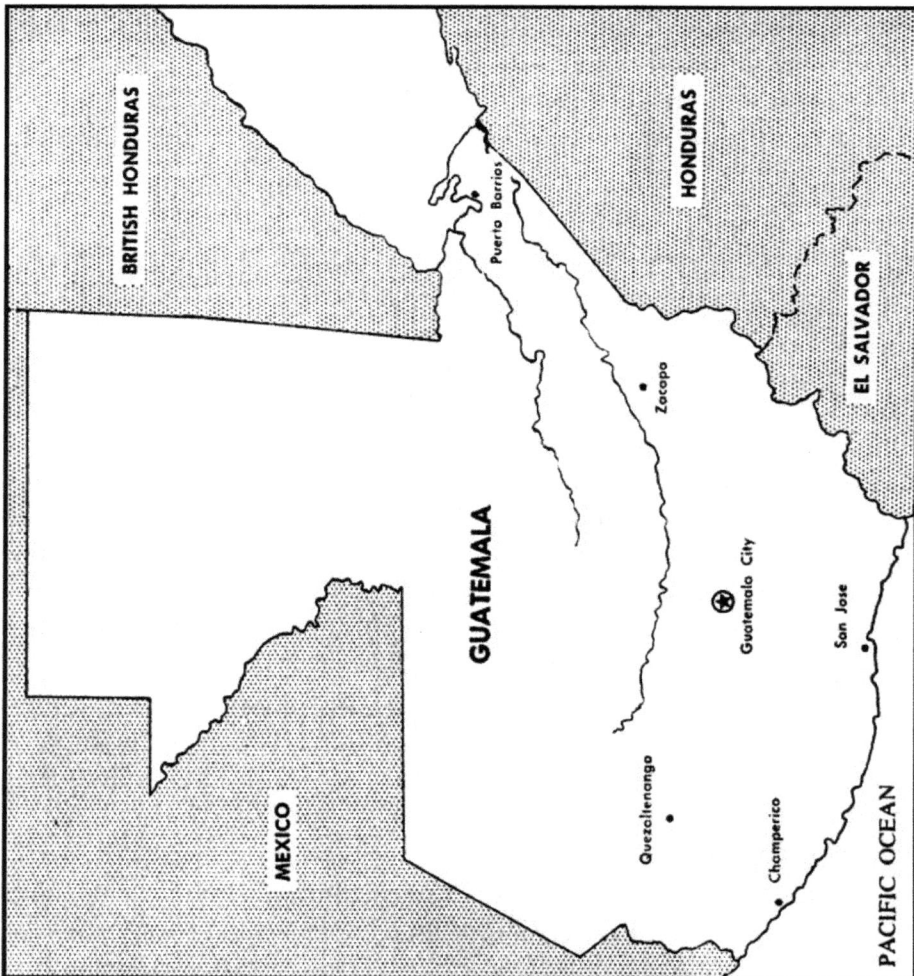

serious repercussions. On June 22 Ubico rejected these demands and suspended civil guarantees, avowedly because of the activities of "Nazi-Fascist agitators." This turned the affair into a major political crisis.

The *Memorial de los 311*, a petition prepared by young lawyers and signed by 311 prominent citizens, was submitted to the government on June 24. It explained the reasons for the unrest and asked that constitutional rights be restored. That same day the National University students in Guatemala City held a peaceful demonstration of protest. That evening, at a second gathering, they demanded Ubico's resignation. They were brutally dispersed by the dictator's strong-arm men; 25 persons were killed.

The following day a procession of middle-class women and children, dressed in mourning to protest the strong-arm methods used in dealing with the students, was fired upon. The death of a young school teacher, Maria Chinchilla, gave the revolution its needed martyr and symbol. Enraged citizens broke off negotiations for a peaceful settlement, which the diplomatic corps had persuaded Ubico to undertake. Businessmen and railway workers joined a protest strike; banks and businesses were closed. Ubico called out the troops and on June 27 placed the employees of the railways and public utilities under military law. These measures did not halt the strikes and violence. A deluge of petitions poured in from people and groups in all walks of life, demanding Ubico's resignation. These, coupled with the prolonging of the general strike, finally persuaded Ubico, and on July 1 he turned authority over to a military triumvirate.

But the revolution did not accomplish the desired end of replacing a tyrannical form of government with a constitutional democracy. General Federico Ponce Vaides forcibly "persuaded" the Legislative Assembly to appoint him Provisional President and set out to pattern himself after his predecessors.

THE ENVIRONMENT OF THE REVOLUTION

DESCRIPTION OF COUNTRY

Physical characteristics

Guatemala, the third largest country in Central America, with an area of 42,364 square miles, is comparable in size to the state of Tennessee.[1] It is located between Mexico on the north, British Honduras on the east, El Salvador and Honduras on the southeast, and the Pacific on the northwest.

It is similar to most other Central American nations in that highlands form the "backbone" of the country. In the south lies a dense jungle, which gives way near the Pacific to a series of mountain ranges which run, spine-like, across the country from Mexico to Honduras. These highlands provide the cool elevated basins where most of the inhabitants live. The coastal valleys and jungles are unpleasantly hot and rainy and consequently are sparsely populated.

The people

Guatemala is the most populous country of Central America, with a 1940 census count of 3,284,269, or 68 persons to the square mile.[2] Most of the people, however, live in the southern highlands; almost half the country is very sparsely settled if at all. The majority of the inhabitants (60 percent) are Indian; Guatemala has the largest proportion of Indian population of any Central American country. The second major group is made up of mestizos, or Indian-white breeds (35 percent). There is a very small minority of whites and just a few Negroes.[3]

The Indians are divided into several major groups. In the west are the Toltecs, to the east the Aztecs, and in the north the Maya-Quiches, the largest group.[4] Also worthy of note are the Cakchiquel, the Kekchi, the Mam and the Tzutuhil.[5] The overwhelming majority of Indians do not speak Spanish; there are 21 distinct Indian dialects.[6]

Most of the Indians live in isolated communities in the highlands, leaving the cities to the "white" man or the Ladino. Guatemala City, the capital and the largest city in Central America, had a 1940 population of 163,826. It was followed by Quetzaltenango (33,538), Puerto Barrios, a major port (15,784), and Zacapa (14,443).[7]

Communications

Transportation facilities in 1944 were inadequate. Those that did exist were centered on the productive and densely populated highland areas. Even there they were so limited and primitive as to slow down economic progress seriously.

In the coastal lowland and jungle areas comprising 46 percent of the country there were only cart roads and some river traffic. The only railroad was on the United Fruit Company plantation in Isabel. The population of Peten, the largest department, had to import food by plane, riverboat, and pack mule.

Three railroads operated within the republic; the largest of these, the International Railway of Central America, was controlled by U.S. interests. The railroad system connected San José with Guatemala

City and Puerto Barrios. A branch led to El Salvador. Along the Pacific lowlands there was a connection with the Mexican railroad system.[8]

Pan American Airways and Central American Airways connected the department capitals and neighboring American countries. Puerto Barrios on the Atlantic and San José and Champerico on the Pacific were the major ports for ocean trade.

Roads of all types totaled less than 5,000 miles;[9] the Pan American Highway and the Highway of the Pacific were the main arteries.

Natural resources

The Guatemalan economy in 1944 was predominantly agricultural, as it is today. Over 90 percent of the working population gained their livelihood from the land. This percentage is probably somewhat lower today. Maize, coffee, bananas, chicle, and sugar are the most important products. A large area devoted to pasturage contributes to the growth of stockraising. Enough is produced to satisfy internal demands for meat, and many hides and some frozen meat are exported.

Guatemala carries on a small number of extractive industries. Silver is still found in some of the old mines, and small amounts of gold are panned from mountain streams. Other mineral resources include sulfur, lead, zinc, tin, copper, chromite, manganese, and salt, but they are not mined in significant quantities.

SOCIO-ECONOMIC STRUCTURE

Economic system

The commercial life of Guatemala had been largely controlled by the small ruling native aristocracy. Until World War II the great coffee plantations were largely in the hands of Germans who had recently migrated to Guatemala. During the war, these properties were confiscated by the state and the industry was nationalized. Foreign penetration from another quarter, the United States, resulted in the development of another very important crop, the banana. The United Fruit Company, which started in 1909, by 1944 virtually controlled the entire banana growing industry, which at that time was still relatively small.

Corn for local consumption utilized approximately 60 percent of Guatemala's land. In 1941 the United States supplied 78.5 percent of the imports and took 92 percent of the exports. Textiles, petroleum products, medicines, and motor vehicles were the major imports.[10]

Industry had not been very significant in the Guatemalan economy. It was concentrated mainly on processing agricultural products and producing light consumer goods. Manufacturing enterprises represented little capital outlay, operated on a small scale, and employed few workers.

Class structure

Since there was little in the way of mineral resources to attract the Spaniards, those who did settle remained as *latifundistas*, gaining their wealth and position from the agricultural produce of their large plantations, or as officials. They settled around Guatemala City and Antigua, leaving the northwestern highlands to the Indians.

The bases of society changed little in the four centuries following the Spanish conquest in 1529. Until after World War II Guatemala remained an underdeveloped agricultural country in which a few large landowners, allied with the officer corps of the army and backed by the representatives of foreign corporations and the hierarchy of the Catholic Church, controlled a large population of illiterate Indians and Ladinos. Political and economic power was concentrated in the hands of the producers of the two major export crops—coffee and bananas. This ruling elite, which constituted slightly over 2 percent of the total number of landowners, held title to more than 60 percent of the cultivated land, while two-thirds of the landowners held only 10 percent of the farmland.[11]

Although Spanish culture had made some penetration, particularly in religion, speech, and dress, the Indian villages continued an ancient pattern of life based largely on local traditions and on subsistence agriculture. The Indians had almost no comprehension of the profit motive and little desire to accumulate land or goods. While they were poor by the standards of the industrial countries, it was not an "abysmal" poverty and for the most part they accepted their lot.

In the social hierarchy, the Ladinos had greater status than the Indians. Broadly speaking, a Ladino is one who knows Spanish and has adopted some of the customs and habits of the white man. This term, therefore, refers primarily not to race but to culture—habits, customs, and patterns of life. Any Indian may become a Ladino merely by moving to the city and adopting as much of the "white" man's culture as he can, by gaining some wealth, or by marrying a Ladino woman.[12]

Literacy and education

About 80 percent of the adult inhabitants of Guatemala are illiterate[13] despite the fact that education is compulsory for children from 7 to 14 years of age.[a]

Teaching is an especially difficult problem because the teachers must either be fluent in the Indian dialects or be Indians able to bridge the cultural gulf between the native and the Ladino. In addition the schools are poorly equipped and inefficiently managed, but probably not more so than in many other parts of Latin America.

The National University in Guatemala City, with an enrollment of 855 and a faculty of 120, was and still is the only institution of higher learning. It offered courses in Law, Social Sciences, Engineering, Natural Sciences, Pharmacy, Medicine, and Economic Sciences.

Major religions and religious institutions

Although the population is predominantly Roman Catholic, all faiths enjoy freedom of worship. The religion of the Indians is a mixture of old Mayan beliefs and Catholicism, as it was taught during the colonial period rather than as it is practiced today in Western Europe and the United States.[15] The Indian is indifferent to much of what the official church says, but he does recognize a certain orthodox sphere in which the Catholic priest is competent to act. The Catholicism of the male Ladino is a matter of social and political advantage rather than of religious conviction; the men attend church only on very special occasions. The women tend to be more strongly attached to the church.

GOVERNMENT AND THE RULING ELITE

Description of form of government

In theory the government of Guatemala in 1944 rested on a democratic constitution dating from 1879. Public powers were divided between three popularly elected branches; the executive had a 6-year term, the legislature a staggered 4-year term and the judiciary a 4-year term.[16] In practice, however, the government was completely dominated by a strong executive supported by the military, the large landowners and foreign interests. Any potentially dangerous opposition had a choice between self-imposed exile, asylum in a foreign embassy, jail or death. Congress met occasionally to "confirm" Ponce's legislation.[17]

---

[a] Another source sets the figure at only 67 percent in 1945.[14]

Description of political process

Political parties and power groups supporting government

The pattern of Guatemalan politics had been set in the 19th century. It was characterized by the rule of strong men who were able to maintain themselves in power for long periods. The most significant change in the 19th century was probably in the relationship between church and state. From 1871 to 1885 Guatemala was under control of Justo Rufino Barrios, a determined "liberal" in the sense that he tried to reduce the political and social power of the Catholic church and succeeded in doing so. The Liberal Party machine became firmly entrenched under his rule and remained the dominant force in politics. The idea of constitutionalism was accepted, but the constitution never became more than a symbol.

The shift from the Conservative to the Liberal period did not mark the advent of democratic practices. The Conservatives had stood for localism, cultural isolation, and political gain through powerful established institutions, especially the church and the army. The creed of the Liberals, with its stress on secularism, internationalism, and "Europeanism," had an economic basis in the large-scale commercial exploitation of coffee. The great care demanded by the high-grade highland coffee led to labor shortages and new problems.[18]

The tradition of the rule of a strong man continued with Ubico, who assumed the presidency in 1930. Ubico rigidly suppressed all opposition and used secret police to guard against possible plots against his regime. Following the revolution in El Salvador in May 1944 he became aware of the growing restlessness of the middle group and declared a state of siege. Yet popular dissatisfaction with his oppressive rule became so great that he was forced to resign. For the first time it appeared as if the traditional ruling groups, composed of conservative army officers, large landowners, and domestic and foreign commercial interests, were losing their grip on the political reins of the country. The revolution in June 1944 was only the first stage in the struggle for power between the traditional ruling class and the emerging, progressive middle-class elements. Gen. Federico Ponce Vaides succeeded to the Presidency, but "Ponce still took orders from Ubico, who remained quietly in the background."[19]

General Ponce followed in the footsteps of his predecessors, demanding and gaining the support of the traditionally conservative-minded Liberal Party elements. A large portion of the army which had forced the Legislative Assembly to name him Provisional President became his personal following. Large landowners and important church elements acknowledged his leadership, unwilling to support what they regarded as a Ladino-dominated attempt to usurp their authority in the June movement. Ponce attempted at first to gain the backing of the middle-class Ladino elements by permitting the organization of labor unions and political parties and granting concessions to workers, students, and teachers. He discontinued this attempt, however, when these elements began to gain too much popular support.

Character of opposition to government and its policies

When Ponce lifted restrictions on the functioning of political parties, the revolutionary elements within the country, together with returned exiles, took immediate action; a number of new political parties sprang up. A small group of professionals and intellectuals organized the *Renovación Nacional* (RN, or National Renovation Party). The *Partido Social Democrática* (PSD, or Social Democratic Party) was formed by older professionals, many of whom had been in exile during the Ubico regime; they tended to gather about Adrian Recinos, a well-known conservative, prochurch, aristocratic scholar. But theirs was a weak party, with little chance for popular support. The June revolution had been carried out by the young, and they were not willing to cede their potential power to older, more conservative leaders who had been in exile or had remained quiet under Ubico's restrictions. These younger elements formed the *Frente Popular Libertador* (FPL, or Popular Liberation Front), which attracted the majority of the center- and leftist-oriented and acknowledged Professor Juan José Arévalo as its standard-bearer. A Communist Party was also organized, but its membership and status at that time were so insignificant that it became lost among the other revolutionary elements. Campaigning for the promised presidential election provided a *raison d'être* for these embryonic organizations, and it soon became clear that Arévalo had the strongest support among those who were politically active, especially the hard-working Communist minority.

When Ponce announced in early August that he would run for President on the old Liberal Party platform, the opposition speeded up its campaign. An attempt to force the legislative assembly to amend the constitution, so that Ponce could continue in office, evoked scathing criticism from the opposition parties. After printing editorials critical of government policies, the publisher of the largest Guatemalan daily, *El Imparcial*, was murdered. Overt opposition was once again outlawed.

Legal procedure for amending constitution or changing government institutions

The pre-Arévalo Constitution provided that changes in the fundamental law could be initiated by a two-thirds vote of the legislative assembly. The assembly then would be dissolved and the President would call an election to choose a constitutional assembly. After this assembly had acted on the proposed reforms new elections would be held to select a new legislature.[20]

Difficult and complicated as this amendment procedure might appear, Ubico and his predecessors used it repeatedly to broaden their powers and to protect themselves from charges of unconstitutionality. Control of the legislative assembly by the dictator's personal followers made the process relatively simple.

Relationship to foreign powers

General Ponce maintained the close ties with the United States that his predecessors had developed. Large American enterprises, such as the United Fruit Company, the Empresa Electrica, and the International Railway of Central America retained their monopolistic advantages and were thus able to exert a strong influence over the nation's economy. The bulk of Guatemalan exports continued to go to, and the majority of imports to come from, the United States.

The role of military and police powers

Many of the more important army officers quickly threw their support to General Ponce after the overthrow of Ubico. As members of the old elite, they naturally tended to support the system that gave them their power; therefore, they were highly instrumental in "persuading" the legislative assembly to appoint Ponce Provisional President. They remained, on the whole, a loyal and obedient force, ready to crush any opposition to the new dictator.

WEAKNESSES OF THE SOCIO-ECONOMIC-POLITICAL STRUCTURE OF THE PREREVOLUTIONARY REGIME

History of revolutions or governmental instabilities

The history of Guatemala since independence had been marked by the successive rise of *caudillos*, each of whom held office as long as he was able. Yet there had been few attempts at revolution, not because of the personal appeal of the dictators, but because Guatemalan society, so accustomed to dictatorship, was unable to envision or unwilling to implement alternatives. The rural Indian majority was completely disinterested in the politics of what for them was a meaningless concept—"Guatemala." In addition, their geographical separation into small isolated groups and the varied conditions under which they lived made any resistance they might have offered local and unfocussed. More important, perhaps, were the forceful and effective suppression of all opposition, the lack of articulate public opinion and the weakness of the middle class.[21]

Events during World War II ushered in important changes. The middle class revolted against an autocratically-managed society in which it found no place, and thus the dictator Ubico was overthrown, even though he was replaced for a short time by a man who attempted to follow in his footsteps.

Economic weaknesses

The basic economic structure was one of monoculture; depending largely on the exportation of coffee and bananas, and the trade in both these products was highly competitive, thus making the economy highly vulnerable to fluctuations in world market prices. The country depended on trade for many of the basic articles of daily life.

Internally, the feudalistic-type landholdings perpetuated the poverty and economic exploitation of the Indian masses. Forced to live on subsistence farms, isolated because of inadequate roads, lacking educational and sanitary facilities, the Indians constituted a rural proletariat so backward and underprivileged that they contributed little to increased productivity or industrial development.

The lower-middle-class laborer and the native businessman were also discriminated against by the foreign monopolies that controlled the economy, and could do nothing to remedy the situation. Rigid censorship kept the public from challenging governmental policy, and labor was not allowed to organize to enforce its demands for higher wages.

Social tensions

The majority of the Indian population lived on isolated subsistence farms under conditions of poverty, ignorance, and lack of sanitation. Nevertheless, they were relatively content and kept to themselves, avoiding most Ladinos and the city.

Discontent existed mainly among the educated middle class and the urban workers. There was no place for them in the semifeudal society that had prevailed since the time of the conquistadores. They did not in fact have the rights which the constitution guaranteed them. Censorship was rigid. They were not allowed to organize. Wealthy foreigners joined with the traditional elite to control the country. It was no wonder that there was antagonism between these two groups: one unwilling to relinquish any power, the other demanding a greater share of it.

Government recognition of and reaction to weaknesses

When Ponce first took over in early July 1944, he revoked Ubico's decree suspending civil rights. A free press sprang up. Refugees returned. Labor was allowed to organize and political parties to form. Most urban Guatemalans joined the labor "guilds," which for the first time ensured them living wages. Specific concessions were granted to students, professors, and workers—three elements that had been most active in bringing about the downfall of Ubico's regime. The tight monopolies on sugar, meat, and tobacco were broken up.[22] Pro-Nazi Foreign Minister Carlos Salazar was forced to retire. German-owned coffee plantations were expropriated. But these measures were only temporary, and soon gave way to severe measures to repress middle class discontent with the continuing authoritarianism.

FORM AND CHARACTERISTICS OF REVOLUTION

ACTORS IN THE REVOLUTION

The revolutionary leadership

The leaders of the revolution were mostly younger army officers, assisted by a few liberal-minded civilians. Four names stand out as significant. Jacobo Arbenz Guzmán was a captain in the Guatemalan army and a political officer of the *Escuela Politécnica*, Guatemala's West Point. Dissatisfaction with the Ponce regime and an administrative shakeup within the school caused him to resign. He was essentially a strong nationalist, bent on breaking the monopoly of foreign interests. To accomplish this, he helped formulate a plan to remove Ponce from office and replace him with a governing junta.[b]

Jorge Toriello was a local businessman, who represented the interests of the middle-class Ladinos. He too was a nationalist whose main interest lay in implementing full civil liberties and social and economic reforms—for labor, professionals, and the university.

Francisco Javier Arana was the third member of the "planning committee." An army tank commander who led the elite *Guardia de Honor* (Honor Guard) corps of the capital, he was representative of the young army officers who had come up through the ranks. Unknown to the civilian leaders and supporters of the movement, he had little knowledge of, or interest in, their aims, and was concerned primarily with personal power and gain. He became the central figure among

---

[b] Several years later, in November 1950, he was elected President and it was during his regime that the Communists reached the peak of their strength in Guatemala.

the three conspirators, both because of the military forces he commanded and because the nonmilitary sector of the anti-Ponce movement had been much weakened by the arrest, exiling, and forced asylum of civilian revolutionary leaders.

Juan José Arévalo was a 40-year-old professor of middle-class background. He had become an Argentine citizen during 15 years residence in that country, where he had taken semivoluntary exile after a dispute with Ubico cost him his position in the Department of Education. He was better suited to appeal to the Ladinos than either a military person or a member of the old aristocracy. With his civilian status, his social position as a middle-class Ladino, his intellectual profession, and his comparative youth, he seemed to embody the major characteristics of the revolutionary elements.

His ideas were decidedly radical in the traditionally conservative environment of Guatemala. Formulated under the imposing title, "spiritual socialism," they affirmed the necessity of achieving "human dignity." On a national scale this implied economic and social transformation. The common man achieved "greater dignity when he was in good physical health, literate, able to provide for himself and his family, when his rights as a worker and as an individual were protected."[23] But Arévalo expressed no intention of transforming the economic and social systems. He merely observed that the semifeudalist state would be "softened with measures for the defense of the workers, land redistribution, the raising of wages, better housing, and the socialization of culture, of hospital services, [and] the popularization of diversions."[24]

The revolutionary following

Middle-class inhabitants of Guatemala City provided both the backbone and the support for the revolution of October 1944. These were largely persons without special ties either to the old landowning aristocracy or the present military and political elite. They were a minority of the Ladinos in the capital. University students and young professionals seemed to have been those most influenced by the democratic victories of the Allies and of the people of El Salvador; they became the most outspoken supporters of the revolution. Their ranks were swelled by older, less active anti-Ubico forces, many of whom had been in exile. These were old enough to have witnessed and perhaps participated in the Unionista movement that overthrew the despotic Estrada Cabrera in 1920 and in the subsequent, relatively liberal, administration.

ORGANIZATION OF REVOLUTIONARY EFFORT

Internal organization

When it became evident that Ponce intended to follow in the footsteps of his predecessors, Capt. Arbenz Guzmán went to El Salvador, supposedly to visit his wife's family, but actually to attempt a revolution from across the border. In the middle of September he was joined by Jorge Toriello, the go-between for the civilian revolutionary leaders and that part of the army loyal to Arbenz. They began to plan the coup. When a third plotter joined the group it came as a surprise, for no one had thought of Maj. Francisco Arana as a possible ally. His powerful military aid was welcomed, for it was clear that the revolution would have to be backed by force. Ponce's repression had caused the most important civilian revolutionary leaders to seek asylum in embassies or go into hiding; even Arévalo had to turn to the sympathetic Mexican Ambassador for political asylum. By October 15 the details of the shaky alliance between Arbenz, Arana, and Arévalo had been settled. In company with a small group of officers, Arbenz planned to attack quickly, before the conspiracy could be discovered.

No details of contacts between the conspiratorial group and civilian party leaders have been found for this study. It seems from the events that such contacts must have existed. There was also a group of younger officers recruited to serve the revolutionary cause who probably had advance knowledge of the plan to oust Ponce.

External organization

The coup was planned in El Salvador, largely because many of the people of that country, having recently rid themselves of a dictator, were sympathetic toward the efforts of the Guatemalans to do likewise. Arbenz, Toriello, and Arana therefore hoped for aid from the new Salvadorean Government. To what extent they succeeded has not been ascertained.

GOALS OF THE REVOLUTION

Concrete political aims of revolutionary leaders

The goals of the revolution reflected the grievances of the middle class that supported it. The young leaders of the revolution wanted civil liberties; freedom of speech, press, and association were particularly stressed, as they had long been prescribed. A parliamentary approach to the solution of political problems was to replace the authoritarian regime of General Ponce.

Social and economic goals of leadership and following

Social and economic goals included recognition of the rights of labor, both organized and unorganized, rural and urban. Foreign interests had long been sacrosanct, enjoying monopolistic privileges; it was now believed time for native and governmental interests to take over. The ownership of too much land by too few people was to be remedied by an agrarian reform. Educational facilities were to be extended to the more than half of the population who had no schools; without education, it was felt, little could be done to improve the socio-economic-political situation.

Arévalo himself characterized the revolution as being directed against the following conditions:

1. A government headed by men who had themselves continually reelected.

2. Officials who represented a social and political minority uninterested in, or even unaware of, the sufferings of the people.

3. Official indifference toward exploitation of resources by foreign or Guatemalan capitalists with no thought except private profits.

4. The absence of freely functioning political parties and a free press, and the lack of autonomy of the judicial and legislative departments.

5. The lack of popular organizations, including labor unions, to represent the interests of the people and to protest abuses.

6. The existence of boss rule by military or political officials who represented miniature dictatorships.

7. General lack of respect among officials for the human personality.

8. The incapacity of the government to inspire the people to use their sovereignty "as though their country had not yet shed its colonial clothes to convert itself into a Republic."[25]

REVOLUTIONARY TECHNIQUES AND GOVERNMENT COUNTERMEASURES

Methods for weakening existing authority and countermeasures by government

When Ponce, contrary to public hope, announced that he would run for President on the old Liberal Party ticket, legal opposition activities speeded up. The enthusiastic reception given Arévalo upon his arrival in the capital after Ubico's fall caused Ponce to initiate a campaign of severe repression. A large number of young lawyers and the leaders of the FPL were arrested; leaders of the Arévalo forces were again forced to seek asylum in foreign embassies to avoid arrest or exile.

In early October, Ponce forced through the first of the constitutional changes that were designed to allow him to continue in office. The opposition leaders subsequently appear to have decided that Ponce must be overthrown to keep the country from entering into another long period of dictatorial government. Meanwhile Ponce attempted to rally support by stirring up racial conflict. He brought groups of Indians into Guatemala City, armed them and waited for disorder to break out. But violence did not break out and the Ladinos had no need to turn to him for protection as he had hoped.

On October 16, Arévalo, Jorge García Granados, and Roberto Arzu issued a *Manifesto del Frente Unido de Partidos Políticos y Asociaciones Cívicas,* in which the newly organized political parties and association of university students proclaimed a *paro político* (political strike). This document was mimeographed and distributed in the streets and among the groups supporting the revolution. It had the effect of stopping work in several sectors of the economy. Two days later, on October 18, the faculty and students of the National University declared a strike.

The next night Guatemala City reverberated with rumors of an imminent coup d'etat. There were three contradictory versions, however, which served to confuse the secret police and keep them from taking effective action. The first stated that "Ponce had decided to take definite action to strengthen his hold on the Presidency," the second that "War Minister Corado was going to take over the executive power to 'restore order'," and the third that "the people were about to start a popular uprising to finish the work begun in June."[26]

Methods for gaining support and countermeasures taken by government

The middle-class revolutionary element primarily composed of Ladinos, had immediately taken advantage of the "privileges" granted them after Ubico's resignation. But, as Ponce's measures became more and more repressive in September and October, the revolutionary ranks were expanded to admit young officers who had been influenced by the democratic propaganda of the Allies while training in the United States and Europe, and who saw little hope for a profitable future under Ponce's regime. A small number of these officers were able to bring into the active revolutionary camp all of their loyal subordinates. These men proved a highly important factor in the plan for ousting the new dictator.

MANNER IN WHICH CONTROL OF GOVERNMENT WAS TRANSFERRED TO REVOLUTIONARIES

The events of mid-October seem to indicate that the revolution had been planned by Arbenz and his fellow-conspirators to include both military and civilian elements. The October 16 manifesto calling for a political strike was followed on October 18 by the strike in the National University and by a multiplicity of rumors circulating through Guatemala City on the night of the 19th. These set the stage for the coup which overthrew the government on the 21st.

On the evening of October 20 a small group of rebel army officers smuggled 70 students and workers into the strategic *Guardia de Honor* Fortress in the capital. The officers of the fortress and prisoners held there killed the commander and seized the fort. Captured lend-lease military equipment stored there was distributed to civilian revolutionary groups from the barracks of the *Guardia de Honor*. The rebels then blew up the San José Fortress, destroying hundreds of dollars worth of explosives. Loyal artillery in the Matamoros Fortress fired into the barracks of the *Guardia de Honor* but a counterbarrage reduced Matamoros to rubble. Within 12 hours the provisional government was overthrown; revolutionary elements laid siege to the National Palace and received the unconditional surrender of Ponce and his Cabinet. 800 to 1,000 casualties were the result of the 12-hour fight. By October 22 peace had been restored, and Ponce and his chief advisers were placed aboard a plane for Mexico.

THE EFFECTS OF THE REVOLUTION

CHANGES IN THE PERSONNEL AND INSTITUTIONS OF GOVERNMENT

Arana, Arbenz, and Toriello made up the junta that was to act as the interim government. But effective political power could come only from cooperation between the military members of the junta and the citizens of Guatemala City. Arévalo's backers extended their support to Arana through the *Frente Popular de Partidos*, a loose conglomeration of Arevalísta forces. In return, the students demanded immediate elections for the legislature to take advantage of the present enthusiasm of the masses. The first decree of the junta dissolved the assembly and announced elections to be held on November 3, 4, and 5.

In what is considered to have been the freest election Guatemala ever experienced Arévalo was elected President by an overwhelming majority. He took office on March 15, 1945. Arbenz became Minister of War, Arana was appointed Chief of the Armed Forces, and Toriello assumed the office of Finance Minister. (He was ousted a year later when he attempted to gain control of the government.)

Simultaneously with Arévalo's assumption of the Presidency a new constitution was promulgated. It was one of the most liberal and progressive political charters in Guatemalan history. It gave generous guarantee for all the basic rights of labor and free institutions and strongly emphasized the responsibility of the state for the economic and social welfare of the underprivileged.

MAJOR POLICY CHANGES

Arévalo was labeled a "Communist" almost immediately, at least partly because his government severed relations with Spain and established them with the Soviet Union. Nationalism found a ready scapegoat for Guatemalan backwardness in charges of foreign "imperialism" and dedicated itself to overthrowing it.[c]

Agrarian reform laws expropriated about 130 coffee *fincas*, most of them belonging to persons of German descent, but nothing was done about the large holdings of the United Fruit Company. That company had been protesting the new labor code and thus created a great deal of antagonism between American and government personnel. It was not until the regime of Arbenz that more radical agrarian reform measures were instituted.

LONG RANGE SOCIAL AND ECONOMIC EFFECTS

Arévalo's government passed a large amount of reform legislation, aimed at the expansion of education, the protection of organized labor, social welfare, industrialization, and agrarian reform. School teachers' salaries were raised, more schools for the Indians were established, education was demilitarized, the university was granted

---

[c] A Latin American writer, Germán Arcienagas, in describing the situation in early 1950 at the end of Arévalo's term, quotes a conversation between U.S. Ambassador Richard C. Patterson and Arévalo. "Unofficially, Mr. President," Patterson told Arévalo one day, "I want you to know that as far as I am concerned personally, your government will never get a dime or a pair of shoes from my government until you cease the persecution of American business." In turn, Arévalo said to reporter Samuel Guy Inman, "You do not have an ambassador of the United States here, but a representative of the United Fruit." The Ambassador was declared *persona non grata* and recalled in early April of 1950.[27]

autonomy, 135 schools were built, and a large number of books by the world's great authors were published at low prices. A health and sanitation campaign was initiated and 17 hospitals were built. A labor code was adopted and a social security system put into effect.

Under the succeeding Arbenz regime, American-owned enterprises became the principal targets of Communist attacks. The Agrarian Law of July 1, 1952, provided for the cultivation of unused lands by small farmers so as to create a body of landholders whose varied crops would contribute to the stability of the economy. Owners were to be gradually recompensed and buyers were to repay the government in small installments. The United Fruit Company was obviously the principal target of this law, for it owned a great deal of land, much of which was held in reserve for future development. Therefore 234,000 acres on the Pacific coast and 174,000 acres on the Atlantic coast were expropriated, leaving the company less than half that much; payment for the land was far less than its true value.

OTHER EFFECTS

Within the decade 1944 to 1954 that marked the gradual disintegration of the former all-powerful oligarchy, Guatemala witnessed the spread of Communist influence until, in 1953, it had become the strongest outpost of Communist political power outside the Iron Curtain. This "banana republic" suddenly found itself the center of a raging international controversy and an important battlefield in the cold war.

A shipment of Soviet bloc arms to Guatemala in June 1954 provoked the U.S. Government, which saw its vital strategic interests in jeopardy, to counter the move by sending armaments to Guatemala's neighbors. Some of these arms quickly found their way into the hands of Col. Castillo Armas, leader of a strong exiled anti-Arbenz faction. When this group crossed the border from Honduras in 1954, the Guatemalan Army, which had little love for the Arbenz regime, joined the "invaders" in ousting Arbenz and the Communists.

NOTES

1. *Britannica Book of the Year* (Chicago: Encyclopaedia Britannica, 1944), 327.
2. William Lytle Schurz, *Latin America* (New York: E. P. Dutton and Company, Inc., 1949), p. 66.
3. Preston E. James, *Latin America* (New York: The Odyssey Press, 1959), p. 658.

4. German Arcienagas, *The State of Latin America* (New York: Alfred A. Knopf, 1952), p. 292.

5. Daniel James, *Red Design for the Americas: Guatemalan Prelude* (New York: The John Day Company, 1954), p. 32.

6. *The Worldmark Encyclopedia of the Nations* (New York: Worldmark Press, Inc., Harper and Brothers, 1960), 399.

7. *Britannica Book of the Year*, loc. cit.

8. James, *Latin America*, p. 669.

9. *New International Yearbook*, (New York: Funk & Wagnalls, 1944), 271.

10. Ibid.

11. Ronald M. Schneider, *Communism in Guatemala: 1944–1954* (New York: Frederick A. Praeger, 1958), pp. 2–3.

12. William S. Stokes, *Latin America Politics* (New York: Thomas Y. Crowell Co., 1959), p. 19.

13. *New International Yearbook*, loc. cit.

14. Samuel Guy Inman, *A New Day in Guatemala, A Study of the Present Social Revolution* (Wilton, Connecticut: Worldover Press, 1951), p. 21.

15. Richard N. Adams, "Social Changes in Guatemala and U.S. Policy," *Social Change in Latin America Today: Its Implications for United States Policy*, Published for the Council on Foreign Relations (New York: Harper and Brothers, 1960), p. 240.

16. Chester Lloyd Jones, *Guatemala Past and Present* (Minneapolis: University of Minnesota Press, 1940), p. 94.

17. K. H. Silvert, *A Study in Government: Guatemala* (New Orleans: Middle American Research Institute, Tulane University, 1954), no. 21, p. 2.

18. Ibid., p. 2.

19. Austin F. MacDonald, *Latin American Politics and Government* (2d ed.; New York: Thomas Y. Crowell Company, 1954), p. 526.

20. Silvert, *Government*, pp. 2–3.

21. Ibid., p. 2.

22. Lee Benson Valentine, "A Comparative Study of Successful Revolutions in Latin America, 1941–1950" (Ann Arbor, Michigan: University Microfilms, Inc., 1958. PhD. Thesis, Stanford University [1958]), 62.

23. Charles William Anderson, *Political Ideology and the Revolution of Rising Expectations in Central America 1944–1958* (Madison: University of Wisconsin Microfilm, 1960), p. 268.

24. Ibid., p. 270.

25. Inman, *A New Day*, pp. 40–41.

26. Valentine, "Study," p. 63.

27. Arcienagas, State, p. 295.

RECOMMENDED READING

BOOKS:

Arcienagas, German. *The State of Latin America.* New York: Alfred A. Knopf, 1952.

Bush, Archer C. *Organized Labor in Guatemala: 1944–1949.* Hamilton: Colgate University Press, 1950.

Holleran, Mary P. *Church and State in Guatemala.* New York: Columbia University Press, 1949.

James, Daniel. *Red Design for the Americas: Guatemalan Prelude.* New York: The John Day Company, 1954.

Martz, J. D. *Communist Infiltration in Guatemala.* New York: Vantage Press, 1956.

Schneider, Ronald M. *Communism in Guatemala: 1944–1954.* New York: Frederick A. Praeger, 1958. This study—as well as those by Arcienagas, James and Martz—provides detailed information on the growth of communism in Guatemala under the succeeding Arévalo and Ponce regimes; analysis as to why it was able to take root and prosper in this Central American country. Generalizations are made about the conditions conducive to the growth of communism in Central and South America.

Silvert, K. H. *A Study in Government: Guatemala.* New Orleans: Middle American Research Institute, Tulane University, 1954. Bush and Silvert provide the best overall studies of the particular period concerned; they give the most complete information available on the events leading up to and culminating in Ponce's overthrow. Both give good background sketches.

Stokes, William S. *Latin American Politics.* New York: Thomas Y. Crowell Co., 1959. A general discussion of Latin America.

Valentine, Lee Benson. "A Comparative Study of Successful Revolutions in Latin America, 1941–1950." Ann Arbor, Michigan: University Microfilms, Inc., 1958. PhD. Thesis, Stanford University [1958]. Good account of the socio-politico-economic background. Gives a full account of the Ubico regime and its overthrow.

THE VENEZUELAN REVOLUTION OF 1945

SYNOPSIS

In October 1945, a coalition between a left-of-center political organization and a dissatisfied group of young army officers staged a weekend revolution against a regime (under Presidents Eleazar López Contreras and Isías Medina Angarita) that had ruled Venezuela since the death of the dictator-President, Juan Vicente Gómez, in 1935. Loyal troops offered some resistance, but within 24 hours the revolutionary forces had won control of the government buildings in Caracas, the President had been forced to resign, and a provisional government had been set up.

BRIEF HISTORY OF EVENTS LEADING UP TO AND CULMINATING IN REVOLUTION

A remarkable period of Venezuelan history came to an end when Juan Vicente Gómez died in 1935. Gómez had given his country 27 years of able though dictatorial rule. He held firmly onto the political reins, thanks to a well-equipped army and an efficient spy network, despite numerous attempts to unseat him. He made deals with foreign oil companies to develop Venezuela's petroleum industry, which became the country's greatest asset.[a]

The new order that succeeded the Gómez rule was led by Eleazar López Contreras, then the favorite of military and civilian groups who had greatly resented the Gómez dictatorship. A new constitution, drawn up in July 1936, showed some trend toward democracy. López reduced presidential powers, evicted a number of strongly established Gómez supporters, and announced a 3-year plan of public works. These measures failed to gain the support of the young intellectuals, however, and opposition mounted.

López named Isías Medina Angarita as his successor in 1940 and through the prevailing system of indirect elections, managed to secure Medina's election. Like his predecessor, Medina governed with moderation and allowed considerable political opposition to develop. Democratic Action (AD), the strongest opposition party, waged a strenuous campaign against the government in the 1944 congressional elections. Medina's candidates won, however, thus strengthening

---

[a] See page 149.

147

his chances of being able to name the government's candidate for the coming presidential elections with little opposition.

Democratic Action, frustrated in its attempts to gain power at the polls, allied itself with a group of young officers who had their own grievances and rose in revolt against Medina in October 1945.

THE ENVIRONMENT OF THE REVOLUTION

DESCRIPTION OF COUNTRY

Physical characteristics

Approximating the combined areas of Texas and Oregon, Venezuela is on the Caribbean coast of South America just above the Equator. It is bordered by Colombia, Brazil, and British Guiana, and has four distinct topographical areas: the Andes highlands along the coast, a lowland area surrounding Lake Maracaibo, the extensive lowlands in the Orinoco River valley, and more highlands south of the Orinoco, along the Colombian border. The climate varies from tropical to temperate according to the elevation.[1]

The people

In 1945 Venezuela had a population of 4 million people, 90 percent of them in the northwest region, in and around the cities of Caracas, Valencia, and Merida, in an area comprising only 25 percent of the total. The majority of Spanish-speaking Venezuelans are mestizos, a combination of white, Negro, and Indian ancestry. The predominantly "white" group constitutes approximately 10 percent of the population, but has provided most of the country's leadership.[2] Aboriginal tribes, such as the *Molitón*,[b] still exist, and a sprinkling of European immigrants is found in the cities. Caracas is the largest city as well as the political center. Maracaibo is the second ranking city.

Communications

Except in the Andes highlands, where most of the railroads and highways are found, Venezuela's communications system was not extensive in 1945. Railroads were built during the late 19th century with foreign capital, and during the Gómez regime road building developed. Shipping over the 10,000 miles of navigable waters was significant: oil, mostly in crude form, was carried by tankers from the

---

[b] "Indian" peoples of the Andean region adjoining Colombia. See Stanley Ross, "They Want to Be Alone," in the *Inter-American*, IV, 5 (May 1945).

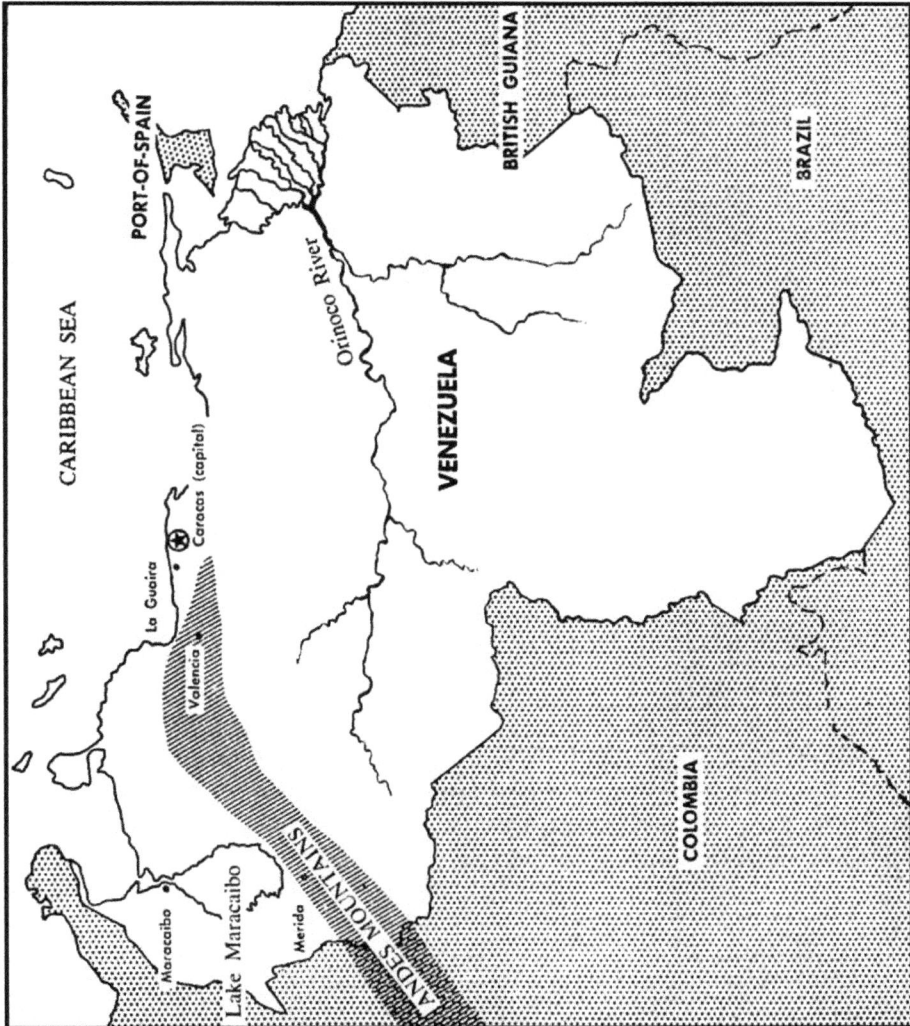

Lake Maracaibo region to the oil refineries on the islands of Curacao and Aruba.

Natural resources

Oil is the most important of Venezuela's resources for international trade. In 1928 Venezuela was the second largest producer and the biggest exporter of petroleum products in the world.[3] Iron, the second most important resource, is mined in large quantities. Gold, industrial diamonds, asbestos, and a limited amount of coal are also found.

SOCIO-ECONOMIC STRUCTURE

Economic system

Venezuela traditionally had been a predominantly agricultural country. However, the petroleum industry expanded so rapidly during and after the Gómez dictatorship that by 1940, 90 percent of export revenues were derived from petroleum products. As Venezuela lacked funds to develop its oil industry, investment capital and technical skill were supplied by foreign interests, which placed Venezuelan oil under foreign control.[4]

Prior to 1945 the oil industry was a boon to the economy, since the royalties from oil products made possible the liquidation of all domestic and foreign debts, and kept the treasury filled. On the other hand, the oil industry created an unbalanced, one-commodity economy, which neglected agriculture[c] and the expansion of heavy industry. Because of this top-heavy economic structure, Venezuela was forced to import large quantities of food and finished products.

Wealth derived from oil production brought increased demands for foods and services which stimulated the growth of small industries. Caracas and Valencia became industrial centers whose factories were engaged in the making of beverages, processing of foods, and manufacturing of textiles for domestic use. These small industries provided much of the increased employment opportunities. Agricultural production included coffee and cacao for export.

Class structure

Material wealth was concentrated in the 3 percent of the landholders who owned 70 percent of the land,[5] and in the industrialists and oil tycoons, who lived in Caracas and profited from oil royalties. Oil profits did not trickle down to the masses to raise the general standard of living. The oil industry paid wages well above the national level; but these wages were not high enough to compensate for the high cost of living. Most of the people were poor. The well-paid managerial and technical staffs were made up mostly of foreigners who lived in the large cities or in the centers of oil production.

A developing middle class, composed of professionals and businessmen, became quite significant after 1936. Presidents López and Medina permitted more freedom of expression, thereby allowing the movement for social reform to grow. The greater demand for foods

---

[c] Agricultural production failed to keep pace with the increasing demands of the large urban population that had developed from urbanization and the growth of population in general.

and services generated by the oil boom increased the importance of the small business class. Both elements of the middle class were relatively well-to-do.

Literacy and education

Illiteracy was widespread; few persons had as much as 6 years of elementary schooling. Students who attended colleges and universities were generally sons and daughters of the landed aristocracy residing close to the few educational centers. Literacy was a prerequisite for voting until the congressional elections of 1946.

Major religions and religious institutions

The majority of the people of Venezuela are Roman Catholics. There was always some degree of union between church and state until the development of anticlericalism in the late 19th century. Because the church supported the Gómez dictatorship, many of the intellectual liberals and the urban lower class who despised the dictatorship became anticlerical. Other religions are tolerated.

GOVERNMENT AND THE RULING ELITE

Description of form of government

Prior to 1945 Venezuelan constitutions had always provided for the election of presidents by Congress, and in practice the selection of the Congress had always been a self-given function of the executive. This system of indirect elections facilitated the establishment of "legal" dictatorships: by controlling the Congress, a President could become the master of the nation by placing puppets of his choice in the Presidency.

The 1936 Constitution, retaining the system of indirect elections provided for congressional selection of the President for a single 5-year term; he was not eligible for immediate reelection. Congress was composed of two houses; the Senate, whose members were selected by the various state legislatures; and the Chamber of Deputies, whose members were named by the councils of municipalities. These councils were elected by local elite groups. Members of the Supreme Court were selected by Congress. The pre-1945 regime was essentially a military government under two generals: López Contreras, from 1936 to 1940, and Medina Angarita, from 1940 to 1945.[6]

Description of political process

Political parties and power groups supporting government

Venezuela had no real functioning political parties before 1936, except for loosely organized bands of politicians. Gómez ruled with an iron hand for almost 30 years and had left no room for opposition and the development of party politics. When he died, therefore, he left an enormous political vacuum for Congress to fill. Lacking the techniques of party politics, Congress faltered. The Cabinet met hastily and chose López to act as provisional President until Congress could choose a successor. Because of López' apparent popularity at the moment, Congress had no alternative other than to select him for a full presidential term. López, in effect, chose Medina to succeed him in 1940. His choice took into account the support of army generals, landowners, and industrialists. Medina, however, alienated conservative elements by instituting liberal policies and received some support from the left.

Several organized groups gave support to these two prerevolutionary governments. The Bolívar Civic Groups, collectively known as the President's Party, were organized solely for the purpose of supporting President López.[7] Medina, who lost the full support of the army during his term of office, formed a group known as the Venezuelan Democratic Party (*Partido Democrático Venezolano*). Its members were mostly public employees who felt compelled to support the man who controlled their political destinies. Medina also received the support of the Communists, who were unable to muster enough strength to name a candidate of their own in the 1945 elections.[8]

Press support for the two prerevolutionary presidents came largely from the conservative newspaper *La Esfera*, although the Communist newspaper, *Últimas Noticias*, also supported Medina in 1945.

Character of opposition to government and its policies

Twenty-seven years of political suppression under Gómez had brought an increasing desire for wider participation in politics. The tactics of crushing opposition with a powerful army, a spy network, and police action which characterized Gómez's rule continued to be feared by the masses even after his death. Anti-Gómez demonstrations were broken up by the police, and some persons thought this action was López' answer to those who dared raise their voices for civil liberties.[9]

The masses quieted after they received a taste of López' moderate policies, however, but leftist organizations, including the Communists, agitated strikes in 1937 and 1938, "demanding constitutional reforms and other measures leading to more freedom of opinion."[10]

López took steps to reduce leftist influence by exiling agitators and instituting restrictive measures, but continued demands for a more widely-based form of government kept the masses aware of their inferior role in their country's politics.

Organized opposition after 1936 came largely from the Revolutionary Organization of Venezuela (*Organización Revolucionaria Venezolana*), a left-oriented precursor of the Democratic Action. Organized by Rómulo Betancourt and supported by labor leaders and intellectuals, it was soon dissolved by López, many of its leaders being forced into exile. Betancourt remained in Venezuela, however, and reorganized his Revolutionary Organization under the name of the National Democratic Party (*Partido Democrático Nacional*), which included most of the original members of the older group.[d] The party continually issued propaganda through a column in the Caracas daily *Ahora*, an irregularly published periodical, and countless handbills. The party was highly critical of the national petroleum policies of the regime and of its alleged corruption.

By 1940 politically conscious Venezuelans readied themselves for a real campaign against López' choice for President, Medina Angarita. The National Democratic Party was reorganized under the name of Democratic Action and chose Rómulo Gallegos to run against Medina. Gallegos lost, but Democratic Action continued its political opposition in preparation for the 1944 congressional election. This 1944 Congress was to select Medina's successor. However, in 1944 the majority of Medina's candidates won seats in the Congress, which meant that Medina's candidate for the Presidency was certain to be elected. Medina's choice was the Venezuelan Ambassador to the United States, Diogenes Escalante, a civilian regarded by all parties as highly capable. Moreover, Escalante gained the full support of Democratic Action leaders after they met with him in Washington.

Escalante returned from Washington to campaign but suffered a nervous breakdown. The new candidate, chosen by Medina, Angel Biaggini, was not acceptable to Democratic Action. Since AD felt that its own candidate would meet certain defeat because of the Medina-controlled majority in Congress, AD approached the President and proposed that all parties agree on a neutral candidate, with the understanding that he would serve for one year. During that time, the proposal continued, a new Constitution providing for direct popular election of the President should be written. Medina rejected the pro-

---

[d] Betancourt's efforts to organize opposition to the government is covered best in Stanely J. Serxner, *Acción Democrática of Venezuela, Its Origins and Development* (Gainesville: University of Florida Press, 1959).

posal, and AD decided that revolution was the only way to institute popular elections.[11] López and his conservative supporters did not endorse the candidacy of Biaggini, but campaigned for the election of López.

Organized labor was still in too embryonic a state to offer much support or opposition. The Confederation of Venezuelan Workers claimed 400,000 members, but it was only in the process of formation in 1945.[12]

Legal procedure for amending constitution or changing government institutions

Constitutional changes could be initiated in either chamber of Congress or in state legislative assemblies. Proposals for constitutional changes needed the approval of the National Congress and ratification by an absolute majority of the state legislative assemblies.[13]

Relationship to foreign powers

Venezuela participated in a number of international gatherings of political character from 1939 to 1945. It took part with other American nations at Chapultepec in 1945 in establishing an inter-American system which called for "continental solidarity" in case of attack by Axis powers. Although this was merely an arrangement for the duration of the war and not a permanent alliance, it provided for reciprocal assistance and cooperation in the event of invasion. It was followed by the Río de Janeiro Treaty of Reciprocal Assistance (1947) to which Venezuela is also a party. Venezuela had declared war on the Axis powers following the attack on Pearl Harbor and in 1945 became a charter member of the United Nations. Abroad, the Medina government was regarded as democratic.

The role of military and police powers

Military rule had been a frequent occurrence in Venezuela. Many of the rulers had been military dictators who used the armed forces to establish themselves. Regimes which lacked popular support had been kept in power by the army. Since the death of Gómez the regime had been more dependent on army support and was notably weakened when that support was withdrawn.[14]

WEAKNESSES OF THE SOCIO-ECONOMIC-POLITICAL STRUCTURE OF THE PREREVOLUTIONARY REGIME

History of revolutions or governmental instabilities

Venezuela had had 52 revolutions of one kind or another in the century and a quarter since its independence.[15] Thus the position of the ruling groups was always precarious. Most of the successful revolutions, however had not brought about any radical changes in society or even in the governmental system but merely substituted new rulers in place of the old. No attempts had been made to overthrow López by force; but he had been faced with a wave of strikes, particularly in 1937.

Economic weaknesses

Venezuela's major economic weakness had been the one-commodity character of its economic system. The lure of good wages in the cities drew peasants away from the farms; agricultural production failed to meet the demands of the people and the government was forced to import large quantities of food. Aside from this, the economic situation was adversely affected by World War II. The country was entirely dependent upon foreign trade for the sale of its principal products and for the manufactured goods necessary to meet domestic needs. War conditions created a scarcity of goods and also made shipping uncertain. Costs and prices rose and the economy was caught in an inflationary spiral.

Social tensions

The traditional social order began to be transformed after World War I. The development of the oil industry brought about radical modifications in the social structure. The attraction of oil refining and other industries in the urban centers prompted peasants to leave the countryside, and created conditions and opportunities which stimulated the growth of a middle class. A new social group, composed of the developing middle class, was added to the elite landholders, the church hierarchy, the insignificant professional and commercial class, and the great mass of illiterate agricultural workers. It was this new social group that demanded fundamental changes in the old order and precipitated a new political movement.

The struggle between the old and the new social groups was duplicated in the armed forces. The young men who entered military academies as cadets continued to come from the middle class as in the past; but the new urban-oriented graduates of these academies, unlike the traditionally urban-oriented officers, owed no special allegiance to the landed elite or the church hierarchy. As a result, they had little enthusiasm for protecting the traditional order. The young officers thought of themselves as members of a new enlightened class of social and economic reformers and often regarded their generals as unimaginative and behind the times.

Government recognition of and reaction to weaknesses

A number of social and economic policies were instituted by López and Medina to counter the weaknesses inherited from the Gómez regime and the economic disorders brought on by the war. López' 3-year plan included the improvement of public health and education, improvement of shipping facilities, encouragement of "desirable" immigration of farmers from Europe, and increased importation of American-made products.[16] He sponsored new labor legislation, but lost the support of the more radical elements when he refused to endorse all the demands of the union leaders. Medina began his Presidency by taking the middle of the political road, but as his term drew to a close, he turned to the left in an attempt to attract the support of labor. Meanwhile, his moderate policies had permitted considerable freedom of the press and allowed opposition to organize even when his policies were under attack. He set up a food production office in collaboration with a U.S. mission, provided funds to stimulate agriculture, stockraising, and industry, and granted cheap long-term credit for sound projects supported in part by Venezuelan capital. His program also included an income tax and legislation to strengthen the government's position vis-a-vis foreign oil companies.[17]

These attempts to strengthen the country's economy had accomplished little, however, by the time the revolution took place. López, blaming the industrial unrest on Communists and alien agitators, took steps to reduce their influence by decreeing measures that restricted the activities of those who were prominent in the trade union movement. Transients and tourists were restricted in the period of time that they might remain in the country, and many were required to obtain identity cards. On the other hand, by the support of labor toward the end of his term, and by favoring the unions in management-labor disputes, Medina had incurred the distrust of many business interests. His proposal for an income tax alienated the landowning families. Workers were dissatisfied because increased earnings did not keep pace with inflationary prices.

FORM AND CHARACTERISTICS OF REVOLUTION

ACTORS IN THE REVOLUTION

The revolutionary leadership

The leadership was a combination of Marxist-oriented, well-educated professionals and businessmen of the middle class, and young, politically-minded army officers. They had a common desire to block what threatened to be the self-perpetuation of an undesirable regime, a common social philosophy, and a common ambition for power. Of the four leaders, two were civilians and two were army officers.

Rómulo Betancourt, who emerged from the revolution as Provisional President, had had a long record of opposition to the government. He was first arrested by Gómez police in 1928 for participating in a student demonstration, and continued his antigovernment activity both at home and in exile by lecturing and writing.

Betancourt was Marxist but not Communist in his political philosophy. He became a member of a Communist Party while in exile, but only for a brief period. His was not an international ideology, but rather a national one: He believed in applying Marxist principles to Venezuelan conditions. His ideology found an instrument in a revolutionary group which later became the Democratic Action. His first public office was the post of Councilman of the Federal District, which he won in the 1944 congressional elections.

The other civilian revolutionary leader was Rómulo Gallegos, one of Venezuela's most highly respected men. A novelist, he also had been an instructor in several Caracas schools. A number of leaders, including Betancourt, had been his students. In 1936 he had been appointed Minister of Education under López, but later resigned to protest a government move against political dissenters. He had run against Medina in 1940 as Democratic Action nominee for the Presidency.

The other two important revolutionaries were members of the armed forces. Lt. Col. Carlos Delgado Chalbaud inherited his revolutionary ideas from his father, who was killed in a 1929 attempt to remove Gómez. Chalbaud was educated as an engineer, and did not enter the army until 2 years prior to the revolution. He taught at a military academy, but had never commanded in the field.[18] Capt. Mario Ricardo Vargas, on the other hand, had a relatively long service record and had been attached to the presidential staffs of both López and Medina. Vargas also had a much closer association with members of Democratic Action.

The revolutionary following

The 1945 revolution was a military coup d'etat and was staged by a military organization in collaboration with leaders of Democratic Action. The exact number of participants is not known. A reporter estimated the number of conscripts to be 10,000, led by 900 officers.[19] However, some troops remained loyal to Medina.[20] Only the younger military officers were aware of the coup. *Time* magazine reported that the army did most of the fighting.[21]

ORGANIZATION OF REVOLUTIONARY EFFORT

The revolutionary organization was a semi-Socialist, semimilitarist coalition formed by the Democratic Action Party and a group of disgruntled young army officers who belonged to the Patriotic Military Union (*Unión Patriótica Militar*). This army group was a conspiratorial clique, drawn from the junior and middle rank army leadership, similar to the Group of United Officers in Argentina. The relationship between the party and the leaders of the Unión is not clear,[e] although Betancourt claims that his party and the leaders of the Union planned the revolution jointly and his leadership was decided beforehand.[25]

GOALS OF THE REVOLUTION

Concrete political aims of revolutionary leaders

The political aims of the revolutionary leaders varied, especially as between the politicians and the military men. Both groups were dissatisfied with the prerevolutionary regime, but for different reasons. The Democratic Action was attempting to prevent the old regime from perpetuating itself; it intended to keep the army subordinated to a civilian government.[26] The army, on the other hand, was dissatisfied with the little influence it had had with President Medina. Officers felt that low army salaries could be increased if graft were eliminated from the budget. The army's aim was to overthrow the Medina government and establish a new government within which it could more successfully press its demands for higher salaries and better military equipment.[27]

---

[e] One authority indicates that the two organizations conspired jointly from the initial planning stage.[22] A reporter writes, however, that the whole plot was a military conspiracy, and that although Betancourt was in on the scheme, he was not a tactical planner, but was merely called upon to join with the army in order to give the revolution political substance.[23] Another source maintains that Medina abdicated to the young officers, and that the Presidency was then handed to Betancourt.[24]

Social and economic goals of leadership and following

Although the Democratic leaders considered themselves Socialists, the solutions they advocated for Venezuela's social and economic problems were similar to measures of the New Deal administration in the United States.[f] They intended to introduce legislation to bolster national production. The national budget, they believed, could support such legislation if Venezuela demanded a larger share of royalties from the oil companies. They also aimed at establishing a productive public works program, and at placing ceilings on rents and prices of certain items. Goals further included subsidies for certain food imports, and a peaceful rapprochement between capital and labor.[28]

REVOLUTIONARY TECHNIQUES AND GOVERNMENT COUNTERMEASURES

Methods for weakening existing authority and countermeasures by government

The revolt began on October 18, 1945. It seems to have been well-planned. For more than a year some of the officers had spoken of a possible revolution.[29]

The uprising had originally been scheduled to take place at the end of 1945—possibly in December. However, President Medina got wind of it and began to arrest suspects. The plan was hastily changed and the date was advanced to October 18.[30]

On that day Gen. Andrónico Rojas, Commander of Caracas Military District, went to the military center of Maracay to deal with a reported conspiracy. The insurgents successfully isolated him, and Gen. N. Ardilla, who commanded the forces at Maracay, was forced to surrender after a 30-minute tear gas skirmish and an exchange of fire between the insurgents and some police units.

At 4 p.m. on the same day, another group of army officers revolted at the San Carlos barracks in Caracas and by 9 o'clock had forced their way into the strategic *Escuela Militar* (Military Academy) and the Miraflores Palace (Presidential Offices), and arrested Medina and a number of government officials, who were held as hostages. The fighting continued in the streets of Caracas on October 19, and there was scattered shooting throughout the city. Just before noon on that day, rebel planes from Maracay dropped bombs on San Carlos barracks and flew menacingly over police headquarters. The police soon

---

[f] Betancourt, although a Socialist, aimed at a moderate reform program and promised to respect existing economic interests.

surrendered. By evening of the 19th, the insurgents had control of Caracas, the port of La Guaira, the Maracay arsenal, the San Carlos barracks, and several interior towns. Loyalist troops showed some resistance, but soon capitulated because they lacked heavy weapons and aircraft.

The insurgents imposed a state of siege in Caracas which was in effect for several days. Newspapers were censored, radio stations closed, and the transport system paralyzed. The electric current was shut off in sections where there was fighting, and communications with the rest of the city were cut off.

Some National Guard units and progovernment Communists showed the stiffest opposition to the insurgents, though for different reasons. Communist units broke into local barracks on the 20th, and confiscated guns and uniforms. They set up a defense post on a six-story housing project and eventually made an attack upon the rebels. They also attempted to recapture the Military Academy, where the hostages were reportedly held. Loyalist cavalry units joined the Communists in the assault, but were eventually defeated by rebel tanks, artillery, and air support.

Other loyal troops were reportedly massing in the interior, but they lacked heavy weapons and aircraft and surrendered one by one. Many army officers in the interior, suspicious of city revolutions, held out for a short period. They soon joined the rebel units, however, as did the military garrisons around the Lake Maracaibo oilfields. It was reported on the 20th that the insurgents had consolidated their hold in eastern, southern, and central Venezuela. By the 21st, although some generals in the western Andean states were still holding out in the hope of massing counter-revolutionary forces in that area, the west Andean states offered their support to Betancourt.

The government of the Revolutionary Junta (*Junta Revolucionaria de Gobierno*) became the *de facto* government on October 27, 1945. News reports indicated 100 to 300 dead, and 300 to 1,000 wounded.[31]

Methods for gaining support and countermeasures taken by government

The insurgents appear to have used two methods to call for the active support of the people: radio broadcasting and dropping leaflets from airplanes. Revolutionary propaganda began early during the fighting, after the insurgents captured five radio stations. Armed civilian volunteers were called upon via radio on the night of October 18 to lend their support to the revolution. However, the thousands of civilians who did answer the call only hindered the revolution: they were undisciplined and reportedly roamed the streets and looted homes. Troops had to be sent from the Maracay military center to restore order. A broadcast late on the 19th declared that the revolution was over and announced the names of a seven-man junta. The broadcast called on the civilians to surrender their arms to members of the armed forces.

Leaflets were dropped over several cities during the first several days. The earlier ones urged the people to go on strike in support of the revolution. Later leaflets explained the aims of the revolution and asked the people to respect the revolutionary decrees.

Progovernment propaganda was issued mostly from the state of Tachira in the western part of the country. A radio in San Cristóbal claimed on the 19th that although the insurgents had succeeded in paralyzing Caracas, pro-Medina forces had control of the rest of the country. The radio also announced that the loyalists were concentrating forces in four states to march on Caracas. A later broadcast announced that Medina was taking over the palace in Caracas with loyalist troops.

In Caracas, progovernment Communists captured a police radio mobile unit and used it to broadcast pro-Medina propaganda.[32]

MANNER IN WHICH CONTROL OF GOVERNMENT WAS TRANSFERRED TO REVOLUTIONARIES

Control of the government was transferred to the revolutionary junta by Decree Number 1 of October 19, 1945, following the arrest of President Medina and other government officials. Medina signed his resignation on that day. He and his principal advisers were hurriedly sent into exile. López was also asked to leave the country.[33]

THE EFFECTS OF THE REVOLUTION

CHANGES IN THE PERSONNEL AND INSTITUTIONS OF GOVERNMENT

Decree Number 1 also made provisions for organizing the junta that replaced the prerevolutionary government. The junta was initially composed of two army officers, 13 members of Democratic Action, and one other civilian. Valmore Rodríquez, who occupied the office of Provisional President for a short time before the official consolidation of the junta, was replaced by Rómulo Betancourt on October 24. Chalbaud was appointed Defense Minister.

MAJOR POLICY CHANGES

The junta immediately began a three-pronged drive: against older conservative army men who still supported Contreras and Medina, against obstacles to the development of democracy, and against economic instability. Defense Minister Chalbaud, in order to prevent a counterrevolution, removed from the army, and thus from positions of influence, older officers sympathetic to López and Medina. To the displeasure of many officers, the army was transformed from a group which traditionally determined the tenure of the President to a nonpolitical body.[34]

The junta continued to rule by decree. It initiated a land reform intended to divide large estates into small farms; started construction of low-cost housing at government expense; developed a publicly subsidized merchant marine. A heavy excess profit tax, directly aimed at the foreign oil companies, was enacted to enrich the public treasury. Ceiling prices were fixed for foodstuffs.[35] Rents were reduced, electric rates were cut, and new arrangements for a "50–50 percent split" in oil royalties with foreign oil companies were negotiated to provide what Venezuelan officials considered a more equitable distribution of oil profits.

LONG RANGE SOCIAL AND ECONOMIC EFFECTS

The new government, elected in December 1946 and headed by Gallegos, was soon faced with military dissatisfaction. Within 2 months a series of rebellions sprang up in various parts of Venezuela. They were suppressed, but some reached serious proportions. The army, allied with leaders of the Committee Organized for Independent Elections, demanded more voice in the Cabinet, but the demand was refused. The government then declared a state of siege and suspended constitutional guarantees. On November 24, 1948, army tanks and trucks rumbled into Caracas, and the army established a military junta, thus curtailing any long range social and economic effects which the 1945 revolution might have produced. The government had failed in its effort to make the army nonpolitical.

OTHER EFFECTS

After a brief democratic interlude under Betancourt and Gallegos, a three-man military junta ruled Venezuela until Marcos Pérez Jiménez, in an attempt to gain personal control, purged the junta of his associates. He decided to hold a presidential election in November 1952. When it became obvious that he was trailing behind the opposition in the first returns, Pérez Jiménez imposed a tight censorship on election returns, announced his victory, and established himself as the Provisional President. In early 1953 the Constituant Assembly named him President for a 5-year term. Students, army officers, businessmen, and the church hierarchy turned against him, and in January 1958 he was forced to flee the country. A short-term provisional government was installed and paved the way for the second democratic election, which chose Rómulo Betancourt as President.

Betancourt faced three economic problems: two were traditional and concerned the unbalanced economy, which favored the oil industry over agricultural production and expansion of other industries; the third was the foreign debt which the graft and the unwise spending of the Jiménez government had rolled up.[g] Betancourt's government drew up plans to bolster the national economy and promote the general welfare. Steps were taken to diversify agriculture, increase agricultural production, to raise industrial production, particularly that of iron ore, and to increase tourist trade. By the fall of 1960, well over 1 million acres of land had been distributed, and more distribution was being considered.[36]

Despite this favorable beginning, opposition rose against Betancourt. Dissension within the government, which depended upon a coalition of the Democratic Action, the Committee Organized for Independent Elections (COPEI), and the Democratic Republican Union (URD), led to a reorganization of the Cabinet. The reorganization excluded URD members, thus creating an important opposition group. In 1961 there were several leftist demonstrations, involving many students, in favor of Castro. To counter these demonstrations, Betancourt had to call on the army and temporarily suspend constitutional guarantees.[37] However, some leaders of the coalition parties, labor unions, and armed forces still support his program.

---

[g] Actually, Venezuela's economy was somewhat prosperous under Jiménez. Oil production was high and made the country the second largest oil producer in the world. Mining was developed by foreign capital and iron ore took second place in the nation's exports. However, the bulk of the national treasury was spent on modern public buildings in Caracas as well as on public housing and other public works; prices were rising and much of the food—even such staples as wheat, corn, and eggs—was imported.

Betancourt's success or failure will be strongly affected by outside forces especially the attitude of the United States. In 1960, Venezuela sent a delegate to the International Petroleum Accord in Baghdad, where five oil-producing countries formed the Organization of Petroleum Exporting Countries and agreed to maintain prices and avoid dumping oil on the international market. In 1961, the United States listed Venezuela as one of the "bright spots" on the continent where the Alliance for Progress, sponsored by the U.S. government, has a good chance for success. U.S. aid is designed to advance the reform movement instituted by Betancourt.

NOTES

1. Austin F. MacDonald, *Latin American Politics and Government* (2d ed.; New York: Thomas Y. Crowell Company, 1954), pp. 419–425.

2. Arthur P. Whitaker, *The United States and South America, The Northern Republics* (Cambridge, Mass.: Harvard University Press, 1948), p. 58.

3. Ibid., p. 60.

4. Ibid.

5. Donald Marquand Dozer, "Roots of Revolution in Latin America," *Foreign Affairs*, XXVII, 2, 275.

6. MacDonald, *Politics*, p. 433; see also Thomas Rourke, *Gómez, Tyrant of the Andes* (New York: William Morrow and Company, 1936), p. 53.

7. MacDonald, *Politics*, p. 435.

8. Stanley J. Serxner, *Acción Democrática of Venezuela, Its Origins and Development* (Gainesville: University of Florida Press, 1959), p. 5.

9. MacDonald, *Politics*, p. 431.

10. Serxner, *Acción Democrática*, p. 3.

11. Ibid., pp. 2–8.

12. Whitaker, *The United States*, p. 82.

13. Amos J. Peaslee, *Constitutions of Nations* (Concord, New Hampshire: The Rumford Press, 1950), II, p. 510.

14. Dozer, "Roots," p. 276.

15. Edwin Lieuwen, *Arms and Politics and Latin America* (New York: Frederick A. Praeger, Inc., 1960), pp. 125–128.

16. MacDonald, *Politics*, p. 434.

17. Whitaker, *The United States*, p. 99; see also MacDonald *Politics*, p. 436.

18. Ray Josephs, *Latin America: Continent in Crisis* (New York: Random House, 1948), p. 90.

19. Ibid.

20. MacDonald, *Politics*, p. 440.

21. *Time* (October 29, 1945), 46–48.

22. MacDonald, *Politics*, pp. 439–440.

23. Josephs, *Latin America*, p. 66.

24. Serxner, *Acción Democrática*, p. 10.

25. Ibid., p. 11.

26. Dozer, "Roots," pp. 280–281.

27. MacDonald, *Politics*, p. 440.

28. Josephs, Latin America, p. 64; and Lee Benson Valentine, "A Comparative Study of Successful Revolutions in Latin America, 1941–1950" (Ann Arbor, Michigan: University Microfilms, Inc., 1958. PhD. Thesis, Stanford University [1958]), 140.

29. MacDonald, *Politics*, p. 440.

30. Ibid.

31. *Time*, pp. 46–48; Josephs, *Latin America*, pp. 66–67; Serxner, *Acción Democrática*, p. 8; and *New York Times*, October 20–25, 1945.

32. Josephs, *Latin America*, pp. 66–67.

33. MacDonald, *Politics*, p. 440.

34. Serxner, *Acción Democrática*, p. 13.

35. Ibid., pp. 13–14.

36. C. A. Hauberg's "Venezuela Under Betancourt," *Current History*, XL, 236 (April 1961), 237–238.

37. Ibid., pp. 238–239.

RECOMMENDED READING

BOOKS:

Davis, Harold E. (ed.). *Government and Politics in Latin America.* New York: The Ronald Press Company, 1958. A general selection of articles written by the editor and other writers on contemporary Latin American politics.

Josephs, Ray. *Latin America: Continent in Crisis.* New York: Random House, 1948. Discusses the 1945 Venezuelan revolution briefly, but is more detailed than most general works.

MacDonald, A. F. *Latin American Politics and Government.* New York: Thomas Y. Crowell Company, 1954. A standard text on Latin American political institutions. Useful for information on prerevolutionary government and postrevolutionary political developments.

Whitaker, Arthur P. *The United States and South America, The Northern Republics.* Cambridge: Harvard University Press, 1948. Discusses wartime economic and political problems of Venezuela and postwar trends.

PERIODICALS:

Dozer, Donald Marquand. "Roots of Revolution in Latin America," *Foreign Affairs,* XXVII, 2, 274–288.

Hauberg, C. A. "Venezuela Under Betancourt," *Current History,* XL, 236 (April 1961), 232–240.

OTHER:

Serxner, Stanley J. *Acción Democrática of Venezuela, Its Origin and Development.* Gainesville: University of Florida Press, 1959. This monograph is not quite complete for the purposes of this study, but contains more pertinent information than any other work concerning the 1945 revolution and the role of the AD.

THE ARGENTINE REVOLUTION OF JUNE 1943

SYNOPSIS

On June 4, 1943, the Castillo government in Argentina was overthrown by army revolutionary forces. The coup d'etat was executed in one day; except for a brief exchange of fire between two military units, there was little fighting.

BRIEF HISTORY OF EVENTS LEADING UP TO AND CULMINATING IN REVOLUTION

Spanish rule made a lasting impression on Argentina. "The land of the Río de la Plata," as Argentina was known, was politically and economically controlled from far-off Lima, and the interior cities were favored to the disadvantage of Buenos Aires. Discriminatory practices isolated Buenos Aires during most of the colonial period, and the city did not begin to prosper until the end of the 18th century. Isolation resulted in a fixed pattern of disunity, characterized by distrust between the *provinciaños*, the people of the provinces, and the *porteños*, the people of the port city of Buenos Aires. Power struggles between Buenos Aires and the rest of the country have had much influence on the course of Argentine history.

From 1853 to 1943, except for a period of 14 years, Argentina was ruled by a conservative oligarchy comprised of wealthy landowners, bankers, and merchants. Oligarchic rule, according to the philosophy of this class, was as "natural" as the medieval "divine right of kings." However, organized opposition developed toward the end of the 19th century, so that in 1916, the Radical Civic Union, a moderate party, was able to take over the government. The moderates ruled Argentina until 1930. In that year a military coup d'etat wrested power from the moderates and restored the selfish and conservative group that had ruled prior to 1916.

For 13 years after the coup a bloc of generals and landowners ruled Argentina, with the support of bankers, merchants, and the high clergy. Gen. José F. Uriburu, leader of the coup, dissolved Congress, installed himself as President, and turned public sentiment against himself by advocating the establishment of an authoritarian Fascist state. Groups of military officers and businessmen repudiated his plan, and within 6 months he found it wise to announce an election for November 1931. Agustín P. Justo, a conservative general, was elected, and over the next 6 years he was able to rehabilitate the country and pull it out of the depths of the world economic depression. He took an active part in international affairs. Dr. Roberto M. Ortiz, Justo's Finance Minister, was elected President in 1937 and continued his predecessor's policies. But Ortiz fell ill and the reins of government fell into the hands of his Vice President, Ramón S. Castillo. Castillo was a reactionary who held pro-Axis sympathies and instituted isolationist policies.

As Argentina readied itself for the presidential elections in 1943, no political party offered an outstanding leader to run against Castillo's hand-picked candidate. Moreover, the opposition was aware that the elections would be rigged in favor of the Conservatives. There was talk of revolution, but the Radicals quarrelled among themselves and could not organize sufficient strength to overthrow the oligarchy. The army, however, had organization and strength. Castillo's Minister of War, Gen. Pedro P. Ramírez, called together a few of his fellow officers and planned a coup. Castillo got wind of the impending revolt, called an emergency Cabinet meeting, and ordered that Ramírez be placed under arrest. However, the revolutionary troops, led by army generals, marched against the government on June 4, 1943, for the second time in 13 years. Public buildings were surrounded and in less than 24 hours military rule was established with a minimum amount of bloodshed.

THE ENVIRONMENT OF THE REVOLUTION

DESCRIPTION OF COUNTRY

Physical characteristics

Argentina has the most tillable and best watered land in Latin America. One-third the size of the United States, it has a north-to-south dimension of 2,300 miles, and a maximum east-to-west distance of 800 miles. It has vast areas of fertile plains, the pampas, and is situated in a temperate zone, though temperatures are high in the north and low in the south. It is bounded by the Atlantic Ocean, Uruguay, Brazil, Paraguay, Bolivia, and Chile.

The people

Argentines are primarily of Spanish and Italian extraction. They claim to be 95 percent white, although, according to one source, there is an intermixture of Indian and Negro blood, perhaps more than Argentine nationalists want to admit.[1] A great tide of immigration in

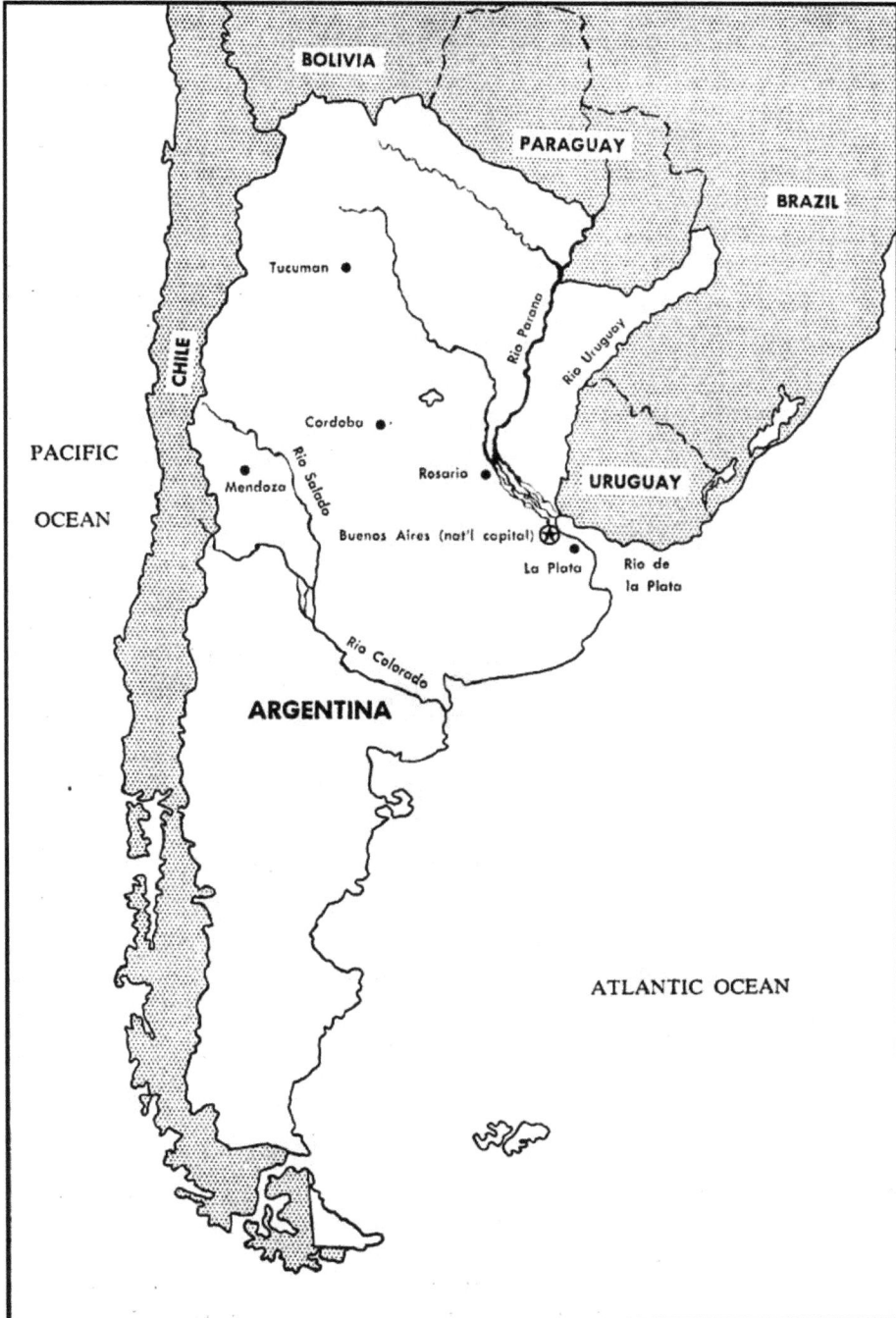

the late 19th and early 20th centuries brought many Europeans to Argentina. Of 16 million Spanish-speaking people in 1947, well over half lived in urban areas.[2] Buenos Aires is the capital; other major cities are Rosario, Córdoba, La Plata, Tucumán, and Mar del Plata.

Communications

Communications prior to 1943 included a highway system that provided transit through most of the provinces, and British- and French-owned railroads. However, the highways were in great need of repair, and the railroads were allowed to deteriorate during the war. Inland and coastal waterways provided additional service for transporting goods.

Natural resources

Argentina has not been a major producer of minerals, but tungsten was produced in large quantities and some lead and zinc were mined in the northwest. Oil has been produced since 1907 and has increased in importance. Argentina also has large natural gas reserves.

SOCIO-ECONOMIC STRUCTURE

Economic system

While predominantly engaged in raising and processing agricultural products, Argentina, nevertheless, was one of the most highly developed countries in the Western Hemisphere in 1943. Agricultural and livestock products, wheat, meat, wool, corn, hides, and skins constituted most of the exports, as traditionally they had done. The development of heavy industry had been retarded by the lack of iron and coal. Most industrial goods had to be imported.

British and "North American" capital investments exercised a major influence. Fifty percent of the owners of industrial establishments, which included railroad transportation, meat packing, textile manufacturing, production of tires, development of electric power, assembling automobiles, operation of subways and streetcars, maintenance of telephone systems, and production of quebrácho extract used in tanning leather, were foreigners. Foreign capital dominated virtually all economic activities except agriculture.[3]

Class structure

Although the traditional social structure was not strictly a plural one, it was nevertheless characterized by schisms, first between the social elite of Buenos Aires and the social elite of the provinces; and second, between the small percentage of landowners and the rest of the population.

Landownership was the main base of power. The landowning class traditionally dominated Argentina socially, economically, and politically. This group was exclusive and had ready access to higher education, a privilege denied most Argentines. Often the landowners lived in Europe and took no personal interest in administering their large estates so long as those continued to make profits. Left under the command of land captains, the Argentine peons were underpaid, badly housed, and politically unrepresented.

Large bankers, merchants, and industrialists in Buenos Aires constituted the urban social elite. They held themselves in high esteem and clung to the belief that they were the only civilized people in Argentina. Argentines living in the provinces were regarded as uncivilized.

A large middle class developed toward the latter part of the 19th century and rapidly grew in importance as well as in number. By 1940 the middle class comprised nearly half the population.[4]

As the concentration of land increased during the century prior to 1943, large numbers of peasants migrated to the cities to seek employment. In 1869, 67 percent of the population was rural. By 1914 the percentage had decreased to 42, and in 1943 over half the population was urban.[5]

Argentina has always been able to feed its people and the poverty of the peons has been less severe there than in most Latin American states. The Argentines were blessed with a comparatively high per capita income, as well as high calorie consumption and high meat consumption because of wheat and beef surpluses.

Literacy and education

Unlike most Latin American states, Argentina has a literacy rate approximating levels in Western Europe. By 1930, 85 percent of the population were literate and that figure has continued to increase.[6] The first of eight national universities, established in the provincial capital of Córdoba, was founded in 1713. The principal university, in Buenos Aires, was founded in 1821. The national government administers all public education. Some schools are administered by the church.

Major religions and religious institutions

An overwhelming majority of Argentines traditionally have been Roman Catholics. Although the church greatly influenced politics under Spanish rule, there was a gradual separation of church and state under the republic. However, Argentina retains "national patronage," a system under which bishops are appointed by the President of the republic.

Attitudes toward the Church differed greatly between Buenos Aires and the provinces. Catholics in Buenos Aires were only moderately religious and attended church on occasion. Provincial Catholics, on the other hand, were known as devout churchgoers. Anticlericalism developed among the liberals in the late 19th century and in some labor unions in the present century.

GOVERNMENT AND THE RULING ELITE

Description of form of government

Constitutional government had existed in Argentina since 1853. The Argentine Constitution provided for a chief executive elected for 6 years and ineligible for immediate reelection. Authority was strongly centralized, with the intent of assuring national unity. Though there was some provincial autonomy, the chief executive had the right to intervene, with Congressional approval, in provincial affairs if he considered the national government threatened by "internal disorders or foreign invasions." The government in power employed this formidable weapon on many occasions to intervene in local elections, in order to weaken the opposition.[7]

Adhering to the doctrine of the separation of powers, the Constitution also provided for legislative and judicial branches. The Congress consisted of an upper house, the Senate, and a lower house, the Chamber of Deputies. The chief federal court was designated the Supreme Court.

Description of political process

Political parties and power groups supporting government

Politics traditionally had been the monopoly of the upper social class. In 1912, however, President Sáenz Peña instituted election reforms which guaranteed to all males over 18 a free and secret vote. Minorities were to be represented, and voting was to be compulsory. Two major factors continued to reduce the democratic flavor of Argentine politics in the 90 years from 1853 to 1943: fraudulent manipulations of both national and local elections from within the political circle, and frequent demonstrations and uprisings, which resulted in the deposition of presidents, from without the political circle.

The Conservative Party, also known as the National Democratic Party, was organized in 1909, when it became evident that the ruling elite was dangerously threatened by opposing groups. Strongly supported by landowners, the party continued to assert that government should

172

be controlled by a select minority of landowners and other wealthy economic groups. Between 1916 and 1930 a more moderate group representing middle-class elements ruled, but a conservative coup in 1930 ended its political control. Following the 1930 coup, the Conservatives immediately instituted reactionary measures which removed the democratic facade: Congress was dissolved, liberals were removed from public office, provincial elections were cancelled, labor unions were suppressed, and newspapers were subjected to strict censorship.

In order to unite the reactionary forces against the democratic groups, General Uriburu had formed the Argentine Civic League. Organized on the pattern of European Fascist organizations, the League intended to introduce Fascist and Nazi ideas to strengthen and perpetuate conservative rule. A system of corporate representation of economic groups in the government was proposed. However, public sentiment turned against General Uriburu and he was persuaded to announce an election for November 1930.

Agustín P. Justo, elected in 1931, became somewhat progressive while holding office, and instituted reforms[a] which satisfied the landowners, the business community, and the industrial workers. He also made generous concessions to foreign capital. Justo led his party in guiding the Argentine economy in the direction of increased prosperity. He allowed political opposition to exist, but electoral frauds were customary, provincial governments were subjected to repeated interventions, and in many cases political dissidents were barred from public employment. Roberto M. Ortiz, Justo's successor, continued the progressive trend. He fought for free and secret elections, and advocated nonintervention in provincial affairs. Ortiz fell ill shortly after being elected in 1937 and was replaced by Vice President Castillo, a pro-Axis reactionary who reverted to political repression, fraudulent elections, and provincial intervention.

This situation continued as the nation prepared for the 1943 elections. A state of siege called by President Castillo in 1941 kept the country in constant tension, as he manipulated political strings in order to install Rubustiano Patrón Costas as his successor. Costas was extremely unpopular, and his nomination aroused angry opposition from Conservatives and Radicals alike. However, no political party offered a really able leader to oppose him. The Radical Civic Union (UCR) opposition party nominated General Ramírez, but the party felt that it had little chance of winning even with army support: UCR knew that the election, dominated by Castillo, would be corrupt. There began to be talk of revolution.

---

[a] See p. 177.

Character of opposition to government and its policies

Opposing the Conservative Party was the Radical Civic Union founded in 1892 and politically more to the center than its name indicates. The UCR originally represented the rising middle class, but very soon extended its popular appeal to worker groups. It won a large number of seats in the Buenos Aires elections of 1912, and in 1916 it won control of the national government. By 1930, however, it had become corrupt and had alienated businessmen, the army, the press, and university students.

In its first quarter century, the UCR was able to maintain party unity. After the 1916 victory, however, a faction led by Marcelo del Alvear and opposed to the policies of President Hipolito Irigoyen formed the Antipersonalist UCR. The Antipersonalist UCR swung slightly to the right and was able to gain support from some conservatives. Presidents Justo and Ortiz had been elected by a coalition of Antipersonalists and Conservatives.[b] UCR, in an attempt to strengthen its support within military circles, named General Ramírez as its candidate for the 1943 elections.

Other opposition parties were the Socialist Party and the Communist Party, but neither carried much weight politically. The Socialist movement, although it was supported by some workers in Buenos Aires, was more an intellectual movement than a popular one. The party immersed itself in doctrinal questions, and kept breaking up into splinter groups. The Communist Party had some support from the labor movement, but had very little influence. The state of siege was partly responsible for preventing labor from organizing on a larger scale.

Newspapers flourished in Argentina prior to 1943, but were generally suppressed whenever they became hostile to a regime. One of the most important newspapers was *La Nación*, highly literary and very energetic. *La Prensa*, an independent daily with an international reputation, had a circulation of over 450,000 by 1943. *La Vanguardia*, a Socialist newspaper, continually opposed the government, and it was suppressed in 1931, 1942, and again in 1943.[8]

Legal procedure for amending constitution or changing government institutions

The Argentine Constitution of 1853 provided for relatively easy amendment, yet it was altered only two times after 1860. The infrequency of constitutional changes may be due to the fact that presidents took it upon themselves to interpret the fundamental law to fit

---

[b] This coalition was called *Concordancia*.

their own ends. Congressmen often objected to this practice, but the courts rarely interfered.

Article 21 of the 1853 Constitution stated that the necessity for constitutional amendment "must be declared by Congress by a vote of at least two-thirds of the members present . . ." A national convention summoned by Congress and composed of representatives from the provinces would then consider the articles recommended for revision by Congress.[9]

Relationship to foreign powers

Although there was a powerful pro-Fascist element in Argentina, the ruling group maintained an official policy of neutrality, and displayed open sympathy for England and her allies. The isolationist policies of UCR President Irigoyen were shelved in 1931 and, under Justo, Argentina's relations with other nations improved. Participation in the League of Nations was revived. During World War II, Argentina continued its trade relations not only with Great Britain and the United States but with Germany and Italy. President Castillo, who had pro-Axis sympathies, continued to advocate neutrality even after the Japanese attack on Pearl Harbor, and was sharply criticized by U.S. officials.[c]

The role of military and police powers

The chief role of the army in the 20th century prior to 1930 was to maintain internal order. The army had not participated in any international conflict since the Paraguayan war (1865–70), and it had not been very active in politics for nearly 50 years. However, general dissatisfaction mounted when the aging Radical leader, Irigoyen, failed to improve the grave economic situation caused by the world crisis in the late 1920's, and to prevent widespread corruption. General Uriburu, acting in behalf of the Conservatives, ousted Irigoyen and established a military government. From 1930 on the military played a key role in politics by keeping in power traditionalist military and civilian presidents.

---

[c] Surrounded by admirers of the Axis, Castillo allowed the Nazi press to issue its propaganda while suppressing any expression of Allied sympathy.

WEAKNESSES OF SOCIO-ECONOMIC-POLITICAL STRUCTURE OF THE PREREVOLUTIONARY REGIME

History of revolutions or governmental instabilities

Argentina is not one of the countries notorious for frequent coups and revolutions. Unlike most Latin American countries, it has had comparatively stable governments. However, since the establishment of constitutional government in 1853, internal disturbances had produced enough pressure to bring about the resignation of two presidents. In 1880 opposition forces made an unsuccessful attempt to overthrow the "conservative oligarchy;" and twice, in 1890 and again in 1893, determined but weak rebellions were attempted. Both of the latter attempts were suppressed but they caused enough unrest to result in presidential resignations. No regime was overthrown until 1930.

The 1930 military coup which overthrew a UCR-controlled government was comparatively peaceful. General Uriburu's revolutionary troops, consisting of a few thousand soldiers and cadets of the Military Academy, moved into Buenos Aires early on September 6 and demanded the President's resignation. There was little violence and little bloodshed. The workers were disinterested: no strikes were called, no demonstrations were held, and no plants or shops were closed. Some fires were set, some homes and shops were looted, there was a lot of milling about and shouting; but all this took place after the coup was over.

Economic weaknesses

The lack of a diversified economy was Argentina's greatest economic weakness. The country lacked heavy industry and had to rely on imports of manufactured goods to satisfy domestic needs. These imports were paid for by exporting agricultural products. Thus the lack of economic diversification caused undue dependence on the fluctuating demand and prices of goods on the international market. The economic depression of the late 1920's and 1930's nearly ruined Argentina, as beef, wool, and hides were selling at extremely low prices. Export prices rose in the late 1930's but the benefits were reduced by a rise in the cost of living.

Social tensions

Social tensions in the post-World War I period grew out of the antagonism between two major social groups: conservatives, who attempted to maintain the traditional order, and a new middle class that attempted to change it. The conservatives were the large landowners, some of them descendants of those who settled Argentina shortly after the "discovery." The new middle class was made up of Spanish and Italian immigrants who came to Argentina in the late 19th and early 20th centuries, and developed from the increased industrialization (mostly food processing) and urbanization of that era. The middle class opposed and threatened to dislodge the traditional and conservative groups from their dominant economic and social positions. It was partly successful between 1916 and 1930.

The military was called upon by both the conservatives and the liberals to help protect their separate vested interests. The military took it upon itself to play the role of moderator between the old agrarian and the new industrial interests, and reconcile the divergent social groups. The generals contended that "social peace" could be achieved only by integrating the two groups and directing their antagonism against a "common enemy": Jews and Communists. Antisemitism and anticommunism as a major policy of the 1930's and 1940's resulted.[10]

The military itself was not free of internal differences. Representatives of the rising middle class appeared in the lower echelons and offered a new ideology that conflicted with the ideology of the older generation of officers. These younger officers did not support the "landed aristocracy," and neither did they favor an old-type military dictatorship supported by the landowners and church hierarchy. Although they were as unconcerned with democratic political institutions as were the older officers, they were the "sponsors of fundamental change and reform, the underminers of traditional institutions, and proponents of public-welfare measures."[11]

Government recognition of and reaction to weaknesses

The four presidents who came to power after the 1930 coup did not respond to the basic social, economic, and political problems of Argentina. General discontent among the masses and the lower-middle class did not concern them so long as these elements did not get out of hand. They were conservatives and they felt it their duty to preserve the vested interests of the upper classes.

Policies instituted after 1930 were primarily enacted to solve the economic ills caused by the worldwide economic crisis. General Uriburu, on his part, removed thousands from the public payroll in order to cut public expenses drastically and instituted a rather ruthless dictatorship. President Justo, on the other hand, concentrated heavily on rehabilitating the economy. His economic policies, which included devaluation of the currency, fixing of commodity prices, a quasi-governmental monopoly on grains, reduction in foreign imports, and

177

increased foreign investments, brought swift economic recovery. He also enacted laws to protect the rights of the industrial workers; agricultural laborers remained at the mercy of the landowners. Ortiz's ill health prevented his effecting major changes; and Castillo's reactionary policies undid much of Justo's progress. A state of siege was declared to cope with political and social unrest.

FORM AND CHARACTERISTICS OF REVOLUTION

ACTORS IN THE REVOLUTION

The revolutionary leadership

The instigators and organizers of the 1943 military coup d'etat were officers of the armed forces with widely divergent political views. Some were extreme rightists and frowned on social and political experimentation. Others were moderate and expressed some desire for progressive reforms. The four outstanding leaders of the coup had been trained by German military missions and were highly influenced by Fascist organizations in Italy and Germany.

Gen. Arturo Rawson, who was the senior officer at the strategic Campo de Mayo military garrison in 1943, was a professional soldier. He served in the army's general staff and held various regimental commands. His father, Gen. Franklin Rawson, led a small revolutionary force against one of the governors of Buenos Aires years before his son became a revolutionary leader.

Gen. Pedro P. Ramírez had both a military and a political background. He had been sent to Germany and Italy on military assignments, and had taken an active part in the 1930 coup d'etat. In 1931 he was the Military and Air Attaché in the Argentine Embassy in Rome. As a ranking army chief, he held the Cabinet post of War Minister under President Castillo. Although he had long been Conservative, he became associated with the Radicals several months before the coup when they offered to make him their presidential candidate.

Gen. Edelmiro Farrell also enjoyed an assignment in Fascist Italy during his military career. He was more inclined than the other actors toward a military dictatorship, but he was overshadowed by Col. Juan Domingo Perón. General Farrell commanded the mountain troops of the Mendoza garrison in 1943.

Colonel Perón emerged as the most important figure of the 1943 revolution. At 16 he was well along his way in military school, and his adult life was basically military. He was military attache to Chile in

1936, but had to be withdrawn in 1938 because he was charged with espionage. Sent to Italy as military attache, he was highly influenced by Fascist ideology. He was elevated to the rank of colonel in 1941.[d]

The revolutionary following

The 1943 revolution was a military coup d'etat executed by a military organization and supported by the middle-class elements of the Radical Party and a faction of the Conservatives. Approximately 8,000 soldiers participated in the march against the capital. The revolutionary following included neither workers nor peasants. Only a small number of military officers knew of the coup.

ORGANIZATION OF REVOLUTIONARY EFFORT

The organization of the revolution consisted of a three-cornered arrangement between Conservatives, Radicals, and a group of young army officers known as the United Officers Group (GOU). The divergent groups were represented in a military committee in charge of the revolution, which was led by Generals Rawson and Ramírez. The committee was formed shortly before the coup was executed. Each group had its own reasons to overthrow the Castillo government.

GOU, which emerged as the dominant group of the revolutionary organization, seems to have been founded in 1940 in a military garrison in Mendoza. Although its founders are not definitely known, it has been suspected that Perón was one of its early leaders. He was stationed at the garrison in 1940. GOU had a membership of approximately 60 percent of the 3,600 active officers on the army list. The group was organized to give political orientation to the army. It professed disgust with civil administrations. It wanted military authoritarianism and strict control over most phases of national life. GOD was patriotic, nationalistic, and antiforeign.[13] The names of the GOD officers became known only after the coup. General Farrell represented GOU on the revolutionary committee.

GOALS OF THE REVOLUTION

Concrete political aims of revolutionary leaders

The Conservative element of the revolutionary leaders represented the landowners, and their primary objective was to prevent the indus-

---

[d] Several sources indicate that Perón was the real "brains" behind the revolution. One source states that Perón presumably preferred to occupy a minor post after the revolution and manipulate the political strings from behind the scenes while consolidating his power.[12]

trialists, represented by the Castillo-Costas clique, from dominating the political arena. Some were reported to have become strongly pro-Allied. General Rawson, for instance, wished to declare war on the Axis powers and benefit from American lend-lease material. General Ramírez, whom the Radicals thought they could influence, wanted to establish a Fascist state. He abandoned both the Conservatives and the Radicals.

General Farrell and Colonel Perón, members of GOU, had an announced political program. On May 13, 1943, GOD circulated a proclamation stating the position of the group. The proclamation was quite frank in its pro-Axis sympathies and its conception of Argentina's "manifest destiny." Under a military dictatorship, the GOD declared, Argentina could unify all the South American states under its guardianship as Germany was then "uniting Europe" under its leadership.[14]

More specifically, the GOD wanted to end the complacent alliance between the generals and the oligarchy. It felt that the army had been neglected by civilian administrators; by taking control of the government by force, it expected to obtain modern weapons and equipment, recruit more men, and receive better pay. GOU appeared primarily interested in power.

Social and economic goals of leadership and following

The older officers, Ramírez especially, envisaged a corporate-type economic system in which the industrial and agrarian sectors could operate harmoniously, responding to the interests of the country and not subject to foreign economic interests. Moreover, Ramírez wanted to industrialize the army by training its members to operate business enterprises. He also wanted to establish "social stability" by removing the causes of discord among the various social groups.

GOU also expressed a desire to bring the Argentine Armed Forces into more sympathetic relations with the rest of society. By establishing a military dictatorship and winning public support, the group explained, the army could carry out an armaments program for the control of the continent. GOU felt that control over radio, press, literature, education, and religion was necessary. Only under these conditions, it argued, could the army guarantee freedom. The sole mission of the people, it felt, was to work for the state and obey its leaders.

REVOLUTIONARY TECHNIQUES AND GOVERNMENT COUNTERMEASURES

Methods for weakening existing authority and countermeasures by government

The revolution was swift and virtually bloodless. General Ramírez has been given credit for plotting the coup, and it appears that the scheme was only 3 or 4 days old when executed. The important events took place within a 3-day period: the 3rd, 4th, and 5th days of June 1943. Argentine officials were aware of the revolutionary ferment. President Castillo had discovered some time earlier that War Minister Ramírez was somehow implicated and had demanded his resignation. Ramírez refused, and on June 3 he in turn, along with a group of officers from Campo de Mayo military garrison, demanded the resignation of the President. Castillo also refused to resign and issued a manifesto ordering one General Rodolfo Marquez to command the "forces of repression,"[15] and Chief of Police General Domingo Martinez to defend the city.

Two events occurred on June 4: the timely escape of the President and his Cabinet, and the march of General Rawson's troops on Buenos Aires. How much preparation was needed for Castillo and his Cabinet to make their escape is not clear. They all drove to the port of Buenos Aires early in the morning, lifting a drawbridge along the route, and boarded a minesweeper. From there President Castillo wired the President of the Supreme Court saying that the seat of the government had been established on the ship.[c]

Meanwhile, 8,000 troops under the command of General Rawson began their march on the capital. At the navy's Mechanical Training School outside the city, they were challenged by "loyalist" troops. The exchange of fire lasted 40 minutes and caused nearly 100 casualties. This fighting appears to have resulted from a misunderstanding between the garrison and school commands. Had the school commander been aware of Rawson's mission, there would have been no fighting in this revolution. After a short explanation of the nature of the march, "loyalist" troops joined the rebels and proceeded to the city.

Not until 10 a.m. did bustling Buenos Aires become aware of the revolution. The troops entered the city and some units immediately surrounded Casa Rosada, the President's mansion. The Central Police Barracks, the Banco de la Nación (National Bank), and other government buildings were occupied before noon. As the people milled

---

[c] One author, describing the events of the revolution, writes that Castillo had hoped to rally the support of the navy behind his government.[16]

about watching, leaflets stating that the revolution was a return to democracy were distributed. By 1 p.m. the troops were conducting mopping-up operations. Troops from other garrisons were reported marching toward the city.

Throughout the morning and part of the afternoon the State Radio Station was broadcasting pro-Castillo bulletins. Junta men ordered the station to cease counterrevolutionary broadcasting and announce the fall of the Castillo regime. By 3 p.m. all stations were broadcasting General Rawson's proclamation that the revolution was purely Argentine and devoid of foreign political inspiration. The General asked the people for confidence and tranquility.

Civilians took no part in the revolution. During the day people went about their business as usual, and when it became known that a revolution was in progress some of the curious came downtown to observe. The crowds became somewhat rowdy, perhaps because the revolution was not exciting enough for their taste, and cars and buses were overturned. General Rawson came out on the balcony of Casa Rosada to make a speech at 5:20 p.m. The people applauded and shouted *"Viva Democracia."* All began to calm down by 7 o'clock. On the next day Castillo resigned and dissolved the Congress.[f]

Methods for gaining support and countermeasures taken by government

The revolution was well under way before any measures to rally the support of the people either to the government or to the revolutionists were undertaken. Castillo managed to have the State Radio broadcast his bulletins for a part of the day, but the bulletins aroused curiosity rather than sympathy. If Castillo did attempt to gain the support of the navy while he was on the minesweeper, he did not succeed. Admiral Saba Sueyro, Commander of the Argentine Navy, was a member of the revolutionary committee.

Neither did the army succeed in rallying the Argentine people behind the revolution. Leaflets, radio broadcasts, and public speeches did stir the public; but for the most part the Argentines and their leaders remained aloof and maintained a "wait and see" attitude.

---

[f] For further information on the revolutionary activities see Robert J. Alexander, *The Perón Era* (New York: Columbia University Press, 1951), pp. 12–14; and Austin F. MacDonald, *Latin American Politics and Government* (New York: Thomas Y. Crowell Company, 1954), pp. 53–54.

MANNER IN WHICH CONTROL OF GOVERNMENT WAS TRANSFERRED TO REVOLUTIONARIES

Control of the government was transferred in a simple ceremony on June 5. Castillo had returned to Buenos Aires and met with the military commandant, Diego Mason, to deliver his resignation and to dissolve the Congress. The military junta was in control of the Argentine Government.

THE EFFECTS OF THE REVOLUTION

CHANGES IN PERSONNEL AND INSTITUTIONS OF GOVERNMENT

The new government was a military dictatorship whose leaders were constantly involved in power struggles. General Rawson was the first to receive an official title. On the day Castillo resigned, Rawson appeared again on the balcony of Casa Rosada and declared himself Provisional President. Whether this was according to plan is not known. Ramírez would have been a more likely choice. In any event, Rawson held his post for no more than 28 hours. Ramírez emerged as the new President. He held the position for 8 months. General Farrell, who was Ramírez's Vice President, succeeded him; and Farrell in turn was followed by his Vice President, Colonel Perón.

Following the 1943 revolution, Perón was appointed Under Secretary of War, and began to consolidate power in the army. He became acting War Minister under Farrell and was soon named permanent Minister. A nonmilitary and non-Cabinet position of Director of National Department of Labor was given him in 1943, and one month later he was raised to the office of Minister of Labor and Welfare. From this vantage point he proceeded to organize labor under his control. When he became Vice President under Farrell, he had virtual control of the Argentine government.[17] Perón was officially elected President in 1946.

The Castillo cabinet was replaced by officers of the Argentine Armed Forces. Of the new Cabinet members, only two were civilians, both professed Naziphiles. As soon as Ramírez was elevated to the Presidency he named an all-military Cabinet. High governmental posts were filled with members of the Armed Forces. Provincial legislatures and city councils ceased to function.

MAJOR POLICY CHANGES

Those Argentines who were optimistic when the coup occurred were soon disappointed. The military dictatorship issued a series of decrees more oppressive than Castillo's undemocratic practices. Four months after Ramírez's accession to the Presidency, protesting citizens published a democratic manifesto deploring their loss of civil liberties. This precipitated a reign of terror. Opposition politicians were locked in concentration camps. Student opposition was suppressed. Organizations of manufacturers and businessmen were threatened. Seventy newspapers were closed or suspended for criticizing the regime.[18]

But the Ramírez regime also had its supporters. Rents were lowered; minimum wages for farm laborers were established, salaries in the government service were increased, and utility charges were greatly reduced. This rallied thousands of unorganized workers behind the dictatorship. The church hierarchy also supported Ramírez. Church and state had been separate since 1884. Under the new government, the fundamentals of the Catholic faith had to be taught to every school child regardless of the wishes of the child or its parents.

Foreign politics also took on a new look. The pro-Axis attitude which the Argentine Government had maintained under its policy of neutrality ended officially in January 1944, when the dictatorship severed diplomatic relations with Germany and Japan. The United States and Great Britain had high-pressured Argentina into a move the regime would not have made of its own accord. Pro-Axis groups were nevertheless very active after the official break.[19]

LONG RANGE SOCIAL AND ECONOMIC EFFECTS

Juan Domingo Perón was heard of only infrequently in the months immediately following the revolution. However, he had already begun building the organization which was to elevate him to power 3 years later. The social and economic changes for which he became responsible gave the 1943 revolution lasting significance. During the years prior to the 1946 elections Perón concentrated on effecting progressive changes within three sectors of Argentine society upon which his administration was to rest: the army, organized labor, and the church.

Perón's official post in the War Ministry enabled him to consolidate his power within the army. He became undisputed leader of the GOU and soon members of the GOU were able to control the army. He was able to promote army officers who supported him and give them choice assignments. By decreeing higher wages for the noncommissioned officers and the common soldiers he won favor with all ranks.

Even more important than the army in his rise to power was labor. The constructive changes which he brought about in social security and the development of labor unions made Argentina the most advanced Latin American country in social insurance and labor organization. Social security institutes covering virtually all of Argentina's workers were established. Extensive construction of low-cost housing was undertaken. Ten days of annual leave with pay was secured for many workers. An increase in real wages raised living standards.[20] Organized labor was either persuaded or coerced into supporting Perón; and the power which he held over labor unions after he was elected made these organizations virtual arms of the government.

The third source of Perón's strength was the Catholic Church. The Argentine Cardinal and several Archbishops endorsed Perón's candidacy in 1946 by signing a pastoral letter which was read in churches throughout Argentina. Perón favored religious instruction in the schools; other candidates opposed it. Although some of the lower clergy spoke out against Perón, there was no doubt that the weight of the church was behind him.

The 1949 Constitution formalized Perón's virtual dictatorship. The constitutional prohibition against two successive terms was abolished, permitting a President to remain in office indefinitely if he controlled the election mechanisms. The Perónista Party and the Labor Party became the agencies of the army and organized labor, strongly supporting Perón. The traditional procedure for confirmation of federal judges was drastically changed. As interpreted, the new clause provided that all judges, even those confirmed years earlier, had to come up for confirmation. In effect, this weeded out all anti-Perón judges.[g]

OTHER EFFECTS

After the fall of Perón in 1955, Argentina was in the throes of an economic crisis which threatened the stability of the nation. Although many of these economic problems were resolved by the administration of Arturo Frondizi between 1958 and 1962, political and social dislocations continued to generate unrest.

---

[g] The immediate effects of the revolution are found in Ray Josephs, *Argentine Diary* (New York: Random House, 1944), passim. The Perón dictatorship is best described in Robert J. Alexander, *The Perón Era* (New York: Columbia University Press, 1951), passim.

The Frondizi administration's austerity program,[h] although unpopular, appeared to have been working successfully toward stabilization and development in 1962. The inflationary trend precipitated by Perón's economic policies had been curbed, foreign credit standing had been restored, and price stability had been achieved. Meaningful wage increases had given the Argentine consumer higher purchasing power. However, there were still many unresolved problems.[21]

The Conservatives—landowners and industrialists—still retained considerable social and economic power, and continued to be the benefactors of Argentine prosperity. President Frondizi's preelection promises of social and economic transformations gave way to a moderate economic development which only strengthened the existing social structure. Except for a few economic restrictions, little has been done to curb the increasing wealth of the upper classes.

That is why Frondizi has lost much of the mass support he had in 1958 when he was elected President. Like Perón, Frondizi promised the lower classes a long overdue agrarian reform; as under Perón, the assault on the Conservatives remained largely verbal. Overwhelmingly popular in 1958, Frondizi had become a minority President 3 years later. Social changes restricting the power of the Conservatives could have relieved the administration's precarious position and produced an identity of interest between the administration and the industrial workers. Opposition from the Conservatives would have been negligible. The split among the Conservatives, i.e., between the landowners and beef barons, and the financiers and industrialists, had weakened the Conservatives.[22]

The Armed Forces, led by Gen. Pedro Arumburu, strongly supported Frondizi, but by 1962 there were signs that this support was wavering. Several military leaders have made attempts to unseat Frondizi. The fate of Argentina may rest on the government's ability to raise economic production still further.

NOTES

1. Robert J. Alexander, *The Perón Era* (New York: Columbia University Press, 1951), p. 8.
2. George I. Blanksten, *Perón's Argentina* (Chicago: The University of Chicago Press, 1953), p. 4.
3. Ibid., p. 239.

---

[h] Frondizi's austerity program included holding down wage levels and restricting imports.

4. Arthur P. Whitaker, "The Argentine Paradox" *The Annals* (March 1961), 105.

5. Blanksten, *Argentina*, p. 250.

6. Whitaker, "Paradox," pp. 104–105.

7. Hubert Herring, *History of Latin America* (New York: Alfred A. Knopf, 1960), pp. 610–611.

8. Blanksten, *Argentina*, pp. 209–216.

9. Austin F. MacDonald, *Latin American Politics and Government* (2d ed.; New York: Thomas Y. Crowell Company, 1954), pp. 97–98.

10. Felix J. Weil, *Argentine Riddle* (New York: John Day Company, 1944), pp. 41–49.

11. Edwin Lieuwen, *Arms and Politics and Latin America* (New York: Frederick A. Praeger, Inc., 1960), pp. 125–127.

12. Weil, Riddle, p. 46; and Lieuwen, *Arms*, p. 67.

13. Blanksten, *Argentina*, pp. 52–53; and Ray Josephs, *Argentine Diary* (New York: Random House, 1944), pp. 136–140.

14. Alexander, *Perón*, pp. 12–13.

15. Josephs, *Diary*, p. 5.

16. Ibid., p. 4.

17. Blanksten, *Argentina*, pp. 53–56.

18. Ruth and Leonard Greenup, *Revolution Before Breakfast, Argentina 1941–1946* (Chapel Hill: University of North Carolina Press, 1947), p. xi.

19. MacDonald, *Politics*, pp. 66–69.

20. Alexander, *Perón*, p. 25.

21. Arthur P. Whitaker, "Social and Economic Crises in Argentina," *Current History* (April 1961), 210.

22. Ibid., p. 213.

RECOMMENDED READING

BOOKS:

Alexander, Robert J. *The Perón Era*. New York: Columbia University Press, 1951. Analytical and general, this book deals extensively with postrevolutionary developments and the Perón institutions.

Blanksten, George I. *Perón's Argentina*. Chicago: University of Chicago Press, 1953. Describes Peron's government as the result of the revolution. The book contains some background material.

Greenup, Ruth and Leonard. *Revolution Before Breakfast, Argentina: 1941–1946*. Chapel Hill: University of North Carolina Press, 1947.

Herring, Hubert. *History of Latin America.* New York: Alfred A. Knopf, 1960. Very general in Latin American studies, but includes good background material, particularly on the political process of Argentina leading to the revolution.

Josephs, Ray. *Argentine Diary.* New York: Random House, 1944. Perhaps the best account of revolutionary activities. Author describes dictatorship of Argentina as Fascist threat to Western Hemisphere.

MacDonald, Austin F. *Latin American Politics.* 2d ed. New York: Thomas Y. Crowell Company, 1954. Has good background of prerevolutionary Argentine politics and sufficient information on immediate results.

Weil, Felix J. *Argentine Riddle.* New York: John Day Company, 1944. An analysis of the underlying causes and the development of the country's political, social, and economic situation in 1944.

PERIODICALS:

Whitaker, Arthur P. "The Argentine Paradox," *The Annals* (March 1961), pp. 103–112.

Whitaker, Arthur P. "Social and Economic Crises in Argentina," *Current History* (April 1961). Both articles deal with the contemporary problems of Arturo Frondizi's administration.

OTHER:

Ashby, Joe Charles. "Labor and the Argentine Revolution," presented as a thesis to the University of Texas, Austin, Texas, August 1950. The dissertation traces significant developments in the social and economic progress of Argentina following the revolution and up to 1949.

THE BOLIVIAN REVOLUTION OF 1952

SYNOPSIS

In April 1952 the *Movimiento Nacionalista Revolucionario* (MNR), a progressive party, staged a revolution which, after a few days of heavy fighting, wrested power from the military junta that had ruled Bolivia for a year. The uprising brought to an end an era of political chaos during which the small group that had dominated the country's political and social structure tried desperately to stave off the growing demands of the people for social progress.

BRIEF HISTORY OF EVENTS LEADING UP TO AND CULMINATING IN REVOLUTION

Until rather recently, Bolivia was governed by a domestic oligarchy much as it had been governed by the Spaniards before independence. The Spanish-speaking "elite" dominated the country, primarily through the army. Most of the Indian population—over 50 percent of the total population still spoke pure Indian dialects—lived completely withdrawn from the rest of society. They worked the land for estate owners and managers and they mined tin for pitifully little compensation.

The seeds of the revolution can probably be traced back to the Chaco War between Bolivia and Paraguay. From 1932 to 1935 the two countries had battled for the possession of an inhospitable area mostly on the western slopes of the Andes in the belief that it contained rich oil deposits. The war left Bolivia exhausted and even poorer than before. Most of the Chaco was lost to Paraguay. One major effect of the war was that it demonstrated to the Indians and mestizos who were pressed into service that their white masters were not as efficient and superior as they had appeared. Once uprooted from the land through service in the army, the Indians developed a social consciousness they had not possessed before.

The revolution was not, however, carried out by the peasant Indians, who were the most suppressed element, despite the existence of a "grass roots" agrarian movement, but by lower-middle-class groups, by workers who had become a power in politics since the Chaco War, and by many dissatisfied intellectuals. It was instigated and led by the MNR, which had competed with the Bolivian Communist Party for the support of dissatisfied elements.

An election in 1951 had shown the MNR to be the most popular party, but the outgoing President refused to allow the MNR to come to power. Instead he turned the government over to a military junta, which ruled Bolivia in authoritarian fashion. The years preceding the revolution had been marked by extreme turmoil and political instability, and by the desperate attempts of the old ruling and elite groups to maintain the *status quo*. After the MNR took over the government in April 1952, a broad social revolution was initiated. Its success is still in the balance.

THE ENVIRONMENT OF THE REVOLUTION

DESCRIPTION OF COUNTRY

Physical characteristics

Bolivia has been called the American Tibet. Its area is roughly the size of California and Texas combined. About two-fifths of the country lies on Andean plateaus 12,000 to 14,000 feet above sea level. This area contains about 80 percent of the population. Mountain peaks rise more than 20,000 feet above sea level. The climate is cold, with little seasonal variation, and the country is barren. Lower down on the Eastern Andean slopes are deep valleys, virgin forests, and prairies. Exploitation of these areas for grain, meat, rubber, and other products is handicapped by lack of labor—only 20 percent of the population live in an area that occupies 60 percent of Bolivia's territory—and lack of transportation facilities.

The people

The inhospitable condition of the country is responsible for its very low population density. Between 3 and 3½ million people live in Bolivia; of these, 54 percent are pure Indians, speaking Indian dialects, between 10 and 15 percent are white, mostly of Spanish descent, and the remainder are mestizos. La Paz is the major city; its population is 321,000. Sucre is the legal capital, but most government offices, including the Congress, are in La Paz. Sucre, and Bolivia's second city, Cochabamba, are situated on the eastern slope of the Andes, where the climate is less severe.

Communications

Transportation is made difficult by the mountainous nature of the country. The total road mileage in 1955 was 12,500 miles, but only a

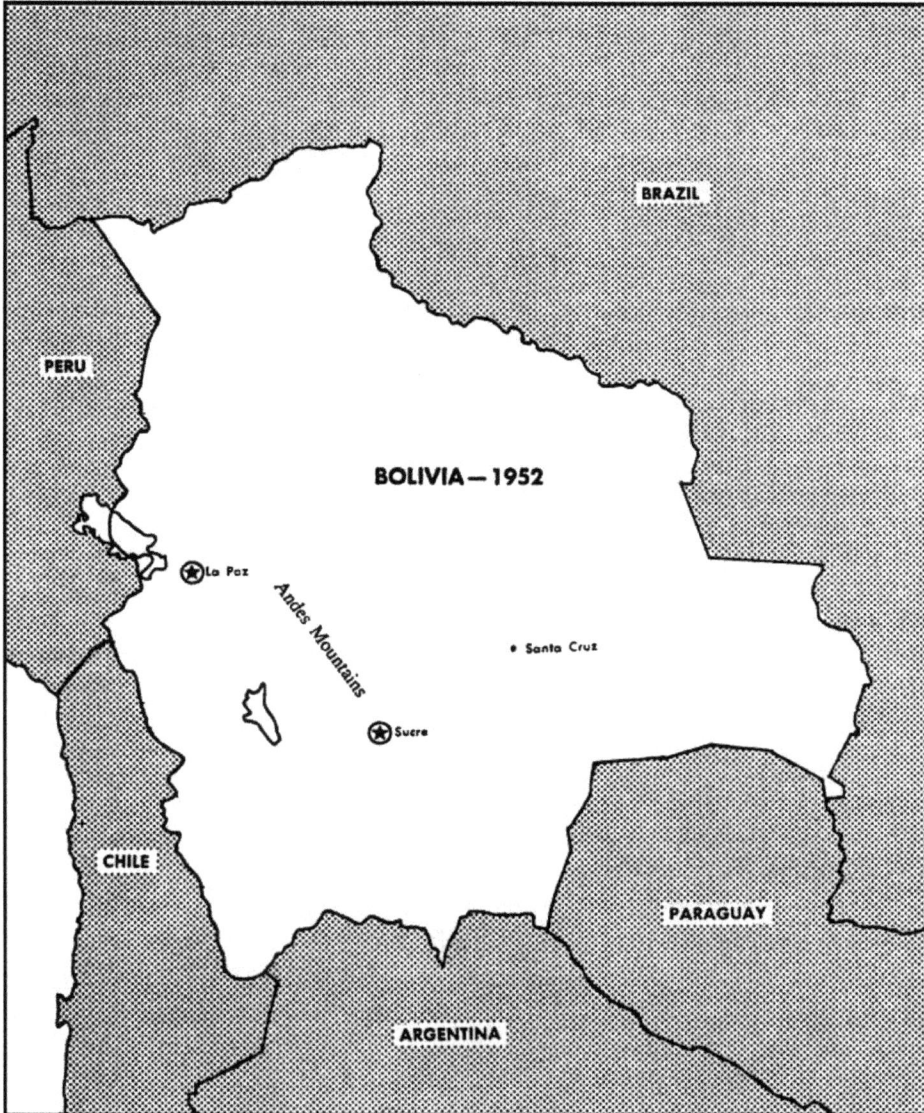

small fraction of that was paved. Only recently, after the 1952 revolu-
tion, was a major highway completed between Cochabamba and Santa
Cruz, facilitating the agricultural development of the eastern lowlands.
The railroad system is somewhat better developed and serves the min-
ing industry. Bolivia is landlocked; it lost its access to the Pacific after
the Chaco War in 1936. Railroads connect the country with ports in
Chile, where Bolivia has a free port, and Peru, and another line from
Santa Cruz leads to Brazil and the Atlantic.

Natural resources

Bolivia has an abundance of natural resources, despite the poverty of its people. It has been called a "beggar on a throne of gold."[1] Tin is the most important mineral resource; some lead and silver are mined. Oil deposits exist in the Chaco area but before 1952 they had not been exploited on a large scale. Indeed, comparatively little of the potential agricultural, forest, and mineral wealth had been developed. The major reasons for this were the lack of transportation, an inadequate supply of labor, political instability, lack of technology, and insufficient capital.

SOCIO-ECONOMIC STRUCTURE

Economic system

Before the revolution, most of the productive land was held under a system of the latifundian type. Large areas were owned by a few people, often residing in the cities. The Indians who worked the land were virtually serfs serving their landlords or the patrons administering the estates. More than 70 percent of the nation's labor force was employed in agriculture, though tin mining constituted a much more important segment of the economy from the standpoint of foreign trade and the money economy. Considerable foodstuffs had to be imported. Tin accounted for almost two-thirds of the nation's exports. Before 1952, tin mining was in private hands and was controlled by a few foreign-owned companies. The mining industry was based on cheap labor. Company stores supplied necessities to the miners at prices "calculated to keep them in perpetual debt."[2] The prosperity of the mine owners and managers, government revenues, and the state of the economy in general fluctuated with the price paid for tin on the international market. World War II brought great demands for tin, and prices soared. After the war, lower prices and reduced demand put Bolivia's economy into a precarious condition, which was only temporarily relieved by the Korean War boom.

Industry did not play a major role in the economy. The growth of manufacturing had been handicapped by lack of manpower, lack of coal, and inadequate transportation facilities. About 80 percent of industry still operated at the small shop or artisan stage. Most industrial plants were in La Paz or Cochabamba.

Class structure

One student of the revolution noted: "Before 1952 Bolivian society and its power structure were cast in the mold of the Spanish colo-

nial institutions which persist in many Latin American republics."[3]
In Bolivia this mold was particularly rigid. The upper crust was composed of the urban, Spanish element, comprising 10 to 15 percent
of the population. Before 1952, most laborers worked on latifundios,
where they barely eked out an existence after paying their dues to
the landowners or managers of the estates. The situation of the tin
miners was not much better. Their pay ranged from 20 to 140 bolivianos per day; a kilo of flour cost 28 bolivianos and a cotton shirt 300.[4]
Despite the overall labor available for economic development, there
was often unemployment in the mining industry when production
was low because of lack of demand on the international market.

Wealth was concentrated in the hands of a few mining magnates
and estate owners. The middle class was small and without great
influence. The problem, however, was not so much that wealth was
unevenly distributed as that there was little wealth to distribute. The
net yearly income per person in 1951 was $86.[5] There was little social
mobility, although Indians did become mestizos. Whites and mestizos controlled the country. The author of a standard book on Latin
American politics characterized the class structure as follows:

> Some have great estates; a few have important financial
> interests in the mines. Many more are the managers of
> the estates, or skilled workers, or minor government offi-
> cials. The Indians are the laboring class. Most of them
> are engaged in agriculture. If they do not have their
> own small plots of ground, they work on the estates,
> where they are virtually serfs. From sun to sun they
> toil for the mere right to exist. They receive almost no
> money wages; their health and education are neglected.
> Yet nothing could induce them to move from the place
> they have known as home.[6]

Literacy and education

Seventy to 85 percent of the population are illiterate. Schools were
concentrated in the urban areas; the Indians remained illiterate and
without knowledge of Spanish. This fact rather than their race made
them "Indians"; when they moved to a city and learned Spanish they
graduated to the class of Cholos (mestizos).

Major religions and religious institutions

Bolivia, like other South American countries, is Catholic, but religious tolerance is practiced, and there are a fairly large number of
Protestants and Protestant missions. However, all but 5 percent of the

population are Catholic. Archbishops and bishops are appointed by the President, after the Senate submits three names for each vacancy. Papal bulls and decrees were approved by the President and the Senate before becoming effective.

GOVERNMENT AND THE RULING ELITE

Description of form of government

Bolivia is a republic. The pre-1952 constitution provided for a strong executive. The President was elected by direct vote for a 6-year term. If one candidate failed to get a majority, Congress decided from among the candidates. The President appointed and dismissed the Cabinet and had almost unlimited power of appointment and dismissal of all administrative officers. This power was not subject to approval by Parliament except for high officers of the army, who had to be confirmed by the Senate. Supreme Court judges, however, were appointed by the Chamber of Deputies. The President also played an important role in the law-making process by initiating legislation. His veto could be overridden.

The legislature was divided into a Senate and a Chamber of Deputies. Senators and Deputies were elected by direct vote, the former for 6 years, the latter for 4 years. Before 1952, there was a literacy requirement for suffrage and women could not vote.

Description of political process

Political reality diverged greatly from the constitutional framework. "Dictators have succeeded one another with monotonous regularity. Many have died violent deaths. Fourteen constitutions have been adopted without substantially changing the political pattern . . . Leaders were more important than principles. The Army dominated the nation."[7] Thus one authority summed up politics in Bolivia. Between 1941 and 1950, there were no less than 25 revolutionary uprisings, ranging from "abortive plots, through brief bloody strikes, to a bitter three weeks civil war. Of these attempts, two resulted in the overthrow of the President."[8]

The political parties were a mixture of nationalism, socialism, authoritarianism, and reformism. Between 1936 and 1943, political life was dominated by the *Razón de Patria* or Reason of the Father Land (RADEPA), a political group of younger army officers, strongly influenced by German and Italian Fascist ideas. This group, relying on secret maneuvering, and aided by the conditions brought about by the Chaco War, never gained popular support, and the governments which it dominated were unable to maintain effective public order. "The political vacuum began to be filled in the early 1940's by the formation of half a dozen parties of varying tendencies, from the Nationalistic Revolutionary Movement (MNR), which was able to reach an understanding with RADEPA, to the Communist Stalinist Party (PIR). The MNR, founded by intellectuals such as the university professor Victor Paz Estenssoro, was able to combine the popular themes of revolution with a program of government which included, in the beginning, the RADEPA's ideas of a powerful centralized state headed by the military."[9]

In 1943, the RADEPA and the MNR organized a rebellion which succeeded in ousting a President who had been supported by more conservative army leaders. The RADEPA installed one of their own, Maj. Gualberto Villarroel, as President while the MNR leader, Paz Estenssoro, became Minister of Finance. A Marxist, radical group (PIR) remained in the opposition, together with more conservative army officers. Villarroel's regime soon became one of Bolivia's most oppressive.

In 1946, a coalition of labor and student groups disposed of Villarroel although mass violence and street mobs hanged him on a lamp post. The MNR was discredited and Paz Estenssoro was forced to leave the country. Other MNR leaders went into exile or underground.

Outside of La Paz, especially among the tin miners, Villarroel's ouster was viewed with strong regrets. The new coalition proved unable to establish a stable government. The MNR continued to enjoy considerable support among the workers and middle class elements, and in elections in 1949 it showed surprising strength in electing congressmen.

In the presidential elections of May 1951 the MNR received, officially, just under 50 percent of the vote. MNR claimed, however, that actually it had polled 79 percent of the vote.[10] The truth may well be somewhere in between. The Constitution provided that if no candidate received an absolute majority Congress should choose the President. MNR leaders threatened to install Paz Estenssoro by force of arms should Congress decide for someone else. Mamerto Urriolagoitia, the outgoing President, unwilling to turn the reins of government over to the MNR, persuaded a group of army officers to form a military junta. (During this time many of the MNR leaders, including Paz Estenssoro, were still in exile.)

The military junta was a conservative group which had little public support and much opposition. The two major opposition parties were the MNR and the POR, a Trotskyite party. The PIR vacillated between supporting the government and siding with the MNR. Opposition to the military junta and support for the MNR came from labor groups, particularly the tin miners, from students and intellectuals, and from the underprivileged agricultural population. The latter, however, was no strong political force despite its large numbers. There was some support for the MNR within the army but, judging from the fact that after the revolution MNR considered 80 percent of the officer corps unreliable, this support could not have exceeded 20 percent of the officers.

Legal procedure for amending constitution or changing government institutions

An amendment to the Constitution could be enacted by Congress without the participation of the President. A two-thirds majority was required in each house, but the amendment did not become effective until after the next congressional elections and a second reading in Congress which also required a two-thirds majority.

The constitutional provisions for amending the basic law and for changing administrations were unimportant, however, since Bolivia has had a long tradition of revolution and violence in accomplishing political change. Neither the governments in power nor the opposition had much faith in orderly political or constitutional processes.

Relationship to foreign powers

Bolivia had no binding alliances that would have influenced internal developments. Its economic dependence on the export of tin made foreign policy an important aspect of internal politics. An important test of the ability of a government was provided by the economic agreements it concluded with other countries, especially the United States, relating to the quantity and price of tin to be exported. Despite Bolivia's economic dependence of the United States and other Western powers, antiforeign and especially anti-"Yankee" propaganda slogans were constantly stressed by the political opposition between 1946 and 1952.

The role of military and police powers

Robert Alexander, the author of a study on the revolution, has described the role of the military as follows: "The Army constituted the backbone of the regimes in power between 1946 and 1952, and was frequently used during this period—and before it—to discipline the turbulent mine workers and coerce the peasants."[11] The officers came primarily from the urban middle class, the men were mostly Indian conscripts who had to serve for 1 year. The army, containing some 18,000 soldiers, was poorly outfitted; it had some artillery but little other heavy equipment. There was also a national police force, the *carabineros*, which until 1952 did not play a significant role in politics.

WEAKNESSES OF THE SOCIO-ECONOMIC-POLITICAL STRUCTURE OF THE PREREVOLUTIONARY REGIME

History of revolutions or governmental instabilities

Bolivia has had few periods of political stability and they were always succeeded by periods of violence. None of the revolutions or coups d'etat before 1952 brought any significant changes in the social or political structure.

Economic weaknesses

The greatest economic weakness was the dependence on the fluctuating international demand for tin. Inadequate roads and other transportation facilities made it difficult to develop the agricultural potential of the country. Petroleum deposits did exist, but before 1952 only a small amount of oil was obtained. The life and economy of the country were controlled by a small upper class for their own benefit rather than the country's as a whole.

Social tensions

The Chaco War between Bolivia and Paraguay, which lasted from 1932 to 1936, sharpened already existing social tensions. As a result of the war, which Bolivia lost, Indian peasants and mine workers were drafted into the army and exposed to the inefficiency of the officer corps. The white upper class thus lost some of its previously unquestioned supremacy. The Indian recruits also broadened their limited social horizon and some of them were attracted to the cities, where they climbed up the social ladder and became Cholos. Social unrest also spread to the countryside, though this process of fermentation was not completed until after the revolution. The whole period between the Chaco War and the national revolution in 1952 was one of unrest and political instability. It differed from previous periods of unrest particularly in that it involved Cholos and Indians.

Government recognition of and reaction to weaknesses

The military junta which governed between 1951 and 1952 tried to aid the economy without altering the basic structure of society. To quote Alexander again: "It obtained technical assistance from the United Nations, signed important agreements with neighboring countries, tried to secure a good price for tin through skillful negotia-

197

tions in Washington, increased imports of foodstuffs to ease shortages and decreed a general amnesty under which all exiles were granted permission to return to Bolivia."[12] Not all exiles took advantage of the offer; Paz Estenssoro remained in Montevideo. Aside from these positive steps, which did little to solve the basic social conflicts, the junta ruled in authoritarian fashion. Its promise of early elections was not fulfilled. Political parties were suppressed and free expression of ideas in the press was interfered with by the authoritarian regime.

FORM AND CHARACTERISTICS OF REVOLUTION

ACTORS IN THE REVOLUTION

The revolutionary leadership

The most important figure in the Bolivian revolution was the acknowledged leader of the MNR, Paz Estenssoro. He was a soft-spoken intellectual and former professor of economics. During the Villarroel regime he had been Minister of France, but he was released from the Cabinet because of his alleged Nazi sympathies. He himself denies having advocated collaboration with the Axis.[13] After 1946 he was exiled and spent time in Argentina and Montevideo, but he continued to be the leader of the party.

The second important figure in the MNR was Hernan Siles Zuaz, the founder of the party (and later President of Bolivia from 1956 to 1961). He is the son of a former President turned rebel. A lawyer and journalist by profession, he had served as an MNR delegate to Congress. He was the principal leader of the revolution in Bolivia at the time.

Juan Lechin, the leader of the tin miners' organization, also played a major role. He, too, had been exiled several times, but had always returned. He had served in Congress and his main loyalty was toward the mine workers. His political views leaned toward those of Trotsky; he represented the left wing of the MNR.

The fourth leading figure was Gen. Antonio Seleme, the head of the *carabineros*, who was instrumental in aligning the police force on the side of the revolutionaries. His importance ceased when, on the second day of the revolution, he took refuge in the Chilean Embassy for fear the revolution had failed.

The revolutionary following

Most of the popular support for the MNR came from workers and the poorer middle classes. Its leadership was composed primarily of intellectuals and labor leaders. In the years before the revolution the MNR had succeeded in recruiting some Indians, particularly among the miners. These Indians became the spokesmen for their people on a national level. The support for the MNR and its revolutionary ideas was greater outside La Paz, especially in the tin mining areas. In the capital the old conservatives, the landed aristocracy, and the army had greater influence.

ORGANIZATION OF REVOLUTIONARY EFFORT

Internal organization

The status of the MNR alternated between that of a regular political party and that of an outlawed party. Between 1946 and 1952 it was severely persecuted. "At one time there were some 5,000 exiles belonging to the MNR and its allies in Argentina alone."[14] The party itself must have been rather disorganized during this time, since many of its leaders were in exile. "If an MNR leader succeeded in winning an election, the government usually managed to send him into exile or dispose of him in some other manner."[15] For the elections in 1951 Paz Estenssoro was nominated as the party's candidate for President and Hernan Siles for Vice President. During the campaign, the MNR received the support of the Marxist parties, including the PIR, which by now had become the official Communist Party.

Details of the conspiratorial organization which engineered the revolution are not available in the sources consulted. It appears that there was a conspiracy between the leadership of the MNR, General Seleme, head of the *carabineros*, and the leaders of the mining areas. Paz Estenssoro remained in exile and the revolution in La Paz was led by Siles and Lechin.

External organization

There is some doubt as to how much support the MNR received from the Perón machine in Argentina. Many of its leaders found exile in Argentina; but Paz Estenssoro left that country for Uruguay because of disputes with Perón. According to one author, Perón did not resume his support of the MNR until after the 1952 revolution. However, some MNR exiles were given jobs in the Argentine Government between 1946 and 1952.[16]

GOALS OF THE REVOLUTION

The goals of the revolution corresponded to the political and social program of the MNR. This program was first issued in 1942 and confirmed in 1946. It recognized the imperative need for land reform, which would give the peasants ownership of the land they tilled, and for better conditions for the workers. More specifically, the program called for the nationalization of the mines and the organization of strong labor unions. In the economic sphere the goals of the revolution were to free Bolivia's economy from its extreme dependence on the international market for tin and to place the means of production of tin in the hands of Bolivians.[17] To accomplish this, the petroleum industry was to be expanded and transportation facilities improved. Political power was to pass into the hands of the people at large, including illiterates and the underprivileged Indian population. Another goal was to eliminate the dominant position of the army in political life.

REVOLUTIONARY TECHNIQUES AND GOVERNMENT COUNTERMEASURES

The revolutionary struggle for the MNR lasted from the fall of Villarroel—in whose government the MNR had an important voice—in 1946 until the revolution of 1952. Almost immediately after the revolution of 1946, the miners began contemplating a counterrevolution to install a government favorable to them. They had been unable to come to the aid of the Villarroel government because events in La Paz had moved so fast that the fighting was over before they could intervene, but they never accepted the new regime. The mine owners, in the meantime, encouraged by the political shift, tried to destroy the union. In 1947, one of the large mining companies, Patino Mines, discharged all its employees and rehired only those it considered "loyal."[18] Early in 1949, there was a strike of the mine workers and, as a result, the mine union was virtually destroyed and its leaders were sent into exile. The union reorganized, however, and further strikes and riots resulted.

In May 1949 the MNR was permitted to participate in elections for Congress. The party made a good showing, but did not win a majority. Encouraged by the results, the party plotted with the mineworkers and some army officers. The plot was discovered, the MNR was again outlawed, and the head of the mineworkers, Juan Lechin, was again exiled.

In August 1949 another revolutionary uprising took place. The government responded by full-scale mobilization. Outside La Paz the revolutionaries were quite successful. For several days they held Cochabamba and Santa Cruz. They captured the radio station in Santa Cruz and announced a new government headed by Paz Estenssoro. Finally, however, loyal government troops crushed the revolution. One observer described the event as a "revolution planned on the idea of civil war, the people of the country against the army, the country against the capital."[19] Several other uprisings followed in 1950; all were suppressed. In one of these, students barricaded themselves in the university and battled government troops for 3 days before they were subdued.[20]

The propaganda themes employed by the MNR and its political allies included a strong appeal to nationalism. This took the form of denunciation of "Yankee imperialism," and of all foreign ownership of Bolivian economic enterprises, particularly the large tin mining companies. This was coupled with demands for social justice for the underprivileged and the removal of the small upper class from political domination.

Between 1946 and 1951, the MNR, unable to muster sufficient political power to take over the government by legal means, made numerous attempts to overthrow it by force. All these attempts failed, but they demonstrated the inability of existing regimes to rule the country in an orderly manner. Before the elections in 1951 MNR decided to try once more to gain power legally. Paz Estenssoro was nominated for President and Hernan Siles for Vice President. As noted previously, the MNR received at least a plurality of votes, but the old political group showed its unwillingness to relinquish its power and turned the government over to a military junta. The MNR's seizure of power came after it had been denied the fruits of the election. The party had become convinced that only a revolution could end the prevailing political anarchy.

The military junta attempted to suppress the revolutionary fervor of the people by forbidding strikes and mass demonstrations which might "be considered acts of sabotage or attempts against the security of the nation." It also attempted to stifle criticism in the press. However, the opposition was too strong to be silenced.[21]

The details of the planning for the revolution are not described in the sources used in this study. It seems to have been clear that the uprising would rely heavily on the support of the workers and miners. Originally the revolt had been scheduled for later in 1952, but the defection to the conspirators of the chief of the *carabineros*, General

Seleme, made it possible and probably necessary to act earlier. Now the MNR could count on an organized armed force, the police, to support and partially arm the workers and miners. Some arms, captured and hidden during previous uprisings, were already available to the miners.

MANNER IN WHICH CONTROL OF GOVERNMENT WAS TRANSFERRED TO REVOLUTIONARIES

On April 9 the armed revolt started simultaneously in all major cities. It was led in La Paz by Hernan Siles and Juan Lechin. The mining centers were quickly brought under control. Little has been learned about the fighting outside the capital. In La Paz, however, loyal government troops battled the workers and the police for 2 days and the issue seemed to hang in the balance;

> On the second day it appeared as if the revolt were going to' be suppressed. General Seleme took refuge in the Chilean Embassy and when he left the Embassy again on the following day, things having taken a turn for the better from his point of view, he was told that he was no longer in charge, and that the revolution was now completely in the hands of the MNR.

> Pro-government troops fought their way from the military school down below the center of La Paz almost to the heart of the city. Besides the police and armed workers, they were opposed by the Indian market women of La Paz, reported to have played an important role during the fighting by going up to the simple Indian soldiers who made up the regiments garrisoning the capital, and seizing their guns from their hands. The army troops were finally turned back, and the victory of the revolt was sealed when batteries mounted on the rim of the plateau above La Paz were dismantled by cadet volunteers of the Police Academy and when armed miners arrived from the camps near Oruro.

> Thus, an armed uprising gave power once more to the MNR. Due to his lack of faith in victory, General Antonio Seleme, who might otherwise have become President, turned over full control of the movement to the civilians of the MNR.[22]

Thus the events are described by Alexander. On April 15, Paz Estenssoro returned triumphantly, and was immediately sworn in as President.

THE EFFECTS OF THE REVOLUTION

CHANGES IN THE PERSONNEL AND INSTITUTIONS OF GOVERNMENT

Paz Estenssoro assumed the Presidency on the basis of the MNR's claim that it had won a majority in the 1951 election, and Hernan Siles became Vice President. Juan Lechin was appointed Secretary of Labor. The new government asserted that it had assumed the role to which it had been legally entitled since the elections of 1951. The basic institutions of government remained the same and so did the Constitution. Democratic processes did not, however, flourish. The new regime was again authoritarian, probably out of necessity; nevertheless two elections have been held at the scheduled time and both times the MNR has been victorious. Siles succeeded Estenssoro as President and Estenssoro was reelected in 1961. One of the important consequences of the revolution was a complete reorganization of the army. Eighty percent of the officer corps and most of the men were discharged and replaced by others considered loyal to the MNR. The military academies were closed and not reopened until much later.

MAJOR POLICY CHANGES

Policy changes were many and far-reaching. The tin mines were nationalized, but the owners were permitted to continue operations. The labor unions, which had been greatly weakened during the revolutionary struggle, were strengthened. The literacy requirement for voting was removed and women were given suffrage. Education was made available to larger numbers of people,[23] including the Indians. The most basic and important change occurred in 1953, when land reform eliminated the latifundian system and gave the land to the Indian peasants. The land reform was the work of the peasants themselves. They had begun to organize after the Chaco War and were taking matters into their own hands after the revolution. The MNR was forced to act quickly to legalize the actions of the peasants and to keep its prerevolutionary promises. In its foreign policy the MNR has repulsed overtures by Argentina for a close alliance. Instead the regime has sought to cooperate with the United States, though without antagonizing Argentina. As a result, Bolivia has received substantial economic aid from the United States.

LONG RANGE SOCIAL AND ECONOMIC EFFECTS

The long range social and economic effects of the revolution have been a leveling of Bolivian society. The poor gained and the old upper class all but disappeared. The Bolivian people, including the Indian peasants, have become a political and social force that can no longer be disregarded. On the whole, the MNR has been successful in maintaining political stability, but unrest persists just below the surface. In the beginning there were counterrevolutionary plots and uprisings, but they were easily suppressed. More recently unrest among the workers, particularly the tin miners, has again been evident. Nor could the economic problems existing before 1952 be corrected overnight.[a] On the whole, the MNR has followed a moderate economic course despite the major social changes which it encouraged. This moderation has cost the party the support of the more radical groups and also of the Communists, who at first had thrown in their lot with the MNR. Recent uprisings (in June 1961) appear to have been Communist-inspired. Progress for the workers has not been as rapid as many had hoped and social tension caused by poverty and ignorance continues.

NOTES

1. *The Worldmark Encyclopedia of the Nations* (New York: Worldmark Press, Inc., Harper and Brothers, 1960), 70–84.

2. Austin F. MacDonald, *Latin American Politics and Government* (2d ed.; New York: Thomas Y. Crowell Company, 1954), p. 526.

3. Richard W. Patch, "Bolivia: The Restrained Revolution," *The Annals of the American Academy of Political and Social Science*, Vol. 334, (March 1961), p. 124.

4. Le Benson Valentine, "A Comparative Study of Successful Revolutions in South America, 1941–1950" (Ann Arbor, Michigan: University Microfilms, Inc., 1958. Ph.D. Thesis, Stanford University [1958]), 175.

5. *Worldmark Encyclopedia.*

6. MacDonald, *Politics*, p. 527.

7. Ibid., p. 531.

8. Valentine, "Study," p. 178.

---

[a] Land reform has had the effect of lowering production and has not solved the problem of poverty.

9. Patch, "Bolivia," p. 127.

10. Robert J. Alexander, *The Bolivian National Revolution* (New Brunswick: Rutgers University Press, 1958), p. 41.

11. Ibid., p. 144.

12. Alexander, *Revolution*, p. 43.

13. Ibid., pp. 31–33.

14. Ibid., p. 38.

15. Ibid., pp. 38–39.

16. Ibid., p. 41.

17. Ibid., p. 9.

18. Ibid., p. 39.

19. Ibid., p. 29.

20. Valentine, "Study," p. 189.

21. Alexander, *Revolution*, p. 42.

22. Ibid., pp. 44–45.

23. Ibid., pp. 47–48.

RECOMMENDED READING

BOOKS:

Alexander, Robert J. *The Bolivian National Revolution.* New Brunswick: Rutgers University Press, 1958. The only book on the Bolivian revolution in English. It deals primarily with postrevolutionary developments and is sympathetic toward the MNR.

Herring, Hubert. *A History of Latin America.* New York: Alfred A. Knopf, 1960. A standard history of Latin American countries. A short section on Bolivia is useful as background information.

MacDonald, Austin F. *Latin American Politics and Government.* 2d ed. New York: Thomas Y. Crowell Co., 1954. A standard textbook on Latin America with a good section on Bolivia and the events preceding and following the revolution.

PERIODICALS:

"Bolivia's Revolutionary Regime," *The World Today,* II (1955). A discussion on the impact of the revolution on Bolivia's economy.

Patch, Richard W. "Bolivia: The Restrained Revolution," *The Annals of the American Academy of Political and Social Science,* 334 (March 1961). A short account of the revolution and its causes. Particularly good for the impact of the revolution on the Indian peasants and the social structure in general.

OTHERS:

Valentine, Lee Benson. "A Comparative Study of Successful Revolutions in South America, 1941–1950." Ann Arbor, Michigan: University Microfilms, Inc., 1958. PhD. Thesis, Stanford University [1958]. The account does not include the revolution, but is useful for developments between 1940 and 1950.

THE CUBAN REVOLUTION: 1953–1959

SYNOPSIS

On July 26, 1953, Fidel Castro and 165 Cuban revolutionaries attacked the Moncada Army Post with its garrison of 1,000 soldiers. The attack failed, but marked the beginning of a revolutionary movement, named after the day of the attack, "the 26th of July Movement." During the ensuing years it constituted the major opposition to the authoritarian regime of Gen. Fulgencio Batista. After a bitter struggle, which during its final phases took the form of a civil war, Fidel Castro and his movement gained complete control over Cuba in January 1959. His success was due to the widespread dissatisfaction with the dictatorship of Batista and the appeal of his program, which concentrated on social reform and freedom from oppression. After Castro became head of state, communism gained the upper hand and, in late 1961, Castro finally admitted that he had always been a Marxist-Leninist and that he had always planned to transform Cuba into a Communist state.

BRIEF HISTORY OF EVENTS LEADING UP TO AND CULMINATING IN REVOLUTION

The rise of Fulgencio Batista in 1934 marked the end of an era for Cuba.[a] Until that time Cuba had been less than fully independent. The Platt Amendment, passed by the U.S. Congress in 1901 and incorporated in the Cuban Constitution under pressure from Washington, granted the United States the right "to intervene for the preservation of Cuban independence," and "for the maintenance of government adequate for the protection of life, property and individual liberty."[1] It also gave the United States the right to ratify Cuban treaties with other nations, and to establish military bases in Cuba. The amendment was replaced by treaty in 1934.

The first phase of the Batista dictatorship lasted from 1933 until 1944. Until 1940 he had controlled Cuba through puppet presidents; from 1940 and 1944 he assumed the Presidency himself. Under Batista the Cuban economy made definite progress. Roads were improved, the educational system was enlarged, and the press enjoyed considerable freedom. Yet the Cuban people were little better off than they

---

[a] See pages 213 and 214 for the discussion of the 1933 revolution and the rise of Batista.

207

had been during the rule of Spain and illiteracy remained widespread, especially in rural areas.

In 1944 Batista retired voluntarily and took up residence in Florida. Under the succeeding presidents, Ramón Grau San Martín and Carlos Prio Socarras, there was a breakdown of police forces, courts, and public administration. In March 1952 Batista returned to campaign for the Presidency, but in the midst of the campaign seized power again in a revolution that lasted 2 hours and cost three lives. While his earlier regime had been relatively moderate, he now suspended the constitution of 1940 and established an absolute and ruthless dictatorship.

Batista's return to power dashed the hopes of a young lawyer, Fidel Castro, for election to Congress. He asked the Supreme Court to declare Batista's rule illegal, but that body failed to act and Fidel Castro determined to overthrow the dictator by force. On July 26, 1953, he led a raid against the Moncada Army Barracks which ended in his arrest and that of the few survivors. Pardoned in May 1955 in a general amnesty for political prisoners, Castro went to Mexico, where he and other members of the July 26th Movement began preparing for another revolution. The rebel forces were trained at a ranch outside Mexico City and an underground organization in the United States obtained money and weapons.

In December 1956 Castro landed in Cuba with 82 men, with disastrous results. Most of them were killed, a few imprisoned, and only 12, including Castro, survived to take refuge in the mountains of the Sierra Maestra. From here the revolution gradually spread as Castro built up his forces and won popular sympathy. Batista fought back with a reign of terror but could not stem the growing opposition. In May 1958 Batista launched a major attack against the rebel forces, estimated to be only 300 strong. The attack failed to eliminate Castro's forces; instead, his support increased everywhere. In July 1958 an agreement between Castro, representing the July 26th Movement, and various other organizations united the opposition. Toward the end of 1958 Castro's forces were strong enough to come down from the Sierra Maestra and capture the city of Santa Clara. On New Year's Day 1959, Batista resigned and fled the country.

THE ENVIRONMENT OF THE REVOLUTION

DESCRIPTION OF COUNTRY

Physical characteristics

Cuba, with its area of 44,217 square miles, including the Isle of Pines, is about the size of Pennsylvania. The coastline is deeply indented and provides many excellent harbors. About half of the island has semitropical flat or rolling terrain; the remainder is hilly and mountainous. The mountains consist of three intermittent clusters rather than one continuous chain. The highest and most rugged range is the Sierra Maestra, in the easternmost providence of Oriente.

The people

Cuba, the most populous member of the Antilles, has a population of 6,627,000 or 150 persons per square mile.[2] In 1953, 57 percent of the inhabitants lived in cities and towns.[3] There has been an increasing trend toward urbanization and industrialization. The major city, and the seat of economic and political power, is Havana, with a population of 785,455.[4] Other large cities are Marianao, a suburb of Havana, Santiago de Cuba, and Camaguey.

The pre-Colombian Indian population was virtually wiped out during the early years of Spanish rule. According to the 1953 census almost three-quarters of the population were white;[5] the remainder were Negroes and Mulattoes. Spanish is the official language.

Communications

No part of Cuba is far from at least one of her many excellent harbors, all of them connected with the interior by roads, railroads, or both. Cuba has an extensive net of railroads, over 5,000 miles of all-weather roads, and 3,200 miles of other roads.[6] Before 1959 Havana's José Marti International Airport was a crossroads for inter-American air transport.[7] Three other international airports were served by large foreign airlines. Interregional air transportation was maintained by several domestic companies.

Natural resources

Cuba is one of the richest nations in the Americas, with highly fertile soil and large quantities of minerals. The economy is predomi-

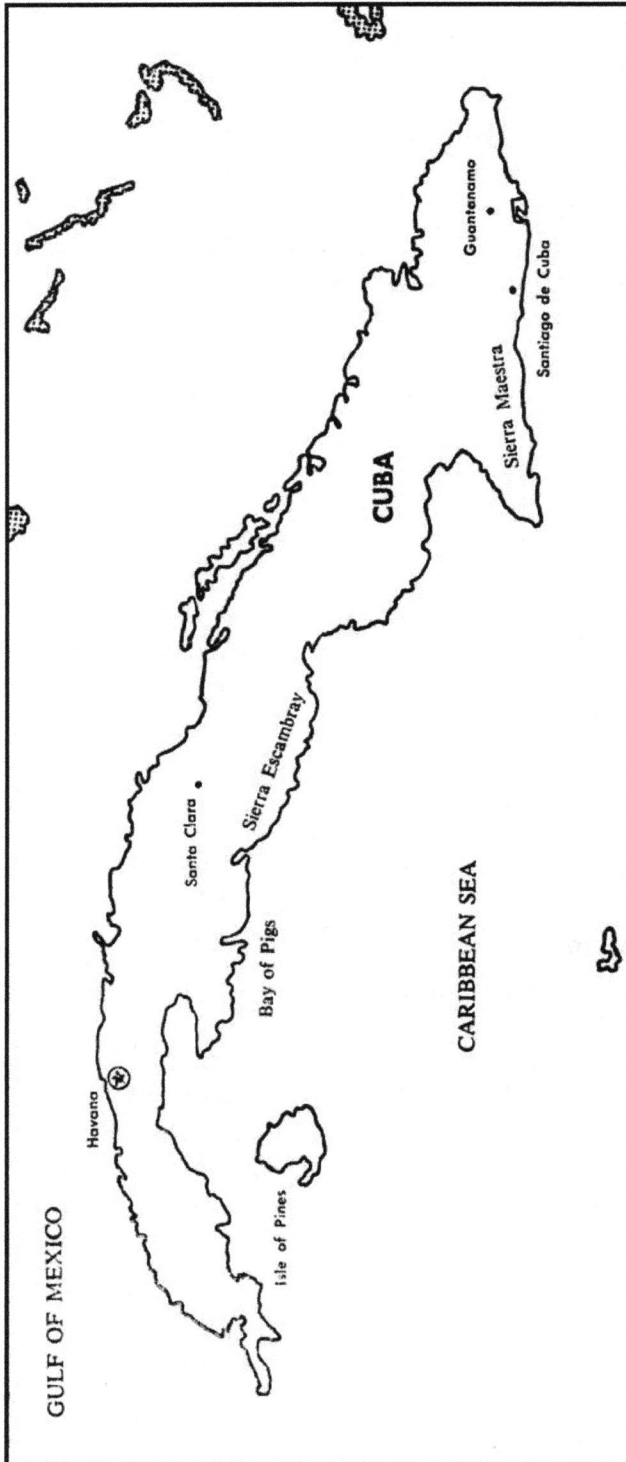

GULF OF MEXICO

Havana

Isle of Pines

Bay of Pigs

Santa Clara

Sierra Escambray

CUBA

Sierra Maestra

Guantanamo

Santiago de Cuba

CARIBBEAN SEA

nantly agricultural; over 40 percent[8] of the laboring population[b] are engaged in working the land and more than 90 percent of the exports are derived directly or indirectly from the soil. Sugar is by far the most important crop; other products are tobacco, coffee, corn, winter vegetables, fruits, and henequen.[9] Animal husbandry was second to sugar in value of output. An estimated 68 percent[10] of the total land area was occupied by stock ranches.

Although the island has no coal and very little oil, mineral deposits are found in abundance in the eastern mountains. The deposits of iron and nickel are considered among the most important potential sources in the world.[11] Manganese, chrome, and copper are also found in large quantities.

SOCIO-ECONOMIC STRUCTURE

Economic system

Cuba's economic life centered largely around the big estates held by sugar companies and the large ranches, many dating back to colonial times. Over the years Cuba became the world's largest producer and exporter of sugar. Sugar accounted for four-fifths of the annual export value, nearly one-third of the industrial income, through the manufacture of rum, sugar, and industrial alcohol, and about one-half of the annual agriculture income. It has been estimated that about two-thirds of Cuba's total income is dependent either directly or indirectly upon the sugar industry.[12] Tobacco, the second main export crop, was grown on small farms requiring intensive cultivation, largely by sharecroppers, responsible to large absentee landowners or latifundistas.

Mineral ore extraction industries made up an increasingly important sector of the economy, and minerals represented the third most valuable export. The government owned all mineral rights; it was the sole purchasing and selling agent in the rich Oriente Province. However, some mining enterprises were foreign owned.

About two-thirds of Cuba's foreign trade was with the United States. These economic ties dated back to the early part of the 20th century, when a reciprocity treaty was signed between the two countries. In exchange for exports of sugar and sugar products, tobacco, and minerals, Cuba imported foodstuffs, fuels, capital goods, raw materials for industry, motor vehicles, and consumer goods. In 1955, 73.4 percent of Cuban imports came from the United States.[13]

---

[b] The labor requirements of sugar production, Cuba's largest industry, vary enormously from one season to another.

Class structure

Cuba had one of the most highly-developed middle classes in Latin America. Its 1957 per capital income was $361, the fourth highest of the 20 Latin American republics.[14] From the mid-1940's a progressively larger share of national income had gone to urban labor as a result of collective bargaining. Yet there were still large differences of income and wealth. Agricultural laborers rarely owned land; and except for those who were employed in other agricultural pursuits such as tobacco, cattle, and food production, they worked only during the 3 to 4 months when sugar was planted, worked, and harvested. During the rest of the year—the "dead time"—they worked for very small wages and lived off their meager savings, or on credit. They were tied to company stores by increasingly large debts, sometimes stretching back to their parents and even grandparents.

Within the cities life was dominated by the middle class. There was social mobility from the lower-middle class to the upper-middle class, which had traditionally aligned itself with aristocratic values. At the peak of the social pyramid were the owners of large sugar plantations and cattle ranches and the wealthy professionals. Social mobility depended largely on university education in one of the traditionally-esteemed fields, such as law and medicine, and on the possession of money. Color was not a strong deterrent against moving ahead.

Literacy and education

Education was theoretically compulsory and free; nevertheless, 42 percent of the population in rural areas and 11.6 percent in urban areas were illiterate.[15] Those who could afford to pay for education had a tremendous advantage, because the government did not supply enough teachers, schools, and equipment to carry out the law providing for free education.

There were a number of universities, the largest of which was the University of Havana. Some received governmental financial support, but enjoyed administrative and educational independence. Most students became teachers, doctors, or lawyers; only a few became scientists or engineers.

Major religions and religious institutions

Roman Catholicism was the principal religion. Church and state were separate, and the church had less influence, especially among the middle class, than in other Latin American countries.

GOVERNMENT AND THE RULING ELITE

Description of form of government

The Constitution of October 10, 1940, provided for a parliamentary-cabinet system. Governmental power was divided between three popularly elected branches; the President had a 4-year term, the two-house legislature a staggered 4-year term, and the terms of members of the judiciary varied. In practice, however, the government was almost completely and continuously dominated by a strong executive. When Batista became "Provisional President" in 1952 he suspended Congress and the Constitution. Lawmaking powers were placed in the hands of Batista and his Cabinet. An 80-member Advisory Council was created to make suggestions on legislation. Congress and the 1940 Constitution were reinstated in 1955, but political parties and their leaders had little confidence that they could operate freely.

Description of political process

Political parties and power groups supporting government

When the dictatorship of Gerardo Machado was overthrown in 1933, political power passed into the hands of army enlisted men under the leadership of Sgt. Fulgencio Batista, who promptly set about building the army into a political power instrument. Two-thirds of the 600 career officers were forced to retire; Batista proclaimed himself a colonel and handed out commissions to sergeants, corporals, privates, and even to a number of civilians. He increased the size of the army by one-third, to 16,000 men, raised the pay of officers and enlisted men, established a new military academy, modernized military installations, and increased the Armed Forces' share of the national budget from 15 percent to 22 percent. High-ranking officers were again placed in high civil service positions, where they quickly became involved in the customary graft.

Although at first he had suppressed the organized labor movement, Batista allowed it to be rebuilt under government tutelage after 1934, thus creating his second important body of support. Labor was instrumental in his election to the Presidency in 1940, after he resigned from the army to run for office. During his regime, civilians ostensibly occupied the top government positions, but real power was held by his former army colleagues. At the end of his constitutional term, in 1944, Batista retired from political life, accepted the electoral defeat of his hand-picked successor, and went to Florida.

In 1952 he decided to return to Cuba; political life there was still notoriously corrupt, violence was increasing, and the Armed Forces were becoming restless. When it became apparent that he would not be likely to win the 1952 election legally, he executed a coup with the help of young officers and established himself as a virtual dictator. During the following years, urban labor continued to back Batista, though the middle classes were agitating for the restoration of constitutional government and the peasants were dissatisfied with their lot. Batista kept support of the army by rewarding faithful officers and weeding out those suspected of opposition.[16]

Character of opposition to government and its policies

Aside from the 26th of July Movement, the opposition from 1952 to 1958 took two forms: political parties and professional civic organizations working for a legal change of government; and clandestine groups offering underground resistance to Batista's extreme measures.

Political parties, which Batista had dissolved on taking power, began to reorganize in 1953, when Batista announced that elections would be held in the near future. There was much maneuvering, but no unified opposition solidified. The elections were postponed several times. In 1954, former President Grau San Martín, leader of the *Auténtico* Party, a coalition of progressives and conservatives that had defeated Batista's candidate in 1944, announced his candidacy but withdrew on the eve of the election, charging that it was rigged. The *Auténticos* then turned to former President Prio Socarras as leader, but their strength had receded. In 1958 they supported the 26th of July Movement.

Senator Eduardo Chibas, one of Grau's strongest supporters in 1944, had broken with him in 1947 to organize the Cuban People's Party (*Ortodoxo*). He rallied most of the younger generation and a few older groups. It was on this party's ticket that Fidel Castro ran for Congress in 1952. The party preached a left-of-center ideology, with strong emphasis on nationalism and reform. It broke into three factions after the 1952 coup. By 1958 most of the party's members had come to support Castro's movement.

The Popular Socialist (Communist) Party was established in 1925, and concentrated mainly on infiltrating large organizations, particularly the labor unions, where it exerted strong influence. The party worked in cooperation with the Machado government in 1933 and with Batista in 1938. It reaped its reward in 1940 when Batista named two Communist Party leaders to his cabinet as Ministers without portfolio, in order to tie labor unions to his government.

Batista and the Communists parted ways in 1954 and the party was declared illegal. However, Batista considered the 26th of July Movement his main enemy, and the Communists continued to work quite openly and to exercise influence in the labor movement.

Deriding Castro as a "bourgeois romantic," the party did nothing at first to further his movement. The Communist line changed in January 1958, and by the middle of the summer the party had established liaison with Castro in the Sierra Maestra and was offering aid.

Professional civic organizations began to oppose Batista shortly after the coup.[c] Opposition was weak at the beginning, but gained momentum after the Moncada incident and especially after the November 1954 elections. As Batista resorted to more and more drastic measures in the vain attempt to crush underground resistance and guerrilla activity in 1956 and 1957, the professionals began to demand his resignation. Some Havana judges, for example, sent a letter to the Chamber of Administration of the Court of Appeals of Havana stating that the administration of justice had been "mocked, ridiculed and abused."[17] The document was signed by 14 magistrates. The Roman Catholic hierarchy issued a document which, although carefully worded, virtually asked for Batista's resignation. The Cuban Medical Association registered a protest in 1957, and early in 1958 a statement signed by the Joint Committee of Cuban Institutions, representing 45 organizations, requested the "termination of the regime and the abdication of those in the executive and dissolution of Congress. . . ."[18]

The most important underground resistance group other than the Castro movement was the *Organización Auténtico* (OA) headed by Manuel Antonio de Varona. The group was small, but well-financed by former President Prio. OA had a revolutionary force, the Second National Front of Escambray, in a mountain range east of Cienfuegos. An American, William A. Morgan, was in charge of training fresh recruits. The organization joined the 26th of July Movement in 1958.

A militant rebel organization of university alumni and students, the *Directorio Revolucionario* (DR), also conducted some underground activity. In March 1957 the group attempted to assassinate Batista in the presidential palace. Faure Chomón, secretary-general of the DR, went into exile, but returned in February 1958 with a small band of armed guerrillas. They landed on the coast of Las Villas province and made their way into the same mountains that harbored the Second National Front of Escambray. DR also joined forces with the 26th of July Movement in the summer of 1958.

---

[c] For instance, the magazine *Bohemia*, edited and published by Miguel Angel Quevedo, censured Batista's overthrow of the constitutional government.

Legal procedure for amending constitution or changing government institutions

The 1940 Constitution provided that amendments could be initiated by either the people or Congress. Under the popular procedure the signatures of at least 100,000 literate people were required on any proposal for change. Congress then decided within a 30-day period whether to call a special constitutional convention or submit the proposal to a plebiscite.

One-fourth of the members of either house could also initiate a proposal for amendment. A vote of two-thirds of all the members of Congress in joint session was required for approval, and the amendment had to be ratified in the same manner in the next two regular legislative sessions.[19] Under the provisional constitution (1952–55), an amendment required a two-thirds vote of the President's Cabinet, officially known as the Council of Ministers.[20]

Relationship to foreign powers

President Batista maintained close economic and military ties with the United States. Large American enterprises played an important role in the economic and political life of the island and trade was primarily with the United States. A U.S. military mission and American arms and ammunition strengthened Batista's attempts to maintain internal stability. In March 1958 American arms were cut off, but the military mission remained until the Castro takeover.

Batista's relations with Gen. Rafael L. Trujillo, dictator of the Dominican Republic, had been less than harmonious before January 1957. The dictators charged each other with harboring conspirators, and in several cases the feud led to the recall of diplomats. Early in January 1957 Trujillo indicated to Batista that he was ready to come to an agreement. The two dictators negotiated a mutual friendship pact which assured Batista of the Dominican Republic's noninterference in Cuban affairs. In effect, the pact assured Batista that revolutionary forces in his country would not be supplied or reinforced by his eastern neighbor.[21]

In April 1952, the Soviet Union broke diplomatic relations with Cuba because the diplomatic immunity of two of its couriers had not been honored.

The role of military and police powers

Batista relied heavily on an efficient secret police and well-equipped Armed Forces to maintain his power. He had taken over Machado's secret police system. As opposition mounted, the police uncovered and put down several plots to overthrow Batista, destroyed underground arsenals, and rounded up suspects.

In both instances when Batista seized power, in 1933 and 1952, he owed his success to the Armed Forces. In turn, they owed their wealth, position, and privileges to him. Under the command of a single family, whose head was the man closest to Batista, Gen. Francisco Tabernilla Dolz,[d] the Armed Forces were a strong and obedient force. They were superior in numbers and weapons to the opposition forces; but during the last 2 years of Batista's regime many military conspiracies occurred and Batista was forced to keep purging suspected elements. Gradually the Armed Forces disintegrated and toward the end many soldiers joined Castro.

WEAKNESSES OF THE SOCIO-ECONOMIC-POLITICAL STRUCTURE OF THE PREREVOLUTIONARY REGIME

History of revolutions or governmental instabilities

In the first quarter century of independence from Spain, governmental "stability" in Cuba was maintained entirely by U.S. intervention. Three times—in 1906, 1912, and again in 1917—certain elements attempted to take things in their own hands and settle political disputes by force of arms. Each time the United States intervened to restore order. After the 1906 episode, the United States remained in control of Cuba for 2½ years.

From the late 1920's the men in power resorted to dictatorial tactics to maintain their positions. President Machado snuffed out opposition by suppressing freedom of speech, imprisoning and deporting hundreds of people, closing down the national university, and declaring a state of siege. A revolt broke out in 1931, but was suppressed. In 1933 a crisis was precipitated by a general strike; and an army revolt 21 days later led to a change in government.

Batista dominated Cuban politics from 1933 to 1944, but the period was not free from unrest. There were a number of riots, university demonstrations, and labor strikes. The same unrest prevailed under Presidents Grau San Martín and Prio Socarras. Batista returned to the political scene, and in March 1952 led a revolt that resulted in his becoming the absolute dictator.[22]

Between Castro's unsuccessful attempt on the Moncada Army Barracks in 1953 and the Cuban invasion of 1956, several conspiracies

---

[d] One son, Carlos, was commander of the air force; another son, Francisco, Jr., was commander of a tank group; a brother-in-law, Brig. Gen. Alberto del Río Chaviano, was commander of Moncada and the theater of operations in Oriente Province.

were suppressed. In April 1956, for example, several military leaders and a civilian group attempted to unseat Batista and establish a provisional government. Only 25 days later a group of men emulating Castro attempted to capture the Gorcuria Army Fort at Matanzas. The leaders of these plots were either killed or imprisoned.

Economic weaknesses

Cuba's economy had several major weaknesses. As noted above (II–B–1) it was precariously dependent on the fluctuating world market price of a single crop, sugar. Though Cuba is one of the richest agricultural countries in the world, it had to import 30 percent of its food. About 40 percent of the profits from raw sugar production went to American-owned enterprises.

Most of the land was concentrated in the hands of a few families; the majority of the peasants owned little or none. Agricultural laborers had regular work only during the short sugar season.[23] Government policies also retarded the economy as ". . . business profits were held down by the continuous wage inflation, by injudicious laws which both hampered efficient management and added to operating costs and by a graft-ridden government which preyed on business enterprises."[24]

Social tensions

In the early 19th century land was distributed widely in the form of separate landholdings or in communal ranches. As sugar production increased toward the end of the 19th and the beginning of the 20th centuries, sugar mill owners acquired land on an immense scale to assure themselves of an adequate supply of sugar cane, and the land became concentrated in the hands of a few. Small farmers were uprooted, and sugar production came to dominate the entire economy. From a country of small proprietorships and "family farms," Cuba became a land of latifundia whose rural population was predominantly composed of renters and wage workers.

The sugar mill became the symbol of relative security to agricultural laborers. Many small landowners who had sold their farms to the mills remained in the area and rented company-owned land. Even those who kept their land were dependent upon the sugar companies, which owned the transportation system, to transport sugar cane to the mills. Renters and owners alike became tied to the sugar mill; they depended on it for renting land, milling sugar, and receiving credit.[25]

The need for agrarian reform was pointed out in the late 1940's by a rural sociologist:

Political unrest, arising from the frustration of the desire of peasants to obtain possession of and security on the land, will be chronic in Cuba until more positive action is taken in this respect. Admittedly the problem is a difficult one, with the existing rights of large land-holders to consider; but it is not a question that can be continually postponed. It is likely that continued delay in carrying out the law may result in serious political consequences.[26]

Government recognition of and reaction to weaknesses

After the overthrow of Machado the new government took a number of steps to allay dissatisfaction. A law passed in 1937 and other supplementary regulations set forth rights and obligations of producers, laborers, and sugar mills. Production quotas, rent controls, and minimum wages were established. The common practice of importing laborers from Haiti and Jamaica at harvest time was abolished. Many of the provisions of this act were included in the 1940 Constitution. Manifestations of displeasure continued, however, but they were suppressed.[27]

One of Batista's first acts after the 1952 coup was to suspend the 1940 Constitution and decree a provisional constitution that modified the old one in many respects. He emphasized that these changes were only temporary. Elections were held in 1954, Congress reconvened, and the 1940 Constitution was reinstated. A campaign for a general amnesty of all political prisoners and exiles was begun and in May 1955 Batista found it wise to sign the amnesty bill Congress had passed.

Labor weathered the 1952 coup and preserved its strength as an institution that the government did not dare to eliminate. In the meantime social protection and insurance were provided by labor legislation, and labor unions continued to make progress through collective bargaining. The living levels of the farmers, agricultural laborers, industrial workers, storekeepers, and others were higher, in comparative terms, than those of people in most Latin American countries.

The sugar economy, despite its inherent instabilities, was bolstered and further American investment was encouraged. No attempt was made to diversify the economy. Little was done to better the lot of the peasant laborers through the construction of houses, schools, or hospitals. Many of the provisions of the comparatively advanced Constitution of 1940 were ignored. Any sign of discontent was quickly suppressed by the military or the police. But the severity of Batista's measures eventually lost him even their support.

FORM AND CHARACTERISTICS OF REVOLUTION

ACTORS IN THE REVOLUTION

The revolutionary leadership

The revolution was instigated and conducted by sons and daughters of the upper and middle classes. The leaders were all intellectuals. They had been graduated from universities or at least had attended a university, and had become, or aspired to become, professionals. They included lawyers, professors, architects, engineers, and doctors. Until the revolution few had combat records. Most of the leaders "gravitated" toward socialism; but the degree of socialism differed. Some were convinced Marxists and others were collectivists of various other kinds.

Fidel Castro, often referred to by his associates as the "Maximum Leader," was born in 1926, the son of a relatively wealthy family. His father, Angel Castro, had emigrated to Cuba from northwestern Spain, and prospered in the sugar and lumber industries. Fidel attended parochial schools in Oriente Province and was graduated in the upper third of his class from a high school in Havana. In 1945 he entered the Law School of the University of Havana. His studies were interrupted in 1947 when he joined an expeditionary force that was training to invade the Dominican Republic to overthrow Trujillo. The plot failed, and Castro returned to the university, where he gained prominence in campus politics. He was arrested a number of times for participating in mass meetings and student demonstrations.

"One of the most controversial episodes of Castro's career was his participation in the *Bogotazo*, the riots of April 9, 1948, in Bogotá, Colombia, during the Ninth Conference of American States. . . ."[28] Castro headed a Cuban delegation that was attending an anticolonialism and anti-imperialist student congress in Bogotá, and became involved in a citywide riot triggered by the assassination of a Colombian Liberal. He managed to slip out of the country and return to Havana.

Castro was graduated from Havana University in 1950 and became a partner in a law firm. He found time for politics and in 1952 he was running for Congress on the *Ortodoxo* Party ticket. After Batista's coup d'etat canceled the elections, Castro became a full-time revolutionist with the sole purpose of overthrowing Batista. His dynamic personality and oratorical skill made him a natural revolutionary leader.

Maj. Ernesto "Che" Guevara, a declared Marxist, distinguished himself in the revolution as a guerrilla tactician and as the theoretician of the 26th of July Movement. The son of an architect and builder, he was born in Argentina, studied medicine there, and participated in several conspiracies against Juan Perón. Later he went to Guatemala, where he held an inconspicuous army post in the pro-Communist government that was overthrown in 1954. He met Castro in Mexico in 1955. Trained there as a guerrilla fighter, he came to Cuba with Castro's original landing party in 1956. He is one of Castro's closest confidants.

Raúl Castro, Fidel's younger brother, also attended parochial schools and the University of Havana. He seems to have been more closely affiliated with Communist organizations than was Fidel. In 1952–53 he was a delegate to an International Congress on the Rights of Youth and visited several Soviet-bloc countries. He participated in the Moncada attack and continued to be one of his brother's closest confidants throughout the revolution.

The revolutionary following

The revolution was instigated, controlled, and developed by middle-class leaders; and most of those who were attracted to the revolutionary cause came from the middle class. Many of the recruits who rallied around the original band of young intellectuals in the mountains were peasants, but even the "influx of the last 4 or 5 months failed to give [the revolution] anything like a mass character."[29] The total number of the revolutionary following in the mountains from beginning to end was estimated at 3,000 men.[30] There is no estimate of the number who took part in the underground resistance movement. "The heaviest losses were suffered by the . . . middle class urban resistance movement."[31]

ORGANIZATION OF REVOLUTIONARY EFFORT

Introduction

Fidel Castro's revolutionary movement began to develop long before guerrilla bases were established in the Sierra Maestra in 1956–57. Soon after Batista's coup d'etat in March 1952, Castro determined to dedicate himself to the dictator's overthrow. After one year of recruiting, training, and plotting, Castro's small revolutionary "army" executed the attack on Fort Moncada, Cuba's second largest military fortress. The revolutionaries planned to seize arms, ammunition, tanks, and trucks from the 1,000 soldiers stationed there, then seize the radio stations and call upon the people of Oriente Province to rise against Batista. The plot failed but Fidel Castro's name had become known, and the 26th of July Movement was born.[32]

Castro was sentenced to 15 years imprisonment, but was released in the general amnesty of May 1955, after serving only 19 months. He went to Mexico to start preparing for another attempt. There he met other exiles of the 26th of July Movement. They enlisted the aid of Col. Alberto Bayo, a former officer of the Spanish Air Force, who had fought against Franco during the Spanish Civil War, and who had considerable experience in guerrilla warfare. Bayo was placed in charge of secretly training the growing volunteer corps on a ranch outside Mexico City.[33]

Weapons, ammunition, equipment, and a yacht were acquired on the black market, with money raised partly by underground branches of the 26th of July Movement in the United States, partly contributed by former Cuban President Carlos Prio Socarras, a refugee in Florida. A mutual friend, Teresa Casuso, had brought Castro and Prio together.[34] On November 25, 1956, Castro's expedition set sail for Cuba in the tiny, damaged, under-stocked yacht, *Granma.*

The 82 members of the expedition landed safely a week later somewhat east of the intended beachhead. The *Granma* foundered, and much of the food and supplies were lost. Word of the invasion reached government officials quickly and aircraft and army patrols were sent to hunt down Castro and his small army. Only 12 of the original expedition reached the Sierra Maestra.[35] There they started to build up their guerrilla organization.

Internal organization

The military organization of the guerrilla units was flexible enough to allow for variations according to the surroundings within which they operated. These surroundings were categorized as either "favorable ground" or "unfavorable ground." Hills, mountains, and jungles were thought of as favorable ground for guerrilla bases, if they were established at a "safe" distance from enemy positions. Flat country, unbroken terrain, junctions, and suburban areas were considered unfavorable ground.

The military organization was headed by a commander in chief, who authorized the commanders of the different bases of action to govern their respective territories.[e] Under the base commanders were several columns of varying sizes, each headed by a column commander named by the base commander. The column commander named the

---

[e] Fidel Castro, presumably outlined strategic objectives after consulting with Che Guevara, Raúl Castro, and Camilo Cienfuegos. They decided, in March 1958, to establish a base in northeastern Oriente Province. Raúl Castro led the march on April 1 and became the first commander to establish a new base. Che Guevara was given the mission to establish another base in the province of Las Villas toward the end of August 1958.[36]

captains and lieutenants under him. The first rank above the soldiers was that of lieutenant.

With few exceptions, the combat units on "favorable ground" numbered less than 150 troops. An ideal number, according to Che Guevara, is 100.[37] These constituted the column headed by a commander. The column was divided into platoons of 30 to 40 men, each headed by a captain. The platoons were in turn divided into squads of 8 to 10 men each led by a lieutenant.

Aside from combat duties, some squads and platoons were assigned various auxiliary functions. For example, each base had an elected Commission of Internal Order, charged with maintaining camp discipline, promoting soliders for valor, and educating recruits. Special teams were assigned to transport, watch, and advance guard and rear-guard duties. An extra-combat utility platoon, referred to by the guerrillas as the "suicide platoon," was composed of guerrillas who had volunteered for this duty.

The military organization for fighting on "unfavorable ground" differed primarily in number—units did not exceed 15 men. Rapid mobility and absolute secrecy were essential.[38] Until the middle of 1958 the number of guerrilla units in action remained very small; according to one source it did not exceed 300 even as late as May 1958.[39] This did not reflect lack of proper support; it was the result of careful screening of potential recruits and of logistical problems. During the last few months the numerical size of the organization increased considerably as a deliberate attempt was made to capture large areas and to set the stage for giving Batista the coup de grace.

The need for internal civil organizations arose when the rebel army expanded its bases and incorporated a large number of Cuban civilians. An administrative council was set up in each base to coordinate military and civilian efforts. This council took the executive, legislative, and judiciary functions, and was usually headed by someone with legal training. The council organized the peasants and the workers, collected taxes and donations, and set up a civil health administration and an accounting department in charge of supplies. It institutionalized and regulated life within the rebel bases by issuing penal codes, civil codes, and agrarian reform laws.

Peasants and workers were organized to support guerrilla forces. Most of the cultivated land and farm animals that could produce meat, milk, butter, and cheese, were collectivized. Guerrilla agencies were set up to distribute the food to the military and civilian populations, and to barter with the peasants. Peasants were also used as supply carriers and their buildings as storage depots for arms.

As military needs increased, the civil organization became more complex. Small industries were established to manufacture items for the revolutionary army. For example, boots and shoes were made from the hides of slaughtered animals; and tobacco was rolled into cigarettes and cigars in small shops. Secret "bonds" were printed for the purpose of exchanging goods outside the guerrilla bases.

Peasants, both men and women, were organized into an efficient postal and intelligence system which maintained direct contact with the enemy forces.

The 26th of July Movement's underground units fell directly under the command of the base in the Sierra Maestra, and were interconnected by an underground postal system. New bases were established in several provinces after August 1958; and the underground units operating in a particular province came under the jurisdiction of the base commander in that province. The movement was composed of three major groups: the Civic Resistance Movement, the National Labor Front, and the National Students Front, all organized in cells. For instance, the Civic Resistance Movement had three sections: propaganda, fund-raising, and supply, each designated by a letter of the alphabet and divided into cells of approximately 10 persons. Each person was urged to enlist another 10 to form another cell. The chief of the propaganda section had 400 persons under him, of whom he knew only 12.[40]

Sabotage and terrorist groups also fell under the civil or clandestine branch of the revolutionary effort. An example was the "Triple A," in Havana, which had an organization of 18 saboteurs during 1957. At meetings the organization was divided into groups of three or four persons, and one leader and one bomb was assigned to each group. Each group was assigned a specific target.[41]

A political organization, the Cuban Civilian Revolutionary Front, was established in the summer of 1958, when various parties and groups opposing Batista finally agreed on a compromise united front program on lines laid down by Castro. José Miró Cardona was named secretary general coordinator. The Communist Party was not invited to sign the agreement.

External organization

The external organization of the revolutionary effort was primarily an underground network that reached out to several Latin American countries as well as to several large cities in the United States. According to one source there were 62 centers of underground activity throughout the Americas.[42] Castro himself established centers in

Miami, Tampa, and Key West in 1955. An agent was planted in the Cuban Embassy in Washington to steal secret documents. Another was officially registered with the State Department as a foreign agent; he was a lobbyist and often protested shipments of war material to Batista. The main functions of the external organization were to obtain secret information from Batista sympathizers and solicit funds and supplies from Castro sympathizers.[43]

Several large ranches in Mexico were used as quasi-military training bases for guerrilla fighters. Field exercises led by Colonel Bayo prepared the revolutionists for their initial attack on Cuba. The residences of Cuban nationals in Mexico were often used as arsenals. The Mexican-trained recruits were transported to Cuba by boat or plane.

GOALS OF THE REVOLUTION

In discussing the revolutionary goals, a distinction must be made between those avowed during the revolutionary struggle and those later admitted to by Castro and other Communist leaders.

Concrete political aims of revolutionary leaders

No coherent program for postrevolutionary reconstruction was presented by the revolutionaries. Castro's passionate but murky statements of political philosophy covered all the freedoms and liberties found in the Magna Carta, the American Constitution, and the Cuban Constitution of 1940, but contained few specifics.[44] The leaders' announced goal was to bring Batista's oppressive regime to an end and replace it with a more democratic form of government chosen by the people in free and honest elections. They promised a government that would guarantee, under its constitution, the rights of habeas corpus, trial by jury, and freedom of speech, press, and assembly.[45]

Social and economic goals of leadership and following

The social and economic goals, as disclosed before 1959, were encompassed in the phrase "social justice." Large plantations were to be confiscated and apportioned among the landless peasants. By requiring all companies and corporations to pay their "fair" share of taxes, the leaders expected to finance a comprehensive program of social and economic reforms: education, medical care, road construction, public works, unemployment compensation. The revolutionaries were also interested in ridding Cuba of gambling, racketeering, prostitution, dope peddling, and other vices.[46] After the revolution had succeeded it became apparent that Castro and some of the Communist leaders had other goals, which they had purposely concealed.

REVOLUTIONARY TECHNIQUES AND GOVERNMENT COUNTERMEASURES

Methods for weakening existing authority and countermeasures by government

When Castro and his "army" of 82 men landed in Cuba in 1956 they expected to link up immediately with other guerrilla forces, storm the government garrison at Manzanillo and seize supplies, weapons, and ammunition, and make their way to a prearranged hideout in the Sierra Maestra. There they would outfit the hundreds of volunteers expected to pour in. Meanwhile sympathizers would create confusion in the cities by bombings and shootings. In a short time, Castro hoped, the country would rally to his cause and a general strike would complete the downfall of Batista.[47]

As has been noted, his plans misfired badly. His new strategy, after he had begun to rebuild his shattered forces, was based on a two-front attack on the Batista forces: guerrilla warfare in the mountains; and propaganda, strikes, riots, sabotage, terrorism, and subversion in the cities. The basic area for armed fighting was to be the countryside.[48] However, Castro hoped that his guerrilla movement in the Sierra Maestra would inspire revolutionary activity in the cities also.[49]

For the first few weeks the 12 men who had survived the landing were continuously on the move, dodging Batista's soldiers. Then recruits began to come in, sympathizers provided arms and supplies, and the guerrilla force grew rapidly.

A permanent base was established in a remote area of the Sierra Maestra. It shortly acquired the characteristics of a village. It had an administration which regulated the life of its inhabitants. It set up small industries, schools, a hospital, a radio station, and a newspaper. It recruited and trained its small army. Detachments were sent out to establish new bases.

The guerrillas had two main objectives: first, to expand their forces into a regular army; second, to inflict increasingly heavy losses on the enemy.

Hit-and-run attacks on enemy columns were usually conducted at night, in areas at least one day's march from the guerrilla camp. Guerrilla units of four or five men would take up stations surrounding the enemy column. One unit would open the attack; when the enemy countered, that unit would retreat to strike from another point, while other units did the same. In this way the guerrilla bands were able to inflict heavy losses without engaging the enemy in a decisive battle.

A special point was made of annihilating the advance guard of an enemy column. The growing awareness of the Batista forces that the advance guard's position was dangerous helped to demoralize Batista's army. "Molotov cocktails" were ideal weapons for encirclements. Their mission completed, the guerrillas swiftly withdrew with the captured equipment. Their intimate knowledge of the countryside gave them speed and maneuverability.

As the revolutionary movement grew and received increasing support from the people, the guerrilla bands became more adventurous. They descended from the mountains, usually in groups of 10 to 20, to raid Cuban villages where enemy detachments were stationed. In most cases food, arms, and ammunition were captured and the guerrillas usually escaped before enemy reinforcements arrived. To cripple Batista's economy, the revolutionaries set fire to sugar plantations, doing enormous damage.

Suburban guerrilla warfare and sabotage in the cities were second front diversionary tactics in the final stage of the revolution. The revolutionary force had by then developed into a compact regular army and engaged the enemy on "defined fronts" in regular warfare. The number of conquered zones increased, and city after city fell into rebel hands. Batista's army had grown weak and demoralized from many defections and offered little resistance.

Demonstrations, strikes, riots, and revolts occurred sporadically throughout the revolutionary period. Not all of these were planned by the revolutionary high command and executed by the revolutionary following. A parade of women demonstrators, organized by underground leaders, set off a general strike in Santiago de Cuba July 31, 1957; it was put down by the army and police. A nationwide strike scheduled by the revolutionary leaders for April 1958, which was to paralyze the whole country and lead to the final overthrow of the Cuban government, was not a complete success. The strike was total in some cities, but it failed to reach national proportions.

Sabotage was employed by the revolutionists to destroy communications: telephone and telegraph lines, bridges, railroads, and public utilities. In a number of cases, cities were completely paralyzed.

The Cuban Government attempted to deal with the revolutionary movement by conducting a military offensive against the guerrilla forces in the mountains, and by suppression, reprisals, and counter-propaganda in the cities.

Batista announced in May 1958 that 12,000 troops were to be committed "to a spare nothing effort to put an end to Fidel Castro" and his revolutionary movement in the Sierra Maestra. "It was the biggest troop movement in Cuba's history, including the war for independence from Spain."[50] The army intended to force the guerrillas into decisive battle. The government troops were grouped into 13 combat teams of about 900 men each, and had tanks, armored cars, trucks, and half-trucks. Air force bombers bombed and strafed the combat areas as a prelude to the ground attack.[51]

The initial attack was unsuccessful. To avoid a head-on encounter, the guerrillas retreated into their mountain redoubt. Batista's forces were not prepared to engage the enemy in mountain and jungle warfare; they remained in the foothills, where they were continuously attacked by guerrilla snipers and lost much of their small arms and equipment.

The revolutionaries' diversionary tactics forced Batista to leave hundreds of soldiers in the cities to help the police maintain order. Terrorists, saboteurs, and young people only suspected of supporting the revolutionary movement were arrested.[52] Many members of the underground were imprisoned.

Batista made effective use of restrictive measures to suppress and localize demonstrations. Communications facilities were closed down, transport systems were halted, curfews were imposed. In a few cases peasant families suspected of aiding the guerrillas were taken from their land and resettled in nearby cities or towns. News of disturbances often did not reach cities in other provinces until after order had been restored. For instance, the demonstrations in Santiago de Cuba and the revolt in Cienfuegos were not made public until everything had quieted down.

Unable to locate and destroy the guerrillas' radio transmitter, the government began newscasts over the same frequency and had some success in countering revolutionary propaganda. False propaganda leaflets also spread confusion among Castro's supporters.

Methods for gaining support and countermeasures taken by government

Revolutionary propaganda originated from two sources: the civic organizations in the cities, and the guerrilla bases in the mountains. Leaflets, pamphlets, and newspapers, as well as word-of-mouth communication, were utilized in distributing information in the cities. The guerrilla army employed the same methods, and in addition set up its own radio transmitter. The general content of the propaganda material included news of guerrilla activities and instructions to non-combatants on how to conduct resistance.

A fundamental part of the guerrilla tactics was to treat the inhabitants of guerrilla-occupied zones decently and reward them for aiding the revolutionists. One of Castro's first acts on reaching the Sierra Maestra was to arrest a landlord who had increased his holdings at the expense of the peasants. "So we tried him and executed him," said Castro, "and won the affection of the peasants."[53] Castro thus became the Robin Hood of the Sierra Maestra.

Posing as social reformers, the guerrillas were able to convince the peasants that this was a peasant revolution and that the peasants would reap the benefits. Time and again Castro denied that he was a Communist. He always talked in terms of "democracy" and "freedom." His personal qualities alone won countless followers.

Castro had little trouble finding recruits. His movement generated enthusiasm especially among young idealists. Following the December 1956 invasion, Batista announced that Castro was dead; and when word spread that the news was false, Castro "was obtaining recruits faster than he could supply them with weapons."[54]

The initial reaction to the revolution among people generally, however, was one of apathy. The news of Castro's landing stirred little interest in Havana at first. "It was considered merely another hare-brained scheme, like the suicidal attack on Moncada post in 1953."[55] However, Castro sent emissaries to meet with secret groups in the cities and urban support began to develop. Segments of the Roman Catholic hierarchy lent their moral support. The Popular Socialist (Communist) Party, which was well organized and had considerable influence in labor unions, withheld its support until the last months of the revolution, however.

The agreement signed by the various opposition groups with Castro on July 20, 1958, resulted in the formation of the Civilian Revolutionary Front, and was issued as a Manifesto to the Cuban people. It received wide circulation by the underground organization in Cuba and the United States. It was broadcast from Caracas and other radio stations inside and outside of Cuba. The agreement called for cooperation in the common cause, continued cooperation after victory, the arming of the people, and the cooperation of labor and business in a general strike to be called to aid the military front when needed. The manifesto called on the soldiers to desert, and on everybody else to support the revolution. Above all, it stressed the theme of unity.[56]

One of Castro's greatest assets was the support of foreign correspondents. An interview with Herbert Matthews of *The New York Times*, published in February 1957 with photographs of Castro in the Sierra Maestra, had tremendous impact, "not only in Cuba, but on the entire hemisphere."[57] Other U.S. newspapers and news agencies followed Matthews' example and gave Castro wide and favorable publicity in the United States. This publicity, more anti-Batista than pro-Castro, may have contributed heavily to Castro's victory. As a result of it, pressure was brought upon the State Department by some members of Congress and by many citizens to withdraw military support from the Cuban Government. The United States stopped selling arms to Batista in March 1958 and persuaded other countries to do likewise. Revolutionary sympathizers continued to smuggle arms, men, and ammunition to the Castro forces.[58]

One source, quoting the London *Intelligence Digest*, maintains that Castro's revolution was also "lavishly financed from behind the iron curtain."[59] Raúl Castro, according to this source, "made several trips behind the iron curtain on fund-raising missions and to arrange supplies of arms."[60] This support is alleged to have come "from Russian and Chinese subsidies, channeled through secret groups in several Latin American countries . . ."[f61] The exact nature and extent of such support cannot be deduced from accounts published so far. Reports from eyewitnesses have spoken of unidentified submarines, which were claimed to have been Russian, landing supplies.

The Batista government conducted an extensive propaganda campaign to discredit Castro and his movement. Radios and newspapers carried false reports that Castro had died during invasion. Later the government announced that the guerrilla bands were being successfully rounded up and that the revolution was over. Batista denounced Castro as a Communist and a tool of Russian conspiracy. He offered rewards ranging from $5,000 to $100,000 for information leading to the capture of the revolutionary leaders.[62]

Batista made several attempts to arrest the rapid decline of his popularity in Cuba and the United States. In the United States his public relations adviser emphasized the support of the State Department and Pentagon officials. Photographs of Cuban officials with U.S. Government and military officials were publicized.

In the spring of 1958 Batista proclaimed that the regular presidential elections would be held in June of that year. The Cuban Government promised to guarantee free and open elections, but the opposition was understandably skeptical. The elections were post-

---

[f] This source also states that Castro received "powerful support" from Rómulo Betancourt, President of Venezuela. It goes on to say that Castro received $50 million from the Democratic Action regime of Venezuela.

poned until November, and Batista, ignoring his pledge, engineered the seating of his own candidate. This further alienated support both at home and in the United States.[g]

MANNER IN WHICH CONTROL OF GOVERNMENT WAS TRANSFERRED TO REVOLUTIONARIES

Batista fled the country on January 1, 1959. He left Cuba in the hands of a junta headed by the senior justice of the Supreme Court, who, because the Vice President and the leader of the Senate had resigned also, was the constitutional successor. The junta was fated to govern only a short time. In Santiago de Cuba, Fidel Castro named Manuel Urrutia, a former judge, President of the Republic in conformity with a previous agreement among opposition party leaders. Urrutia designated two of his selected ministers to negotiate with the junta. The junta agreed to evacuate the presidential palace upon the arrival of Urrutia, and control of the government was transferred to the revolutionaries.

THE EFFECTS OF THE REVOLUTION

CHANGES IN PERSONNEL AND INSTITUTIONS OF GOVERNMENT

It is doubtful whether anyone, including Fidel Castro, knew exactly what course to take after Batista fled the country. President Urrutia and the Cabinet members, handpicked by Castro, were able and dedicated Cubans of mixed political views who had been active during the revolution. Urrutia appointed Castro as head of the Armed Forces. However, several members of the new government were removed after a short time. In February 1959 Castro himself took the office of Prime Minister, and in July he forced Urrutia to resign, after the President had raised the issue of communism and placed him under

---

[g] Nathaniel Weyl writes that the election was undermined by Castro and postponed because Castro intimidated the liberal candidates who were trying to fight Batista by legal means and that he "was determined to prevent the elections at all costs." Castro's intentions were made public in a manifesto promulgated on March 17, 1958, and by Law #2, which was promulgated approximately a week later.[63] However, Jules Dubois indicates that American Ambassador Earl T. Smith already knew that the elections "would be postponed to the month of October or November" on March 15, 1958, 2 days before Castro's manifesto was made public.[64] It would seem more probable, therefore, that the election was postponed to give the candidates and officials more time to make preparations.

house arrest. Most of the original Cabinet members were forced out; many went into exile.

These shifts were a result of Castro's radical policies. Moderate politicians were eliminated from important governmental posts, and power was increasingly concentrated in the hands of three radical extremists, Fidel Castro, Che Guevara, and Raúl Castro. Fidel Castro emerged as the undisputed leader of the revolution and has continued to maintain a dominant position.

Che Guevara was away on a trip to the Middle and Far East during the summer of 1959, and many observers thought he was on the way out. However, after he returned in September he was named to an important position in the industrialization department of the National Institute of Agrarian Reform described below. Later he became President of the National Bank and tsar of the Cuban economy, and was firmly entrenched as second man in the hierarchy.

The third member of the ruling elite is Raúl Castro. He has been credited with having done an excellent job in organizing the Armed Forces of which he was given charge after October 1959. Politically, he is the most radical of the three and is considered most likely to succeed his brother as head of the government. He has a strong personal following in the Cuban Army, which may become a decisive factor in any future power struggle.

MAJOR POLICY CHANGES

Castro established headquarters in the Havana-Hilton Hotel, and from his suite poured out hundreds of decrees designed to place the new regime in direct control of all aspects of Cuban society. Government measures became increasingly excessive and irresponsible— excessive in expropriation of wealth, seizure of private commercial enterprises, arbitrary arrest, trial by revolutionary tribunals, and elimination of moderate supporters as well as opponents; irresponsible in the many costly projects undertaken with little consideration for sound administrative practices, financing, and technical advice.

A law for agrarian reform was published in May 1959 and the National Institute for Agrarian Reform (INRA) was set up to administer it. The general idea was to limit the size of landholdings by expropriating land from large landholders, and to distribute a small acreage to each landless peasant. The law also provided for the establishment of cooperatives, whereby the peasants could share collectively-owned farm equipment.

However, INRA set about nationalizing land, stock, and machinery on a scale far beyond that authorized by the law. On its own initiative INRA expropriated cattle and machinery as well as land. Although the law provided that landowners should be compensated with government bonds, INRA gave no bonds and no receipts in exchange for the land, and no hearings to those who tried to appeal. It has been estimated that by mid-1960 INRA had possession of over one-half of Cuba's land area.

The new government also undertook a crash program to raise living standards by providing low-income housing, schools, health services, and public recreation facilities. New construction projects erupted all over Cuba. To finance these costly projects the government reorganized the old National Lottery into the new National Institute for Savings and Housing (INAV). The organization sold numbered bonds for prizes and financed INAV projects with the profits. However, profits were only marginal and many projects were abandoned for lack of funds.

The government took steps immediately to eliminate political and military leaders of the old regime. Approximately 600 persons were executed by revolutionary tribunals. Thousands of suspected counter-revolutionaries were imprisoned, scores lost their property, and many were driven abroad. There was an unprecedented exodus, not only of rich and middle-class Cubans, but also of peasants. Three of the most important newspapers, *Prensa Libre, El Diario de la Marina*, and *El Villareno*, were suppressed when they showed signs of opposition. Radio and television communications came under full control of the government.

International policy was altered just as radically. There had always been a close, if not always friendly, relationship between Cuba and the United States. This relationship deteriorated rapidly. Castro made an unofficial visit to the United States soon after his victory, apparently to convince the North Americans that he was not a Communist and desired continued business ties with this country. Public opinion, as reflected by the press and the popular welcome, was still disposed to be friendly, although it had been shocked by the courts-martial and executions. American business interests received Castro's announcements with skepticism, however. He returned home and took steps to strengthen Cuba's economic ties with the Sino-Soviet bloc. Engineers, agronomists, plant managers and technicians, and military supplies were imported from the Soviet Union, Czechoslovakia, and Red China.

LONG RANGE SOCIAL AND ECONOMIC EFFECTS

Socialization on Communist Party lines is far advanced in Cuba. The government has taken over not only most of the land, but almost all of the large enterprises. In the process Castro and his associates have established a totalitarian system. The army has been disbanded and replaced by a highly indoctrinated armed militia and youth groups. Opposition parties no longer exist. Labor unions have been deprived of their freedom, to remove the danger of strikes and to make wage freezes possible. The role of the Catholic Church is uncertain and it is questionable whether the masses would choose the church in preference to Castro if a choice were offered. On December 2, 1961, Castro told the world that he was "a Marxist-Leninist and . . . [would] be a Marxist-Leninist until the last days of [his] life." "On July 26, 1961 the fusion between the Communist Party and the 26th of July Movement was completed with the announced formation of the *Partido de la Revolución Socialista.*"[65]

Other radical changes have been made in the social structure, the judicial system, and the system of education. Now that the landowners have been dispossessed, and the middle class, which helped Castro to power, has been persecuted and alienated, the peasants and workers hold the predominant position in the class scale. The Supreme Court was purged in December 1960 and its membership, faithful to Castro, reduced to less than half. Lesser judges were also purged. Education has been remodeled. Textbooks have been rewritten to teach Castro's political, social, and economic doctrines.

Some constructive social and economic reforms have been accomplished. More schools have been established, and a decided attempt has been made to wipe out illiteracy. New and improved housing for the poorer elements has been built. Public health measures and new hospitals have wiped out sicknesses prevalent in the countryside. Vacation and recreational facilities have been provided for the masses. It is likely that these improvements are not all in the nature of a carefully displayed showcase but it is as yet difficult to assess the true nature of progress.

Most of the underprivileged still look to Fidel Castro as their "Maximum Leader" and as the "instrument of their deliverance."[66] Many of them, for the first time in their lives, feel that they have political power. Many have benefited greatly, and many more are hopeful because they see new homes, new schools, and the possibility of better opportunities in the future.

OTHER EFFECTS

Part of the Cuban middle class that supported Castro's revolution is now plotting to overthrow the regime it helped to establish. Many of its members have fled to the United States, and joined with some pro-Batista elements to establish anti-Castro revolutionary organizations. The most important are: the Revolutionary Movement of the People, led by Manuel Ray, which also has an underground organization in Cuba; and the Democratic Revolutionary Front, led by Dr. Manuel Antonio de Varona y Loredo.

The two revolutionary organizations met in Miami on March 21, 1961, and set up a seven-man revolutionary council with Castro's former Prime Minister, Dr. José Miró Cardona, as President. All the members of the council are exiles. The council became known as the Revolutionary Council of Cuba.

The primary purpose of the Council was to provide a political base for a military organization of exiles formed a year earlier. The objective was to gain a beachhead on Cuban soil, establish a government-in-arms, and request diplomatic recognition.

On April 17, 1961, the attempt was made and failed disastrously. Twelve hundred exiles, armed with small arms and bazookas, landed at Cochinos Bay, south of Matanzas Province, hoping to ignite rebellion and mutiny among Castro's forces. The invasion was easily crushed by a well-armed and skillfully led Cuban militia. Many prisoners were taken, though some of the rebel forces escaped to the mountains of Escambray to reinforce an anti-Castro guerrilla base there.

There are at least six main reasons for the failure. (1) Cochinos Bay was a bad location for the landing, since it is swampy and swift deployment was impossible. Castro's main force in Havana, only a few hours away by good roads, was able to move in on the invaders on short notice. (2) Those who planned the invasion did not rely on well-known guerrilla leaders but on the sons of the middle- and upper-class rich, including some Batista men. (3) The planners did not forewarn the anti-Castro underground in Cuba. The plan was probably withheld from the underground for security reasons, but at least one council member said afterward that lack of information was the chief cause of the failure. (4) The invaders lost communication with their own high command because the ship that carried the transmitting equipment was sunk.[67] (5) Although many Cubans were disturbed by Castro's methods and his "betrayal" of the revolution, it does not follow that they were "ripe for revolution."[68] (6) Finally, Castro seems to have been well informed of all stages of the invasion by an excellent intelligence system.[69]

NOTES

1. Hubert Herring, *A History of Latin America* (New York: Alfred A. Knopf, 1960), p. 404.

2. *Rand McNally Indexed Classroom Atlas* (Chicago: Rand McNally and Company, 1961), ii.

3. Theodore Draper, "Castro's Cuba: A Revolution Betrayed?", *The New Leader* XLIV, 2 (March 27, 1961), 12.

4. *The Worldmark Encyclopedia of the Nations* (New York: Harper and Brothers, 1960), 230.

5. Ibid.

6. *Britannica Book of the Year: 1958* (Chicago: Encyclopaedia Britannica, Inc., 1960), 230.

7. *Worldmark* , p. 231.

8. *Britannica*, p. 839.

9. *Worldmark*, p. 233.

10. Ibid., p. 234.

11. Leo Huberman and Paul M. Sweezy, *Cuba: Anatomy of a Revolution* (New York: Monthly Review Press, 1960), p. 2.

12. *Worldmark*, p. 232.

13. Robert F. Smith, *The United States and Cuba, Business and Diplomacy 1917–1960* (New York: Bookman Associates, 1960), p. 166.

14. Nathaniel Weyl, *Red Star Over Cuba: The Russian Assault on the Western Hemisphere* (New York: The Devin-Adair Company, 1960), pp. 185–186.

15. Draper, "Castro's Cuba," p. 12.

16. Edward Lieuwen, *Arms and Politics in Latin America* (New York: Frederick A. Praeger, Inc., 1960), pp. 97–100.

17. Jules Dubois, *Fidel Castro: Rebel-Liberator or Dictator* (New York: The New Bobbs-Merrill Company, Inc., 1959), pp. 221–222.

18. Ibid., p. 227.

19. Amos J. Peaslee, *Constitutions of Nations* I (The Hague: Martinus Nijhoff, 1956), 668.

20. Austin F. MacDonald, *Latin American Politics and Governments* (2d ed.; New York: Thomas Y. Crowell Company, 1954), p. 570.

21. Dubois, *Fidel Castro*, pp. 148–149.

22. MacDonald, *Politics*, pp. 557–568.

23. Huberman, *Cuba*, pp. 6–10, 22.

24. Weyl, *Red Star*, p. 189.

25. Lowry Nelson, *Rural Cuba* (Minneapolis: The University of Minnesota Press, 1950), pp. 92–97.

26. Ibid., p. 255.

27. Ibid., pp. 100–101.

28. Dubois, *Fidel Castro*, p. 17.

29. Draper, "Castro's Cuba," p. 8.

30. Ibid.

31. Ibid.

32. Dubois, *Fidel Castro*, pp. 26–83.

33. Ibid., pp. 111–113.

34. Teresa Casuso, *Cuba and Castro* (New York: Random House, 1961), pp. 111–112.

35. Dubois, *Fidel Castro*, pp. 139–140.

36. Ibid., pp. 245, 302–303.

37. Che Guevara, *Guerrilla Warfare* (New York: Monthly Review Press, 1961).

38. Ibid., p. 34.

39. Huberman, *Cuba*, p. 63.

40. Dubois, *Fidel Castro*, pp. 249–250.

41. Ray Brennan, *Castro, Cuba and Justice* (Garden City: Doubleday and Company, 1959), p. 135.

42. Dubois, *Fidel Castro*, p. 212.

43. Ibid., pp. 99, 271; and Brennan, *Castro*, pp. 194–195.

44. Brennan, *Castro*, pp. 82–83.

45. Dubois, *Fidel Castro*, pp. 166–172; 262–263.

46. Brennan, *Castro*, pp. 82–83.

47. Huberman, *Cuba*, pp. 52–53.

48. Guevara, *Warfare*, p. 15.

49. Theodore Draper, "The Runaway Revolution," *The Reporter*, XXII (May 12, 1960), 15.

50. Brennan, *Castro*, p. 227.

51. Ibid., pp. 227–229.

52. Ruby Hart Phillips, *Cuba: Island of Paradox* (New York: McDowell, Oblensky, 1957), p. 291.

53. Dubois, *Fidel Castro*, p. 145.

54. Phillips, *Cuba*, p. 297.

55. Ibid., p. 291.

56. Dubois, *Fidel Castro*, pp. 280–283.

57. Phillips, *Cuba*, p. 240.

58. U.S. Senate, *Hearings before the Subcommittee to Investigate the Administration of the Internal Security Act and Other Internal Security Laws of the Committee on the Judiciary*, Part 9, 86th Cong. 2nd Sess. (Washington, D.C., United States Government Printing Office, 1960), pp. 688–689.

59. Weyl, *Red Star*, p. 142.

60. Ibid., p. 149.

61. Ibid., pp. 141–142.

62. Dubois, *Fidel Castro*, p. 149.

63. Weyl, *Red Star*, pp. 181–182.

64. Dubois, *Fidel Castro*, pp. 223–224.

65. Robert F. Smith, "The United States and Latin American Revolutions," *Journal of Inter-American Studies*, IV (January 1962), 101.

66. Irving Peter Pflaum, *Tragic Island: How Communism Came to Cuba* (Englewood Cliffs, New Jersey: Prentice-Hall, Inc., 1961), p. 19.

67. Ibid., pp. 103, 139, 187, 184.

68. Reinhold Niebuhr, "Mistaken Venture," *The New Leader*, XLIV, 18 (May 1, 1961), 3.

69. Tristram Coffin, "Probing the CIA," *The New Leader*, XLIV, 20 (May 15, 1961), 4.

RECOMMENDED READING

BOOKS:

Brennan, Ray. *Castro, Cuba and Justice*. Garden City: Doubleday and Company, 1959. An account of the Cuban revolution. Although the book is pro-Castro, most of its descriptions check with other sources.

Dubois, Jules. *Fidel Castro: Rebel-Liberator or Dictator*. New York: The New Bobbs-Merrill Company, Inc., 1959. A well-documented and detailed study of Castro's revolution and one of the more reliable sources.

Guevara, Che. *Guerrilla Warfare*. New York: Monthly Review Press, 1961. Written by one of the Cuban revolutionary leaders, the book is a theoretical and practical treatise on the conduct of revolutionary warfare, based on the Cuban experience.

Huberman, Leo, and Paul M. Sweezy. *Cuba: Anatomy of a Revolution*. New York: Monthly Review Press, 1960. The authors, editors of the magazine, *Monthly Review*, see the Cuban experience as a peasant revolution whose success was based on the support of the peasants as a class.

Matthews, Herbert L. *The Cuban Story*. New York: George Braziller, 1961. The author, the first correspondent to interview Fidel Castro in the Sierra Maestra, maintains that the Cuban experience was not a Communist revolution although it became a totalitarian dictatorship. He concludes that the revolution's importance is unequalled by any event in Latin America since the wars for independence.

Pflaum, Irving Peter. *Tragic Island: How Communism Came to Cuba.* Englewood Cliffs, New Jersey: Prentice-Hall, Inc., 1961. An eyewitness report and appraisal of postrevolutionary Cuba.

Phillips, Ruby Hart. *Cuba: Island of Paradox.* New York: McDowell, Obolensky, 1959. The book, written by a correspondent of *The New York Times,* is a good but romanticized analysis of Fidel Castro and his movement.

Weyl, Nathaniel. *Red Star Over Cuba: The Russian Assault on the Western Hemisphere.* New York: The Devin-Adair Company, 1960. An extreme version of the Cuban revolution, described as part of the international Communist movement. The author contends that Fidel Castro has been "a trusted Soviet agent," since 1948.

PERIODICALS:

Draper, Theodore. "The Runaway Revolution," *The Reporter,* XXII (May 12, 1960), 14–20. An analysis of Castro's relationship with Communists and communism during and shortly after the revolution. A good summary of the revolution.

Draper, Theodore. "Castro's Cuba: A Revolution Betrayed?" *The New Leader,* XLIV 2 (March 27, 1961). This article is primarily a review of what has been written about the Cuban revolution. The author argues that the revolution was a middle class one, first in the name of the entire Cuban population, then of the peasants, and finally of the peasants and workers; and that the revolution promised by Castro before taking power was betrayed.

Shapiro, Samuel. "Cuba: A Dissenting Report," *The New Republic,* CXLIII (September 12, 1960), 8–26. The author concludes that the revolution and the land reform which followed were essentially Marxist. Postrevolutionary events are interpreted rather sympathetically.

OTHERS:

Special Warfare Area Handbook for Cuba. Prepared by the Foreign Areas Studies Division, Special Operations Research Office, The American University, Washington, D.C., for the Department of the Army. June 1961. The most complete and general source available for the study of the Cuban revolution.

SECTION III

··················

NORTH AFRICA

General Discussion of Area and Revolutionary Developments
The Tunisian Revolution: 1950–1954
The Algerian Revolution: 1954–1962

TABLE OF CONTENTS

GENERAL DISCUSSION OF AREA AND REVOLUTIONARY DEVELOPMENTS

THE TUNISIAN REVOLUTION: 1950–1954

THE ALGERIAN REVOLUTION: 1954–1962

TABLE OF CONTENTS

NORTH AFRICA

GENERAL DISCUSSION OF AREA AND REVOLUTIONARY DEVELOPMENTS

DESCRIPTION OF AREA

The countries comprising North Africa, as defined in this study, are Morocco, Algeria, Tunisia, and Libya. The entire region, which the Arabs refer to as the *Maghreb* (West), shares a common cultural heritage, although Tunisia and Morocco have distinct historical traditions of their own. Islam is the dominant religion, and Arabic is the language of three-quarters of the population; the others speak Berber dialects or European languages. French is the dominant European language, although Spanish is widely spoken in the western part of North Africa and Italian in the eastern sector.

The economic, cultural, and political life of North Africa is concentrated in the densely populated coastal strip which extends roughly 2,600 miles from southwestern Morocco to northern Tunisia and reaches inland for 50 to 100 miles. Here the terrain is mountainous and the climate comparable to that of Italy, Spain, and the south of France. This part of North Africa produces grain, citrus fruits, cork, olives, wines, and other agricultural products native to the Mediterranean basin. Beyond the Atlas Mountains lies the Sahara Desert, where extensive deposits of iron ore and petroleum have recently been discovered but are not yet being fully exploited.

The population of North Africa, which is one of the world's most rapidly growing, was estimated in 1958 to be around 26 million. Of these, 10 million were in Morocco, 10 million in Algeria, 4 million in Tunisia, 1 million in Libya, and 1 million scattered throughout the Sahara Desert regions. Around 1½ million North Africans are of European descent, and two-thirds of these live in Algeria. However, the ancestors of many of these "European" Algerians settled in North Africa over a hundred years ago. They are a vital factor in the economy of the area, owning the most productive farms and most of the leading business enterprises. There are about half a million Jews in North Africa, many of whom identify more with the Arab population than with the European.[1]

The economy of North Africa is based primarily on agriculture. Apart from mining and building industries, there has been only limited industrialization. Most of the best farming lands came under the control of European settlers during the colonial period. Consequently, there is an Arab-Berber peasantry and a European landowning elite in many rural districts. Rapid urbanization and the introduction of modern hygiene have caused the Arab population in the cities to expand faster than the limited industrial development warranted. The result has been a decline in the living standard of most urban Arabs in the midst of general prosperity among the European residents. Around every large city have sprung up shanty-towns, called *Bidonvilles* in French, where the Arab urban proletariat live. These Arabs provide an abundant source of inefficient, by Western standards, but cheap labor for European business enterprises. They also provide the main body of popular support for the nationalist and revolutionary movements led by the Western-educated Arab middle class and traditional elite. The antagonisms arising out of North Africa's socio-economic structure have been the most significant factor in the development of Arab nationalism in the region during the postwar period.

HISTORICAL BACKGROUND

North Africa has experienced a succession of foreign invasions and colonizations: Phoenician, Roman, Arab, Turkish, and European. The modern period of North African history opened with the French occupation of Algeria in 1830. Turkish control over the area had long been nominal; in actual fact, the coasts of Algeria were controlled by autonomous Turkish administrators and an oligarchy of traders and pirates, and the interior by tribal chieftains. Morocco, the most isolated part of North Africa, was governed by a native sultan and a similar commercial and tribal oligarchy. Tunisia and Libya were nominally parts of the Turkish Empire but were in practice governed by local oligarchies. Tunisia was added to France's North African empire as a "protectorate" in 1881, and in 1912 France and Spain divided Morocco into two "protectorates." In 1911, Italy "liberated" Libya from Turkey, and that North African country became an Italian colony.

European rule in North Africa varied according to whether the European power was France, Spain, or Italy; according to the historical traditions and institutions of the area; and according to the way European rule came about. Algeria, for instance, was conquered militarily and was administered directly by France as a colonial possession. There were extensive military operations, confiscation of land, European immigration, and destruction of religious and political institutions. In Tunisia and Morocco, on the other hand, the impact of European control was less severe. As protectorates, these countries preserved their traditional monarchs and the form, if not the substance, of indigenous administration. Here European settlement was much less extensive than in Algeria and was largely confined to the cities, in the case of Morocco, and to a few fertile agricultural districts in Tunisia.

Tunisia, with a historical tradition of its own which looked back to ancient Carthage and several illustrious periods of Arab history, had the advantage of a well-established tradition of receptiveness toward European culture. Morocco had a similarly strong historical tradition, which Algeria lacked, but Morocco was not as homogeneous ethnically and culturally as either Tunisia or Algeria and was the most isolated and culturally primitive of the French possessions in North Africa. While French rule came to Tunisia by relatively peaceful means, in Morocco the French had to contend with a succession of Arab and Berber trial rebellions, the most important of which was the Rif War of 1921–27, originating in the northern Spanish sector. Morocco was not fully "pacified" until 1934.

France's colonial policy was, in theory, to "civilize" the Arab-Berber native population and assimilate them into French society and culture; however, in practice, all non-Europeans, whether educated in French ways or not, were subjected to racial discrimination. This contradiction between theory and practice was a major irritant in Franco-Arab relations. Spanish colonial policy in the small portion of North Africa that Spain controlled (the northern sector of Morocco) was one of coexistence with the Arabs rather than assimilation of them into European civilization, as both French and Italian colonial policy envisaged. To most North African Arabs, Spain's colonial policies were the least objectionable form of European control. Italy, especially after the rise of Fascism in the 1920's, pursued repressive colonial policies against the Arab population of Libya. The Mussolini regime's theories of racial and cultural superiority and its promotion of Italian migration to North Africa made Italian rule particularly objectionable to the Arabs.

The positive aspect of European colonialism consisted primarily of its enormous material achievements, although it may also be said that Europe's most significant long range effect in the region has been the creation of a Western-educated Arab elite. The economy of the coastal areas, especially around the larger cities of Casablanca, Rabat, Tangier, Oran, Algiers, Tunis, and Tripoli, is comparable to that of Western Europe. The negative side of European colonialism in North Africa, as elsewhere in the colonial world, was its neglect of the human factor. Economic development and prosperity were largely limited to the politically dominant European minority, while the majority of the

Arab population existed at a standard of living greatly inferior to that enjoyed by Europeans in the area.

An Arab elite educated in modern European ways but prevented from enjoying the fruits of European civilization became a revolutionary element with which European colonial authorities found it difficult to deal. From this educated elite came the leadership of the Arab nationalist movements which, in Morocco, Tunisia, and Algeria, have challenged European authority since World War II. Libya, which became an independent kingdom in 1951, gained its national independence more as a result of the defeat of Fascist Italy in World War II than through the efforts of Libyan nationalists.

REVOLUTIONS IN THE AREA

Since World War II there have been three nationalist revolutions in North Africa: the Tunisian and Moroccan revolutions, which ended in 1956, and the Algerian revolution, which began in 1954 and, in 1961, was in its seventh year. These Arab movements, all directed against French rule, have taken markedly different courses, although all three shared a similar revolutionary goal—national independence and socio-economic reform. Such factors as the historical background, the institutional relationship with the French, the cultural development, and the personalities of the nationalist leadership were among the determinants in the course each of these revolutions has taken. All three revolutions started out with relatively moderate demands for local political autonomy, or at least political equality with the resident European minority, and the removal of social and economic discrimination against the Muslim population.

TUNISIA

The distinguishing characteristic of the Tunisian revolution was the fact that it was accomplished with a minimum of violence under the leadership of a broadly-based, comprehensive nationalist political party, the *Neo-Dastour*, which had been founded as early as the 1930's and was under the firm control of a nationally recognized popular political leader, Habib Bourguiba.[a]

Most of the Tunisian nationalist leaders come from the Western-educated intellectual and semi-intellectual groups who comprise an emerging middle class social element, many of whom are professional

---

[a] For a detailed discussion of the Tunisian revolution see the summary following this discussion.

people. Neither Communists nor military personnel played an important part in the revolution; the military did not participate because North African Muslims were never allowed to serve in the French military service in Tunisia. When their earlier, more moderate demands were rejected by the French, the nationalists demanded complete political independence. Tunisian nationalists are mainly secularists and have tended since independence to pursue left-of-center economic policies.

The political objectives of the revolution were ultimately obtained by peaceful means, although *fellaghas*, as the guerrilla-terrorists are called in North Africa, were used to give emphasis to nationalist political demands. Headquarters were established by the *Neo-Destour* in major capitals and cities throughout the world, in order to "internationalize" the revolution. Organizations for North African Muslim students in Paris have performed an important recruitment function for the nationalists, and have provided an additional measure of foreign publicity. The practical effect of international pressure on the final outcome of the revolution is subject to wide disagreement; however, the *Neo-Destour*, like the *Istiqlal* (Independence) Party in Morocco and the FLN (*Front de Libération Nationale*) in Algeria, has devoted considerable effort to international activity.

MOROCCO

The *Istiqlal*, the nationalist political party founded in the 1940's, which led the revolution, was never a comprehensive national movement, and lacked a single leader of great national popularity like Bourguiba in Tunisia. However, Morocco had a valuable national symbol in the person of its traditional ruler, the late Sultan Mohammed V, who was both the religious and political head of Moroccan society. The Sultan was converted to the nationalist cause during World War II, and when the French deposed him in 1953 he became inseparably identified with Moroccan nationalism.

More isolated from European cultural influences and less ethnically homogeneous than Tunisia, Morocco's nationalist movement was not characterized by Tunisia's moderation in political orientation and high degree of centralization in political organization. The presence of a large minority of Berber tribesmen, who had traditionally looked to the French for protection of their interests against the urban-based Arab majority, posed a special problem; Moroccan nationalists were required to devote considerable time and effort to winning over the Berbers to the nationalist cause. Since there is no outstanding Islamic institution of higher education in Morocco, as there is in Tunisia, Moroccans desiring religious training have traditionally gone to Al Azhar in Cairo, and this became a means of injecting Egyptian political influences into the Moroccan nationalist movement. Generally speaking, Moroccan nationalism was more anti-Western and more influenced by Egyptian ultranationalist attitudes than was Tunisian nationalism.

The actors in the revolution were drawn from the same groups as in Tunisia, with the exception that traditionalist religious leaders played a more active role in Morocco, and as a result the movement has tended to be economically more conservative.

The techniques used in the Moroccan revolution were very similar to those used in Tunisia, and as in Tunisia, its objectives were finally attained peacefully. Before granting full independence to either protectorate, the French vacillated between periods of administrative reform and negotiation with the nationalists and periods of political repression and military operations against the guerrilla bands. Eventually, between 1954 and 1956, France decided to settle the account by negotiation rather than by military force. France was under great pressure at this time from events in Indochina and was faced with an outbreak of terrorist and guerrilla activities in Algeria, which had hitherto been considered relatively "safe" from Arab nationalist activity, because of that area's legal connections with France and the long presence of the French in Algeria. Algerian nationalists have perhaps rightly concluded that their military operations against the French in 1954–56 won independence for the Tunisians and Moroccans.

ALGERIA

Arab nationalism developed more slowly in Algeria than in either of the neighboring French protectorates, due to the fact that the influence of France was felt more strongly there. Legally and administratively a part of France and integrated into the economy of France, Algeria had over a million European settlers, many of them belonging to families that had lived there for well over a century. Moreover, France had intentionally destroyed the few systems of Algerian nationalism that existed at the time of the French conquest. By imposing European standards of education and culture on the educated Muslim population, the French had made many Algerian Muslims think of themselves as French Muslims.

Ferhat Abbas, who was to serve as Prime Minister of the Algerian "Provisional Government in Exile" from 1958 to August 1961, expressed the feelings of most educated Algerian Muslims when he wrote in 1936 that he had consulted both present and past history but had not found "the Algerian nation" and so could not become a nationalist. The condition for Abbas' continued acceptance of French identification was, however, "the political and economic emancipation of the natives"; without that he could not envisage a "lasting France in Algeria."[2] When the contradiction between theory and practice in French colonial policy became abundantly clear in the postwar period, Ferhat Abbas and the sizable portion of the Muslim community which he represented turned to nationalism as the last resort.

The distinguishing characteristics of the Algerian nationalist movement have thus been its late appearance and, prior to 1954, its factionalism and internal rivalries. There was neither a nationalist political leader of Bourguiba's stature nor a traditional ruler and national symbol comparable to Sultan Mohammed V to give unity and centrality to the Algerian movement. This situation was changed in the course of 1954, when a group of young militants, who had been strongly influenced by the Nasserist revolution in Egypt and by events in Tunisia and Morocco, where guerrilla warfare was leading to negotiations for independence, organized the *Front de Libération Nationale*, with headquarters in Cairo. By 1956, almost all leading Algerian nationalists had joined this militant nationalist organization, and in September 1958 the FLN proclaimed the formation of a "Provisional Government-in-Exile" for Algeria.

The Algerian FLN is considerably different in social background from both the Tunisian *Neo-Destour* and the Moroccan *Istiqlal*. The *Istiqlal's* social base was the Muslim political-religious elite in Morocco, the *Neo-Destour's* was the Tunisian bourgeoisie, but the FLN's social base has been the Muslim lower-middle class. The Algerian nationalist organization came into being as a resistance movement, organized by activists who had despaired of political agitation. The FLN, while founded by intellectuals, combined in its leadership practical men of action, "people whose experience of life [has] been as noncommissioned officers in the French Army during the Second World War or in the fighting in Indochina, and who are mainly mechanics, small farmers, and the like."[3] It was only after months of guerrilla fighting that the FLN absorbed virtually every political and intellectual leader in the Algerian nationalist movement.

The present phase of the Algerian revolution started abruptly on the night of November 1, 1954, when armed bands of guerrillas made concerted attacks on French *gendarmerie* posts throughout Algeria. The resort to armed rebellion came after a decade of intense political agitation, during which Algerian Muslims at first made moderate demands, and later insisted on complete independence.

The FLN has conducted a vigorous military and diplomatic offensive against the French since 1954. Its leaders abroad have procured arms and war materiel for the combatants in Algeria and have maintained active liaison with the rebel leaders operating in the mountains and the native quarters of Algerian cities. They have instituted some forms of governmental control over rebel-held territory, have been able to win diplomatic recognition from some Afro-Asian countries, and have developed close military and diplomatic ties with the Soviet bloc and especially with China. The Algerian nationalists have also employed terroristic methods in Paris to exert pressure on the French Government and to arouse foreign interest in their cause. In 1961, the French Government agreed to negotiate with FLN representatives; however, these negotiations had made little headway toward a peaceful settlement of the Algerian rebellion by the late summer of 1961.

RESULTS AND OUTLOOK

For the two North African countries in which the nationalist revolution has attained its political goal, the post-independence period has been fraught with difficulties, many of which were foreshadowed during the revolutionary period. As noted above, the nationalist movement embraced elements which were widely divergent in ideological views and in socio-economic and cultural background. In the postrevolutionary period these differences have come to the fore, especially in Morocco, where Berber-Arab conflicts are a threat to national unity. Even in highly centralized and homogeneous Tunisia, the leadership of the *Neo-Destour* has had to deal with dissident "leftist" tendencies in the labor organization, and more serious than that has been the continued pressure on Bourguiba exerted by former guerrilla terrorists for a militant, anti-French foreign policy. Many of the *fellagha* bands in Morocco did not disband when independence came in 1956 and these continue to operate in remote mountain sections. Thus, the main domestic political problem in Tunisia and Morocco in the post-independence period has been to devise ways of controlling the social forces stirred up during the revolutionary period.

The Algerian Revolution has greatly aggravated this problem of control in Tunisia and Morocco. Tunisia had been especially subject to involvement in the Algerian conflict, since the Bourguiba government has been compelled by popular sentiment to permit Algerian guerrillas to use Tunisian territory as a base of operations against the French. This has aggravated Franco-Tunisian relations at a time when Tunisia desperately needs France's economic assistance. Tunisia's national interests conflict with the interests of the Algerian nationalists at several points, most notably over disputed territory in the oil-rich Sahara regions. However, Pan-Arabist popular sentiment acts as a constant and compelling motivation for a more militantly anti-French and more pro-Algerian foreign policy than Tunisia's leaders might otherwise adopt. The pacification of Algeria would probably ease many of Tunisia's domestic problems.

Recent events suggest that the hitherto politically inactive and pro-western kingdom of Libya may no longer be immune to revolutionary activity. A sparsely populated area and one which is economically less developed and culturally more backward than its neighbors, Libya is currently experiencing an oil boom which is having a definite effect on its political and social structure. Foreign technicians and a group of native Libyans connected with the oil economy live in air-conditioned luxury apartment houses and villas in the midst of an urban proletariat living in shanty-town conditions; moreover, civil servants are feeling the pinch of inflation induced by the oil boom.

Libya is without a political party system and strongly under the influence of a traditional sectarian elite. During the past decade the major portion of Libya's national budget has been derived from U.S. payments in exchange for the use of American-built air bases.[4] By the summer of 1961, the kind of Pan-Arabist and anti-Western political agitation that had long been current in the rest of the Arab world was beginning to be stirred up in Libya by the Syrian-based *Ba'thist* Party.[b] Most of the Libyan *Ba'thists* are Palestinian Arab refugees employed by the Libyan Government as teachers and by the oil companies as technicians. Future political conflicts in this country are likely to revolve around such questions as the use of oil revenues and the presence of foreign bases.

For 7 years the Algerian nationalists have waged a bitter guerrilla war against the French Army, the European settlers, and Algerians whom the FLN accused of collaborating with the French. Whatever peaceful settlement may eventually be arrived at must take into consideration such problems as Algeria's economic dependence on France and on the disputed Sahara region for the large-scale industrialization that the country's expanding population requires, and Algeria's considerable non-Muslim minority population. Post-independence difficulties like those being experienced today in Morocco and Tunisia will still lie ahead of an independent Algeria, although pacification of North Africa may contribute to the solution of these problems.

A final factor in the future of North Africa is the role which France and the French will play. On one hand, there are the metropolitan

---

[b] See pp. 408–409 in Middle East section.

French and the Government in Paris, which at times have been willing to meet some of the political demands of the nationalists. French people in metropolitan France have generally not practiced racial discrimination against the North African Muslims who have come among them, and many humanitarian and liberal Frenchmen have actively supported the cause of the nationalists. On the other hand, the French settlers (*colons*) and the French military hierarchy stationed in the area have been singularly unable to deal with the demands of the Muslim population, whom they generally regard as their social inferiors.

The *colons* and the French Army personnel, who have adopted the European settler attitudes, have been unable to accept the changes imposed by conditions in the postwar world in their personal relations with the Muslim natives and in France's international position. Psychologically these groups are ill-prepared for the sudden decline in French power in the area. There is a tendency to explain the rise of nationalism as the work of the international Communist movement or even as the result of Anglo-American intrigues against France. The French in North Africa find it difficult to recognize any faults in the workings of the French colonial system. *Colon* influence in French governing circles, directed by the powerful French Algerian lobby in Paris, enjoyed a high degree of success in checking the liberal tendencies of the Paris government during the Fourth Republic (1945–58). Although the *colons* were unable to prevent the adoption of some liberal colonial policies and pronouncements, they were often able to sabotage the application of these reforms.

In May 1958, the French Army in Algeria, supported by the *colons*, revolted against the government in Paris and threatened military action against metropolitan France to install a government which would maintain French authority in Algeria. As a result, General de Gaulle came to power and the Fifth French Republic was established. The ultraconservatives were disappointed by de Gaulle's liberal colonial policies, and the same army-*colon* revolutionary coalition that had helped him to power made repeated attempts to overthrow him in 1960 and 1961.

The intransigent attitude of the *colons* and parts of the French Army in Algeria greatly contributes to revolutionary unrest. If the conservative French desires are even partly satisfied, the anti-French rebellion is bound to continue. If, on the other hand, the demands of the Algerian nationalists were to be met, the French Army itself, supported by the *colons*, might revolt against the government. It is difficult to see how the area could become pacified in the near future, no matter what solutions are attempted.

NOTES

1. Nevill Barbour (ed.), *A Survey of North West Africa (The Maghrib)* (London; Oxford University Press, 1959), pp. 1–6.
2. Lorna Hahn, *North Africa: Nationalism to Nationhood* (Washington: Public Affairs Press, 1960), pp. 139–140.
3. Barbour, *Survey of North West Africa*, p. 59.
4. Roger Owen, "Libyan Note-Book," *Africa South in Exile* (London), V, 3, pp. 100–106.

RECOMMENDED READING

BOOKS:

Barbour, Nevill (ed.). *A Survey of North West Africa (The Maghrib).* London: Oxford University Press, 1959. An excellent general work on the area. Contains material on geography, socio-economic structure, historical background, and contemporary political developments.

Gillespie, Joan. *Algeria: Rebellion and Revolution.* New York: Praeger, 1960. A study of the Algerian revolution based on author's Ph.D. thesis at Fletcher School of Law and Diplomacy. Contains material on military aspects of revolution, maps, bibliography, etc.

Hahn, Lorna. *North Africa: Nationalism to Nationhood.* Washington: Public Affairs Press, 1960. A scholarly study of Tunisian, Moroccan, and Algerian nationalism and the revolutions. Contains bibliography, biographical notes, glossary of North African and Arab terms, etc.

Ziadeh, Nicola. *Whither North Africa.* New York: Gregory Lounz, 1959. An excellent survey of political, social, and economic developments in the area.

THE TUNISIAN REVOLUTION: 1950–1954

SYNOPSIS

The Tunisian revolution was a nationalist revolution directed against the French, who were occupying Tunisia as a French protectorate. The *Neo-Destour* Party took the initiative in the nationalist movement from the time of its formation in 1934. Through a combination of intensive domestic and foreign propaganda, underground and guerrilla operations, and a willingness to negotiate with liberal French governments, the party won important constitutional concessions in the years between 1950 and 1956. On March 20, 1956, the French Government ended its protectorate and granted Tunisia complete political independence. The impetus for the Tunisian revolution came primarily from middle-class elements. It received its mass support through the dynamic leadership of Habib Bourguiba. The revolution was accomplished with a minimum of violence and stands in sharp contrast to France's other colonial revolutions.

BRIEF HISTORY OF EVENTS LEADING UP TO AND CULMINATING IN REVOLUTION

The French presence in Tunisia began in 1881 when the Bey of Tunis, under pressure from his European creditors, signed the Treaty of Bardo and thereby transformed the Beylic (Kingdom) of Tunisia into a protectorate of France. Similar arrangements were established between Britain and Egypt in 1882 and France and Morocco in 1912. These "protectorates," established during the last of the 19th century and the beginning of the present century, were little different from the "colonies" that European powers had established in Africa, Asia, and America in earlier centuries. The one significant difference is that a protectorate maintains its traditional ruler. The French governed the country in the name of the Bey, who retained the form if not the substance of national sovereignty, and the institutional machinery for a Tunisian national state remained intact. But for all practical purposes, Tunisia, along with Algeria—a colony—and Morocco—another protectorate—became virtually a colonial possession of France. French, Italian, Corsican, Maltese, and other southern European settlers migrated to Tunisia.

The first stirrings of nationalist discontent occurred before World War I, especially among the younger generation. Their organization, known as "Young Tunisia," conspired actively with the Ottoman Turkish Empire, of which Tunisia had once been a part, in an attempt to drive out the French. These aspirations were doomed to failure. The call for national self-determination associated with President Wilson brought forth a second wave of nationalist agitation. In 1920 the *Destour* Party was formed to press the French for administrative reforms if not complete independence. In 1934 the *Neo-Destour* Party was formed out of the less aggressive parent organization. The defeat of France by Germany in 1940, and the liberal policy that the Germans compelled the Vichy regime to follow, gave Tunisian nationalists their first real opportunity. In the postwar period the French were not able to pacify the country completely and return it to its prewar status as a docile protectorate. The events which preceded the peaceful transfer of authority from the French Ministry of Foreign Affairs to the Bey's Prime Minister in March 1956 included 6 years of violence by the Tunisians and repression by the French, followed by negotiations and reforms carried out in an atmosphere heavy with threats of renewed violence and international repercussions.

THE ENVIRONMENT OF THE REVOLUTION

DESCRIPTION OF COUNTRY

Physical characteristics

Located midway between Gibraltar and Suez on the northern extremity of Africa, Tunisia is equivalent in size to the state of Louisiana.

The northern quarter of this territory, together with a narrow coastal strip down the eastern side of the country, comprises "functioning Tunisia," in which the overwhelming majority of the population live and where the economic life of the country is concentrated. The southern two-fifths of Tunisia lies wholly within the Sahara Desert.

People

In 1956 the population of Tunisia was officially estimated at 3.8 million. The Tunisians are predominantly Arab-Berber in racial composition and cultural identification. Approximately 90 percent of the population is Arab or Berber Muslim; 8 percent is European and 2 percent is Jewish. Arabic is the language of the Tunisian population and French is that of the European sector of society. Tunisia's population is concentrated in the northern quarter, especially in the vicinity of Tunis, and along the eastern coastline. Tunis and its suburbs had a population of almost a million inhabitants by 1956. At that time, other

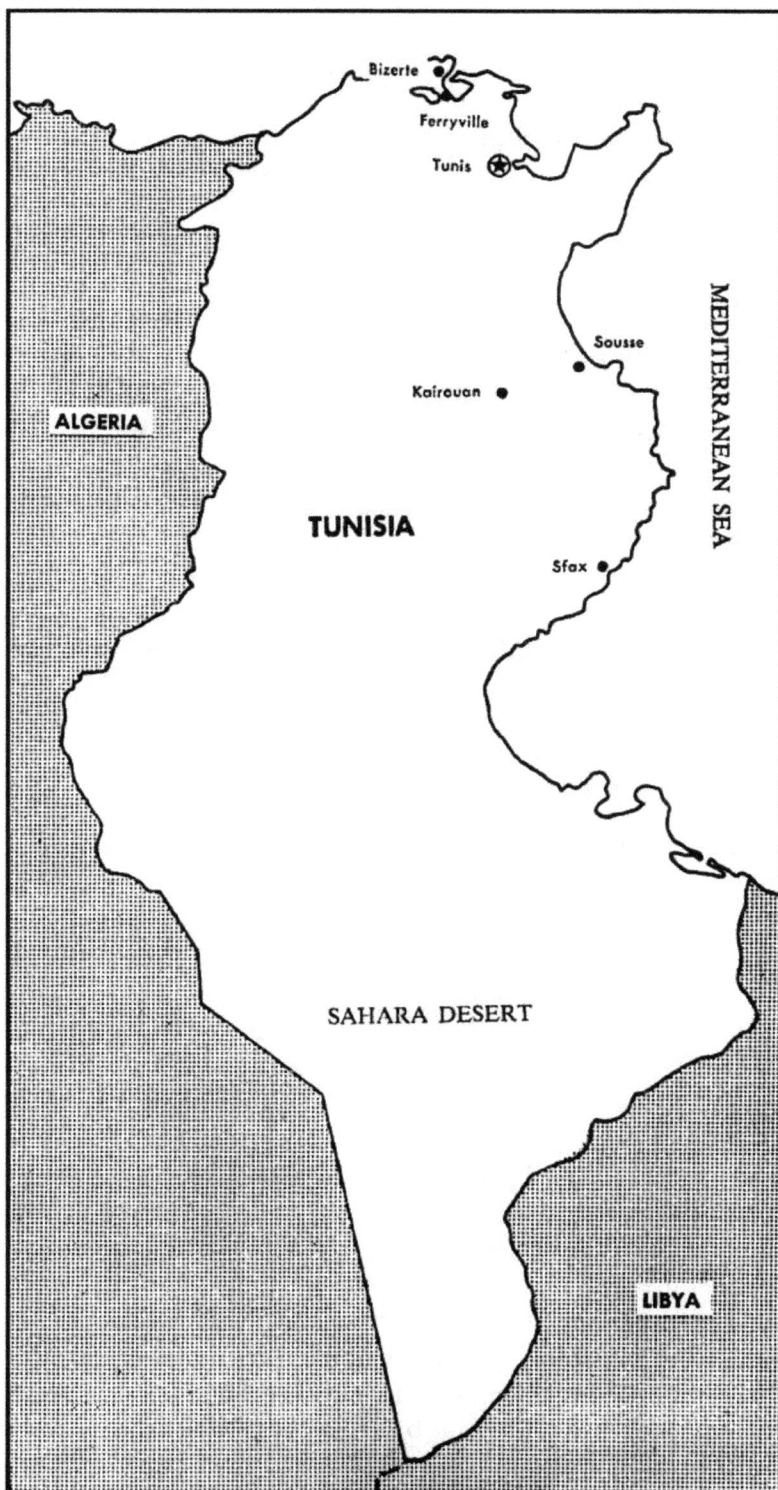

towns, such as Bizerte, Sfax, Sousse, Kairouan, and Ferryville, had less than 75,000 inhabitants.[1]

Communications

Lines of communication generally run from the Algerian border through the economically developed northern part of the country and continue down the eastern coastline to the Libyan frontier. Tunis is linked by rail and highway with Libya and Algeria; the railroad runs south from Tunis only as far as Gabes and transit beyond that point is by highway. The country has four major ports: Bizerte and Tunis on the north coast and Sousse and Sfax on the east. Internal communications radiate from these seaports. Tunis, the nodal point of internal communication, is the site of the country's only airfield of international importance.[2]

Natural resources

Tunisia's agricultural endowments have been relatively meager in modern times. Lack of water is the country's overriding problem and this limits Tunisian agriculture to the northern and eastern edges of the country. Only about 12 percent of the land area is cultivated. Considerable mineral wealth in phosphates, iron, lead, and zinc exists, but there is no coal, oil, or hydroelectric potential within the country.[3]

SOCIO-ECONOMIC STRUCTURE

Economic system

Tunisia is primarily an agricultural country. Under French rule it was self-sufficient in foodstuffs and 60 percent of its exports were agricultural products. Large-scale European-owned farms, located for the most part on the best land, produced yields several times higher than those of Tunisian peasant farms. There is very little industry, and 80 percent of this is concentrated in and around Tunis. Before independence almost all modern industry was European-owned and the French Government played a significant role in its development. Tunisian plants are mostly light industries engaged in processing agricultural and mineral products. Before 1956, the country derived 35 percent of its exports from minerals. The French, who completely dominated the economy of the country, "engaged in considerable economic exploitation. Tax structures were favorable to them, and much of the best land was taken by French colonists. The country was developed primarily as a market for French industry and as a source of raw materials."[4]

Class structure

As a French colony in every respect except for its formal status as a "protectorate," Tunisia had a social structure exhibiting the usual divisions of a colonial society: the European residents (250,000 in 1956) ranging from upper- to lower-class elements, stood apart from—and well above—"native" society. The French *colons* and the higher French civil servants and professional people were the real social, economic and political elite. A European middle- and lower-class group, composed of government clerks, shopkeepers, and skilled workers, stood immediately below the European upper classes. They were conspicuously better off economically than were the vast majority of Tunisians, whom they regarded as their social inferiors, and they identified completely with the local French upper class. More than half of the civil servants in the protectorate were French.[5] Europeans jealously guarded their social status against any encroachments by Tunisians.

Tunisian indigenous society was composed of a traditional elite which overlapped with the upper-middle class; a strong lower-middle class, centered in Tunis; a rural village element; and a nomadic Bedouin element, in the southern desert areas. While the Tunisians found it impossible to achieve social equality with the dominant European element, French rule created a middle class which had not previously existed. The upper-middle-class Tunisian benefited by the opportunity to go to French schools either in Tunisia or in Paris and tended to become Westernized. The leadership of the *Neo-Destour* was composed largely of such people. Unfortunately, education and political sophistication did not carry with it the right to hold high offices in the French administration. The Tunisian found some escape from this impasse in the organizational structure of the *Neo-Destour* Party. The middle and lower-middle classes consisted of shopkeepers, civil servants, skilled workers, and employees of the French-sponsored industries. It was among this group that the *Neo-Destour* found some of its most important support and following. Historically, the Tunisians, and particularly the Tunisian-educated elite, have been more inclined to identify with the West and to adopt Western standards than other Arab groups such as the Egyptians or Iraqis. The political leaders do not identify with the Arab movement.[6]

Literacy and education

According to U.N. figures, Tunisia's literacy rate in 1951 was between 10 percent and 20 percent. Male urban dwellers were the most literate group in the population. Tunisia's educational system was a combination of French secular schools, Arabic religious schools, and a number of French and Arabic mixed institutions which offered instruction in both languages and in both secular and religious sub-jects. Al Zaitouna Mosque in Tunis was the leading Islamic theological institution in North Africa. In 1945 a Tunis branch of the University of Paris was established.[7]

Major religions and religious institutions

The overwhelming majority of native Tunisians were adherents to Islam. Tunisian Muslims are noted for their religious tolerance and moderation. There was never any organized religious fanaticism, as in Algeria, Egypt, and Iran. Although many of the Westernized elite paid only lip service to Islam, the religious sensitivities of the people were still real, and Islamic traditions contributed to the development of Tunisian nationalism.[8] The 250,000 European residents were nomi-nally Christians and 2 percent of the population was Jewish.

GOVERNMENT AND THE RULING ELITE

Description of form of government

Tunisia, as a French protectorate, was governed in theory by its tra-ditional ruler, the Bey of Tunis, but in practice by the French Resident General, who was responsible to the Foreign Ministry in Paris. Prior to the reforms of 1951, the Resident General acted as Prime Minister of the Bey's Franco-Tunisian Cabinet of Ministers, and from 1951 to 1954 he retained veto power over all ministerial action. During the terminal stage of the protectorate (1954–56), the Resident General was replaced by a High Commissioner, who represented and coordi-nated French administrative policies but was without powers of direct participation in the government of the country.

Apart from the French presence, the Tunisian state before 1956 was an absolute monarchy. There was no constitution and, in theory, the Bey had full power to appoint ministers of his own choosing and to enact any law. In practice, however, all important ministerial admin-istrative posts in the Beylical government were held by French person-nel; a few minor posts were given to "reliable" members of the Tunisian traditional elite. Operating under a system of parallel administration, the Bey and the traditional administrative hierarchy retained the sem-blance of power while the Resident General and French supervisors at all levels governed Tunisia. The Bey's district governors, called *Caids*, served as window-dressing for the local French *Controlleurs Civiles*.

In such a governmental setup, there could be no question of executive responsibility to the legislature (which did not exist as an institution) or to the judiciary, which was a mixture of Muslim *Sharia* courts and French civil courts. The independence of the judiciary was generally below the standards of metropolitan France. The Resident General was responsible only to the French Government in Paris, and his tenure of office depended on it. He was more likely to be influenced by the French settlers whom he met socially than by the Tunisians. There was no government institution for the articulation of Tunisian interests. The *Grand Counseil*, composed of both French and Tunisians, drawn from the "reliable" native elite, had only budgetary powers, and the municipal councils permitted towards the end of the protectorate were effectively dominated by French settler interests. The tenure of office of the Bey, who was technically a hereditary ruler, depended ultimately on the will of the French authorities.[9]

Description of political process

Political parties and power groups supporting government

Political activity in Tunisia centered around the question of the continuation of French rule. Nationalist political agitation began before World War I and gained momentum during the 1930's. In the period after World War II nationalism was a fully grown mass movement receiving its impetus and direction primarily from middle-class elements. Opposing the nationalist movement and supporting the French protectorate were a few members of the elite who benefited from the presence of the French. These were joined by the quarter-million European residents, who feared for their personal status in a Muslim-dominated independent Tunisia in which they would be an alien minority instead of a privileged elite. The few Muslim supporters of French rule were not organized as a political group, but the French *colons* were represented by a well-organized and articulate pressure group, in the local branch of the *Rassemblement du Peuple Français* (RPF). The Tunisian branch of the RPF was controlled principally by a very articulate group of French civil servants and supported by powerful financial and agricultural interests. The organization sought to perpetuate France's economic and political dominance. It advocated equal political representation of French and Tunisian inhabitants in a country where 90 percent of the population was Tunisian; it was the first to demand that repressive measures be taken against the Tunisian population in times of disorder, and in general the RPF represented the ultracolonialist position. "Its militancy and privileged position in the political, economic, and social life of the country . . . led it to practically advocate secession from the French Union when it [found] itself at variance with the expressed policy of the mother country."[10]

267

Character of opposition to government and its policies

The nationalist movement found its organizational expression in the *Neo-Destour* Party led by Habib Bourguiba. The old *Destour* Party, the parent organization from which the *Neo-Destour* group broke away in 1934, also opposed French rule but being "less aggressive and catering to the needs of fewer individuals, it did not have much influence on political events in the country."[11] In 1951, the old *Destourists* attempted to recapture the lead in the nationalist movement by accusing the *Neo-Destourists* of collaboration with the French; there was no evidence of any success in this endeavor. The small Communist Party, composed almost entirely of a few thousand Europeans, opposed French rule as a matter of course in keeping with the dictates of the international Communist Party line concerning nationalist revolutions. However, Communist relations with the *Neo-Destourists* were never friendly; the Communists accused the *Neo-Destourists* of collaborating with the French because of their policy of moderation and gradualism. The Communists were never a mass party and they had even less influence than the old *Destourists* in the final success of the Tunisian revolution.

Since the middle 1930's, the *Neo-Destour* Party had been the most active nationalist political organization in Tunisia. Although its leadership and support were primarily middle class, the *Neo-Destour* maintained effective liaison with the mass of the Tunisian population through an intricate system of local branches of the national party structure in all geographic areas and among all social classes. For example, the *Union Général des Travailleurs Tunisiens* (UGTT), the largest labor union, was an affiliate of the *Neo-Destour*, and through it the nationalists were able to enlist the active support of Tunisian urban and mine workers. The *Union Général de l'Agriculture Tunisienne* (UGAT) gained the *Neo-Destour* the support of peasants and small landowners. The Tunisian Chamber of Commerce, composed exclusively of non-French merchants, and the Feminist Movement added additional support to the *Neo-Destour* nationalists.[12]

In sum, the *Neo-Destour* Party was a comprehensive, broadly based nationalist political organization which consistently took the lead in the Tunisian nationalist movement from its inception in 1934 until the achievement of national independence in 1956.

In addition to the party organ, *Mission*, founded during World War II by the able theoretician and Deputy Secretary General of the *Neo-Destour*, Hedi Nouira, there was a wide range of French and Arabic language journals which were sympathetic to the nationalist cause and were willing to join the *Neo-Destourists* in expressing guarded criticism of the French authorities.[13] The *Neo-Destour* Party was illegal most of the time before 1956, and its leader, Habib Bourguiba, was in exile, but the organization continued and flourished as an underground movement. The party was able to operate openly only for brief periods: in 1937–38 during the left-oriented French government of Léon Blum and the Popular Front; in the winter of 1942–43 when the Germans occupied Tunisia; in 1951–52 during the "Government of Negotiations"[a] headed by Tunisian Prime Minister Chenik; and in the final stage of French rule, 1954–56.

Legal procedure for amending constitution or changing government institutions

There was no Tunisian constitution and in theory the Beylical system of government was an absolute monarchy. In practice, Tunisia's governmental institutions could be changed by administrative decrees handed down by the French Government in Paris to the French Resident General.

Relationship to foreign powers

Tunisia was entirely dependent on France from the 1880's until the end of the French protectorate in 1956. French hegemony was interrupted briefly in 1942–43 when the country was occupied by the Germans.

The role of military and police powers

Tunisia's military and police powers were in the hands of the French. One of the constant complaints of the Tunisian Nationalists concerned the presence of French army units in Tunisia and the maintenance of a French *gendarmerie* whose members were primarily French. French military and police forces were frequently used to repress nationalist political agitation, especially during the 1952–54 period of crisis in Franco-Tunisian relations. In the 1954–56 terminal period of the protectorate, the French agreed to a gradual transfer of internal security forces. Some of the French *gendarmerie*, which by this time contained a higher percentage of Tunisians, were turned over to Tunisian control.

---

[a] The "Government of Negotiations" was "formed within the framework of the protectorate treaties to negotiate . . . institutional modifications which would lead Tunisia by stages to internal autonomy."[14]

WEAKNESSES OF THE SOCIO-ECONOMIC-POLITICAL STRUC-TURE OF THE PREREVOLUTIONARY REGIME

History of revolutions or governmental instabilities

During the 60 years of French rule in Tunisia prior to 1940 there had been a few isolated cases of tribal warfare directed against the French, and during World War I and in the 1930's there had been scattered instances of terrorism, but these were never a serious threat to French authority. France's military defeat in 1939–40 severely impaired both her moral authority and her administrative control over the country. The Vichy regime was never free to handle Franco-Tunisian relations without reference to Germany and Italian policies in the area. The Vichy French Resident General was unable to prevent the formation of a predominantly Tunisian cabinet under the pronationalist Bey, Sidi Mohammed al Moncef, who came to the throne in 1942. During the 6 months of German occupation (1942–43) the "Moncefist experiment" flourished and when the Allies abruptly ended this brief respite from direct French rule it left memories which were not to be quickly forgotten.[15] The liberal and popular Moncef Bey was deposed by the Free French in 1943, on the grounds that he was an Axis collaborator, but the Tunisians felt that it was done because of his pronationalist leanings. The French had made a serious tactical error. This action not only alienated a large segment of the Tunisian public, but made a mockery of the treaty on which French rule was based, since the treaty was theoretically designed to "protect" the Bey.

Economic weaknesses

Tunisia experienced the economic difficulties usual in an underdeveloped area which has suddenly come into contact with modern economic activities. For example, European-controlled enterprises mushroomed, while indigenous economic institutions stagnated.[16] As an added difficulty, the country experienced extensive European immigration. The influx of European settlers displaced native farmers from the best farm lands in the northern quarter of the country; mechanization of large-scale agricultural enterprises sent large numbers of farm laborers into the cities in search of work and the excitement associated with urban living. Finally, the economic boom resulting from the wartime activities of, first, the occupying Germans and then the American, British, and Free French forces brought additional Tunisians into the cities. There was not enough industrial and commercial activity after the war to absorb more than a small portion of these people; moreover, European workers had more skill and social influence than the Tunisian natives. The result was the creation of a large poverty-stricken urban proletariat, consisting of many unemployed or marginally employed unskilled workers. The problem of unemployment, ironically, was intensified by improvements in health and sanitation, which lowered the death rate, particularly the infant and child mortality rate, and by the 1950's had increased the proportion of young people in the population. There also developed a greatly expanded urban middle class, composed of such groups as skilled workers, professional and commercial people, students and intellectuals, who were to a large degree Westernized in values and social habits, but for whom there was little room for advancement in the French-dominated economy.

Social tensions

In such an economic situation, social tensions between the Tunisian natives and the resident Europeans were bound to develop. In many ways Tunisian society had benefited from French rule. Western education had been introduced, though it was limited to a small minority; medical and sanitation facilities had been established; light industry enlarged the economic base; and harbor and communication facilities were expanded. However, the French failed to integrate the Tunisians socially and economically with those Europeans of comparable socio-economic standing, despite the feeling of many Frenchmen that assimilation of Muslim and French society and culture was desirable. In restaurants, schools, and many public places Tunisians were either segregated or totally excluded from facilities enjoyed by Europeans. It was customary among the French to address all Tunisians, regardless of their social status, not as *vous*, but *tu*—a familiarity normally reserved for children, animals, or intimate friends, but in this case, used to denote social inferiority. In general, the European's attitude toward the Tunisian assumed a master-servant relationship, whether or not such a relationship existed in fact, and was in many respects comparable to the social attitudes of whites toward nonwhites in certain parts of the United States. The attitudes of cultural superiority, snobbery, and often contempt manifested by many Europeans in their day-to-day contacts with the Tunisians proved to be a major social irritant, especially among educated Tunisians.[17]

Government recognition of and reaction to weaknesses

By the 1930's the French were aware of an organized nationalist movement among the Tunisians. Their reaction vacillated over the next 20 years from repressive police measures to liberal reforms in the governmental structure of the protectorate. The initial reaction in 1934 was to impose stringent controls over the nationalist activities of the *Neo-Destour* Party and to send Bourguiba and other leaders to concentration camps in the Sahara. The Blum Government of 1936–37 brought a brief respite from police repression of political activities, as did the Vichy Government from 1940 to 1943. In 1951 the French Government initiated major governmental reforms; a predominantly Tunisian cabinet was permitted and French influence, although still decisive, was now exercised more indirectly. When the Tunisians pressed for the elimination of even this indirect influence, the French returned to repressive police action for a period of 2 years. The reform in 1951 did not have a lasting effect and French attitudes stiffened again, largely as a result of the French *colon* attitude. Senator Colonna, representing the French in Tunisia and one of the heads of the "North Africa Lobby," sent the following communication to the French Government in October 1951:

> The Tunisians cannot be much trusted and they are incapable of either administering or governing their country . . . in our own interest and in the interest of France; for once Tunisia is free, she will, whatever her promises, join the other camp—maybe the Arab League, or the U.S.A. or the U.S.S.R. (sic!)[18]

The reforms in 1954 gave Tunisia internal autonomy and introduced the terminal stage of the French protectorate. A government was formed that truly represented Tunisian interests and was not a "puppet" government as previous ones had been. It was with this government that the final negotiations for complete independence were conducted. The French had been unsuccessful in luring either the native traditional elite or the Jewish population away from the nationalist cause; the Bey was responsive to, if not enthusiastic about, the nationalist demands of the *Neo-Destour*, and there were several Jewish leaders in the nationalist movement. The French were not able to correct the major weaknesses, partly because they were the direct result of their policy and partly because the political situation and the attitudes of the *colons* did not permit any steps to alleviate them.

FORM AND CHARACTERISTICS OF REVOLUTION

ACTORS IN THE REVOLUTION

The revolutionary leadership

The leader of the *Neo-Destour* Party and the dominant personality in the Tunisian revolution was Habib Bourguiba. Born in 1903 of middle-class parents (the son of a former Tunisian Army officer), Bourguiba studied first in Tunis and then at the University of Paris. When he returned to Tunis in 1927 he reaffiliated with the *Destour* nationalist organization. He became a member of its executive committee and of the editorial staff of its newspaper, *La Voix du Tunisien*. He soon concluded that this party, with its narrow social base and its platonic protestations of anticolonialism, was dying, both from lack of contact with the people and from lack of courage. Forming a group of young militants within the conservative *Destour* organization, Bourguiba and his followers in 1932 started a newspaper significantly called *L'Action Tunisienne*. They urged the *Destour* to become a mass organization holding frequent open congresses instead of meeting in the exclusive social clubs of Tunis.[19]

Bourguiba's faction formed the *Neo-Destour* organization in March 1934; Dr. Mohammed Materi, an older member of the group, was made President and Bourguiba became Secretary General. Arrested and exiled to the Sahara in September 1934, Bourguiba returned in 1936 wearing the halo of martyrdom, and, having kept in close touch with the organization, was elected President of the *Neo-Destour*. He has remained in control of the party organization, despite long absences from the country, spent either in prisons or in traveling abroad to negotiate with the French and to present to the world Tunisia's case for independence. During the ascendancy of Axis power, Bourguiba steered the party clear of any pro-Fascist alignment, although he spent some time in Rome and was actively courted by the Mussolini Government.

Bourguiba combined in his personality both the rational traits of a Western statesman and the emotional appeal of an Arab nationalist leader. Concerning his attitude toward France, a sympathetic French student of North Africa wrote in 1960:

> Bourguiba, probably more than any other North African, symbolizes the emotional attachment to France which his famous political formula "I do not hate France—only colonialism" only begins to describe. In Paris he learned to speak French better than his native Arabic . . . . He also acquired a French wife . . . . Finally, he developed that feeling which, admittedly or not, has colored his every action vis-a-vis the French: a desire to be respected by them and to be accepted as an equal.[20]

Bourguiba's blending of French and Muslim cultures in his personal life was in keeping with the Tunisian tradition of cultural accommodation and assimilation. His colleagues in the nationalist movement shared a similar background of Franco-Tunisian culture and society, and this fact was the major determinant in the political and ideological course that Tunisian nationalism took.

Tunisian nationalism was fortunate in having available a large body of capable secondary leaders. There was Bahi Ladgham, Tunisian-educated and 10 years younger than Bourguiba, an economist and head of the Tunisian Chamber of Commerce, a charter member of the *Neo-Destour* and as such arrested many times, and in 1951 political advisor to the Chenik "Government of Negotiations." There was Salah ben Youssef, 5 years Bourguiba's junior, also a charter member of the movement, arrested and exiled along with Bourguiba in 1934, Secretary General of the *Neo-Destour* during Bourguiba's Presidency, Minister of Interior in the Chenik Cabinet in 1951–52, and after 1954 leader of a dissident ultranationalist faction which opposed Bourguiba as a collaborator. There was Mongi Slim, member of one of Tunis' leading families, educated in France in mathematics and law, active in the *Neo-Destour* after 1936, in charge of party activities during the 1952–54 period of French repression and police actions, and Minister of Interior in the ben Ammar Cabinet (1954–56) which carried on final negotiations for independence. There were many other leading participants in the nationalist organization whose biographies read similarly: they were born of middle-class parents during the World War I period, were educated in French schools (often in Paris), joined the nationalist movement before the Second World War, and shared the arrests and deportations which were the fate of *Neo-Destourists* before 1954.

The political and ideological orientation of the *Neo-Destour* leaders can be described as moderate nationalism. Tunisian nationalism was notably different from Arab nationalism in general in that the Tunisians appeared to be moderates who were willing to move toward independence gradually and through negotiations rather than by violence and terrorism. The only opposition to Bourguiba's moderate nationalism within the *Neo-Destour* organization developed during the terminal period of the protectorate (1954–56), when Salah ben Youssef led a group of ultranationalist dissidents. Under the influence of militant, ultranationalist propaganda emanating from the Nasser regime in Cairo, and inspired by personal antagonism to Bourguiba, the "Youssefists" attacked Bourguiba for "collaborating" with the French by negotiating for independence instead of using guerrilla tactics. When ben Youssef proved intractable, he was drummed out of the party and the dissident faction completely collapsed when independence came in 1956.

The goals and aspirations of the Tunisian nationalist leaders were similar to those of nationalists elsewhere. They aspired above all to achieve national independence. Secondly, they wanted to improve the standard of living, and this they expected to accomplish through large-scale industrialization, irrigation projects, and land reform—measures which they felt the French had been blocking. National independence, the nationalist leaders believed, would open the way for the Tunisian middle classes to play the dominant role in the economy and society that the French *colons* were then playing. Tunisian nationalist leaders were more secular and Westernized and much less tradition-oriented than is usual in colonial societies and their aims reflected this fact.

The revolutionary following

The *Neo-Destour* Party was a mass party which attracted support ranging from the traditional elite around the Bey and the wealthy commercial *bourgeoisie* to the unemployed and unskilled illiterate urban masses and the peasants in the villages. The bulk of the party's membership came from the lower-middle class, which was economically and socially on the rise. The *Neo-Destour* organization brought people with many kinds of social and religious backgrounds together in social intercourse to work for the common cause of independence. The broad base of the *Neo-Destour* and its lack of identification with other Arab movements are shown by the fact that there were Jews among the rank and file in its leadership. In 1949, the party had around a half million official members; there were 200,000 members of labor unions, chambers of commerce and other organizations which were affiliated with the *Neo-Destour*. Through its grassroots organization, on the village and city district levels, and through its control of the labor unions and peasant unions and the local chambers of commerce, the *Neo-Destour* could boast the support of virtually the entire Tunisian nation. In the country's first national election on March 25, 1956, in which 85 percent of the eligible voters participated, the nationalist party obtained 95 percent of the total vote.[21]

ORGANIZATION OF REVOLUTIONARY EFFORT

Internal organization

The *Neo-Destour* began its operations in 1934 by fomenting popular agitations in an attempt to secure internal autonomy—the first step toward independence. The attempt failed because the *Neo-Destour*, at this early stage, was still essentially a cadre-type party. Recognition of this weakness led to the reorganization of the party along mass-type party lines. In September 1934, the top *Neo-Destour* leaders were banished to the Sahara, but were able to maintain contact with the party.

On the highest level, a Central Committee was established as the policy forming body of the party. A Political Bureau was created as the executive branch, with commissions to supervise the implementation of party directives at the lower levels. The basic units of the party—the cells—were incorporated into regional federations. The cells were charged with the responsibility of recruiting members, creating new cells, and implementing party directives. Special attention was given to the creation of auxiliary or parallel organizations which would enable the *Neo-Destour* to increase and maintain party control over all phases of social stratification. These auxiliary or parallel organizations blossomed after World War II into the General Union of Tunisian Workers (UGTT), the General Union of Tunisian Agriculture (UGAT), the Tunisian Association of Chambers of Commerce, and the Feminist Movement. When the French outlawed the party in 1938, 1943, and 1952 it went underground but its organization remained intact. It was able to function perfectly in 1942–43, under the German occupation, and in 1951–52 and 1954–56, when its activities were sanctioned by the French Resident General.

In 1952, a Special Commission within the Political Bureau of the *Neo-Destour* was charged with the conduct of war. Two agencies were created by this special commission:

—A military organization, based in Tripoli, had the task of training cadres and specialists, and giving technical and tactical instructions to leaders of guerrilla bands.

—A civil organization, based in Tunis, was charged with the responsibility of recruiting, supplying, financing, and other aspects of liaison.

The *fellagha* bands (guerrillas) were reorganized in 1952 into two main groups—Urban and Rural. The Urban group formed part of the *Neo-Destour* Party cells and auxiliary or parallel organization. In all cases, *fellagha* leaders took their orders from the special commission within the Political Bureau, and their ranks, at the regional level, were constantly increased by volunteers recruited by the federations or by the cells. By November 1954, their number had reached 2,500.[b]

---

[b] This description of the organization of the *fellaghas*, and their relationship to the *Neo-Destour*, is based on the analysis of French officers. This analysis has not been substantiated elsewhere.[22]

External organization

Tunisian nationalists often set up headquarters in foreign cities, from which they directed nationalist activities at home and conducted their diplomatic offensive against the French: Bourguiba was in Cairo from 1945 to 1949, and in 1952 Ladgham established the Tunisian Information Office in New York. Foreign governments attempted several times to use the Tunisian nationalist organization for their own national interests, as when the Axis powers tried to win Bourguiba's support; the Arab nations, led by the Nasser regime in Egypt, consistently tried to dominate the leadership of the *Neo-Destour*. Only in the case of Salah ben Youssef were the Arabs successful in influencing the course of *Neo-Destour* politics. Bourguiba became completely disillusioned with the Pan-Arabist movement during the 4 years he spent in Cairo. The Soviet bloc never supported the *Neo-Destour*, which the Communists considered too conservative, except by voting against France whenever the Tunisian question came up in the United Nations.

Foreign diplomatic support of the Tunisian revolution was a particularly important factor during the final stages of French rule. Bourguiba's world tour in 1951, the establishment of Tunisian Information Offices in several major foreign capitals in 1951–52, and the appeal to the United Nations from the Tunisian "Government of Negotiations" in January 1952 were significant steps which won support for Tunisian nationalism in the United Nations. In 1952, the Afro-Asian bloc supported a resolution to place the Tunisian question on the agenda of the Security Council; however, this body voted down the proposal. In favor of the proposal were Pakistan, Nationalist China, the Soviet Union, Chile, and Brazil; only the British and French voted no, but the United States, the Netherlands, Greece, and Turkey abstained and thereby prevented passage of the proposal. The reason given by the United States delegation for its abstention was that a discussion of the Tunisian question in the world organization would jeopardize the progress of negotiations between the French authorities and the Tunisian nationalists. Critics of the United States position argued that U.N. action would have forced the French to make greater concessions to the Tunisian negotiators. In metropolitan France there was a growing body of French support for the nationalists, and as early as May 1950, the French Socialist Party had come out in support of Tunisian independence.

GOALS OF THE REVOLUTION

Concrete political aims of revolutionary leaders

The primary political goal of the *Neo-Destour* leaders was always the achievement of national independence. However, for tactical reasons, the nationalist organization limited its demands at first to greater Tunisian participation in the protectorate government and the civil service. By 1945, however, the entire nationalist leadership was openly demanding internal autonomy, with a constitutional and parliamentary monarchy tied to France in the realm of foreign affairs and by economic, cultural, and military relations. In 1946 the same nationalist leaders called for complete independence. The clearest statement of the *Neo-Destour's* concrete political aims was contained in Bourguiba's program announced in 1950. Its most important provisions called for: the transfer of the Resident General's executive and legislative powers to a Tunisian Prime Minister and Cabinet selected by a nationally elected National Assembly; Tunisian control of the civil and administrative services; the suppression of the French *gendarmerie;* the transfer of local authority from French inspectors to the Bey's *Caids* (governors); and the election of municipal bodies, with French interests represented in all areas where there were French minorities. The nationalist leaders were careful to declare their loyalty to the Bey as the constitutional head of state; however, as events were to prove, this loyalty was more a matter of strategy and tactics than of deep conviction.

Social and economic goals of leadership and following

The *Neo-Destour's* social and economic goals were not radical for a nationalist movement in an economically underdeveloped and socially backward country. Nevertheless, Tunisian nationalists intended that the government they planned to set up play an active part in the economy of the country. At a party congress held in Sfax in 1955, the *Neo-Destour* leaders agreed on a comprehensive program which called for the development of light industry and power sources to cope with unemployment, the nationalization of land belonging to religious orders, the modernization of peasant agriculture, and the establishment of village cooperative stores. Nothing was said about the large-scale agricultural enterprises controlled by Europeans, probably because the Tunisian leaders realized that French *colon* influence in the government in Paris could jeopardize progress towards national independence. Also, Tunisia's economy depended heavily on the high output of these European farms, which additionally could serve as models of efficiency for their Tunisian neighbors.

REVOLUTIONARY TECHNIQUES AND GOVERNMENT COUNTERMEASURES

Methods for weakening existing authority and countermeasures by government

The *Neo-Destour* approached the French with demands for total independence, but on a gradual basis. Bourguiba's policy is best outlined as follows:

> Ours is a weak and tiny country, concluded Bourguiba and his friends. If we demand immediate, total independence, we won't get it—or anything else. But if we have a step-by-step program, seeking first a decent voice for Tunisia in the protectorate government, then internal autonomy, and then complete independence, we will have a basis for negotiations with France, and perhaps be able to gain some concrete concessions in the meantime.[23]

Tunisian nationalists had recourse to both legal and illegal techniques in the course of their successful revolution. A salient characteristic was this alternation of legal and illegal methods, corresponding to the vacillation of France's Tunisian policy from liberal to repressive. Legal techniques were available to the nationalists only during the brief life of the Blum government (1936–38), the period of the Vichy government (1940–42), and the German Occupation (1940–43), Tunisian "Government of Negotiations" in 1951–52, and again in 1954–56. At these times, the *Neo-Destour* openly issued manifestoes, published regular organs of party propaganda, and held public congresses at which *Neo-Destourist* speakers attacked French domination.

Another important legal technique was participation by individual *Neo-Destourists* in the Tunisian "government" of Prime Minister Chenik during the German occupation and later in the "Governments of Negotiations" headed by Prime Ministers Chenik and ben Ammar. The decision of the *Neo-Destour* to permit, and even openly encourage, the participation of active members of the *Neo-Destour* organization in these pseudo-nationalist "Governments of Negotiations" represented a major break with the traditions of radical Arab nationalism. By being members of these "governments," composed largely of Tunisians whose nationalism was minimal, *Neo-Destour* nationalists were able to prevent their becoming French puppet organizations and often were able to exercise a decisive influence over the other Tunisian ministers, who might otherwise have been amenable to French influence. The most important legal method of weakening French rule came about as a result of *Neo-Destour* influence over the ben Ammar government. This was the national election of a legislative assembly, Tunisia's first, which the ben Ammar Cabinet announced was to be held in March 1956. The French were reasonably assured that after March 25, 1956, there would be a legally constituted, *Neo-Destour*-dominated national assembly ranged against them.

More in keeping with the traditions of Arab nationalism were the various illegal revolutionary techniques which the *Neo-Destour* employed. It circulated its anti-French manifestoes and propaganda organs clandestinely when the French returned to police repression, and it organized mass protest demonstrations, riots, strikes, and sabotage against the telegraph system and other public works. For example, on April 10, 1938, Bourguiba called a mass rally to protest the banning of the *Neo-Destour* Party, and 10,000 Tunisians turned out to take part in the demonstration. The French *gendarmerie* shot 200 of the demonstrators and arrested 3,000 *Neo-Destourists*, including Bourguiba, who was sent to prison for the second time.

In the postwar period, the *Neo-Destour* underground began large-scale terrorist and guerrilla activities against the French. Known as *fellaghas*, the *Neo-Destour* terrorists made a significant contribution to the atmosphere of tension and civil discord that prevailed in Tunisia after 1945. The *fellagha* bands suspended operations during France's liberal Tunisian policy of 1950–52, but when French repression returned in the 1952–54 period, the *fellaghas* returned.

The *fellaghas* achieved immediate notable success. Organized into small bands, which operated in the familiarity of their mountainous regions, they quickly proved themselves a match for the superiorly equipped French forces. The rigid security which prevailed among these bands denied the French forces of another advantage. Ordinary recruits seldom knew the names of their leaders, and could thus provide their captors with little, if any, information. These disadvantages were further complicated by the ambiguity of the relationship of the *fellaghas* to the *Neo-Destour*.

> The Neo-Destour never claimed any official connection with the *fellaghas*, trying to explain that they had been spontaneously formed by people who had been oppressed. But Taieb Mehiri, Director of the Neo-Destour Policy Committee, was completely in favor of them. Party headquarters in each city began busily to enroll new recruits, and to provide them with the proper contacts.

After their reorganization in 1952, the urban bands of the *fellaghas* were held back. Though thoroughly trained and ready to attack whenever ordered, they were to be used only after the rural bands had captured the countryside. This plan was adopted in order to make the French spread their forces throughout Tunisia, away from the big cities, where the *Neo-Destour* was busy recruiting, financing, and training. If the *fellaghas* could have secured the countryside and forced the French back to the big cities, then the story of Dien Bien Phu might have been repeated all over.

When the Mendes-France government adopted a liberal Tunisian policy in the summer of 1954, the *Neo-Destour* joined the second Tunisian "Government of Negotiations" and *fellagha* operations were no longer countenanced by the leadership of the nationalist organization. However, pro-Nasser ultranationalists led by Salah ben Youssef continued terrorist activities against both the French and the Tunisian regime, which the Youssefists considered "collaborationist," until the achievement of complete independence in 1956.

French countermeasures included police and military operations against mass demonstrations and terrorist activities, arrests, and deportations of *Neo-Destourist* activists, censorship of the press, and other repressive measures, as well as a variety of administrative "reforms" designed to appease nationalistic aspirations. Significant countermeasures were also taken by European *colons* acting unofficially but receiving varying degrees of tacit support from the local French authorities. For example, a group of extreme *colons* calling themselves the *Main Rouge* (Red Hand) began "to employ threats and even physical terror, through such means as exploding bombs near meeting places of suspected nationalists, in order to snuff out Tunisian resistance."[25] French civil servants often sabotaged the working of their administrative departments by not following orders from their Tunisian superiors.[26]

Methods for gaining support and countermeasures taken by government

The success of the Tunisian revolution depended to a large extent on the nationalist organization's use of words, both written and spoken. The *Neo-Destourists* early realized "that the strength of the colonial regime rested less on its rifles than on the resignation of the Tunisian people."[27] For this reason, the *Neo-Destour* sent its intellectuals into the villages to teach the peasants first to read and then to think politically. "Once people became accustomed to attending the Party's 'schools,' it was not too difficult to explain to them the political facts of their lives, to interest them in doing something about the situation, and to get them to attend regular meetings where they could discuss plans of action."[28] The agitation against French domination attracted broad support among the peoples, and the frequent arrests and deportations of *Neo-Destour* activists had the effect of making martyrs of these nationalists.

As a comprehensive nationalist organization, the *Neo-Destour* Party could draw on a wide range of popular support. Its affiliate among business and professional people, the Tunisian Chamber of Commerce, provided considerable financial assistance. With this the *Neo-Destourists* were able to carry on social work among the poor, and this operation added greatly to the popularity of the organization. Through its connections with the labor unions, particularly the UGTT, the *Neo-Destour* had the support of the working class and could call nationwide strikes to support its political demands.[29]

The support of Tunisia's traditional ruler, the Bey of Tunis, was of decisive importance. The first Bey to be active in the nationalist movement was Sidi Mohammed al Moncef Bey, who came to the throne during the Vichy regime. Under his rule, the first predominantly Tunisian Cabinet was formed—without prior consultation with the French Resident General—and when the Allied military forces returned the French to their former position of power in Tunisia, the Bey was quickly exiled. He was a popular hero and a focal point of nationalist sentiment until his death in 1947. The next Bey, Sidi Lamine Pasha, was not enthusiastic about Tunisian nationalism; however, on several crucial occasions, he was pressured into voicing the demands of the *Neo-Destour*. Since the French presence in Tunisia was ostensibly to "protect" the Bey, they could ill afford to ignore demands originating with the Bey. When the Bey hesitated early in 1956 to ask the French to revoke the Treaty of Bardo (which guaranteed both the Beylical monarchy and the French presence), the *Neo-Destour* threatened to denounce the Bey publicly as a French puppet, realizing that he could be easily overthrown if it could be proved that he was working contrary to the nationalists. The aged sovereign signed the document prepared by the *Neo-Destour*. This served as an important legal device in the hands of the *Neo-Destour*—a request by the sovereign, whom the treaty was designed to protect, demanding that this arrangement be ended.

The countermeasures employed by the French to undermine the domestic and international support enjoyed by the *Neo-Destour* organization included, in addition to the combination of repressive measures and liberal reforms discussed in the above section, some of the following techniques: 1) a "smear campaign" against Bourguiba before World War II which alleged that he was an Italian agent; 2) a similar campaign against the *Neo-Destour* movement after the war charging it with having collaborated with the Axis powers; 3) appointments to the United Nations and the conferring of high honors on Tunisians who were regarded by their countrymen as French puppets; 4) subsidization of Arab-language newspapers in opposition to the *Neo-Destour;* and 5) an elaborate propaganda campaign, aimed especially at foreign supporters of Tunisian nationalism, which attempted to show that "Bourguiba and his party were simultaneously Fascists, Nazi collaborators, and Communist sympathizers."[30] The various organs of the European settlers were more vitriolic than the official French press in their attacks on the *Neo-Destour.* According to some observers, the Youssefist opposition to Bourguiba's more moderate nationalists received secret financial support from ultraconservative European settlers.

> . . . Anxious to pursue here the *politique du pire,* or policy of encouraging the worst possible opponents in order to discredit the opposition as a whole, many French were willing to back ben Youssef in the possibility that he would indeed replace Bourguiba the moderate. They hoped that he would then denounce Tunisia's official agreements with France and create such chaos and trouble that French troops would have to intervene and put Tunisia back into the old colonial status.[31]

MANNER IN WHICH CONTROL OF GOVERNMENT WAS TRANSFERRED TO REVOLUTIONARIES

The *Neo-Destour* revolutionaries gained full control of the Tunisian Government in March 1956. They came to power legally and constitutionally, but the constitutional framework which permitted the Nationalist party to achieve political power was the direct result of a combination of negotiations for legal political change, threats, and periods of violence which the *Neo-Destour* had employed to undermine French rule during the 22 years of the party's existence. Although France granted Tunisia its independence unilaterally, in a step-by-step process lasting from 1951 to 1956, it was the *Neo-Destour* organization which was really responsible for these emancipating moves.

The liberal government of Mendes-France granted Tunisia home rule in 1954. In the Tunisian Government set up under Prime Minister Tahar ben Ammar, the *Neo-Destour* was represented by Mongi Slim (Minister of Interior) and Bahi Ladgham who, though not an official member of the ben Ammar government, was in charge of the negotiations with the French. Armed with the Bey's repudiation of

French "protection" early in 1956 and the forthcoming election of a Tunisian national assembly on March 25th, Ladgham was able to threaten France with the prospect of a unilateral declaration of Tunisian independence by this national assembly, which all agreed would be controlled by the *Neo-Destour*. Under these conditions the French agreed on March 20th to an official protocol granting Tunisian independence. In the elections held a few days later the *Neo-Destour* received a resounding majority of 95 percent. Bourguiba became the Bey's Prime Minister, with Foreign Affairs and Defense Portfolios as well, and Ladgham became Vice Premier.

THE EFFECTS OF THE REVOLUTION

CHANGES IN THE PERSONNEL AND INSTITUTIONS OF GOVERNMENT

The most striking institutional change occurred a year after the *Neo-Destour* came to power. On July 25, 1957, the constitutional monarchy was replaced by a republican form of government. Prime Minister Bourguiba became President of the Tunisian Republic. The Bey was deposed "with a minimum of excitement or celebration."[32] He had never been a popular figure; his private misgivings about the nationalist cause were generally known and this deprived his few, though significant, acts in behalf of Tunisian independence of any positive effect of his popularity. The nationalists regarded the Beylical institution as a useful legal weapon in the fight for independence, and after independence had been won they maintained the institution temporarily because "some people felt that [the Bey's] continued presence added a note of stability, or continuity, that was necessary during the early period of organizing the government."[33] For example, some nationalists, such as Mongi Slim, Tunisia's first ambassador to the United States and the United Nations, felt that the United States would be alarmed if the monarchy were overthrown. Others felt that the Bey and the large Beylical family were a luxury the national budget could ill afford, and when the Bey's oldest son became involved in a police scandal the regime felt justified in declaring Tunisia a republic.

MAJOR POLICY CHANGES

Tunisian foreign policy was decidedly pro-Western, and until 1958, the Bourguiba government maintained close relations with France. After the bombardment of the Tunisian border village of Sakiet

Sidi Youssef in February 1958, by French military units in pursuit of Algerian rebel forces who were using Tunisian territory as a base of operations, relations between Tunis and Paris worsened. The bases of contention were the continued presence of French military forces at the naval station and air base near Bizerte, which the French occupied under treaty arrangements agreed to in 1956, and the continued dominance of the Tunisian economy by French agricultural and industrial interests. The Algerian conflict continually poisoned the atmosphere of Franco-Tunisian relations, and in the summer of 1961, Tunisia launched a military attack on the French military base at Bizerte and against French outposts in disputed territory in the Sahara.

Tunisia has consistently opposed Nasser's Pan-Arabist policies, and has appeared anxious to identify itself more closely with the emerging states of Black Africa and the Arab states of North Africa than with the Arab states of the Middle East. Tunisia was admitted to the Arab League in October 1958, but boycotted the meetings until February 1961, when it attended the meetings of the Arab League in the wake of worsening Franco-Tunisian relations. In the United Nations, the Tunisian delegation has participated actively in the Afro-Asian bloc and the Arab bloc, but always as an influence for moderation and pro-Western policies.[34]

LONG RANGE SOCIAL AND ECONOMIC EFFECTS

In social and economic matters the Tunisian revolutionaries proceeded in the same moderate but effective manner characteristic of their political revolution and their postrevolutionary foreign policy. One recent observer wrote: "The very manner of obtaining independence without undue bloodshed and through tortuous negotiation was the predecessor of post-independence Bourguibist diplomacy which has since been extended to the whole field of Tunisian domestic and foreign relations . . . Constant pressure, but with tactical retreats and readvances, persuasion, and nonviolence are its trademarks."[35] There was no nationalization of European property after 1956, although thousands of Europeans left the country as a matter of personal choice, and Tunisians stood ready to buy out the holdings of the departing *colons*. To accomplish its program of light industrialization and improvement of native agriculture, the Tunisian Government actively attempted to attract foreign capital, from both French and other foreign sources. But in so doing, the Tunisians did not wish to have undue foreign influence over their economy. The *Neo-Destour* moved further toward secularizing Tunisian society than the French had ever dared to go. The government nationalized the land of the religious foundations and abolished the Muslim *Sharia* judicial system. Tunisia became the first Muslim country to prohibit polygamy. Many of the reforms and changes that had been goals of the *Neo-Destour* remain to be carried out.

OTHER EFFECTS

The other significant aspect of the Tunisian revolution was its effective institutionalization in the postrevolutionary period of the political forces which brought about the revolution. A recent student of Tunisia described it as "a homogeneous, nationalist, and republican state, with strong tendencies toward secularization, and governed by a single party with highly effective mechanisms for influencing and controlling opinion, and led by a man with equally effective methods of mass appeal, charm and persuasion."[36] On the role of Bourguiba and the *Neo-Destour* in postrevolutionary Tunisia, another observer wrote:

> Bourguiba's strength is based not only on his charismatic personality, but on his gift for organization. His party has covered the whole country with a network of a thousand cells . . . .
>
> No other Arab leader besides Bourguiba has at his disposal such a modern "capillary" organization and power machine. But it all stands and falls with the leader.[37]

It is, therefore, important to note that the recent challenge to Bourguiba's pro-Western orientation, and continued large-scale unemployment in Tunisia, seem to have led to a slight decrease in his personal prestige, both within the *Neo-Destour* and among the people in general.

NOTES

1. Fahin I. Qubain and Milton D. Graham, *Tunisia* (Subcontractor's Monograph, HRAF–63, Johns Hopkins–5), (New Haven: Human Relations Area Files, 1956), pp. 56–58; see also Charles F. Gallagher, "Ramadan in Tunisia; Aspects and Problems of the Tunisian Republic," American Universities Field Staff Report, *North Africa Series*, VI, 1, 8–9.
2. Nevill Barbour (ed.), *A Survey of North West Africa (The Maghrib)* (London: Oxford University Press, 1959), p. 291.
3. *American Geographical Society, Focus*, VII, 1, passim.

4. Ibid., p. 5.

5. Leon Laitman, *Tunisia Today*, (New York: Citadel Press, 1954), pp. 51–56.

6. Qubain and Graham, *Tunisia*, p. 292.

7. Ibid., pp. 227–230.

8. Benjamin Rivlin, "The Tunisian Nationalist Movement: Four Decades of Evolution," *The Middle East Journal*, VI, 2 (1952), 190–191.

9. Keith Callard, "The Republic of Bourguiba," *International Journal*, XVI, 1 (1960–61), 19–20.

10. Laitman, *Tunisia Today*, pp. 196–197.

11. Ibid., p. 195.

12. Ibid., pp. 194–195.

13. Lorna Hahn, *North Africa: Nationalism to Nationhood* (Washington: Public Affairs Press, 1960), pp. 39–40.

14. Alexander Werth, *France 1940–1955* (New York: Henry Holt and Company, 1956), p. 568.

15. Hahn, *North Africa*, p. 26.

16. Qubain and Graham, *Tunisia*, p. 303.

17. Hahn, *North Africa*, p. 8; Laitman, *Tunisia Today*, pp. 44–56, 64–66, et passim.

18. Werth, *France*, p. 569.

19. Hahn, *North Africa*, pp. 8–16.

20. Ibid., p. 16.

21. Ibid., pp. 39–41, 210.

22. Captain A. Souyris, "Concrete Cases of Revolutionary War," *Revue Militaire D'Information* No. 287 (Paris: Ministère des Armées, 1957) tr. by Department of the Army, Office of the Assistant Chief of Staff, Intelligence. Intelligence Translation No. H–2060, pp. 66, 70–72.

23. Hahn, *North Africa*, p. 18.

24. Ibid., pp. 125–126.

25. Ibid., p. 102.

26. Ibid.

27. Ibid., p. 18.

28. Ibid., p. 19.

29. Ibid., pp. 39, 41.

30. Ibid., p. 104.

31. Ibid., p. 172.

32. Ibid., p. 213.

33. Ibid.

34. Gallagher, "Ramadan in Tunisia," passim.

35. Ibid., p. 15.
36. Ibid.
37. Hans Tutsch, "Bourguiba's Tunisia—I," *The New Leader*, XLIII, 9 (1960), 7.

RECOMMENDED READING

BOOKS:

Barbour, Nevill (ed.). *A Survey of North West Africa (The Maghrib)*. London: Oxford University Press, 1959. A comprehensive and factual survey of the countries of North Africa.

Hahn, Lorna. *North Africa: Nationalism to Nationhood*. Washington, D.C.: Public Affairs Press, 1960. A detailed account of nationalist movements in Tunisia, Morocco, and Algeria sympathetic to nationalist cause.

Laitman, Leon. *Tunisia Today*. New York: Citadel Press, 1954. Good source for social and economic background of Tunisian revolution.

Qubain, Fahim I. and Milton D. Graham. *Tunisia*. (HRAF Subcontractor's monograph—63, Johns Hopkins—5) New Haven: Human Relations Area Files, 1956. Contains detailed survey of geographic regions, economy, and social stratification of Tunisia, little on politics.

PERIODICALS:

Callard, Keith. "The Republic of Bourguiba," *International Journal*, XVI, 1 (1960–61), 17–36. Contains a brief survey of the revolution and a detailed discussion of political institutions and dynamics of Bourguiba's government.

Gallagher, Charles F. "Ramadan in Tunisia; Aspects and Problems of the Tunisian Republic," *American Universities Field Staff Report, North Africa Series*, VI, 1. An impressionistic but analytic account of contemporary Tunisian society by a political scientist and student of the Arab world who was in Tunis during the Muslim month of fasting, the Ramadan, in 1960. Discusses secularization of Tunisian society and international attitudes.

Neville-Bagot, Geoffrey. "Bourguiba, Prime Minister of Tunisia: A Portrait of North Africa's Greatest Statesman," *The Islamic Review* (December, 1956), 18–23. A detailed and sympathetic account of Tunisian nationalism from the beginning of the century down to

1956, with particular emphasis and focus on Bourguiba's life and role in the nationalist movement.

Rivlin, Benjamin. "The Tunisian Nationalist Movement: Four Decades of Evolution," *The Middle East Journal*, VI, 2 (1952), 167–193. An analytical and detailed study of the *Neo-Destour* Party down to 1952.

Romeril, Paul. "Tunisian Nationalism: A Bibliographical Outline," *The Middle East Journal*, XIV, 2 (1960), 206–215.

Souyris, Captain A. "Concrete Cases of Revolutionary War," *Revue Militaire D'Information No. 287*. Paris: Minestère des Armées, 1957.

Tutsch, Hans. "Bourguiba's Tunisia—I," *The New Leader*, XLIII, 9 (1960), 6–7. A brief account of Bourguiba's personal and organizational strength in contemporary Tunisian politics.

THE ALGERIAN REVOLUTION: 1954–1962

SYNOPSIS

During the night of October 31 and the early hours of November 1, 1954, bomb explosions and attacks on French military posts were reported throughout Algeria, while the Voice of the Arabs from Radio Cairo announced that the Algerian war of independence had been launched by the Algerian Front of National Liberation (FLN). In more than 7½ years of bitter and bloody fighting, with casualties estimated at over 1,000,000, the FLN developed from a small band of 2,000–3,000 militant nationalists to a revolutionary force of about 130,000. The vast majority of the Algerian people had rallied to its cause, and opposing nationalist leaders had joined its ranks. It was able to create what the French claimed did not exist: a separate and distinct Algerian entity. On March 18, 1962, the French Government of General Charles de Gaulle formally accepted this political fact, despite a favorable military stalemate, and recognized the right of the Algerian people to self-determination. On July 1, 1962, Algeria became an independent nation.

BRIEF HISTORY OF EVENTS LEADING UP TO AND CULMINATING IN REVOLUTION

In 1830 a French military expedition began the limited occupation of the coastal zone of what is today the country of Algeria. Fierce resistance by the indigenous population delayed French expansion, and by 1837 only Bône, Oran, Mostaganem, Arsew, Bougie, and Constantine had fallen. In 1840 the policy of limited expansion was abandoned when it became apparent that the safety of these cities depended on the pacification of the interior. In 1857 Algeria was finally subjugated. Pacification, however, was not achieved until 1881, and much of the unrest of this period resulted from the influx of settlers and the process of colonization. Friction between settlers and the indigenous population over land rights caused uprisings in 1871, 1872, and 1881. By the beginning of the 20th century, however, the settlers had acquired all of Algeria's most fertile lands.

The indecision of French policy also produced friction. Paris wavered between two alternatives: outright annexation or the possibility of granting Algeria some degree of autonomy. The ordinances of 1833 and 1834 proclaimed Algeria an "extension" of France. This implied that French law, without major modifications, would be applied. French citizenship, however, was not extended to the Arabs and Berbers of Algeria at this time. They continued to be subject to special police and military regulations. In 1871 a Governor General was appointed, but Paris maintained its control over Algerian affairs until 1896, when the Governor General assumed major responsibility for administration, with the exception of education and justice. In 1898 a measure of self-determination was granted: Algerians were given a direct vote in the financial, fiscal, and economic affairs of the country through the establishment of the *Délégations Financières*. The indigenous population, however, benefited very little. Denunciation of the new policy by some 500,000 settlers, who by now had become the entrenched political, economic, and social elite, forced the French Government to limit Muslim participation.

Indigenous political ferment began at the end of World War I. A small number of French-educated Muslims and former Muslim officers of the French Army demanded political equality. The disparity between the rights of a French citizen and those of a French subject became their focal issue. Returning Muslim soldiers and factory workers, on the other hand, focused their attention on economic disparity. The demand for political and economic equality influenced the development of three major movements within the Muslim community during the interwar period.

The first movement, the *Fédération des Élus Musulmans d'Algérie*, was composed of French-educated intellectuals. It sought total assimilation into France and political equality within Algeria. It never developed mass support, but such members as Ferhat Abbas and Dr. Ben Djelloul achieved influencing stature.

The second movement, the *Étoile Norde Africaine* (ENA), under the leadership of Messali Ahmed ben Hadj, sought complete independence from France while advocating Islamic-proletarian economic and social reforms. It developed more of a popular base than the movement of the intellectuals.

Alongside the ENA there developed a religious organization, the Association of Ulemas. This third movement was made up of orthodox Muslims who were offended by French controls over their religion. They shared three points in common with the ENA: independence from France, opposition to French culture, and making Arabic the official language of Algeria.

Pressure from these movements met with resistance from the settlers. The acceptance of any program, or even part of any program, sponsored by any of the three movements would have upset the special status of the European community. Response in Paris was divided.

The Conservatives and the business lobbies opposed any concession to the Muslim community, while the Liberals and the Left supported Muslim demands for equality within the French political system. With the advent of the Popular Front Government of Léon Blum in 1936, a reform proposal (the Blum-Violette Plan) was introduced in the French National Assembly to extend French citizenship to some 25,000 Muslims. But the resignation of all the French mayors of Algeria prevented the bill from being implemented.

Under the Vichy Regime, which came to power after the fall of France in 1940, the Muslim community lost many of the small benefits which it had acquired over a period of years. The settlers were given a free hand. Muslim leaders were jailed, and all of the nationalist movements were banned and members persecuted. After the Allied landings in November 1942, attempts by the free French to enlist the support of the Muslim community in the war were met by the Algerian Manifesto. In this Manifesto the nationalist leaders demanded self-determination and agrarian reforms (to solve the crisis in the rural areas where unemployment and food shortage were rampant) as a precondition to their full participation in the war. These demands were brushed aside with vague promises. In 1944 Ferhat Abbas organized the *Amis du Manifeste Algérien* to press for social reform within the French political framework, while the *Parti du Peuple Algérien* (PPA), the newly reconstituted ENA, advocated direct action in the countryside as the only way of achieving improvements. On May 8, 1945, "Liberation Day," the settlers reacted swiftly to nationalist pressure and attempted reforms by the French Government. Provoked by Muslim extremists and fearing that the violence which had hitherto marked the celebration was the signal for an uprising, the European community resorted to massive repression. Police, citizens' militia, and army invaded the Muslim sections of the major cities and at the end of the blood bath an estimate of more than 4,000 Muslims had perished. The PPA, a prime suspect, was hit hardest by the authorities. Its activity as a whole was paralyzed and its organization in the Constantine Department, the immediate area of the uprising, was all but dismantled. In the latter part of 1946 the PPA was reconstituted as a legal party, *Mouvement Pour Le Triomphe des Libertés Démocratiques* (MTLD). Ferhat Abbas, on the other hand, recreated his party as the *Union Démocratique du Manifeste Algérien* (UDMA).

At the MTLD's first congress, held in March 1947, a disagreement arose between the moderate and radical wings of the party. The former advocated abandonment of direct action in favor of cautious reform more in line with the UDMA, while the latter favored the creation of paramilitary organizations and direct action. The congress voted in favor of a policy of political activity only, and postponed the consideration of creating a paramilitary organization. In 1948, after the passage of the Algerian Statute of 1947 and the rigged elections of April 1948, a number of militants proceeded to create, with the support of Messali Hadj, an armed organization—*Organization Secrète* (OS)—within the organizational structure of the party. Its discovery by the authorities, in March 1950, split the party, and the crisis which it precipitated paralyzed party activity. When it became apparent that the unity of the party could not be restored, nine members of the OS created the *Comité Révolutionnaire Pour L'Unité et L'Action* (CRUA) in July 1954.

The considerations that motivated the creation of the OS were based on: (1) fear that political action alone would immobilize the party in legalism, when armed resistance in Morocco and Tunisia was beginning to prove effective; (2) belief that the time for resistance was at hand; (3) the failure of legal methods and the belief that armed resistance alone could dramatize the political problems sufficiently; and (4) the vulnerable position in which the party had placed itself by seeking the legal approval of an authority which denounced it as illegitimate, thus causing it to lose among the masses the benefits which only an intransigent attitude could procure.

In the latter part of October 1954 the members of the CRUA met for the last time and set November 1 as the date for the uprising. On the morning of that day they adopted a new name: *Front de Libération Nationale*. The revolution was on.

THE ENVIRONMENT OF THE REVOLUTION

DESCRIPTION OF COUNTRY

Physical characteristics

Stretching some 650 miles along the coastline of North Africa, between Morocco to the west and Tunisia and Libya to the east, Algeria occupies an approximate area of 850,000 square miles—four times the size of France and one-fourth that of the United States—of which more than 700,000 square miles form part of the Sahara Desert. Two mountain ranges, the Maritime or Tellian Atlas and the Saharan Atlas, which include the almost inaccessible Kabyle and Aures Mountains respectively and which run parallel to the Mediterranean Sea from east to west and divide the country into four broad ecological areas: the Coastal Zone, the Maritime Atlas and the Tell, the Steppe and the Saharan Atlas, and the Sahara Desert. Algeria has no major rivers, and

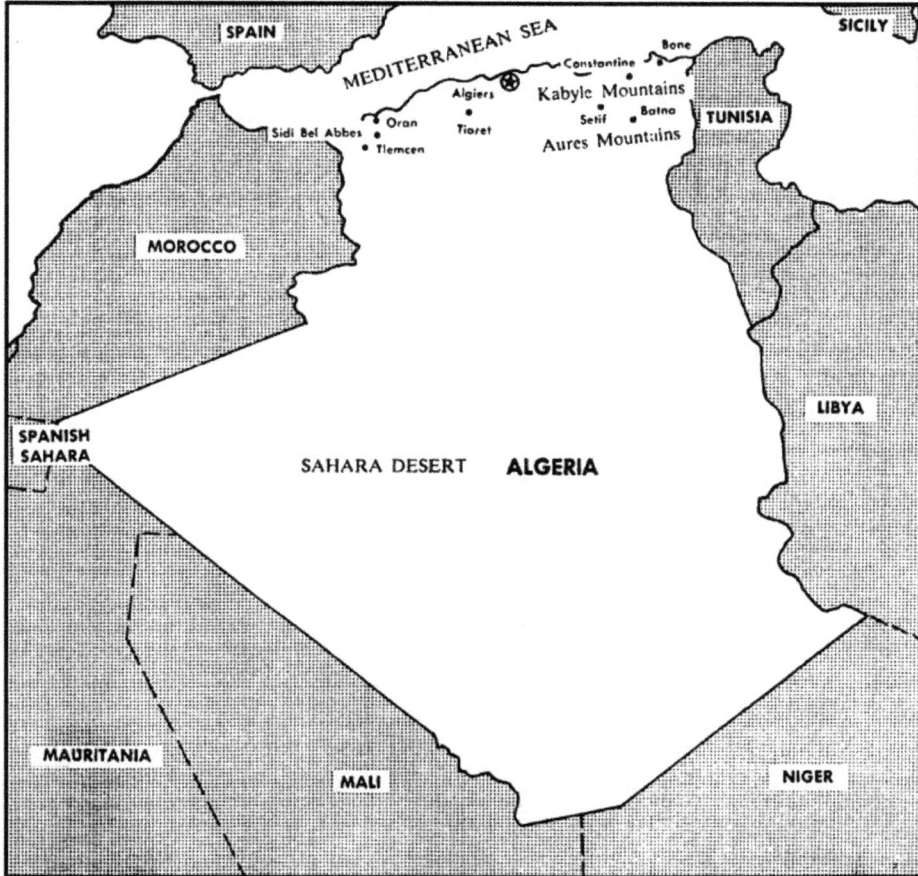

the topographical conditions of the coastal zone have deprived the country of natural harbors, although large manmade harbors have been developed in the gulfs that dot the Algerian seacoast.[1]

The people

The official census of 1954 placed the total population of Algeria at 9,528,670 inhabitants, of which 1,042,426 were Europeans and 8,486,244 were Muslims of Arab-Berber stock.[2] The European inhabitants included about 450,000 of French stock; 140,000–150,000 Jews of North African, Spanish, and Italian origin; 325,000 of Spanish ancestry; 100,000 of Italian descent; 50,000 of Maltese lineage; and a still smaller minority from Corsica. Of the Europeans, however, 89 percent were born in Algeria.[3]

French is the predominant language of the Europeans, although a small percentage speak Arabic or Berber as a second language. Arabic is the predominant language of the majority of the Muslim inhabitants; however, most of the Muslim city dwellers have achieved varying degrees of fluency in French. Berber is still spoken by a small minority.

The overall population density of Algeria was 11.3 per square mile in 1954. However, because of favorable climatic and topographical conditions in the first three ecological areas, population density has tended to vary from 138 in the Coastal Zone, where most of the important cities are located, to .8 in the Sahara Desert. Twenty-two percent of the population lived in the major cities of Algiers (the capital), Bône, Oran, Philippeville, Maison-Carree, Blida, Biskra, Sidi-bel-Abbès, Mostaganem, Sétif, and Tlemcen. The trend towards urbanization showed a marked increase over the 1948 census, but it should be noted that whereas 80 percent of the European community was urbanized, only 18 percent of the Muslims were city dwellers.

Communications

In 1954 Algeria possessed 50,000 miles of roads, 2,700 miles of railroads, no inland waterways, four major ports, and a large number of airports of varying importance. The network of roads, of which 27,000 miles were considered to be first-class routes (national highways and departmental roads), consisted of three west-east highways: "a coastal route from Nemours to LaCalle; a northern transversal from Ujda to the Tunisian Frontier, and a southern transversal from Berguent to Tebessa"[4]—running parallel to the Mediterranean Sea; and three north-south highways : a western transversal from Arzew to Colomb-Bechar; a central transversal from Algiers to Djelfa and southward, and an eastern transversal from Bône to Hassi Messaoud. The better roads, as a whole, tended to be concentrated in the densely populated areas, while most of the interior was serviced by secondary and practically impassable roads.

The railroad network, of which more than half was normal gauge, paralleled the northern transversal and all three north-south transversals, with trunk lines to service important centers located in the vicinity.

The ports of Algiers, Oran, Bône, and Philippeville handled most of Algeria's shipping, which in 1960 amounted to 22.5 million tons. With the discovery of oil at Hassi Messaoud and gas at R'Mel Hassi, the respective terminal ports of Bougie and Arzew are being enlarged to facilitate the storage and shipment of these carburents. Maritime traffic between Algeria and France, however, accounts for most of the tonnage shipped.

Three of Algeria's airports—Algiers-Maison Blanche, Bône-Les Salines, and Oran-La Senia—receive international flights, while 20 regional airports handle most of the domestic air traffic between some of Algeria's largest cities. Military airports service outlying cities as well as military outposts.[5]

Natural resources

Algeria is not richly endowed with natural resources. Adequate deposits of high-grade iron ore near the Tunisian border, rock phosphate in the Constantine region, and medium to low grade coal in the Colomb-Bechar area constitute her major mineral resources. Lead, zinc, antimony, copper, tungsten, barium sulphate, iron pyrites, and salt are also found, but in minor quantities and poorer in quality. The absence of major rivers as sources of electricity deprives Algeria of cheap sources of industrial power, but the recent discovery of oil and gas deposits in the Sahara may yet compensate for this deficiency.[6]

SOCIO-ECONOMIC STRUCTURE

Economic system[a]

The economy of Algeria is primarily and predominantly agricultural. In 1954, 32 million acres of land were considered to be arable, although only 17.5 million acres were actually fit for modern-type cultivation due to irrigation problems. Of the last, 5–7 million acres of the most fertile land belonged to some 22,000 European settlers—an average holding of 250 acres per settler—while 6,300,000 Muslim peasants, practicing subsistence-type farming, lived on the rest of the land, broken up into 600,000 holdings of about 10–12½ acres each. Thus, the agricultural output, which accounted in exports for over one-third of the national income, could support only two-thirds of the population.[7]

With the exception of a small steel-producing plant, Algeria had no heavy industry. The light industry which existed was related mostly to food processing and was owned in its entirety by European settlers or French (metropolitan) concerns with branches in Algeria. The French Government, with monopoly control over the match, alcohol, and tobacco industries, owned the railroads, air services, electricity and gas, telephone and telegraph systems, and the Oran Coal Company. The government has, however, been a prime factor in the further development of industry by providing the necessary financial support. But this support has, in turn, given the government further controls over industrial development.

---

[a] Prior to the French occupation, Algeria was an underdeveloped country. The modern economic aspects of the Algerian economy of today were almost entirely developed by Frenchmen with financial aid from France.

. . . The granting of these benefits . . . gives the government a say in the overall direction of industrial development through the selection for financial aid of the industries most suitable for integration with the French national economy.[8]

Algeria's principal imports include foodstuffs, metal manufactures, electrical apparatus, automobiles, machinery, and wood. Her major exports include wine, iron ore, citrus fruits, vegetables, cork, vegetable oils, potatoes, and esparto grass.[9]

Class structure

Although Algeria became legally and administratively a part of France, her social structure continued to reflect the stereotype divisions and lack of cohesion of a colonial society: Algerians of European origin standing apart from, and well above, the indigenous Algerian population.

The upper and middle classes of Algeria were predominantly European in composition. Nearly 7,500 *Gros Colons* (large landowners and big businessmen), high administrators, and civil servants constituted the effective social, economic, and political elite of the country, while more than 700,000 clerks, teachers, shopkeepers, and skilled laborers formed the middle class. Only about 7,500 Europeans, mainly unskilled agricultural laborers, could be classified as lower class. The Algerian indigenous population, on the other hand, could be classified as lower class, with the exception of a small number of wealthy landowners and a relatively larger lower-middle class.

The 50,000 wealthy Muslims, sometimes referred to as *Beni Oui Ouis* (yes men)[b] had practically no influence. They were completely servile to the dicta of the French administration and were not accepted in either *Gros Colons* or Muslim circles. The traditional Muslim middle class, eliminated during the 19th century, began to reconstitute itself in the early part of the 20th century. Its present average income, however, places it at a level lower than that of the European middle class; hence it is in fact a lower-middle class. In efforts to better its social standing, the indigenous society ran into the opposition of the European settlers who sought vigorously to maintain their advantageous social status.

---

[b] The word *Beni Oui Oui* is actually half-French, half-Arabic: *Beni* is the Arabic for *people* or *folks*, while *oui* is French for *yes*—hence, yes men.

Literacy and education

In 1954 more than 90 percent of all Algerians were illiterate. According to the 1948 census, most of the Europeans were literate, but only 9 percent of the Muslim male population and 2.1 percent of the females could read or write.

The educational system combined French secular schools, Arabic *Madrasas* (religious schools) and *Lycées Franco-Musulmans* (mixed institutions which offered instruction in Arabic and French). The University of Algiers was the only institution of higher learning.

Major religions and religious institutions

All of the indigenous population of Algeria professed the Islamic faith, and the vast majority belonged to the Sunni (orthodox) sect. The Europeans belonged to the Roman Catholic or Judaic faiths, and constituted the two major religious minorities. Although most of the Algerian Muslims ignored the basic tenets of their religion and adulterated it with local pagan beliefs, Islam was instrumental in creating Algerian nationalism by introducing the interchangeable concepts of Pan-Islamic and Pan-Arabic solidarity.

GOVERNMENT AND THE RULING ELITE

Description of form of government

General

The 1947 Organic Statute attempted to strike a balance between the interests of France in Algeria and the demands of the Algerians. It recognized the special political status of the country, and at the same time sought to integrate it with metropolitan France.

Algerians received some measure of self-determination with the creation of an Algerian assembly composed of two colleges—one elected by Europeans and certain special categories of Muslims, and the other elected by the indigenous population. Also, the presence of Algerian deputies in the French National Assembly along with other representatives in the French Council of the Republic and the Assembly of the French Union proportedly guaranteed Algerian interests at the national level. French interests, on the other hand, were safeguarded by the Paris appointment of a French Governor General, endowed with extensive executive powers, to head the French Administration in Algeria.

Responsibility of executive to legislature and judiciary

The Governor General, appointed "by decree of the Cabinet upon the nomination of the Minister of the Interior" as "the general administrative head of the government and representative of the French"[10] interests in Algeria was solely responsible to the French Government. His authority extended over all aspects of defense, security, and civil service activities, except for justice and education, which came under the control of the respective Ministries in Paris. The Governor General was also responsible for the implementation of all legislation enacted by the Algerian Assembly. However, by invoking Articles 39 and/or 45 of the Statute, he could veto any decision which he judged to be detrimental to French interests or beyond the competence of the Assembly.[11]

The powers of the Assembly were limited. Articles 9–12 of the Statute expressly excluded deliberation of "all laws guaranteeing constitutional liberties, all laws of property, marriage, and personal status . . . , treaties made by France with foreign powers, and in general all laws applying to military and civilian departments or posts. . . ."[12] In the financial field, however, all legislation proposed by the Assembly's Finance and General Commissions, the budget, and all fiscal modifications and new governmental expenditures, were contingent on the approval of the Assembly.

Legal procedure for changing executive and legislature

The term of office of a Governor General was indefinite, but tenure depended on the adoption, by successive French Governments, of the Algerian policy which led to his appointment, for only the French Cabinet could effect his recall. The Algerian Assembly was elected for a period of 6 years, with half of the members of each college coming up for reelection every 3 years. It met in three annual sessions of not more than 6 weeks. The Governor General, however, was empowered to cancel its sessions, and the French Government could dissolve the Assembly by decree.

Description of political process

Political parties and power groups supporting government

Most of the Frenchmen of Algeria belonged to the political parties of metropolitan France, and the vast majority of them belonged to the parties of the Center or the Right, such as the *Parti Républicain Indépendent* (R.I), *Mouvement Républicain Populaire* (M.R.P.), *Parti Républicain Radical et Radical Socialiste* (R.R.S.), and the *Parti Socialiste, Section Française de l'International Ouvrier* (S.F.I.O.). Support of the Governor General, or the lack of it, was therefore reflected in the voting record of these parties in the French National Assembly, and the appointment or removal of a Governor General depended, to a great extent, on the approval of these French Algerians. Thus, in 1947, the settlers in collusion with Finance Minister René Mayer, deputy from Constantine, and a member of the R.R.S., were able to effect the removal of reform-minded Governor General Yves Châtaigneau, and secure the appointment in his stead of Socialist Marcel-Edmond Naegelen—a man well-liked by the settlers for his antinationalist tendencies.[c]

Character of opposition to government and its policies

Opposition to the Administration and to France came from three main sources: (1) the Movement for the Triumph of Democratic Liberties (MTLD); (2) the Democratic Union for the Algerian Manifesto (UDMA); and (3) the Association of Ulemas.[d]

Opposition parties and other major groups

The MTLD was founded in the latter part of 1946 by Messali Ahmed Ben Hadj—former member of the French Communist Party and "father" of Algerian nationalism—to replace the outlawed Algerian People's Party (PPA) and its predecessor, the North African Star (ENA).[e] The structural organization of the MTLD was patterned along Communist lines. Cells formed the basic unit. These were then grouped into a *Fawdj* (group), the lowest territorial designation. Other territorial designations included the *Fara's* (section),

---

[c] For further information on the affiliation of French Algerians see Charles-Henri Favrod, *La Révolution Algérienne* (Paris: Librairie Plon, 1959), p. 106.

[d] The policies of the Algerian and French Communist Parties were, at best, ambiguous and contradictory. In the late twenties and early thirties the Communists supported the claims of the Algerian nationalists. In 1936, with the advent of the Popular Front Government of Léon Blum, the Communists, in an about-face, supported the assimilationist Blum-Violette proposal. The French Government which ordered the repressions in the wake of the 1945 Constantine uprising included two Communists: Maurice Thorez and Charles Tillon. It was the latter who, as Air Minister, ordered the aerial bombing of native villages. The Algerian Communist Party, on the other hand, denounced the uprising as Fascist-inspired, and its members participated actively in its suppression. At the outbreak of the revolution in 1954, the Communists once more denounced the nationalists. Less than 2 years later, the Algerian Communists sought to join the FLN, while the French Communist Party supported the FLN in the French National Assembly.

[e] The ENA movement was founded in France in 1925 by Hadj Abdel Kader, a member of the Central Committee of the French Communist Party, as an adjunct to that party. In 1927 Messali Hadj assumed the leadership of the movement and his Communist background left a deep imprint on the structural organization of the ENA. It is not, therefore, surprising that the PPA and the MTLD should have had structural organizations patterned along Communist lines.

Kasma (locality), *Djiha* (region), and *Willaya* (province).[f] Leadership of the party resided in a Central Committee and a Political Bureau; the commissions named by the former dealt directly with leaders at the local level through the *Kasma*, where officials for Local Organization (ROL), Propaganda and Information (RPI), Local Assemblies (ARL), Trade Union Affairs, and Finances were to be found. A General Assembly, which met on an *ad hoc* basis and represented the different sections, was convened whenever it was deemed necessary to define and approve the policy of the MTLD.[13]

The type of following which the MTLD attracted gave it its proletarian and revolutionary character. Membership consisted mostly of poor and disgruntled Algerian workers—some of whom had emigrated to France in search of work—who were always ready to resort to violence and direct action. This brought about severe repressions in the form of arrest, incarceration, and banishment—which, by necessity, imposed on the MTLD a cloak of clandestineness and secrecy. By 1954 its membership was estimated at over 14,000.

The MTLD program demanded: (1) the election by universal suffrage, without racial or religious distinction, of a sovereign Algerian constituent assembly; (2) the evacuation of Algeria by French troops; (3) the return of expropriated land; (4) Arabization of all secondary education; and (5) abandonment of French control over the Muslim religion and religious institutions.

Although this platform represented a more cautious and prudent approach to practical Algerian politics than that of the MTLD's predecessor, the PPA, it was by far the most radical of all the postwar opposition platforms, for it demanded in essence full self-determination and proletarian-Islamic social reforms.[14]

In 1946 also, Ferhat Abbas, a pharmacist from Setif and an intellectual in his own right, founded the UDMA. In 1921 he had founded the Young Algeria Movement and in 1944 he had joined in creating the Friends of the Algerian Manifesto. The UDMA was a cadre-type party, with little mass support. Its membership was chiefly drawn from French-educated Muslim intellectuals and from the professional class.

---

[f] The territorial organization of the MTLD extended to and divided France into the seven Provinces of Marseille, Lyon-Saint-Etienne, Western France, Paris and suburbs, Lille-North, and Ardennes and Strasbourg-East.

In general, the UDMA program sought federation of a free Algeria with France. Ferhat Abbas stated his position[g] in the following terms:

> Neither assimilation, nor a new master, nor separatism. A young people undertaking its social and democratic education, realizing its scientific and industrial development, carrying out its moral and intellectual renewal, associated with a great liberal nation; a young democracy in birth guided by the great French democracy: such is the image and the clearest expression of our movement for Algerian renovation.[16]

The Association of Ulemas was founded in the 1930's by Sheik Abdel Hamid ben Badis, a graduate of the Islamic Zeitouna University in Tunis. The objective of the Association was religious—the revival of Islam in Algeria through a modernization of its practices—but it took on political overtones when the teachings of Ben Badis and two of his principal assistants, Sheiks El Okbi and Ibrahimi, came into conflict with the assimilationist efforts of the French administration. These teachings, based on the Wahabi reform movement of Saudi Arabia and the doctrines of the 19th century Egyptian reformist Mohammed Abdo, tended to generate nationalist feelings by stressing the unity of the Islamic world, brought about by a common religion, language, and history, and the impossibility of unity between Algeria and non-Muslim France. Ben Badis violently attacked the Blum-Violette proposal by stating that:

> The Algerian people is not France, and does not wish to be France, and even if it wished, it could not be, for it is a people very far from France by its language, its customs, its origin and its religion.[17]

Degree and type of criticism in press and other news media

Broadcasting in general, and the 12 radio stations of Algeria in particular, were controlled by the French Ministry of Information, and therefore unavailable to the opposition parties. However, printed matter and word-of-mouth communication were used extensively, especially by the MTLD. The MTLD published two French-language newspapers—*L'Algérie Libre* and *La Nation Algérienne*, one Arabic-language newspaper, *Sawt El-Ahrar* (Voice of Free Men), and numerous

---

[g] In the 1930's Ferhat Abbas had supported direct assimilation of Algeria with France. He declared that he was French and that there was no foundation for Algerian nationalism since a historical Algerian fatherland had never existed. "We are," he concluded, "children of a new world, born of the French spirit and French efforts."[15] The defeat of France in 1940, the anti-French repressions which the Vichy Government initiated, and the Atlantic Charter influenced the adoption of this new position toward France.

pamphlets and tracts. The efficacy of this media, however, was seriously limited by (1) the high rate of illiteracy among workers, and (2) the frequent banning and censuring of the newspapers for their revolutionary tones. The use of word-of-mouth communication, in the form of weekly meetings in cafes, banquet halls, and suburban city halls, conferences, national congresses, and rallies proved more effective. The French could retaliate only by frequently imprisoning the principal leaders of that party. The UDMA published one newspaper: *La République Algérienne.* Its tone was generally moderate, and only occasionally was it censured or banned. Ferhat Abbas contributed articles and speeches.

Other overt opposition

The Association of Ulemas used their *Madrassas* to oppose French assimilation and integration plans. By stressing the Arabic language as the mother-tongue of all Algerians and Islam as a uniting and distinct religion, the Ulemas maintained the elements of a traditional society. By further providing for the free education of their promising students in Middle Eastern universities, the Ulemas sought to imbue future Algerian leaders with Pan-Arabic and Pan-Islamic spirit.

Legal procedure for amending constitution or changing government institutions

The Algerian Assembly was not empowered to amend the 1947 Statute or make changes in government institutions. These were the responsibility of the French National Assembly, in which Algeria was represented by 30 Deputies.

Relationship to foreign powers

Because Algeria was legally and administratively part of France, all treaties and agreements entered into by France with foreign powers were binding on Algeria. Thus, Algeria was integrated into NATO and formed part of its southern flank.

The role of military and police powers

Because of its special status, Algeria formed part of the 10th Military Region of the French defense establishment, and the Algerian police formed an "integral part of the French police"[18] system. The commanding general of the 10th Military Region and the director of the Algerian police were answerable to the Governor General, but were directly responsible to the Chief of Staff of the Armed Forces and the Minister of the Interior respectively. The role of the military and the police overlapped in many instances, and at times the country came "alternatively under military and civilian control as the situation . . . fluctuated between comparative quiet and downright disorder and rebellion."[19]

Prior to the 1954 revolution, the task of the military consisted of defending Algeria from external invasion, quelling internal revolts and uprisings, and administering the southern remote areas, including the Sahara. In the small cities and towns of these areas, the police force, though not a part of the army, nonetheless took its orders from the army. To deal with the outlying oases and nomadic tribes the army created the *M'khazniya* (camel corps) as an integral police force. After 1954, however, the role of the army was expanded to include the enforcement of law in conjunction with the police.

The Algerian police system came under the direction of the *Sûreté Générale* (Public Security), and was headed by a *Directeur de Sûreté*. Its mission was the enforcement of law in the cities. The police force in these cities was headed by a *Commissaire Central* (chief of police) and included a "Criminal and Investigation Department, Narcotics and Vice Squads, Traffic Police, Information contingent, Antisubversive detail and the Harbor Police." The *Direction de la Surveillance du Territoire* (secret police) formed the other part of the *Sûreté Générale*. Its members were recruited from police ranks and the Armed Forces, and its mission was to "keep watch on subversive activities and conduct counter-espionage."[20]

The *Gendarmerie* (constabulary) assumed the function of the police in rural areas. However, the *Gendarmerie* differed from the police in that it formed part of the army administration and took its orders from the Director of Public Security only in times of peace. During crises, the *Gendarmerie*, composed mainly of "ex-soldiers or serving soldiers who have exceeded their minimum required term with the colors," would be placed on a war footing and have its command unified with that of the army.[21]

WEAKNESSES OF THE SOCIO-ECONOMIC-POLITICAL STRUCTURE OF THE PREREVOLUTIONARY REGIME

History of revolutions or governmental instabilities

Revolts, resistance, and uprisings have plagued Algeria from the inception of French rule in 1830, and the resulting influx of European settlers. Insurrection in the east and west of Algeria, headed by Ahma Bey of Constantine and the Emir Abdel Kader respectively, greeted the fall of Algiers. It became widespread when Kabyle and Aures tribesmen joined, and this full-scale insurrection was not subdued until 1850. In 1864 an insurrection which was to last for 7 years broke out in the southern Oran region, and was followed in 1871 by another in the Kabyle Mountains. The uprising of 1881 was the last of the 19th century. In the early 1900's resistance to French presence took the form of large-scale emigration of native Algerians to the Middle East. In the 1920's and 1930's, resistance turned to political agitation, generally sporadic, disorganized, and lacking in focus. A number of indigenous parties and movements were created, ranging in their demands from full independence to equal status with the European inhabitants of the country. The abortive uprising of May 1945 climaxed more than 25 years of political agitation, and the severity with which this incipient uprising was suppressed widened the gulf between the indigenous population and the French administration and France. However, many of the causes of the 1954–62 revolution can be traced to the *immobilisme* (systematic opposition to progress) of French rule, and the continued attachment of France to its traditional colonial policy, best described by Charles-Andre Julien as the "Politics of Lost Occasions."[22]

In Algeria the political weaknesses included a low degree of political integration, discontinuities in political communication, and uneven "reach" of political power. The Organic Statute, which purported to give the Algerian Muslims some measure of self-determination, was never fully applied. Meaningful application was circumvented by fraudulent elections, frequent suspensions, and disparity in representation, whereby nine million natives equalled one million Europeans. The outcome of the 1945 uprising had estranged the Muslims. The urgent passage of the Statute was, in itself, an attempt to redress the situation. But the establishment of two colleges perpetrated the estrangement, whereas the creation of a single house elected on the basis of universal suffrage without ethnic and religious distinction would have led to greater political integration. The fraudulent election of *Beni Oui Ouis,* who obviously did not represent the Muslim masses, prevented the emergence and understanding of Muslim aspirations, and placed the attainment of political power beyond the reach of the true Muslim political elite. Thus the Muslim masses were not fully represented in the Algerian Assembly and had, in reality, no voice in the administration of their country.

Economic weaknesses

Algeria suffered from low productivity, large-scale unemployment and under-employment among the Muslims, and a tremendous population growth. It produced only enough foodstuffs to feed two-thirds of its population. This was due, in part, to the scarcity of fertile land and to the failure of the administration to integrate the Muslim farmers fully into the modern economic system. Increases in agricultural productivity did not keep step with the tremendous population growth—estimated in 1954 at more than 250,000 births per year, and the majority of the Muslim farmers, using archaic tools and outdated methods, worked plots of land too small and too poor to yield even minimum basic food requirements. Thus a great majority of them were forced to seek employment as agricultural laborers on the large European combines or as industrial laborers in Algeria or France. These, however, proved to be partial solutions only: employment on combines was seasonal, giving rise to under-employment, while the combined industries of Algeria and France could absorb only 600,000 Algerians. In 1954 unemployment was estimated at 500,000, but the figure exceeded 2,000,000 when the under-employed of both sexes was added.[23]

Social tensions

Social tensions in Algeria arose from two main factors: (1) the superiority complex of the settler and his contempt for the Muslim; and (2) the settlers' fear of the future revenge of the Muslim. The settlers have "remained, through atavism, that which their fathers were at the beginning of their settlement in Africa, pioneers, men of action and isolated men . . ."[24] In 1892 Jules Ferry, the French Prime Minister, described the settler in the following terms:

> . . . He is not wanting in virtues; he has all those of the hard worker and patriot; but he does not possess what one might call the virtue of the conqueror, that equity of spirit and of heart, and that feeling for the right of the weak . . . It is hard to make the European *colon* understand that other rights exist besides his own, in an Arab country, and that the native is not a race to be enslaved and indentured at his whim.[25]

The settlers derived their sense of racial superiority from the fact that they were able to achieve, in the span of a few years, what the indigenous population had been unable to achieve in centuries: a modernized country.[h] This ability to develop Algeria gave them economic, political, and social preeminence—a special status which they have sought to preserve at all costs. They were, therefore, interested in preserving the *status quo*, and effectively blocked major legislation which tended to improve the lot of the Muslims and thus upset the *sta-*

---

[h] Modernization in Algeria is limited to a few urban centers. The great bulk of the country is still underdeveloped.

tus quo. The propaganda aspect of nationalistic agitation selected the settlers as a prime target. Reports received by Governor General Naegelen from French administrators and prefects, late in 1948, revealed that nationalist slogans—the suitcase or the coffin; the French will be thrown into the sea; we will divide the lands of the *Colons*, every one of us getting his share—had not only created a feeling of insecurity, but had engendered a civil war psychosis among the Europeans which manifested itself in tendencies to leave Algeria for France, or to arm and fortify their communities.[26]

Government recognition of and reaction to weaknesses

The French Governments initiated numerous legislative programs which attempted to remove these inherent weaknesses, with particular emphasis on the economic aspects. To increase agricultural productivity, the system of large plantations was extended to cover more than 50 percent of the arable land. Modern agricultural equipment and cultivation techniques were utilized, and increased irrigational and land reclamation projects were introduced to expand the amount of acreage under cultivation. To better the economic conditions of the indigenous population, wages of agricultural laborers were increased, and new employment outlets were created in the urban centers to handle surplus labor from the rural areas. Unemployment compensation was also introduced. However, these programs were offset, in the main, by a galloping population growth, exhaustion of the available arable land through intensive cultivation, and the fact that the fertile areas remained in the hands of the settlers.[27]

In the political field practically all of the programs met with dismal failure. In 1865 Napoleon III began a policy of *rattachement* (assimilation)[i] by granting all Muslims and European Algerians French citizenship. In 1892 it became apparent that this policy had failed. The Muslim population became French subjects, but not citizens; ". . . they retained their personal status, had no political rights, and continued to be subject to special police regulations."[28] In 1898 the French Government created the *Délégations Financières* in an attempt to give the indigenous Algerians a voice in the financial, fiscal, and economic policy-making. Settler denunciation forced the government to limit Muslim participation to a handpicked one-third of the total number of delegates. Under the Algerian Charter of 1919 the French Premier Georges Clemenceau attempted to give the Muslim full voting rights. It was defeated by a coalition of rightist deputies and

---

[i] The policy of *rattachement* aimed at making Algeria equal to any of the provinces of France. This implied the application of the French Administrative system in all its aspects and without modification, to Algeria.

settler lobbyists, and a watered-down version of the charter gave the Muslim the right to elect the members of the Financial Delegations only. In 1936 the implementation of the assimilationist Blum-Violette proposal, which would have given some 25,000 Muslims French citizenship, was blocked by the resignation of all the European mayors of Algeria. The MTLD won a sizable victory in the municipal elections of Algiers in October 1947. Fearful of a similar victory in the January 1948 national elections—the first national election to be held, by virtue of the Organic Statute, for the purpose of electing delegates to the Algerian Assembly—the administration postponed the election until April, and then proceeded to manipulate the returns so as to ensure the overwhelming election of its handpicked candidates. Full voting rights and representation were, therefore, continually denied the indigenous population, and this denial prevented a lessening of social tensions, which were intertwined with political weakness. Political dominance of the settlers represented one way of maintaining their special social status.

FORM AND CHARACTERISTICS OF REVOLUTION

ACTORS IN THE REVOLUTION

The revolutionary leadership

The initial leadership of the revolution was composed of Hussein Ait Ahmed, Mohammed Ben Bella, Mohammed Larbi Ben M'Hidi, Mohammed Boudaif, Mustapha Ben Boulaid, Rabat Bitat, Mourad Didouche, Mohammed Khider, and Belkacem Krim. This group was essentially drawn from the lower-middle and lower classes, and did not include intellectuals or politicians of statute. They were all in their late twenties or early thirties, had been militants in the PPA and MTLD, and a number of them had gained experience in warfare while serving in the French Army (Ben Bella, Boudaif, and Krim had risen to the rank of sergeant, and had served in various European campaigns). Apart from the various positions of importance which they had held in the MTLD or its covert paramilitary branch, the *Organization Secrète* (OS), Khider was the only one of the group to hold the official position of deputy from Algiers to the French National Assembly. During the course of the revolution new members were to be added to the revolutionary leadership, many of whom, such as Ferhat Abbas and Ahmed Francis, came from other political parties.

Politically, this group believed in a one-party system as the best way of achieving unity of action and purpose. They favored democratic centralization and collective leadership within the party, and rejected the idea of a single party ruler or charismatic leader. The unity of the party was to be "sought at the base rather than the higher echelons of leadership." However, "unity was not required before seizing initiative; on the contrary, action might be the best way to bring all patriotic Algerians together."[29] Ideologically they adopted some of Mao Tse-Tung's theories on revolutionary warfare—a revolution of the broad masses of the nation was the only way in which a nation might be regenerated—and expressed their belief in land redistribution, state direction, and nationalization of public utilities.

The revolutionary following

The revolution was launched by 2,000–3,000 ex-members of the MTLD and the OS, and with practically no popular support. By 1962 the ranks of the revolutionaries had swelled to an effective force of some 40,000–60,000 regulars,[j] and the FLN enjoyed tremendous popular support among the Muslims. Political parties such as the UDMA and the Association of Ulemas, which had disassociated themselves from the revolutionary movement in 1954, voluntarily disbanded, and urged their members to join the FLN as individuals. The population, which had been apathetic at the beginning, enthusiastically supported the FLN directives to strike and demonstrate.

ORGANIZATION OF REVOLUTIONARY EFFORT

Internal organization

The OS, the precursor of the FLN, was founded in 1947 by Ait Ahmed, in conjunction with Ben Bella, Krim, Oumrane, Boudaif, Boulaid, Ben M'Hidi, Ben Tobbal, Bitat, and Zirout. At this stage it was headed by a national chief (Ait Ahmed: 1947–1949; Ben Bella: 1949–1950), who was assisted by three regional chiefs, entrusted with the task of directing and supervising the Organization in their respective areas. These men, in cooperation with Mohammed Khider of the MTLD Central Committee and OS liaison with that party, had the responsibility of defining and expanding the structural organization of the OS. Territorially, the Administrative Departments of Algiers, Constantine, and Oran formed the three main regions of the OS. The region of Algiers was subdivided into three zones, while the regions of Constantine and Oran were allocated only two. These zones were

---

[j] The FLN put the number at over 130,000. Though exaggerated, this number could have included the auxiliary and irregular fighters and the members of the terrorist groups in the urban centers.

equally subdivided into sections or localities, sections into groups, and groups into half-groups. The half-groups, composed of two men and a leader, formed the basic unit of the OS paramilitary force, which by 1950 totaled 1,800 men organized mainly as infantry. This force, moreover, included materiel, transmission, pyrotechnics, and medical-support sections.[30]

In 1950 the French authorities discovered and crippled the OS. Ben Bella, Ben Boulaid and Zirout were arrested. Ait Ahmed fled to Cairo and was subsequently followed by Khider and Ben Bella, after the latter had escaped from jail. Most of the other leaders took to the hills, where they were rejoined by Ben Boulaid and Zirout, who had also succeeded in escaping from custody. In July 1954 these men reconstituted the OS as the Revolutionary Committee for Unity and Action (CRUA)[k] with modification in territorial organization and leadership. The territorial organization of the MTLD was adopted and a system of collective leadership introduced. Ben Bella, Khider, Boudaif, and Ait Ahmed formed the External Delegation, or political leadership, with headquarters in Cairo, Egypt. Ben Boulaid, Mourad Didouche, Bitat, M'Hidi, and Krim, in the *willayas* of the Aures, Constantine, Algiers, Oran, and the Kabyle respectively, formed the Internal Regional Delegation, or military leadership, and were jointly responsible for the future conduct of the revolution. In the latter part of October 1954, the internal delegation of *willaya* chiefs met for the last time in Algiers and set the date for the revolution: November 1954. With the outbreak of the revolution, the CRUA changed its name to FLN.

In August 1956 important military and political modifications resulted from the FLN Soutmmam Valley Congress, which was held at the request of the Internal Delegation. The rebel forces were formally designated as the Army of National Liberation (ALN), and a regular

---

[k] The Central Committee of the MTLD, under the chairmanship of Hussein Lahouel, had dissolved the OS in 1950, and had publicly renounced the use of direct action. To underscore this decision the Central Committee concluded an alliance with the Algerian Front, composed of the UDMA, the Algerian Communist Party, and the Association of Ulemas. Sensing a threat in the action of the Central Committee to his absolute leadership, Messali Hadj chose to take an opposite stand. Over the protest of the Central Committee, Messali Hadj launched a personal tour of Algeria which resulted in demonstrations and clashes. Deported to France in 1952 for subversive activity, Messali Hadj then proceeded to convene at Hornu, Belgium on July 15, 1954, an MTLD congress which voted him full powers and excluded the Central Committee. The latter retaliated by convening another congress in Algiers, on August 13, 1954, which invested it with full powers and declared the unique leadership of Messali Hadj to be outmoded. When it became clear that the partisanship between Messalists and Centralists had immobilized the activities of the MTLD, Mohammed Boudaif attempted to effect a reconciliation between Messali Hadj and the Central Committee. Messali Hadj remained adamant in his demands for full powers and a vote of absolute confidence. Disgusted with both factions, Boudaif called a meeting of old OS members in Berne, Switzerland in July 1954, from which the CRUA resulted.

command structure was established. Six theaters of operations, or *willayas*, were created, commanded by colonels and corresponding to the territorial divisions of the CRUA. The *willayas* were then divided into zones, regions, and sectors, "in which operated battalions of 350 men, companies of 110 men, sections of 35 men, and groups of 11 men."[31]

A five-man general staff, the Committee for Coordination and Execution (CCE), composed of Ramdane Abane, Krim, Zirout, Benyoussef ben Khedda, and ben M'Hidi headed this military structure. A year later the CCE membership was enlarged to include members of the External Delegation, and the Committee was given broad executive powers.

The creation of the National Council for the Algerian Revolution (CNRA) represented the major political decision reached at Soummam. Membership consisted of 17 full members and 17 associates who represented all of the factions within the FLN, and included the nine members of the old CRUA, three leaders from the Internal Delegation—Zirout, Abane, and Oumrane—and three leaders from the External Delegation—Ferhat Abbas, Benyoussef ben Khedda, and Mohammed Yazid. Originally, it was given authority over the CCE, but the inclusion of a number of CCE members in its ranks, as part of the general expansion of its membership in 1957, tended to diminish its powers.

On September 19, 1958, the FLN created the Provisional Government of the Algerian Republic (GPRA), which included the Cabinet posts of Premier, several Vice-Premiers, and the Ministries of the Armed Forces, Interior, Communications and Liaison, Arms and Supplies, Finance, North African Affairs, Foreign Affairs, Information, Social Affairs, and Cultural Affairs. Decisions reached by the GPRA were binding on all members of the FLN in and out of Algeria.[32]

External organization

The External Delegation was entrusted with the task of supporting the revolutionary movement in all its aspects. Located in Cairo, it established bases of operations in Libya, Tunisia, and Morocco for the purpose of supplying and training rebel fighters, and sent roving representatives to "friendly" states to secure financial, military, and diplomatic aid. The cause of Algerian nationalism had initially been espoused by the Arab League as far back as 1948. Algerian nationalists received financial aid and military training in Egypt, Syria, and Iraq, prior to and during the revolution. Libya, Tunisia, and Morocco provided the FLN with safe bases and acted as sponsors for the Front on the international scene and in the United Nations, while attempting to mediate FLN-French differences. Yugoslavian aid in military and medical equipment, along with the financial help of Communist China and some of the "neutral" African nations began in 1957. The United States allowed the establishment of an Algerian Office of Information on its territory, granted GPRA representatives visas to attend the debates on Algeria in the United Nations, and, in 1961, undertook the medical rehabilitation of a number of ALN soldiers.

GOALS OF THE REVOLUTION

Concrete political aims of revolutionary leaders

The main political goal of the revolution itself was national independence and the "restoration of the sovereign, democratic and social Algerian state within the framework of Islamic principles." The internal objectives of the political program of the FLN called for: (1) "Political reorganization by restoring the national revolutionary movement to its rightful course and by wiping out every vestige of corruption . . . ; and (2) The rallying and organization of all the sound forces of the Algerian people in order to liquidate the colonial system." The external objectives called for: (1) "The internationalization of the Algerian problem; (2) The fulfillment of North African unity within the natural Arab-Muslim framework; and (3) Within the framework of the United Nations, the affirmation of active sympathy with regard to all nations supporting [the] liberation movement."[33]

The political aims of the FLN leadership, however, were: (1) To gain the support of the Algerian masses and that of influential Algerian leaders; (2) create a cleavage between the Algerians and the French, thus establishing the concept of an Algerian nation as a separate and distinct entity; (3) to become the only interlocuteur valable (contracting party) for the Algerian nation; and (4) to force France to recognize the separateness of the Algerian nation—hence the emphasis on sovereignty rather than independence in the political program.

Social and economic goals of leadership and following

During the revolution the FLN leadership refrained from committing itself on economic or social goals. Emphasis was placed on the political aspect, with the understanding that economic and social programs would be initiated at the successful conclusion of the revolution. From personal statements made by different FLN leaders, however, it became apparent that no accord could be reached between them. Demands within the FLN ranged from proletarian socialism to ultraconservatism in both the economic and social fields.

REVOLUTIONARY TECHNIQUES AND GOVERNMENT COUNTERMEASURES

Methods for weakening existing authority and countermeasures by government

Objectives

The primary objective of the FLN was to keep the uprising alive, and develop it from mere rebellion to full-scale civil war. The uprising, launched by a small number of ill-equipped and isolated Algerians, scattered in small bands over the Algerian expanse, yielded to the FLN very little materially. However, it signalled a decisive turn of events in Franco-Algerian relations by bringing the Algerian nationalist movement out of its paralysis. If the uprising could be kept alive, it would ultimately leave the Algerian nationalists with two choices; side with France, or actively support the FLN. Attacks, therefore, of the November 1 type were not continued. With the element of surprise gone, such attacks against an alerted French Army and other security units, stationed in and around the urban Algerian centers, would have resulted in the annihilation of the ALN, and the FLN movement with it. The task of the ALN, in that stage, was to fall back on the practically inaccessible rural areas, where French influence was virtually nonexistent, engage in guerrilla warfare to give effective demonstrations of its continued existence, organize and develop its structures, and recruit the local population in the cause of the FLN.

Illegal techniques

Guerrilla warfare

ALN strategy appears to have drawn on—and reversed—the *tache de l'huile* (grease spot) strategy of Marshal Lyautey of France. Lyautey had succeeded in pacifying Morocco in 1925 by massing his troops in settled areas and then spreading "in widening circles a French peace."[34] The ALN, on the other hand, planted a few rebels in remote villages outside the French peace. The task of these rebels was to win over and indoctrinate the villagers, thus enabling the ALN to acquire recruits, food, and hiding places. From these initial "grease spots" the penetration would move, in widening circles, to neighboring areas, eventually reaching the settled outposts under French control. Here the French hold on the Muslim population would be broken (1) by rebel guerrilla and terrorist action (direct attacks on French troops; assassinations; bomb throwing; strikes; boycott of French goods, settlers, and Francophil Muslims; and economic sabotage) which would force the French administration to evacuate the area; or (2) as a result of the severity of French repression which generally followed such attacks and which tended to cast the population with the rebel camp. In the urban centers and cities, which were mainly inhabited by settlers and which were strongly defended by French forces, terrorism alone was used.[35] Terrorism aimed at Francophil Muslims, rural Muslim politicians, and settlers, created an atmosphere of anxiety favorable to the FLN. It silenced the Francophil Muslims, drove a wedge between settlers and Algerians, and forced the French administration to adopt sterner security measures, which meant that more troops were tied down defending the cities and maintaining order.

In the early stages of the revolution, guerrilla action was generally uncoordinated. Small units of less than 10 men, armed mostly with shotguns and obsolete rifles, engaged isolated patrols only for the purpose of seizing their weapons. By 1956, when the ranks of the ALN had swelled and armaments had become more readily available, these guerrilla units were organized into sections of 35 men and light companies of 80 men. In 1957 and early 1958 coordinated major engagements at the battalion level were reported in the Collo Peninsula, El Milia, and Kabyle, Ouarsenis, and the northers section of the Department of Oran. In the second half of 1958, however, when these major engagements proved too costly to the ALN, the large units, heavy and vulnerable, were broken up and reorganized into light, self-sufficient, and highly mobile commando units.

Hit-and-run tactics "against the shifting fringe of French strength,"[36] as by units seldom larger than a company, characterized ALN action. Daylight and fair weather combat in which the overall superiority of the French forces could be brought to bear, were avoided. Guerrilla ambushes, attacks on French units, convoys, and outposts, and acts of sabotage took place, in most instances, at night and during bad weather; and in almost every engagement they were able to achieve surprise due to the help of the civilian population.

> In every operation, the Algerians enjoy a basic advantage: their seemingly omnipresent civilian auxiliary, who serve as 'human radar,' scouts, intelligence agents, and guides.[37]

These guerrilla units acted on the population through persuasion and the use of terror. Their main targets were the recalcitrant rural communities that refused to submit. In these communities the local leadership, generally French-appointed and therefore presumed to be hostile, was eliminated by assassination, and the population was then forced to pay taxes, provide recruits and supplies, and participate in acts of terrorism and sabotage. Once compromised, these communities had no alternative but to make common cause with the FLN, thus providing the ALN units with safe bases of operations and the necessary lookouts and informants. However, these units ingratiated themselves with the local population, thereby gaining their continued confidence, by providing them with an efficient administration which settled their feuds, gave them protection from neighboring raiders, and, in many cases, established elementary schools and medical clinics.[38]

Terrorism

Terrorism took the form of intimidation, assassination, and indiscriminate bombing. Francophil Muslims and rival nationalist leaders were at first warned by letters, bearing ALN letterhead and crest, to desist from cooperating with the French administration, or cease political activity not in conformity with FLN directives. Those who persisted were assassinated, and the order of execution, bearing ALN letterhead and crest, was left attached to the victim. This method, in effect, silenced opposition to the FLN and weakened the position of the French administration by depriving it of the support of some of the Muslim population and leaders. At the same time, it added to the prestige of the FLN among the masses because it tended to prove the effectiveness of the organization.

Indiscriminate bombing—lobbing of hand grenades into crowds, and the placing of delayed-action bombs in streetcars, cafes, stadiums, etc.—was aimed at the European population. It created an atmosphere of anxiety and suspicion which deepened the cleavage between the two main communities and made cooperation almost impossible. It also brought about violent measures of repression, which further antagonized the Muslim population and necessitated the stationing of more troops in the cities, thus relieving some of the pressure put on the ALN forces in the rural areas.

Propaganda

Propaganda was given special consideration by the FLN in and out of Algeria. Political officers were attached at all levels of command in the ALN to indoctrinate the soldiers and the public. Special broadcasts, beamed from Tunis, Cairo, and Damascus, and the constant distribution of leaflets, tracts, and the weekly FLN newspaper *El Moujahid* (The Fighter) constituted some of the other techniques. Propaganda appeals differed. To the educated Muslim the FLN attempted to explain the causes of the revolution with historical, economic, social, and political references. To the masses the FLN represented the revolution as a holy war in defense of Islam and the Arab heritage. In France the FLN relied on sympathetic journalists and writers to write books and pamphlets supporting its cause. It made use of such existentialist writers as Jean-Paul Sartre and a number of church groups who were horrified at the repressionary measures of the French Army. The main target of FLN propaganda were the French liberals. Propaganda appeals attempted to convince them that the war in Algeria was an unjust war and that the FLN action in Algeria arose from the same principles and aspirations that had led to the French Revolution.

The FLN opened offices of information in the major capitals of the world. Through them propaganda literature was made available to the public, while the staffs of these offices made every attempt to establish contacts with the press and important officials, and took every opportunity to expound the FLN cause in public speeches and debates. It was hoped that these influential groups would be able to exert pressure on their respective governments either to support the FLN directly or persuade France to negotiate.

Methods for gaining support and countermeasures taken by government

Counterinsurgency

The Algerian revolution caught France totally by surprise. The quiet years that had followed the May 1945 uprising had lulled the French administration into a false sense of security. Accordingly, the extent of the new revolt and the number of troops required to quell it were grossly underestimated. The French administration was convinced that it was facing another tribal uprising which could be crushed in a matter of a few months. When it became apparent that this was indeed a revolution, the French Army found itself unprepared. It lacked units in France suitable for this kind of warfare. The veterans of the war in Indochina had not yet returned, and the units that had been initially sent to Algeria were unable to cope with the situation. They were NATO-type divisions, created for a European war. Heavy and massive, equipped to fight a frontal war, they proved to be unadaptable to the geographic conditions of combat in Algeria, and ineffectual against the extremely flexible techniques of guerrilla warfare.[39] In February 1955 Jacques Soustelle, the newly appointed Governor General, described the military situation in these terms:

> . . . the resistance to terrorist aggression disposed of very feeble means: regular troops were few and poorly trained for the purpose; little or no extra means; practically no helicopters, few light aircraft, almost no radio equipment . . .[40]

During the first 15 months of the revolution, the French Army resorted to small-scale combing operations. Several battalions were noisily massed to encircle and search a given area where guerrilla action had taken place, while the *Gendarmerie* arrested all known nationalists, regardless of political affiliation, and disarmed all of the clans and tribes, leaving the pro-French defenseless and at the mercy of the ALN. These classical methods of fighting a tribal uprising yielded almost nothing, and served to alienate more and more Muslims.

In April 1956 the French Army adopted new countermeasures. The pacification of Algeria was to be achieved by applying *quadrillage* tactics—"a grid operation garrisoning in strength all major cities and, in diminishing force, all towns, villages, and farms of Algeria.[41] Accordingly, French effectives were increased to 400,000 men; supersonic jet fighters were replaced with slower ground-support planes and helicopters; the Tunisian and Moroccan borders were thoroughly fenced off to cut the supply lines of the ALN; and areas of heavy ALN concentration were declared security zones. The inhabitants of these security zones were moved to resettlement camps; all villages and hamlets were burned; and only French troops were allowed in the security zones, with orders to shoot anything that moved. The tracking down of ALN units was then left to small and mobile handpicked units, generally paratroopers, whose total number never exceeded 50,000 men—a number roughly equal to the effectives of the ALN. The telling effect of these tactics soon prompted the FLN to step up terrorism in the cities, in an obvious attempt to relieve the pressure on ALN units, and in an effort to break French attempts to pacify the population.[42]

Antiterrorist techniques

The "Battle of Algiers" represents the best example of antiterrorist techniques. In January 1957 a census of the inhabitants of Algiers was taken, and new identity cards were issued which gave name, address, occupation, and place of business. Shortly thereafter, a regiment of the 10th Paratroop Division sealed off the Casbah—the Muslim sector of the city—and checkpoints were established at the gates and at all strategic intersections within, with telecommunications between checkpoints. All persons wishing to leave the Casbah had to clear these checkpoints, while paratroop units conducted, at random, *ratissages* (raking operations), in which "homes were cordoned off and searched for suspicious persons, literature, and arms stores." The paratroopers then proceeded to establish a network of *îlots* (small islands). "Under this system, one person in each family group was designated as responsible for the whereabouts of all other members of the family. The responsible family man on every floor in every building was in turn responsible to a floor chief. Each floor chief was responsible to a building chief; all building men to a block leader, and so forth, until there was a chain of command stretching down from French headquarters to the bosom of every family. With this system in operation, the French command could lay hands on any Moslem in the Casbah within a matter of minutes." By September 1957 Algiers was completely under French control, and this system was applied to all major cities until terrorism had come to a halt.[43]

Civic action

It was not until 1958, when the psychological advantage had passed to the FLN, that the French Army was able to stalemate the military situation and proceed to gain the support of the Muslim population through civic action.

In order to win over the population, the French Army began a large-scale reform movement in the rural areas. It created the Specialized Administrative Sections (SAS) and the *Centre d'Instruction à la Guerre Psychologique* in Paris, which was followed by the formation of *Bureaux Psychologiques* in Algeria.

The task of the Specialized Administrative Sections was to bring the French administration into the remote rural areas with the *quadrillage* system, "engaging in all sorts of rehabilitation and construction projects."[44] French Arabic-speaking officers were specially trained in Algerian affairs and then sent to SAS installations. There they built schools, supervised the educational system, provided the population with medical and dental services, initiated work projects to provide employment for the inhabitants, and arbitrated local complaints. Another important aspect of SAS officers' duty was the raising of self-defense (*Harka*) units. These units, recruited from the population, were designed to give their respective installation protection, but they occasionally participated in large anti-guerrilla operations. In pacified areas SAS installations came under civilian authority, but in borderline zones military control prevailed. As a whole, the effectiveness of these installations depended on the continued presence of French authority. When this authority was withdrawn, many returned to the FLN camp.

The Psychological Bureaus were designed to support military and SAS operations. Three Loudspeaker and Leaflet companies were stationed in Algeria, and were responsible for loudspeaker appeals and leaflets that extolled the greatness of France, France's respect for Algerian Muslim institutions, and France's genuine desire to create a new integrated Algeria.

MANNER IN WHICH CONTROL OF GOVERNMENT WAS TRANSFERRED TO REVOLUTIONARIES

On March 18, 1962, after protracted negotiations, an agreement was concluded between the French Government and the FLN at Evian-les-Baines, France. Both parties agreed to order a cease-fire on March 19, and France recognized the right of the Algerian people to self-determination. By virtue of this agreement, a transitional period was to follow the immediate conclusion of the treaty, at the termination of which a national referendum was to be held to determine the future of Algeria. Until self-determination was realized, a provisional government and a court of public law were to be set up to administer and maintain law and order. A High Commissioner was to represent France and was to be responsible for the defense of the country and the maintenance of law and order in the last resort. Also included in the terms were provisions for a general amnesty, guarantees for individual rights and liberties, and clauses concerning future cooperation between France and Algeria, and settlement of litigations and military questions.

THE EFFECTS OF THE REVOLUTION

CHANGES IN PERSONNEL, GOVERNMENTAL INSTITUTIONS, POLICY, AND LONG RANGE SOCIAL AND ECONOMIC EFFECTS

On July 1, 1962, the national referendum was held in Algeria and the overwhelming majority of the population voted for independence. Shortly thereafter, members of the GPRA began to return to Algeria. As of July 15, however, the internal situation remains unclear. A split in the ranks of the FLN has paralyzed all aspects of governmental activity. Only civil order continues to be maintained. No new governmental institutions have been created, and no major changes in personnel and policy have occurred. In the absence of well-developed associational groups, the role of force might very well become paramount in determining future political developments. Outside of the Armed Forces, one can hardly detect the existence of significant institutional groupings. This state of flux makes it extremely difficult to assess, in any way, the long range social and economic effects of the revolution.

OTHER EFFECTS

On May 13, 1958, the French Army of Algeria, long embittered by the ineffectiveness of the French Government, staged a virtual coup d'etat which toppled the Fourth Republic. De Gaulle was returned to power on the assumption that he would keep Algeria French. When, however, the new President of the Fifth Republic gave indications that he was considering independence for Algeria, the army generals who had been instrumental in putting de Gaulle in office felt that they had been betrayed once more.[1] On April 22, 1961, elements of the army of Algeria, led by Generals Raoul Salan, Maurice Challe, Edmond Jouhaud and Andre Zeller, and a number of colonels, attempted another coup. "Their plan was to seize Algiers, to rally the armed forces in the name of army unity and French Algeria, and then to seize Paris, driving De Gaulle from Office."[45] Algiers was seized, but the coup fizzled when the navy, the air force, and draftee units failed to support the generals. Challe and Zeller surrendered to the authorities, but the rest went into hiding, and subsequently created the terrorist Secret Army Organization (OAS). The aim of the OAS was to keep Algeria French despite de Gaulle by starting a counterrevolution. This, however, was never accomplished because the OAS failed to establish a foothold outside the large Algerian cities. Terrorism in the Algerian cities and in France was used instead. Assassinations and bombings became daily occurrences. After the conclusion of the Evian Agreement, the OAS stepped up its terrorist campaign, singling out the Muslim population, in an attempt to provoke communal strife and the intervention of the French Army on the side of the settlers. With the apprehension of Jouhaud and Salan, the movement lost much of its impetus, and final secret agreements with the FLN put an end to OAS activity.

It is noteworthy that the Algerian revolution established the French Army as a political factor. In France the OAS had the sympathy of a large number of French officers, including Marshal Alphonse Juin. Many officers, most of them veterans of the war in Indochina, defected and joined the OAS. And although the navy, air force, and the vast majority of the army did not support the generals in April 1962, they showed no willingness to crush the insurgent movement by force. The army had become so suspect that the defense of Paris

---

[1] The French Army has blamed the politicians in general, and the French democratic system in particular, for the fall of France in 1940, their defeat in Indochina, and the loss of Morocco and Tunisia. After accepting defeat at Dienbienphu, the French Army looked upon Algeria as the battlefield on which it could vindicate its honor.

against a possible invasion from Algiers was entrusted to *Gendarmerie* and *Garde Mobile* units.

NOTES

1. *Algeria*, Subcontractor's Monograph, HRAF–54, J, Hop–1, (New Haven: Human Relations Area Filas, Inc., 1956), pp. 10–18.
2. "Algeria," *Encyclopaedia Britannica* (1958), I, 618.
3. Joan Gillespie, *Algerian Rebellion and Revolution* (New York: Praeger, Inc., 1960), pp. 29–31.
4. Nevill Barbour, *A Survey of North West Africa (The Maghrib)* (London: Oxford University Press, 1959), pp. 202–203.
5. *Ambassade de France, Service de Press et d'Information*, "The Constantine Plan for Algeria," (May, 1961), pp. 14–16.
6. *Algeria*, pp. 494–498.
7. Barbour, *A Survey*, pp. 241–242.
8. *Algeria*, pp. 494–512.
9. *UN Yearbook of International Trade Statistics* (1956), as reproduced in Barbour, *A Survey*, p. 248.
10. L. Gray Gowan, "New Face for Algeria," *Political Science Quarterly*, LXVI (1951), 347–351.
11. Ibid., p. 352.
12. Ibid.
13. Ibid.
14. Gillespie, *Algerian Rebellion*, p. 66.
15. Rolf Italiaander, *The New Leaders of Africa* (Englewood Cliffs, New Jersey: Prentice-Hall, Inc., 1961), p. 18.
16. Gillespie, *Algerian Rebellion*, p. 63.
17. Ibid., p. 45.
18. *Algeria*, p. 455.
19. Ibid., p. 454.
20. Ibid.
21. Ibid., p. 456.
22. Charles-André Julien, *L'Afrique Du Nord en Marcle* (Paris: René Julliard, 1952), pp. 398–408.
23. Barbour, *A Survey*, p. 250.
24. General Georges Catroux, as quoted in Gillespie, *Algerian Rebellion*, p. 13.
25. Jules Ferry, as quoted in Gillespie, *Algerian Rebellion*, p. 12.
26. Julien, *L'Afrique*, pp. 322–323.
27. *Algeria*, pp. 466–467.

28. Barbour, *A Survey*, p. 227.

29. Gillespie, *Algerian Rebellion*, p. 95.

30. Charles-Henri Favrod, *La Revolution Algerienne* (Paris: Librairie Plon, 1959), pp. 107–108.

31. *Underground Case Study*, (Special Operations Research Office, unpublished manuscript), see Chapter on Algeria.

32. Ibid., Chapter on Algeria, passim.

33. Gillespie, *Algerian Rebellion*, pp. 112–113.

34. Joseph Kraft, *The Struggle for Algeria* (New York: Doubleday and Company, Inc., 1961), p. 70.

35. Ibid.

36. Peter Braestrup, "Partisan Tactics—Algerian Style," *Readings in Guerrilla Warfare* (U.S. Army Special Warfare School, Fort Bragg, North Carolina).

37. Ibid.

38. Gabriel G. M. Bonnet, *Les Guerres Insurrectionnelles et Révolutionnaires* (Paris: Payot, 1958), pp. 250–251.

39. Ibid., p. 254.

40. Richard and Joan Brace, *Ordeal in Algeria* (Princeton: D. Van Nostrand Company, Inc., 1960), p. 254.

41. Kraft, *The Struggle*, p. 94.

42. Ibid.

43. Ibid., pp. 104–105.

44. Brace, *Ordeal*, p. 118.

45. Kraft, *The Struggle*, p. 247.

RECOMMENDED READING

Barbour, Nevill. *A Survey of North West Africa (The Maghrib)*. London: Oxford University Press, 1959. A comprehensive and factual survey of the countries of North Africa.

Brace, Richard, and Joan Brace. *Ordeal in Algeria*. Princeton: D. Van Nostrand Company, Inc., 1960. A factual historical presentation of the developments in Algeria after 1945.

Favrod, Charles-Henri. *La Révolution Algérienne*. Paris: Librairie Plon, 1959. Concentrates primarily on Algeria during the interwar period and up to 1954. Gives excellent details of the organization of the nationalist parties. Critical of French policy and administration.

Gillespie, Joan. *Algerian Rebellion and Revolution*. New York: Praeger, Inc., 1960. A detailed account of the Algerian revolution and its causes. Sympathetic to the nationalists.

Julien, Charles-André. *L'Afrique Du Nord en Marcle*. Paris: René Julliard, 1952. The classic on North Africa prior to 1953. Contains critical chapters on Algeria which trace the development of French policy and genesis of Algerian nationalism.

Kraft, Joseph. *The Struggle for Algeria*. New York: Doubleday and Company, Inc., 1961. An impartial analytical study of the Algerian revolution. Studies in detail all parties to the struggle.

SECTION IV

·················

AFRICA SOUTH OF THE SAHARA

General Discussion of Area and Revolutionary Developments
The Revolution in French Cameroun: 1956–1960
The Congolese Coup of 1960

TABLE OF CONTENTS

GENERAL DISCUSSION OF AREA AND REVOLUTIONARY DEVELOPMENTS

THE REVOLUTION IN FRENCH CAMEROUN: 1956–1960

THE CONGOLESE COUP OF 1960

AFRICA SOUTH OF THE SAHARA

GENERAL DISCUSSION OF AREA AND REVOLUTIONARY DEVELOPMENTS

BACKGROUND

DEFINITION OF AREA

As defined in this study, Africa south of the Sahara refers to that part of Africa lying between Dakar on the west coast, the Gulf of Aden on the east, and the Cape of Good Hope at the southern extremity of the continent. Although much of the territory of the Sudan lies within this area, that country, for historical reasons, is not here considered part of sub-Saharan Africa. The island of Madagascar, officially known since 1960 as the Malagasy Republic, is included because of its historical association with Africa. Sub-Saharan Africa is generally referred to in French sources as *Afrique Noire* (Black Africa), because its most distinguishing characteristic is its predominantly Negroid population. Although there are wide ethnic, cultural, and even racial variations among Black Africans—and there are several million "Africans" of pure European or Asian descent who are permanently resident in south, central, and east Africa—the dominant characteristic of the area has, in recent decades, come to be its conscious identification with the Negroid elements.

GEOGRAPHY

In size, sub-Saharan Africa is comparable to the United States west of the Mississippi. The climate throughout most of the area is tropical and rainy with very little seasonal variation. Around the Cape of Good Hope, however, it resembles that of the Mediterranean countries, and a temperate climate not very different from that of the United States prevails in the rest of the Union of South Africa. Southern Rhodesia, and Kenya, though located at the Equator, also have a moderate climate owing to their high elevation. In west Africa and the Congo basin of central Africa dense tropical rain forests alternating with swamps, and savannahs situated on a plateau, punctuated by an occasional mountain range, are the prevailing geographical pattern. There is an arid desert region in southwest Africa, between Angola, Southern Rhodesia, and South Africa. The terrain of south and east

Africa is more mountainous than the rest of sub-Saharan Africa. Here are located the great peaks Mount Kenya and Kilimanjaro and a series of large lakes. One of these is the source of the Nile. The Nile, the Congo, the Niger, and the Zambezi Rivers have all been of great commercial and historical significance in the development of Africa south of the Sahara. Climate, terrain, and disease-carrying insects, such as the tsetse fly, render large portions of Africa's territory undesirable for human habitation.

PEOPLE

Sub-Saharan Africa's more than 140 million inhabitants[1] present a greater racial and ethnic diversity than one finds in any other region of the world of comparable size. Over 3 million are of European descent, and almost as many are of Asian descent, largely from the western part of India. The Black African majority are usually classified as belonging to three major linguistic groups: the Nilo-Hamitic branch, centered in northeast Africa; the Sudanic branch, centered in west Africa; and the Bantu branch, centered in south and east Africa. These ethnic groups are further subdivided into hundreds of tribes and clans, which, through lack of communication for hundreds of years, have developed separate languages and social customs. Only the Bantu tribes retain a degree of linguistic and cultural unity roughly comparable to that of the Indo-European "tribes" of Europe. In countries such as Kenya, Congo, and Cameroun, the three branches overlap and in some instances have intermingled. In the Congo basin and in the deserts of southwest Africa, there are pygmy tribes and Bushman-Hottentot tribes, generally believed to be the earliest inhabitants of Africa, which are culturally more primitive than the Bantu tribes in the area.[2]

Sub-Saharan Africa has not experienced population pressures over wide areas comparable to those confronting most of Asia in recent decades. Nevertheless, where European settlement has been particularly heavy, as in Kenya and parts of South Africa, or where very large-scale economic activity has taken place, overcrowding has resulted. Rapid urbanization has been one of the most striking demographic characteristics of Africa since World War II. The population of the Belgian Congo, for instance, was approximately one-third urbanized at the time of independence in 1960. In most cases, tribal diversity has been maintained in urban areas, and residential segregation along tribal as well as racial lines has been the general rule. Political and social groupings have likewise followed tribal lines.

RELIGION

Africa south of the Sahara has no common religion, and this is another source of the region's cultural diversity. Both Islam and Christianity have penetrated the region, coming from opposite directions and meeting in many places. In western Africa, the northern tribes in the "upcountry" are traditionally Muslim, while the southern tribes along the coast are often Christian. In eastern Africa, the coastal tribes have been widely converted to Islam by the Arab settlers, while many of the "upcountry" tribes have been converted to Christianity. Christian converts, Catholic and Protestant, are found in large numbers wherever European influence has been strongly felt. Black Africa's first Christians were the Ethiopians, who were converted early in the Christian era and have maintained their Coptic Orthodox Church ever since, although completely surrounded by the Muslims since the rise of Islam in the seventh century. The vast majority of Africans are pagan animists, conforming to traditional tribal religious practices and belief patterns.

SOCIO-ECONOMICS

While millions of Black Africans continue to practice subsistence agriculture or to lead seminomadic lives in the hinterlands, Africa south of the Sahara is dotted with enclaves of modern agricultural and industrial activity. Large-scale tropical plantations produce cocoa, coffee, peanuts, palm oil, bananas, cotton, and a variety of exotic spices and seasonings, and modern mining enterprises yield large quantities of copper, uranium, diamonds, gold, manganese, and bauxite. South Africa, the most highly industrialized country, is responsible for over 20 percent of the entire continent's gross product, although its population is only 6 percent of the continent's population.[3] South Africa's 3 million or so European residents derive most of the benefit. The standard of living among the European population of South Africa, the Rhodesias, and Kenya is comparable to that of the United States. The division of wealth has varied considerably from territory to territory. One generalization which can be made is that in tropical Africa, where Europeans did not settle permanently in large numbers, and where an African middle class has developed, the division of the wealth has not followed racial lines to the same extent as in southern and eastern Africa. The coastal cities of west Africa, such as Dakar, Conakry, Freetown, Monrovia, Abidjan, Lagos, and Douala, have a more highly developed indigenous middle class than any other part of sub-Saharan Africa. The mining cities which have mushroomed in the former Belgian Congo, the Rhodesias, and South Africa during the post-war period display considerably less social stability and are characterized by their large urban proletariat of partially detribalized, semieducated African laborers.[4]

WESTERN IMPACT ON THE AREA

The impact of the West on Africa south of the Sahara has varied from one territory to another as a consequence of such factors as the duration and intensity of European influence, the number of Europeans permanently settled in the area, and the nature of the indigenous African culture.

COLONIAL RULE

Although European contacts with Black Africa began as early as the 15th century, the colonial period did not really begin until the last of the 19th century. By comparison with north Africa and Asia, most of sub-Saharan Africa has experienced a relatively recent and brief encounter with European colonialism. Only the coastal settlements have had continuous contacts with Europe for more than a century, and many parts of the interior were not touched by European influence until the beginning of the present century.

The Portuguese, Dutch, British, and other European traders who frequented the west coast of Africa after 1500 came primarily in search of slaves, spices, ivory, and gold. They were not colonial administrators or permanent settlers, and their political influence was confined to the immediate vicinity of the forts and trading posts they built along the coast. The social effects of their coming, however, were disastrous for the indigenous culture, which was one of the most highly developed in Africa south of the Sahara. Slavery, which had long been practiced in west Africa, was internationalized, and a large-scale slave trade soon developed with the newly discovered Americas. Deprived of millions of their ablest inhabitants, west Africa and the Congo regions were weakened materially and socially at a time when European technology and material power were already far superior to those of sub-Saharan Africa. The slave trade also paved the way for colonial rule; one of the justifications for European intervention in the 19th century was the suppression of the slave trade.

In 1885, at the Congress of Berlin, the European powers agreed among themselves on the partitioning of Africa's remaining "unclaimed" territory. During the last quarter of the 19th century, in what has been referred to as the "scramble for Africa," European colonial rule was extended to the remotest hinterlands of the "Dark Continent." Portugal and Spain retained their original footholds, but Britain, France, Germany, and Belgium were the major European colonial powers.

After World War I, Germany's west African colonies, Togoland and Kamerun, were partitioned between France and Britain. German South West Africa was awarded to South Africa, then a part of the British Empire, and German East Africa became Tanganyika, administered by the British, and Ruanda-Urundi, administered by the Belgians. Although in practice they were treated like colonies of the administering European power, the former German colonies were technically not colonial territories but League of Nations mandate territories. After World War II they became U.N. trusteeship territories.

In the period between World Wars I and II, British, French, Belgian, and Italian colonial administrators, business interests, missionaries, and others sought actively to consolidate their countries' political, economic, and cultural influences in Africa. At this time all of sub-Saharan Africa was under colonial administration *except Liberia and Ethiopia;* the former existed in a semicolonial state of economic and diplomatic dependence on the United States. Liberia, founded in the 1830's by emancipated American Negro slaves, was heavily dependent on American business interests, such as Firestone Rubber and certain shipping lines. Ethiopia had retained its independence partly because of its physical isolation, partly because this ancient Christian country enjoyed the economic and diplomatic support of the British, who were the dominant European power in east Africa and Egypt. In 1935 Ethiopia was invaded by Mussolini's army and until its liberation by the British in 1941 was administered as a part of Italy's short-lived African empire. Ethiopia's brief experience with European colonialism is illustrative of the positive content of colonialism throughout sub-Saharan Africa; the Italians built schools, roads, and other public facilities which very likely would not have been built by the Ethiopians.

EUROPEAN SETTLEMENT

European influences have naturally been felt most strongly in those parts of Africa where there have been large concentrations of Europeans permanently resident over a long time. In tropical Africa, Europeans did not settle permanently, primarily because of the climate, but in many parts of south and east Africa, where the climate is generally temperate, Europeans settled in large numbers. They first came in the 17th century, when Dutch and French Protestants landed in the Capetown region of South Africa. South Africa was a Dutch colony until the 19th century, when as a result of the Napoleonic Wars the British occupied the country. During the 19th century, English settlers came to South Africa in large numbers, and thus began the intense rivalry between the two European peoples which culminated in the Boer War at the end of the century. The descendents of the Dutch settlers are called "Afrikaners" and speak a language derived from 17th century Dutch.

In the 1880's English and Afrikaner settlers pushed northward from South Africa into the Rhodesias. South of the Zambezi River, in Southern Rhodesia, there is a large concentration of European settlers which dates from this time. The only other large concentration of European settlers in sub-Saharan Africa is in the British crown colony of Kenya, which was settled during the early decades of this century, primarily by the British. Southern Rhodesia's European population is around 200,000 in a total of almost 3 million, while Kenya has 60,000 Europeans in a population of over 6 million.

In all three areas of permanent European settlement, the indigenous African population and the large Indian minority have been subjected to political, social, and economic discrimination. The non-Europeans have not shared in the high standard of living enjoyed by the European minority, nor, except in Kenya, have they shared in the political rights which the European elite have gradually won from the British. Africans in these areas, and especially in South Africa, are more influenced by European ways than in other parts of Black Africa; however, most Africans in South Africa regard European culture as an alien and hostile culture imposed on them by white Europeans, and there is little of the conscious identification with Europe that one finds, for instance, in the former French protectorate of Tunisia.

INTERACTION OF EUROPEAN AND AFRICAN CULTURES

The impact of European colonialism and settlement on the indigenous peoples of sub-Saharan Africa has been conditioned to a large extent by the nature of the African culture in each area. For instance, west African tribes were more advanced culturally than the Bantu tribes of south and east Africa. West Africans had long been in contact with the Arab world and many tribes had been converted to Islam. Several Negro-Muslim empires had risen and fallen during Europe's Middle Ages, and some coastal tribes had developed a complex system of tropical agriculture, with a plantation economy based on slave labor and an elite ruling and leisure class which patronized the arts. West African artists worked in bronze and indigenous hard woods. While these political and societal relationships were upset by the arrival of the European slave traders after 1500, the west African was better prepared than the Bantu and other African groups for the cultural impact of Europe. Some of the traditional elite maintained their identity throughout the colonial period and have been active in the contemporary nationalist movement.

The more isolated Bantu tribes were culturally less advanced when Europeans began settling in their areas in large numbers. Most of the Bantu were seminomadic peoples who had developed a pastoral economy based on cattle. Their encounter with Europeans in South Africa, Rhodesia, and Kenya resulted in a breakdown in their tribal cultures and, in most instances, in the destruction of their traditional elite. In the 19th century, the European settlers carried on a series of small wars with the Zulus, the major Bantu people of South Africa. Comparable in many respects to the frontier wars with the American Indians during this period, these military encounters led to the destruction of the Zulus' traditional cultural patterns and left a heritage of considerable bitterness.

One Bantu group which has displayed remarkable ability to adapt to European hegemony and preserve its traditional culture and tribal elite under colonial rule is the Baganda tribe in the British protectorate of Uganda. A highly developed group of settled agriculturists, the Baganda have maintained their "parliamentary" system of government and hereditary monarchy with a dynasty dating back for many centuries. The Baganda benefited from the fact that few Europeans settled in their area of east Africa, although it is one of the most fertile regions of the continent. One of the positive contributions of European colonialism has been the development of Makerere University in Uganda.

The European impact has varied widely among the various tribal groups. For instance, the coastal tribes of west Africa, which have had the longest and most intensive contacts with Europe, have become the most Western-educated and technically proficient element in sub-Saharan Africa, while the tribes farther inland have lagged behind in education and technical know-how. In many cases, the tribes which have become most Westernized were small tribes which traditionally had been inferior to others in the area. Thus, traditional power relationships have been reversed by the process of Westernization. This disparate spread of European influences explains much of the recent intertribal conflict in Black Africa. The Ashanti in Ghana, the Fulani in Nigeria and Cameroun, the Lulua in the Congo, and the Masai in

Kenya are all examples of traditionally dominant tribes which have been placed at a political disadvantage by the Westernization of other tribes in the area. The Kikuyu in Kenya and the Baluba in the Congo became Westernized through their association with European settlers, and these tribes have been the most active in the nationalist movement. The tradition-oriented Masai and Lulua, on the other hand, have looked to the British and Belgian authorities respectively for protection of their tribal interests.[5]

Another important effect of European colonialism was the partitioning of tribal areas between two or more European powers. The Ewe, a west African coastal tribe, was partitioned into the German colony of Togoland, the French colony of Dahomey, and the British colony of Gold Coast, and the Hausa and Fulani tribes of the west African "upcountry" were divided by colonial borders separating French and British Africa. Colonial boundaries in west Africa generally ran inland from coastal settlements, while ethnic boundaries tended to run parallel to the coast. Another example of tribal partition occurred around the mouth of the Congo River, where the historically powerful Bakongo tribe live in Portuguese Angola, French Equatorial Africa, and the former Belgian Congo. The Somali tribesmen in east Africa are scattered through Ethiopia, Kenya, and their newly independent homeland of Somalia. These examples of conflict between political and ethnic boundaries have already been the source of considerable political unrest, and they promise to be a continuing irritant.[6]

REVOLUTIONS IN SUB-SAHARAN AFRICA

Although active nationalist movements have developed in Africa south of the Sahara since World War II, there have been no colonial revolutions on the order of those in north Africa and Asia. Instead, nationalism has assumed a variety of forms, ranging from the violence of Mau Mau terrorists in Kenya to negotiated transfers of sovereignty from European colonial administrations to African nationalist political cadres. Since Ghana obtained its independence by peaceful means in 1957, there have been numerous similar instances of nonviolent decolonialization.

As defined in this study of revolutions, some examples of revolutionary activity in sub-Saharan Africa have been the political riots in the French Ivory Coast in 1948–50, the Mau Mau revolt in Kenya in 1952–53, the guerrilla war in French Cameroun in 1955–60, the Angolan revolt begun in 1960, the military coup d'etat in the Congo in September 1960, and the abortive coup d'etat in Ethiopia in December 1960 against its own indigenous government.

Revolutionary activity by African nationalists in Kenya, the Ivory Coast, and Cameroun has displayed a number of common characteristics, such as violent tactics, a tribal-based organization, and Western-educated leadership. It is definitely known that there was considerable Communist influence in both the *Rassamblement Démocratique Africaine* (RDA) organization, which was responsible for the riots in Ivory Coast, and the Camerounian nationalist organization, the *Union des Populations Camérounaises* (UPC). However, despite the claims of European settlers in Kenya, no connection between the Mau Mau organization and the Communist movement has ever been proved to exist.

French Communists who had infiltrated the RDA leadership at the time of its formation in 1946 urged the African leaders to adopt a militant nationalist line. By organizing mass meetings and distributing slogans and propaganda leaflets attacking European "exploitation" and demanding equal rights for the African in all spheres of life, the RDA built up a membership of several hundred thousand in the first 2 years. Although it was an interterritorial and comprehensive nationalist party with local branches throughout French Africa, the RDA's mass support was based primarily in the Ivory Coast. In 1948–49, the RDA instigated a wave of violence in the form of political strikes, boycotts, and riots at political rallies. The French authorities vigorously suppressed these disturbances and were supported by conservative elements of the African population. In February 1950 African "moderates" called on France to send in troops to suppress the RDA as a "foreign party," and in October 1950 the RDA's African leadership broke completely with the French Communists.[7] The UPC in Cameroun retained its connections with the Communists and left the RDA in 1951. The revolutionary activities of the UPC are summarized in another part of this section.[a]

The Mau Mau revolt broke out in 1952, when a secret organization among the Kikuyu, Kenya's most westernized African tribe, began a campaign of organized terror, directed at first against Africans employed by the British colonial administration, and later against European settlers. Like the UPC in French Cameroun, the Mau Mau was tribal-based, and at least some of its leaders were Western-educated. The leaders made extensive use of tribal superstitions and ritual oaths to maintain the loyalty of their followers and to exert psychological pressure on Africans who sided with the Europeans. Little is known definitely about the instigators and organizers of the Mau

---

[a] See pp. 359 ff.

Mau, although European settlers have alleged that its leader was Jomo Kenyatta, a leading Kikuyu intellectual and nationalist politician who, in his 50's, was head of the Kenya African Union, a welfare and political organization for urbanized Africans. Kenyatta was tried and convicted of complicity in the "Mau Mau" movement, largely on the basis of such circumstantial evidence as his earlier writing and the fact that he had been to Moscow in the 1930's.[8]

Little information is available on the revolt which has been brewing in the Portuguese colony of Angola since the end of 1960. However, it is known that a guerrilla organization has been conducting paramilitary operations against Portuguese settlers and Africans whom the nationalists accuse of collaborating with the Portuguese authorities. The leader of the guerrilla bands, which probably number in the thousands, is said to be Roberto Holden, a Protestant mission-educated African nationalist leader of the illegal *União das Populacoes de Angola* (UPA). The Angolan rebels are believed to be receiving considerable material and political support from nationalist groups based in the former Belgian Congo, which borders Angola on a lengthy and poorly guarded frontier. Portuguese authorities have reacted by sending additional troops and war materiel and by intensifying an already rigid censorship and other repressive police measures in the colony.[9]

Of the two military coups d'etat which occurred in sub-Saharan Africa in 1960, the one which overthrew the Lumumba regime in the newly independent and politically unstable former Belgian Congo has been summarized in a later part of this section.[b] The abortive coup which took place in the ancient kingdom of Ethiopia on December 14, 1960, has been characterized as "a palace revolution organized by some of the Emperor's most trusted officials rather than by Ethiopia's considerable body of 'angry young men'."[10] The leaders were two brothers, one the commander of the Imperial Guard and the other a provincial governor. Taking advantage of Emperor Haile Selassie's temporary absence from the country, the rebels took control of the palace and the Addis Ababa radio station and announced that Haile Selassie had been deposed in favor of his son, Crown Prince Asfaw Wossen. Although the rebels proclaimed an end to "3,000 years of tyranny and oppression," the coup was totally lacking in popular support, and Haile Selassie returned a few days later, after loyal elements, particularly in the Ethiopian Army, had regained control of the situation. The two leaders of the coup were hanged publicly, and other officers of the Imperial Guard who were actively involved have been tried and punished. The Crown Prince has been absolved of any com-

---

[b] See pp. 371 ff.

plicity. The significance of the short-lived revolution was that it demonstrated the vulnerability of Ethiopia's Westernizing oligarchy to the type of political unrest which has been sweeping the rest of Africa and the Middle East.

PROSPECTS AND OUTLOOK

The examples of revolutionary activity cited above and those discussed in the following parts of this section suggest the pattern of revolutionary activity which may be expected to develop in Black Africa in the coming decade. With the end of European colonial rule, revolutionary activity will be directed against the indigenous African regimes which take the place of colonial administrations, and in those parts of south and east Africa in which European settlers dominate the political scene, a protracted campaign of guerrilla warfare and terrorist operations by an African nationalist underground organization may develop. The post-independence military coup in Congo, the unsuccessful palace revolt in Ethiopia, and the continued guerrilla war against the newly independent government of Cameroun illustrate the forms of revolutionary activity with which the emerging nations of Africa can expect to have to deal. The Mau Mau activity in Kenya and the Angolan revolt are significant portents for South Africa, Southern Rhodesia, and Kenya.

INDEPENDENT AFRICAN TERRITORIES

The two oldest independent states in Black Africa, Ethiopia and Liberia, are particularly vulnerable to revolution because of their conservative political structures and narrowly based ruling elite. Ethiopia's form of government is an autocracy embodied in the person of the emperor and political participation is strictly limited to a small group. In an attempt to modernize his tradition-oriented and economically underdeveloped country, Emperor Haile Selassie has sent a number of young Ethiopians from the Amharic oligarchy abroad to study. On returning to Ethiopia these Western-educated cadres represent a possible source of danger to the traditional elite, which includes the hierarchy of the Coptic Church, the palace functionaries, and the emperor himself. Frequent administrative transfers about the country are thought to be one of the emperor's ways of preventing the formation of centers of opposition to his regime.

Liberia's Government is patterned after that of the United States, but in practice the political process is almost exclusively in the hands of a few thousand English-speaking and coastal-dwelling descendants of the American Negro slaves who founded the country. The interior is inhabited by tribal peoples who long have been treated as colonial subjects by the coastal population. The development of nationalism in west Africa has forced the ruling elite, headed by President William Tubman, to devote more attention to the political and social integration of the tribal interior with the Westernized coastal elements. In September 1961 the Liberian Government was faced with widespread strikes and demonstrations, which government sources attributed to Communist influences among the tribesmen in the interior.

The states of sub-Saharan Africa which have recently emerged from what was British, French, Belgian, and Italian Africa are particularly vulnerable to revolutionary activity. In the first place, these countries lack national unity. Secondly, they are confronted with what has aptly been called "the revolution of rising expectations." Moreover, as time passes, a "generational cleavage" will develop between the younger generation and the generation which came to power at the time of independence. Finally, many of the political elite came to power virtually without a struggle and through very little effort of their own; consequently, they lack the organizational support, the political experience, and the emotional appeal among the masses that a prolonged nationalist struggle confers on its leaders. The Congo's political elite, for example, which came to power precipitately, has demonstrated extreme instability in the post-independence period.

All these states face severe economic difficulties; all of them are underdeveloped and heavily dependent on a few key agricultural exports for the revenue to pay for the technical equipment and manufactured goods they must import in order to modernize their economies. They face inflation in the midst of poverty, which seems to be the price of the rapid economic development to which Black Africa's political leaders are irrevocably committed. An already narrow range of economic alternatives is further limited by political and psychological factors. For instance, economic logic would dictate the retention of financial and commercial ties with the West; however, for psychological reasons, many of these links with the former colonial powers have been broken. The political elite of these countries must find ways of satisfying the social and economic aspirations of the masses if they are to remain in power.

National independence has invariably widened the scope of popular participation in the political process; consequently, the "revolution of rising expectations" is likely to result in political revolution.

Unfortunately, the new states of Africa are peculiarly lacking in the kind of national unity which older states can usually invoke for political stability in times of crisis. Tribal boundaries seldom correspond to the frontiers inherited from the colonial period, yet tribal ties are usually the strongest loyalties. Rival political factions may identify with rival tribal groups and thereby gain the advantage of a definite geographic base of operations. Also, in a society where even the urban population is less conscious of belonging to socio-economic groups (i.e., in Marxist terminology "lacks class consciousness") than of belonging to different tribal groups, politics naturally tend to be less ideological and national-based and more personal and tribal-based.

The advantages of a geographic base of political support range from the conventional advantages typified in our country by the "Solid South" to the availability of such a territorial base for guerrilla operations against the central government. For example, in Cameroun the revolutionary movement has been centered in the Bamileke tribal district, and in Kenya the Mau Mau movement was exclusively based on the Kikuyu tribe and centered in their rural districts and urban quarters. In the Congo rival political factions have taken geographic positions which have attracted both tribal popular support and international diplomatic support. Thus, the Lumumba faction is based in Stanleyville and Orientale Province, has wide support among local tribes, and has received diplomatic backing from Guinea, Ghana, and the Soviet bloc. Tshombe is based in the rich mining province of Katanga and is supported by the Baluba and other local tribes and by the European settlers. The central government under President Kasavubu is based in Léopoldville among the Bakongo tribe and has the diplomatic support of the United Nations and the Western bloc.

EUROPEAN-DOMINATED AREAS

Colonial revolutions on the order of those in French North Africa may be expected in the coming decade throughout the European-dominated areas of south and east Africa. They may be violent, as in Algeria, or they may be accomplished with a minimum of violence, as in Tunisia. It is difficult to imagine, however, that the political, economic, and social domination of the European minority over the non-European majority can continue without changes in the power relationships among these ethnic groups. These, in turn, may take on the proportions of a political revolution. The situation in the more important of these areas is briefly discussed next.

In Kenya, such changes have been going on since 1960, when the emergency proclaimed during the Mau Mau revolt was officially ended and African politics returned to legal channels. In 1961, the British colonial authorities permitted a nationally-elected legislative assembly with an African majority; however, the African majority was not so large as it would have been under the formula of "one man, one vote," which is the unanimous demand of African politicians. A delicate balance prevails in Kenya between conservative European settlers; a moderate European element organized as the New Kenya Group and led by Michael Blundell; the Indian minority, which is larger than the European but considerably smaller than the African population; and the African nationalists, who are divided chiefly into the Kenya African Nationalist Union (KANU), led by Tom Mboya and supported by Kikuyu and Luo tribesmen, and the Kenya African Democratic Union (KADU), led by Ronald Ngala and supported by the Masai and several minor tribes. The British Government has played the role of "honest broker" and intermediary between these factions in the interest of achieving a peaceful transition from a European-dominated society to a multiracial one. Since 1958 there have been reports of a clandestine movement organized as a successor to the Mau Mau, known as the Kiama Kia Muingi (KKM); its relationship, if any, to African nationalist political organizations is not known.[11]

Southern Rhodesia, a self-governing dominion in the British Commonwealth, is another area dominated by European settlers. Along with the British protectorates of Northern Rhodesia and Nyasaland, Southern Rhodesia is part of the Central African Federation. Although the three territories are economically complementary, African nationalists in Nyasaland and the Rhodesias have been unanimously opposed to the federation, because its practical effect has been to extend the strong European settler influences of Southern Rhodesia over the other two territories. Since 1960 there has been increased political unrest and violence between African nationalists and white authorities in the Federation. African nationalists in Nyasaland, led by Dr. Hastings Banda and the Malawi Congress Party, have been most successful in obtaining political concessions, and this country seems to be taking the course of other British African territories toward eventual independence. African nationalists in Northern Rhodesia have had more difficulty, since this territory is the center of large-scale copper mining and borders on the politically unsettled Katanga Province of the Congo. Southern Rhodesia may either follow the multiracial pattern which seems to be emerging in Kenyan politics, or the dominant whites may turn to South Africa's racial policies. In any event, Southern Rhodesia is highly vulnerable to the type of nationalist revolution which has developed in Angola.

Since the Sharpesville riots in April 1960 and the police repressions which ensued, South Africa's vulnerability to guerrilla warfare and terrorist activity led by African and Indian nationalists has become evident. South African authorities have reacted to non-European opposition by increased police repression and surveillance. The official doctrine of *apartheid* (separateness), which calls for separate economic, social, and political development for each ethnic group in South Africa, has been further implemented by introducing segregation along racial and even tribal lines in institutions of higher education. The development of the "Bantustans," as they are called, is regarded by critics of *apartheid* as economically unfeasible and certain to lead to eventual political fragmentation of the country along tribal lines. Although both English-speaking and Afrikaner-speaking whites are overwhelmingly opposed to altering the dominant role of the European in South African society, the Afrikaner element is particularly unprepared to accept changes in the power relationships among the European and non-European elements of the population. Afrikaner resentment against criticism of *apartheid* by other members of the British Commonwealth led in 1961 to the withdrawal of South Africa from the Commonwealth and the proclamation of a republic. Independent Black African states have added stimulus to the growing opposition of world opinion to South Africa's racial policies. At present there is no independent Black African territory on the borders of South Africa, but an independent Angola, for example, would provide a base of operations for South African non-European guerrilla bands.

Portuguese Angola on the west coast of southern Africa and Mozambique on the east coast both have sizable European minorities. There are around 200,000 Europeans permanently resident in Angola, and around 60,000 in Mozambique; Angola has a total population of about 4.5 million and Mozambique of 6.2 million. Illegal African nationalist organizations are active in both countries, and the Portuguese have had to resort to large-scale military operations against guerrilla bands in northern Angola. A prolonged struggle similar to the Algerian conflict might be expected in Angola if Portugal were militarily and economically prepared for it. The Government of Portugal is a dictatorship, headed since the 1930's by Dr. Antonio de Oliveira Salazar, and the Salazar regime has recently lost some support in Portugal and among Portuguese settlers in Africa. Portuguese settlers have traditionally been more ready than other Europeans to accept Africans on equal terms, and for this reason the chances of arriving at a workable multiracial system would seem to be much better in Portuguese Africa than in the other European-dominated territories of south and east Africa.

NOTES

1. James S. Coleman, "The Politics of Sub-Saharan Africa," in *The Politics of the Developing Areas*, eds. Gabriel A. Almond and James S. Coleman (Princeton: Princeton University Press, 1960), p. 250.
2. Joseph H. Greenberg, "Africa as a Linguistic Area," in *Continuity and Change in African Cultures*, eds. William R. Bascom and Melville J. Herskovits (Chicago: University of Chicago Press, 1959), pp. 15–27.
3. John Scott, "Africa: World's Last Frontier," *Headline Series*, May 20, 1959, p. 3.
4. Peter Ritner, *The Death of Africa* (New York: Macmillan, 1960), pp. 132–134, et passim.
5. Coleman, "Politics," p. 282.
6. Christopher Bird, "Africa's Diverse Peoples Offer Openings for Soviet Agitation," *Africa Special Report*, II, 10 (December 1957), 11–12.
7. Virginia Thompson and Richard Adloff, *French West Africa* (Stanford: Stanford University Press, 1957), pp. 86–92.
8. Ritner, *Death*, pp. 185–192; see also Sir Philip Mitchell, "Mau Mau," in *Africa Today*, ed. C. Grove Haines (Baltimore: Johns Hopkins University Press, 1955), pp. 485–493.
9. *The New York Times*, August 27, 1951, p. 22.
10. "Ethiopian Coup Fails, But Problems Remain," *Africa Report*, VI, 1 (January 1961), 11–12.
11. Ritner, *Death*, p. 194.

RECOMMENDED READING

Bascom, William R. and Melville J. Herskovits (eds.). *Continuity and Change in African Cultures*. Chicago: University of Chicago Press, 1958. A collection of monographs on African linguistics, art, music, economics, and other aspects of culture by leading anthropologists.

Coleman, James S. "The Politics of Sub-Saharan Africa," *Politics of the Developing Areas*. Eds. Gabriel Almond and James S. Coleman. Princeton, N.J.: Princeton University Press, 1960. A scholarly analysis of political processes, social environment, and governmental institutions of contemporary Africa. Author, an American political scientist and African specialist, is Director of African Studies at UCLA.

Hailey, William M., Baron. *An African Survey*, Revised 1956. London: Oxford University Press, 1957. An encyclopedic work on sub-Saharan Africa. Includes reliable data on geography, languages, ethnic groups, political structures, economic activities, etc. Maps and statistics. Lord Hailey is one of the world's foremost authorities on Africa.

Hempstone, Smith. *Africa—Angry Young Giant*. New York: Praeger, 1961. (Published in London by Faber & Faber, 1961, under the title, *The New Africa*.) An account of the author's impressions of political and social developments in North East Africa (Sudan, Ethiopia and Somali), French Equatorial Africa, and British and French West African territories. Author is an American journalist with wide experience in political reporting from Africa.

Herskovits, Melville J. *The Myth of the Negro Past*. Boston: Beacon Press, 1958. A scholarly analysis of the American Negro's African antecedents and his retention of certain "Africanisms." An excellent introduction to the history of pre-colonial Africa. Especially relevant for students of African affairs for its treatment of certain misconceptions surrounding African civilization. Author, an anthropology professor at Northwestern University, has long been the foremost American authority on Africa.

Hodgkin, Thomas. *African Political Parties*. Baltimore: Penguin, 1962. A comprehensive survey of political parties and movements in contemporary Africa. Author, a history professor at Oxford University, is an excellent observer of African political affairs.

Wallerstein, Immanual. *Africa—The Politics of Independence*. New York: Random-Knopf Vintage, 1961. An analysis of social conflicts in colonial Africa which generated the nationalist movement and a discussion of certain problems of nation-building and maintenance after independence. Author is a sociology professor at Columbia.

THE REVOLUTION IN FRENCH CAMEROUN: 1956–1960

SYNOPSIS

In 1955, the *Union des Populations Camérounaises* (UPC), a militant nationalist political organization in French Cameroun, was outlawed for its part in the nationalist agitation and violent anti-European demonstrations and riots which occurred in May of that year. The UPC's openly admitted connection with the French Communist Party was cited as a reason for banning it. Leaders of the UPC either went underground or into exile, and from 1956 to 1960 they conducted an intense campaign of terror and guerrilla warfare against both the French colonial administration and the African leadership to which the French were gradually transferring governmental authority. That the UPC revolutionaries were not successful was largely due to France's decision to relinquish control over the country voluntarily, leaving behind an indigenous administration strong enough to contain the revolution by a combination of military force and political concessions. After French rule came to an end early in 1960 there were sporadic outbreaks of violence led by UPC factions still operating in one tribal area; however, a major part of the UPC had returned to constitutional politics and was functioning as a legal opposition party which solved the demands of the southern tribes and opposed continued reliance on French military support.

BRIEF HISTORY OF EVENTS LEADING UP TO AND CULMINATING IN REVOLUTION

At the end of World War I, Germany's West African colony of Kamerun was divided into British and French zones, which were administered by those powers as mandate territories under the League of Nations. In practice, these territories became colonial possessions, British Cameroons being integrated into the British colony of Nigeria and French Cameroun being integrated into French Equatorial Africa.

French Cameroun, comprising about four-fifths of the former German colony, was more highly developed economically and culturally and more politically conscious than other parts of French Equatorial Africa. The memory of their past association with German, British, Dutch, Portuguese, and other European traders, missionaries, and adventurers stimulated an awareness of the outside world among the tribal peoples of the southern part of the territory that was not shared by the more isolated peoples of the northern interior regions.

After World War II a nationalist movement developed in this southern region, and Camerounian trade union leaders and intellectuals established ties with the nationalist leadership of other African territories and with French leftists, who were then the only European political group sympathetic to African nationalism. The most celebrated expression of this nationalism was the Bamako conference held in French West Africa in 1946. From this meeting of African nationalists and European leftists, many of whom were Communists, came the *Reassemblement Démocratique Africaine* (RDA), a comprehensive interterritorial nationalist organization with local affiliates in every French African colonial territory. The Camerounian UPC was such an affiliate until 1951, when it broke with the RDA over the expulsion of Communist members and influence from the RDA.

THE ENVIRONMENT OF THE REVOLUTION

DESCRIPTION OF COUNTRY

Physical characteristics

Located on the west coast of Africa, just north of the Equator, the territory which prior to 1960 was known as French Cameroun was bounded on the west by the Atlantic Ocean, the British colonial territories of Nigeria and the British Cameroons, on the north by Lake Chad, and on the east and south by French Equatorial Africa and the small Spanish colony of Rio Muni. Its triangular-shaped territory is somewhat larger than the state of California. The terrain is characterized in the southern and coastal regions by lowlands and a gradually rising plateau, which gives way in the northern and inland regions to mountain ranges up to 8,000 feet in elevation. The climate in the southern and central parts of Cameroun is tropical, and the annual rainfall is well over 100 inches; there are dense forests with heavy tropical undergrowth. Farther inland and in the northern region, generally known as the "savannahs," rainfall is considerably lighter, and the characteristic vegetation is tall grasses and small trees. There are four river systems, emptying into the Atlantic, of which the largest is the Sanaga River.[1]

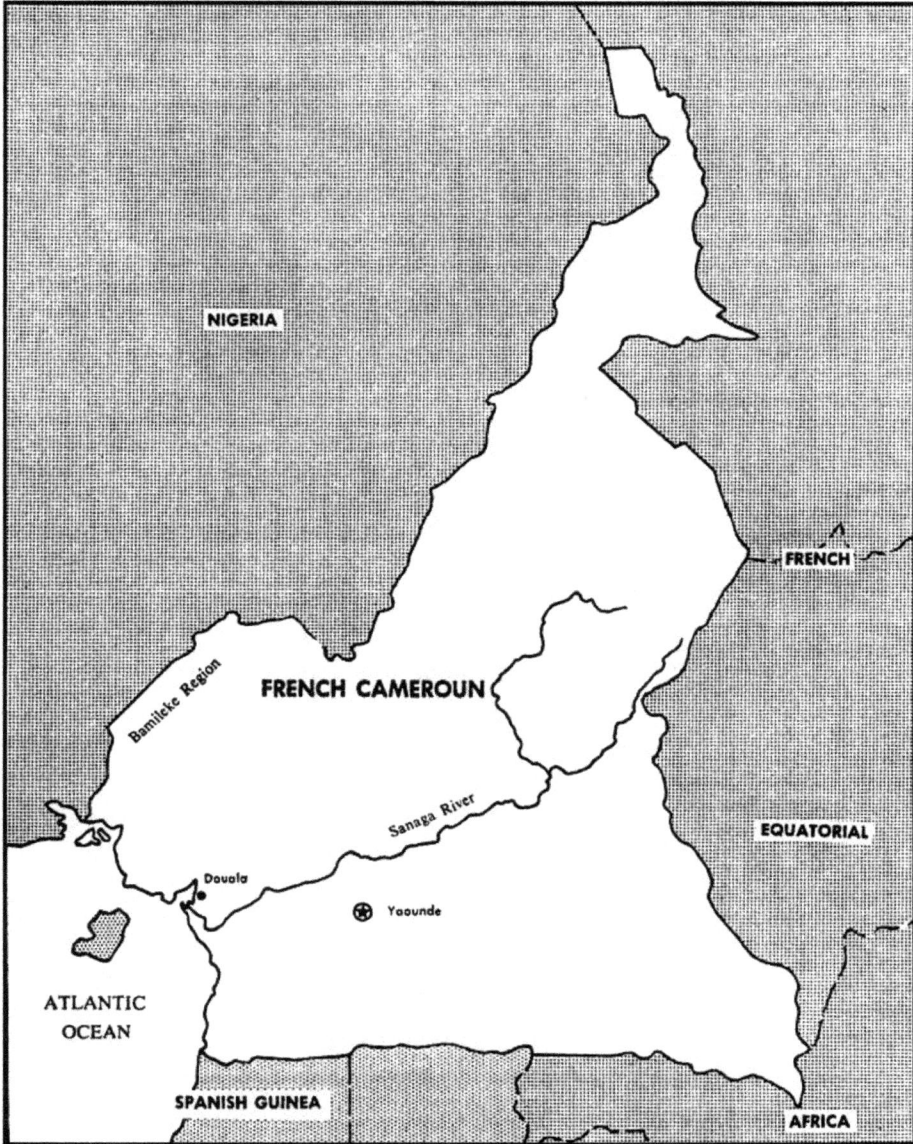

NIGERIA

FRENCH

Bamileke Region

FRENCH CAMEROUN

Sanaga River

EQUATORIAL

Douala

⊕ Yaounde

ATLANTIC
OCEAN

SPANISH GUINEA

AFRICA

The people

The population of French Cameroun was estimated in 1951 to be just over 3 million; of these, only about 12,000 were Europeans. Population density was greatest in the coastal region around Douala, Cameroun's leading seaport and commercial center. Douala had a population in the 1950's of almost 100,000, of whom about 5,000 were Europeans. The administrative capital, farther inland, is Yaounde, a city numbering around 30,000 in 1951, of whom some 3,000 were European. The native population is divided into more than 200 separate tribes speaking as many separate languages and dialects. The majority of the Camerounians speak languages belonging to the Bantu family; for instance, the important Douala, Basa, and Bamileke tribes in the southern region are Bantu. The Bamileke, numbering a half million, are the leading tribe in the southern regions and their language is often referred to as "Bantoid" because it varies so markedly from other Bantu dialects. The northern tribes speak languages related to the Sudanic family; for instance, the important Hausa and the Fulani tribes, whose kinsmen are found throughout West Africa, are Sudanic. There are also a few thousand pygmies in Cameroun, but these people have not taken an active part in the sociopolitical life of the country.[2]

Communications

Although primitive by European and American standards, French Cameroun's communication facilities were not inferior by West African standards. Transportation and communication facilities were most highly developed in the coastal region between Douala and Yaounde. The German-built railway system in the Douala area, designed to serve European banana plantations, was extended by the French to Yaounde in the interior. In 1956 there were 323 miles of railway. The French constructed over 6,000 miles of roads, of which 343 were paved. Internal river transport was negligible; however, the harbor facilities at Douala were well equipped for external water transport. There were large airports at Douala and Yaounde, and a number of smaller fields (22) throughout the country.[3]

Natural resources

Other than fertile soil and tropical climate, the country's only important natural resources are its extensive bauxite deposits, in the coastal region along the Sanaga River, and large but unexploited iron deposits on the coast near Kribi.[4]

SOCIO-ECONOMIC STRUCTURE

Economic system

The economy of French Cameroun was based primarily on agriculture. The country was self-sufficient in foodstuffs. The leading agricultural export crops were cocoa, coffee, and bananas, followed by timber, tobacco, rubber, and peanuts. Most of the banana production came from European-owned plantations in the Douala area. Industry was limited, although at Edea, on the Sanaga River, there was a modern aluminum plant, with a capacity of 45,000 tons per year, and at Eseka there was a large modern sawmill. Electricity for the Edea aluminum plant was provided by a large hydroelectric dam and power plant on the Sanaga. There were also such small industries as a brewery, a cigarette factory, a small factory producing aluminum pots, soap factories, and palm and peanut oil mills.[5]

After World War II, the Camerounian economy became heavily dependent on French public and private financial assistance. The costs of industrialization and modernization of agriculture could not be met by the country's agricultural and mineral exports, which before World War II had netted a surplus of exports over imports. During the period from 1947 to 1959 French financial assistance to Cameroun amounted to about 77 billion francs ($1 equals 247 francs, *Colonies Françaises Afriques*).

Class structure

There was a small European elite, composed of plantation owners and French administrators. Of greater social significance was the native traditional elite, composed of tribal chieftains and their families, many of whom had profited by European rule by becoming paid civil servants of the French colonial administration or by engaging in agricultural and commercial enterprises. European settlement in West Africa never reached the proportions attained in Algeria, South Africa, and Kenya, due to the severity of the climate and the prevalence of tropical diseases. As a result, the native traditional elite was not destroyed as a social group, although individual members were often replaced by others more acceptable to the European authorities. Rather, the West African tribal elite survived under colonial rule as an economic and social, if not political, elite. Tribal chiefs became, in many cases, the leading planters and merchants of their areas, particularly in the coastal regions.

European influence led to the creation of a Western-educated and oriented middle class, antagonistic toward both the foreign colonial administration and the native tribal structure. In French Cameroun this emerging middle class was composed of small traders, small farmers, professional people, skilled industrial workers, minor civil servants, and students and intellectuals. This middle class was, to varying degrees, detribalized and Westernized in terms of education, cultural pattern, and political consciousness. The mass of the population remained in a tribal or semi-tribal state and was not politically conscious, although detribalization and recruitment into the emerging middle class was proceeding rapidly in the southern coastal region, where economic activity was concentrated.

The socio-economic structure was characterized by an unequal distribution of wealth which greatly favored the European and indigenous traditional elite groups. Although "Africanization" of the administration and economy had proceeded further in French Cameroun than in other territories of French Equatorial Africa, the emerging Camerounian middle class found their upward social mobility impeded by the indigenous traditional elite, who, together with the Europeans, dominated the commanding heights of the country's socio-economic and political structure.

Literacy and education

French Cameroun had a higher level of literacy and education than most parts of French Africa. Education was the largest single item in the territory's budget during the terminal period of French rule. In the southern coastal section nearly every child of school age was attending either a mission or government primary school; in the northern part of the country, however, the number of children in school was negligible. In the south the literacy rate was high by African standards—about 50 percent of the population below the age of 40. There were over 1,000 Camerounians studying abroad, mostly in France, in institutions of higher education.[6]

Major religions and religious institutions

In 1956 it was estimated that there were 500,000 Protestants and 622,500 Catholics in French Cameroun, concentrated primarily in the southern coastal regions, and 600,000 Muslims, living for the most part in the northern interior. The remainder of the population, about 1½ million, adhered to various primitive tribal religions and were classified as pagans or animists. Protestant (British Baptist) missionaries were active along the coast of Cameroun as early as the 1840's, and after 1890 Catholic missions were established in the southern region. The dominant Fulani and Hausa tribes of the north have traditionally been adherents of Islam, although the subject tribes in the north have generally not been converted to Islam. The dominant tribes and most outstanding public figures in the country have been for the most part, either Muslim or Christian, despite the relative numerical inferiority of these world religions.[7]

GOVERNMENT AND THE RULING ELITE

Description of form of government

French Cameroun was governed directly by France as a colonial territory from 1916 to 1957 and indirectly as a semicolonial territory with local autonomy from 1957 to 1960. In 1957, France permitted the Camerounian National Assembly, an elected legislative body established in 1946, to select a prime minister and cabinet. During the terminal stage of French rule, Cameroun had a parliamentary system of government patterned after the French Government. This Camerounian "government" was responsible to Parliament and had charge of a large share of the country's domestic affairs, although France retained control of foreign affairs and, through the French High Commissioner and colonial administration, continued to exert a decisive influence over the affairs of the country.

The fully independent Camerounian Government which was set up early in 1960 is likewise patterned on the French governmental system. The Camerounian constitution adopted February 21, 1961, provides for a nationally elected legislative body and President. The President appoints the Prime Minister and Cabinet, who are then responsible to the legislature. The judiciary is independent of the executive and the legislature.

Description of political process

French Camerounian politics have been guided more by personalities and ethnic identifications than by ideological, economic, or social issues. This is typical of sub-Saharan African politics. However, French Cameroun is unique among the countries of French Africa in not having developed at least superficial and formal ties with political groupings elsewhere in French Africa. By 1959, there were 84 legally constituted political parties, all of which advocated complete independence and the reunification of the British Cameroons with Cameroun.[8]

The first Camerounian Government, under Prime Minister M'Bida (May, 1957-February, 1958) and the succeeding government of Prime Minister Ahidjo, which effected the transition to full independence in 1960, were little more than personal coalitions of legislators loyal to the Prime Minister. Most of Prime Minister Ahidjo's supporters were organized into the *Union Camérounaise* (UC), but the Ahidjo government also drew support from an unorganized and unstable bloc of independents and tribal-based politicians.

Opposition to the government, under both M'Bida and Ahidjo, came primarily from the *Union des Population Camérounaises* (UPC).

The UPC has been described as the only real political party in the country. Although it too was tribally and regionally based, its leaders evidenced definite ideological and social motivations.

The UPC was outlawed by the French authorities in July 1955, and for the next three years it operated illegally as a terrorist organization. Between the fall of 1958 and the spring of 1960 the UPC passed gradually from terrorism to legal political opposition to the now fully independent government of Prime Minister Ahidjo. Another source of opposition to the Ahidjo regime after 1958 was former Prime Minister M'Bida, who went into voluntary exile in Guinea and opposed the government on the grounds that it refused to hold new elections before the country became fully independent.

Legal procedure for amending constitution or changing government institutions

Prior to 1960 Cameroun's governmental institutions were creations of the French Government in Paris and as such could be changed only through the regular workings of the French political system. During the terminal stage of French rule (1957–60) Cameroun was governed in accordance with the French *Loi-Cadre* of June 23, 1956; although this document served the usual functions of a constitution, no amending procedure was provided.

The 1960 Constitution recognized the amendment and revision initiative as belonging, concurrently, to the President of the Republic, the Council of Ministers, and the National Assembly.

Every revision and amendment proposal had to be endorsed by at least one-third of the members of the National Assembly before being considered. To become law, the proposal had to be ratified by at least a two-thirds majority of the members of the National Assembly. If it failed to receive the required two-thirds majority, but received a majority of the votes in the National Assembly, it was then submitted to a national referendum.

Amendments and revisions intended to alter the republican form of government, Cameroun's territorial integrity, or the democratic principles that govern the Republic could not be considered.

Relationship to foreign powers

French Cameroun was entirely dependent on France with respect to the political, military, and economic control of its territory. Even after Cameroun became fully independent on January 1, 1960, France retained close economic and financial ties with the country, and French

African military forces were called upon to help the small Camerounian militia restore order during the early months of independence.

Camerounians have been very conscious of their German affiliation prior to World War I. There were many German planters and merchants in the country during the interwar period, and many Camerounians received a German education, some of them in Germany. Particularly in the southern region, ties with the British Cameroons remained close throughout the period of French colonial rule. Consequently, Camerounians tended to be more aware than other French Africans of non-French foreign influences.

The role of military and police powers

During French rule there was a police force (*gendarmerie*) of not more than 2,000 officers and men. The men and some of the officers were African, but the senior police officials were French. In addition to this local force, the French had access to French African troops from neighboring colonial territories of French Equatorial Africa. Several thousand of these African troops from neighboring Chad were called in during the 1956–60 military campaigns against the UPC guerrillas; the Ahidjo regime likewise made use of two battalions of these French African troops, since the Camerounian Army proved insufficient to deal with remnants of the UPC guerrilla bands still holding out in Bamileke district after independence.

WEAKNESSES OF THE SOCIO-ECONOMIC-POLITICAL STRUCTURE OF THE PREREVOLUTIONARY REGIME

History of revolutions or governmental instabilities

There were no serious tribal uprisings and no breakdown in French colonial authority prior to the outbreak of violence in 1955–58. After the military defeat of the German colonial authorities in 1916 the French occupation proceeded without serious incident. In 1940 the French general in command of the colony declared his allegiance to General de Gaulle and the Free French Forces. No military engagements took place in Cameroun during World War II.

Economic weaknesses

In common with other countries in tropical Africa, French Cameroun has been increasingly dependent on foreign technical and economic assistance for its industrialization and modernization programs. For this reason, Cameroun has had to maintain a close political relationship with France. The distribution of wealth within the country has posed a serious problem and has exacerbated social tensions.

Social tensions

A major source of social tension is the traditional antagonism between the Muslim-dominated non-Bantu tribes of the north and the Christian and pagan Bantu tribes living along the coast in the south. The coastal peoples, through their long association with the Europeans, have become the most literate and politically conscious group in Cameroun. This part of the country is more developed economically and, in general, is more advanced culturally than the northern interior regions. The north has long been dominated by the historically powerful Fulani tribal elite. The French left this group in local control, ruling "indirectly" through them. The masses of northern Cameroun were not politically conscious enough to become actively involved in the nationalist movement, which originated and was based primarily in the southern coastal region. The conservative Fulani elite tended to side with the French authorities.

In the southern and western regions, there are social tensions within and among the various Bantu and Bantoid tribal communities. The most serious social situation is in the overpopulated Bamileke tribal area on the western border, where the UPC has been able to exploit social and economic grievances "directed not against the government but against a traditional authoritarian system within the tribe which concentrates wealth, power, and eligible women in the hands of the chiefs."[9] The Bamileke district is the only area in the south which is still ruled by powerful hereditary chiefs, and "some of their administration is by all accounts harsh and corrupt."[10] Attempts to democratize local government have been halfhearted and unsuccessful, partly because the dangers of the situation were not fully realized, but partly because the Bamileke themselves have been ambivalent as to the changes they would like to effect. For example, most of the Bamileke are attached to the institution of chieftainship, but they oppose chiefs who abuse their authority. Thus the essential problem among the Bamileke has been that of a disintegrating tribal structure. "Land hunger and an autocratic and sometimes corrupt administration have shackled an energetic people and produced a simmering discontent which needed very little to spark it into explosion."[11]

Government recognition of and reaction to weaknesses

The French certainly recognized the north-south conflict in Cameroun, and the charge has often been made that they intentionally did nothing to discourage this regional antagonism in the hope that it would delay the spread of Camerounian nationalism from the southern coastal section. Camerounian nationalists, on the other hand, attempted to draw progressive young northerners into the nationalist movement. The intratribal social antagonisms of the south and west, if recognized at all, were approached by both the French authorities and the Camerounian nationalists as problems to be solved primarily by industrialization and economic development. Political solutions were advanced only reluctantly, as in 1957–60, and there was frequent resort to military force and police action, both by the French and the Camerounian authorities. Ahidjo made a serious effort in the spring of 1960 to attract mass political support from the Bamileke tribal areas by organizing a "People's Front for Peace and Unity" composed of deputies from those areas. The Front was given three ministries in the government.

FORM AND CHARACTERISTICS OF REVOLUTION

ACTORS IN THE REVOLUTION

The revolutionary leadership

The leading actors in the UPC-directed revolution were Ruban Um Nyobe and Dr. Félix-Roland Moumie. The UPC's first leader when it was still a legal political party, Ruben Um Nyobe, was an African trade union official, who allegedly had received paramilitary training and political indoctrination in Prague at a Communist center for Afro-Asian trade unionists. In 1952, he went to the United Nations to make certain nationalist demands on the French administration before that world body. After the UPC was banned in 1955, Ruben Um Nyobe, who had become Secretary-General of the revolutionary organization when Dr. Moumie became President, hid out in the jungles of Sanaga-Maritime District. From there he directed guerrilla operations against the French and Camerounian authorities from July 12, 1956 until September 1958, when he was killed by a French military patrol.[12] The son of a Bassa witch doctor and educated by Protestant missionaries, Ruben Um Nyobe combined traditional tribal rituals with modern guerrilla warfare tactics. He was probably not a doctrinaire Communist, but he was a militant nationalist and was willing to utilize Communist material and moral support to the fullest in the fight for independence.[13]

The most widely known leader of the UPC was Dr. Félix-Roland Moumie, who served as President and leader-in-exile of the organization from the time it was banned in 1955 until his death in 1960. Born in 1926, Moumie became a Marxist as a young medical student in France. When he returned to Cameroun he came under the influence of Ruben Um Nyobe and the nationalist movement. Along with several other leaders of the UPC, Moumie fled across the border into British Cameroons in 1955. Later he headed the UPC-in-exile organization, first in Cairo and later in Conakry, Guinea. Moumie became the leading international spokesman for the UPC, writing in leftwing international journals and appearing at Afro-Asian conferences and in Soviet bloc capitals. He maintained that he was a Marxist and a nationalist, but not a Communist; his reliance on Soviet support and material assistance was probably dictated by political expediency rather than by ideological conviction. He died in a Geneva hospital, in October 1960, of thallium poisoning, administered, according to Moumie, by a French ultraconservative terrorist organization known as the "Red Hand."[a][14]

The revolutionary following

The UPC had wide mass support both in the urban centers of Douala and Yaounde, among the newly emerging African middle class and the large African proletariat, and in the rural districts of Sanaga-Maritime and Bamileke. The Basa and Bamileke tribes indigenous to these districts were well represented in the urban areas, and this provided a link between rural and urban branches of the UPC organization. French authorities estimated in 1958 that as much as 80 percent of the population in Sanaga-Maritime and Bamileke districts supported the UPC.[15] In these rural areas all elements of the population except those closely identified with the French, such as the Bamileke tribal hierarchy, supported the nationalist organization. From these elements were drawn the several thousand combat personnel whom the UPC organized into guerrilla bands.

ORGANIZATION OF REVOLUTIONARY EFFORT

Internal organization

The UPC was first organized in 1947 as an affiliate of the *Rassemblement Démocratique Africain* (RDA), an interterritorial nationalist political organization with branches throughout French Africa. The Camerounian branch left the RDA in 1951 in protest against the nationalist organization's decision to break its links with the French

---

[a] See p. 281 in the North Africa section.

Communist Party.[b] The UPC remained a militant but legal nationalist political party until the spring of 1955, when it convened a mass meeting of trade unions affiliated with the French Communist Trade Union Federation, the *Union Démocratique des Femmes Kamérunaises* (UDFK), the *Jeunesse Démocratique du Kamérun* (JDK), and the UPC. The militant nationalist speeches and propaganda that emanated from this mass rally, held on April 22, 1955, set off a wave of anti-French demonstrations and violence to life and property. French countermeasures caused several thousand deaths. Finally, in July 1955, the French outlawed the UPC and its affiliated feminist and youth organizations, the UDFK and the JDK.

From 1955 to 1958, the UPC existed as an illegal underground organization. Based in the districts of Bamileke and Sanaga-Maritime, guerrilla bands carried on sabotage, terrorist, and paramilitary operations against both the French and the Camerounian officials whom the nationalist militants accused of collaborating with the French. In September 1958, in response to the Ahidjo government's offer of amnesty, Mayi Matip, one of Ruben Um Nyobe's lieutenants in the Sanaga-Maritime District, returned to constitutional politics. Since 1958 Matip's wing of the UPC has operated as a legal opposition party. A militant wing of the UPC remained loyal to Moumie, however, and guerrilla operations continued in the Bamileke district. Since Moumie's death this wing has reportedly split into two or more guerrilla bands, led by Ernest Ouandi and Abel Kingue, both former lieutenants of Ruben Um Nyobe.[16]

These splits within the UPC reflect the impact which the environment of the revolution had on its leadership. First of all, there were always conflicts between the exiled leadership abroad and the guerrilla leaders in the jungles at home. Secondly, the countermeasures of the government, which combined military repression with liberal offers of amnesty, the rapid progress of the country toward complete independence, and the death of the most capable guerrilla leader, Ruben Um Nyobe, contributed to the disintengration of the UPC organization in 1958–60.

External organization

After going underground in 1955, the UPC depended heavily on foreign bases of operation and foreign moral and material support. After remaining for a few months in British Cameroons, the UPC shifted its headquarters-in-exile to Cairo. There Moumie and his staff were in contact with both Egyptian and Soviet representatives, and

---

[b] See General Discussion, p. 331.

the UPC received wide acclaim in both the Nasserist and the Soviet press. According to French reports, UPC guerrillas have received Czech arms and war materiel, as well as propaganda material published in the Soviet bloc. When French Guinea became independent in October 1952, Moumie moved UPC headquarters to Conakry in order to be closer to Cameroun.

GOALS OF THE REVOLUTION

Concrete political aims of revolutionary leaders

In 1952, the demands of the UPC leadership included: the reunification of British Cameroons with French Cameroun; the establishment of an elected legislative assembly for the reunified territory; the establishment of a governing council with an African majority in the proportion of five to four; and the fixing of a date for full independence. The UPC's major political demand, which precipitated the nationalist riots in the spring of 1955, was a call for U.N.-supervised elections to a constituent assembly by December of that year. The agitation, in turn, led to the outlawing of the UPC. When the UPC began guerrilla warfare in 1956 the leadership announced four minimal demands: reunification, an elected constituent assembly, amnesty for the guerrillas, and freedom of expression. When the French finally held elections in December 1956, the UPC called for a boycott of these elections, claiming that they would be rigged. Refusing to accept the results of the elections, the leaders of all factions of the UPC—Moumie, Matip, and the guerrilla leaders—demanded that the United Nations hold "free elections" in the country prior to the granting of complete independence on January 1, 1960.[17] The United Nations refused to hold new elections, and the UPC refused to recognize the legitimacy of the Ahidjo regime.

Social and economic goals of leadership and following

Although Moumie frequently expressed the desire to found a "socialist state"[18] patterned after those of the Soviet bloc, once the French had been expelled, the UPC was much too concerned with merely winning independence to prepare a detailed program of social and economic reform. The UPC's following probably expected land reform, rapid Africanization of the civil service, and a higher standard of living after the political objectives had been won.

REVOLUTIONARY TECHNIQUES AND GOVERNMENT COUNTERMEASURES

Methods for weakening existing authority and countermeasures by government

In the period when the UPC was a legal political party it organized mass demonstrations and distributed inflammatory nationalist propaganda protesting French rule and articulating its political demands. This phase of operations reached a crescendo in April and May 1955, when the UPC's nationalist agitation set off large-scale riots in Douala and Yaounde in which a number of Europeans, as well as many Africans, lost their lives. The French reacted with severe police repression in the African quarters of these cities, and on July 13, 1955, the UPC was outlawed. Thereafter, the UPC operated as a clandestine organization and distributed propaganda advocating its political demands, such as the holding of national elections to an All-Cameroun Constituent Assembly by the end of the year.[19]

The next phase of UPC operations began in 1956, when guerrilla warfare broke out in two provinces: Sanaga-Maritime, between Douala and Yaounde, and Bamileke, on the western border of the country adjoining British Cameroons. The UPC leaders, some of whom were abroad while others stayed on in the country, were apparently in complete control of the paramilitary operations, which continued during the next few years. The UPC guerrillas, to whom the French conceded the name "Maquis,"[c] sabotaged transportation and communication lines, burned plantations, and attacked villages believed friendly to the French, often killing unpopular tribal chieftains in the Bamileke areas. The heavy concentration of Bamileke tribesmen in Douala and Yaounde provided the UPC with urban bases for terrorist operations in these cities.[20]

The UPC's guerrilla operations slackened off considerably toward the end of 1958. This was due, first of all, to the loss of their capable underground leader, Ruben Um Nyobe, on September 13, 1958, and, secondly, to the effective combination of countermeasures adopted by the French and Camerounian political authorities in 1958. In February 1958, the first African Prime Minister of Cameroun, André-Marie M'Bida, was succeeded by Prime Minister Ahmadou Ahidjo, who adopted a markedly different policy towards the UPC and the entire nationalist movement. Apparently with French approval, Ahidjo came out strongly for complete independence and reunification with British Cameroons, thus stealing the UPC's thunder. Moreover, Ahidjo

---

[c] Name given to the French guerrillas during World War II.

offered amnesty to all members of the UPC who "returned to legality"; for those who continued to resist he threatened vigorous military countermeasures. After Ruben Um Nyobe's death in September, Mayi Matip, one of his trusted lieutenants, returned to constitutional politics and guerrilla activities virtually ceased in the Sanaga-Maritime District, leaving only the Bamileke area still unpacified. Matip's followers, calling themselves at first the *Force de Réconciliation* (as the UPC label was then still illegal in Cameroun) have functioned ever since as a legal opposition party. In 1948–49, Ahidjo called in additional military forces from Chad and other French African territories and undertook a determined military campaign against the terrorists and guerrillas of the various UPC factions. By December 1959, over 2,000 guerrillas had laid down their arms and accepted amnesty, and the French were sufficiently confident of the situation to withdraw most of their troops.[21] Two or more guerrilla bands continued to operate in the Bamileke district after 1960. These are apparently under the command of two of Moumie's lieutenants, Ernest Orandie and Abel Kingoe. The strength of these Bamileke guerrillas has been estimated to be around 2,000.[22]

Methods for gaining support and countermeasures taken by government

The UPC rapidly gained a mass following, especially in the southern coastal region, during its legal phase from 1947 to 1955. It was the first political organization in French Cameroun to espouse nationalist demands which, although in retrospect they appear moderate, were considered quite revolutionary at the time. The UPC also drew support from labor, youth, and women's organizations affiliated with it. When the UPC went underground in 1955 it retained most of its mass following and, as an illegal guerrilla organization, it may have attracted certain subversive and outlaw elements.

During this phase the nationalist organization actively enlisted the support of dissatisfied elements among the Bamileke and Basa tribes, both in the cities and in their tribal areas. Ruben Um Nyobe, himself a Basa and personally familiar with tribal rituals and superstitutions,[d] made extensive use of traditional tribal beliefs to insure the loyalty and security necessary to the functioning of a modern guerrilla organization. For instance, UPC guerrillas were, according to some sources, required to take an oath of secrecy, known among the Basa as the "Turtle Oath," and were told that Ruben Um Nyobe had direct communion with certain animistic spirits of tribal lore.[23]

---

[d] See pp. 359–360 of this summary.

When the French permitted the Ahidjo government to promise complete independence by 1960, the UPC sought to attract support by coming out for immediate independence and new elections. The northern-based Ahidjo regime was widely unpopular among southerners, and the UPC was able to attract considerable mass support in the south by these demands. Also, the UPC in-exile was able to claim the support of Ahidjo's parliamentary opposition, when former Prime Minister M'Bida joined Moumie in Conakry and also demanded new elections before independence. In the United Nations the UPC had the support of the Soviet bloc and an Afro-Asian bloc of new states led by Guinea. The Western states and the more conservative Afro-Asian delegations, however, upheld the Ahidjo regime's refusal to hold elections until after independence. UPC influence on Guinea, Ghana, and Liberia led these countries to withhold diplomatic recognition of the Ahidjo government for several months. Since independence the Ahidjo regime has taken effective measures to counteract the UPC's appeal to nationalist and southern sentiment. These measures are discussed in Part IV as institutional effects of the revolution.

MANNER IN WHICH CONTROL OF GOVERNMENT WAS TRANSFERRED TO REVOLUTIONARIES

Although UPC's revolutionary goals were achieved in substance by the accession to full sovereignty of the Ahidjo government on January 1, 1960, the revolutionaries did not themselves come to power, and in that sense the 5-year-old revolution against both the French and the Camerounian authorities was not successful.

THE EFFECTS OF THE REVOLUTION

CHANGES IN THE PERSONNEL AND INSTITUTIONS OF GOVERNMENT

The UPC-led nationalist revolution resulted in the establishment of a sovereign state, patterned after the French political system. In February 1960, in order to accommodate the demands of antigovernment political forces and provide a wider base of popular support, the Ahidjo regime replaced the French *Loi-Cadre* of June 23, 1956, with a Camerounian Constitution, under which Ahidjo became President and Charles Assale Prime Minister of a national coalition Cabinet. A southerner and a supporter of Ahidjo, Assale has been described as a "middle-of-the-road uncontroversial figure [who] gives Ahidjo, whose support is almost entirely in the North, a useful foothold in the South of the country."[24] In April 1960 new elections were held, in which Matip's legal faction of the UPC was permitted to participate freely. It won 22 seats in a national legislature of 100 and Matip was offered a Cabinet post. He refused on the grounds that one ministerial post did not constitute adequate recognition of the UPC's political strength.[25]

MAJOR POLICY CHANGES

Owing to the Ahidjo regime's close ties with France and its vigorous suppression of the UPC, which enjoyed wide popularity among radical elements in several of the newly independent African states, the new republic of Cameroun at first had the reputation of being a French puppet. It was not until after the Ahidjo regime had legalized the UPC that Cameroun entered into diplomatic relations with all of its West African neighbors. Since that time, Cameroun has pursued a moderate and generally pro-Western foreign policy. It is significant that as late as the summer of 1961 the Soviet bloc had not sent any diplomatic representatives to Yaounde, apparently expressing the bloc's continued contempt for the Camerounian Government.[26]

LONG RANGE SOCIAL AND ECONOMIC EFFECTS

It is too early to discern any long range socio-economic effects of independence. The country has continued to receive French capital and technical assistance, and the economy is gradually being industrialized and modernized. There has been no serious attempt to solve the problems of overpopulation in the Bamileke District, other than to provide additional economic opportunities for those who migrate to urban areas. Urban problems will probably be among the most serious social weaknesses facing the country in the next few decades.

OTHER EFFECTS

The immediate problem facing the Ahidjo-Assale regime concerns the continued tribal unrest in the Bamileke District. An encouraging factor has been the formation of the *Front Populaire pour L'Unité et la Paix* (FPUP), composed of 18 members of Parliament from Bamileke and other southern constituencies, which cuts across party lines and seeks to articulate Bamileke and other tribal demands by constitutional processes. The FPUP, working closely with many members of the Matip faction of the UPC, stands for the removal of all foreign troops and the formation of a strong national army; the surrender of rebel bands and their acceptance of the government's offer of amnesty; and participation in the national coalition government of Prime Minister Assale. The FPUP was given three ministerial posts in this government.[27] After October 1, 1961, the southern part of British Cameroons will be reunified with Cameroun as a federal region with a certain degree of local autonomy to accommodate the historical differences that separate this formerly British-administered territory from the former French Cameroun. The Political effect will be to increase greatly the political strength of the southern region relative to the north of the country.

NOTES

1. "Cameroons," *The Encyclopaedia Britannica* (Encyclopaedia Britannica, Inc., 1958), IV, 662–663.
2. Ibid., pp. 664–665.
3. "Cameroun," *The Worldmark Encyclopedia of the Nations* (New York: Harper and Brothers, 1960), 123.
4. Michael Crowder, "Background to Independence: The Cameroons," *Africa Special Report*, IV, 3 (March 1959), 8.
5. Ibid.
6. Ibid.
7. "Cameroun," *Worldmark Encyclopedia*, p. 122.
8. Ibid., p. 124.
9. Helen Kitchen, "Cameroun Faces Troubled Future," *Africa Special Report*, V, 1 (January 1960), 2.
10. Ibid.
11. Margaret Roberts, "Political Prospects for the Cameroun," *World Today*, XVI, 7 (July 1960), 307.
12. Ibid.
13. Ronald Segal, *Political Africa* (London: Stevens, 1961), p. 312; see also Walter Z. Laqueur, "Communism and Nationalism in Tropical Africa," *Foreign Affairs*, XXXIX, 4 (July 1961), 618.
14. *Newsweek*, (December 23, 1957), 34.
15. *Time*, (November 14, 1960), 32.
16. Dr. Felix Moumie, "The Cameroons—A Question or Exclamation Mark," Review of International Affairs, IX, 194, pp. 11–13.
17. Segal, Africa, pp. 312–315.
18. Ibid.
19. *Time*, p. 32.
20. Roberts, "Prospects," pp. 306–309.
21. Crowder, "Background," pp. 7–9.

22. Kitchen, "Cameroun," p. 14.

23. *Newsweek*, p. 34.

24. Segal, *Africa*, p. 18.

25. Roberts, "Prospects," pp. 310–313.

26. Rolf Italiaander, *The New Leaders of Africa* (Englewood Cliffs, New Jersey: Prentice-Hall, Inc., 1961), p. 190.

27. Roberts, "Prospects," p. 312.

RECOMMENDED READING

BOOKS:

Hempstone, Smith. *Africa—Angry Young Giant*. New York: Praeger, 1961. Chapter 9 contains one of the best accounts available in English of the UPC uprising, including biographical data on Ruben Um Nyobe and Moumie, and a discussion of Nyobe's guerrilla tactics and operations. Author, an ex-Marine, is the African correspondent for the *Washington Evening Star* and the *Chicago Daily News*.

Italiaander, Rolf, *The New Leaders of Africa*. Englewood Cliffs, New Jersey: Prentice-Hall, Inc., 1961. Contains biographical discussions of 29 African leaders, including Ahmadou Ahidjo of French Cameroun. Quotes Ahidjo's statements before the United Nations and describes UPC activities. Author is a Dutch historian, explorer, and Africanist with wide personal experience in French Africa.

Segal, Ronald, *Political Africa*. London: Stevens, 1961. A who's who of African political leaders and parties. Contains biographical sketches of Ahidjo, Assale, Matip, M'Bida, and brief but factual sketches of the UPC and its factions and other Camerounian political groupings. Author is a South African journalist and political observer now resident in London.

The Worldmark Encyclopedia of the Nations. New York: Harper and Brothers, 1961. Contains an excellent factual survey of French Cameroun's history, economy, social structure, and politics.

PERIODICALS:

"Cameroun Vote Gives Ahidjo Firm Control," *Africa Special Report*, V, 5 (May 1960), 5. Brief account of independent Cameroun's first election results.

"Coalition Regime Sought in Cameroun," *Africa Special Report*, V, 6 (June 1960), 8. Brief account of political groups in and out of Camerounian Government.

Crowder, Michael, "Background to Independence: The Camerouns," *Africa Special Report*, IV, 3 (March 1959), 7–8, 15. General discussion of French Cameroun with emphasis on UPC guerrilla activity and French economic assistance in the territory.

"Dr. Félix Moumie Dies in Geneva Hospital," *Africa Report,* V, 12 (December 1960), 12. Brief account of circumstances of Moumie's death.

Kitchen, Helen, "UN Approval of Cameroons Independence Marked by Controversy over Elections," *Africa Special Report,* IV, 4 (April 1959), 4, 10. Background article on the preindependence election controversy.

Kitchen, Helen, "Cameroun Faces Troubled Future," *Africa Special Report,* V, 1 (January 1960), 23, 7, 14. This article and others in this issue contain excellent background information on the UPC revolution. Author is a journalist and an acknowledged authority on contemporary African politics.

Laqueur, Walter Z., "Communism and Nationalism in Tropical Africa," *Foreign Affairs* XXXIX, 4 (July 1961), 615, 618. Discusses UPC leaders' connections with communism and the Soviet bloc. Author is a well known authority on Soviet relations with Africa and the Middle East.

Le Vine, Victor T., "The 'Other Cameroons'," *Africa Report,* VI, 2 (February 1961), 5–6, 12. General discussion of British Cameroons with respect to reunification with French Cameroun in 1961. Author is an American political scientist who was conducting research in Cameroun in 1961 with a Ford grant.

Roberts, Margaret, "Political Prospects for the Cameroun Today," *World Today,* XVI, 7 (July 1960), 305–313. Describes the extent of guerrilla and terrorist activities in the southern region with particular emphasis on UPC bands in Bamileke District and discusses Ahidjo's attempts to solve tribal unrest among the Bamilekes.

THE CONGOLESE COUP OF 1960

SYNOPSIS

On September 15, 1960, the Congolese Army Chief of Staff, dissatisfied with the political disintegration of the Congo, staged a bloodless coup d'etat against the regime of Premier Patrice Lumumba with the help of some elements of the national army stationed in the capital city of Léopoldville. Congolese troops seized control of the country's only radio broadcasting station and surrounded the Parliament building, preventing all members of Parliament from entering. The President of the Republic was not affected by the coup. The Prime Minister was placed under house arrest, but was able to escape. There was no resistance, and the action was completed in less than 24 hours.

BRIEF HISTORY OF EVENTS LEADING UP TO AND CULMINATING IN REVOLUTION

The relative stability which had prevailed in the Congo during more than 70 years of Belgian rule gave way, in January 1959, to a wave of civil unrest. On January 5, 1959, 30,000 unemployed Congolese, led by some of the political elite, rioted in the capital city of Léopoldville. Although these riots were precipitated by economic conditions, they soon took political overtones, when the Congolese political parties united in their demands for economic and social reform, and eventual independence. This pattern of violence was to mark the last one-and-a-half years of the Belgian administration.

Faced with mounting internal and external pressures, resulting from the inability of the Belgian administration to cope with the agitations of the Congolese elite, and the increasing criticisms of independent African states and other countries, King Baudouin of Belgium expressed, on January 13, 1959, his resolution to lead the Congo to independence. In October 1959 the Belgian Government indicated its desire to grant the Congo some measure of independence in 1960. It called for the election of Congolese political party representatives in December 1959. These delegates would then represent the Congo at subsequent talks with the Belgian Government.

At the Brussels Round Table Conference on January-February 1960, the Congolese delegates presented a unanimous demand for immediate and total independence. An agreement was concluded providing for national elections in April and May 1960 to choose a legislature, the formation of a Congolese Cabinet by June 10, 1960, and independence on June 30, 1960.

It is too early to assess fully the motives of the Belgian Government in precipitately granting the Congo full independence. Fear of the consequences of not acceding to Congolese demands undoubtedly contributed to the decision. There was great pressure everywhere against the continuation of "colonialism" in any form and the Belgians were unwilling to face the bloodshed and high cost of a war which would make them very unpopular in international circles and which would still not allow them to maintain a secure hold on their possessions. Belgian business too may have felt that there was a better chance for salvaging some of its interests by granting independence and thereby gaining some good will from certain Congolese leaders.

In the elections, none of the political parties achieved a plurality. Thus, the Congolese Government which was formed on June 24, 1960, was, essentially, a coalition of the Congo's two major political groupings: the federalists, represented by the Association of the Bakongo (ABAKO) party, and its allies, and led by Joseph Kasavubu; and the centralists, represented by the National Congolese Movement (MNC) party and its allies, and led by Patrice Lumumba.

On July 8, 1960, elements of the Congolese National Army in Léopoldville mutinied against their white officers, and took to mob violence. The mutiny spread throughout the Congo in less than three days. On July 11, Belgian paratroopers occupied Léopoldville and other major Congolese cities. On that same day, Katanga Province proclaimed its independence; Kasai Province followed suit in August. On July 12, 1960, Patrice Lumumba called on the United Nations to eject the Belgian paratroopers, and to help restore order.

In the weeks that followed the arrival of U.N. troops, Lumumba's followers made repeated attempts to reimpose the Government's rule on Katanga and Kasai. These attempts precipitated a power struggle within the central government, between the centralists and federalists, culminating in a reciprocal dismissal by Joseph Kasavubu and Patrice Lumumba of one another. The legislature refused to accept either dismissal, and the jostling for power continued until the coup d'etat of September 15, 1960.

On that date, the Chief of Staff of the Congolese National Army, Col. Joseph Mobutu, announced that the army was taking over the government. He placed Premier Lumumba under house arrest, dismissed the Cabinet and suspended Parliament, and set up a caretaker regime, to be known later as the Council of Commissioners-General. No attempt was made to unseat the President of the Republic, Joseph

Kasavubu; and it appeared that Colonel Mobutu had thrown his support to the federalists. Little is known about the prior planning of the coup; spontaneity seems to have been its major characteristic.

THE ENVIRONMENT OF THE REVOLUTION

DESCRIPTION OF COUNTRY

Physical characteristics

The Republic of the Congo is in central Africa, and is crossed by the Equator in its north central region. It has an area of 905,380 square miles, one-third of the size of the United States, and is divided into six provinces: Léopoldville; Equateur, Orientale, Kivu, Katanga, and Kasai. Built around the basin of the Congo River and its tributaries, the Republic of the Congo has a low-lying central plateau surrounded by

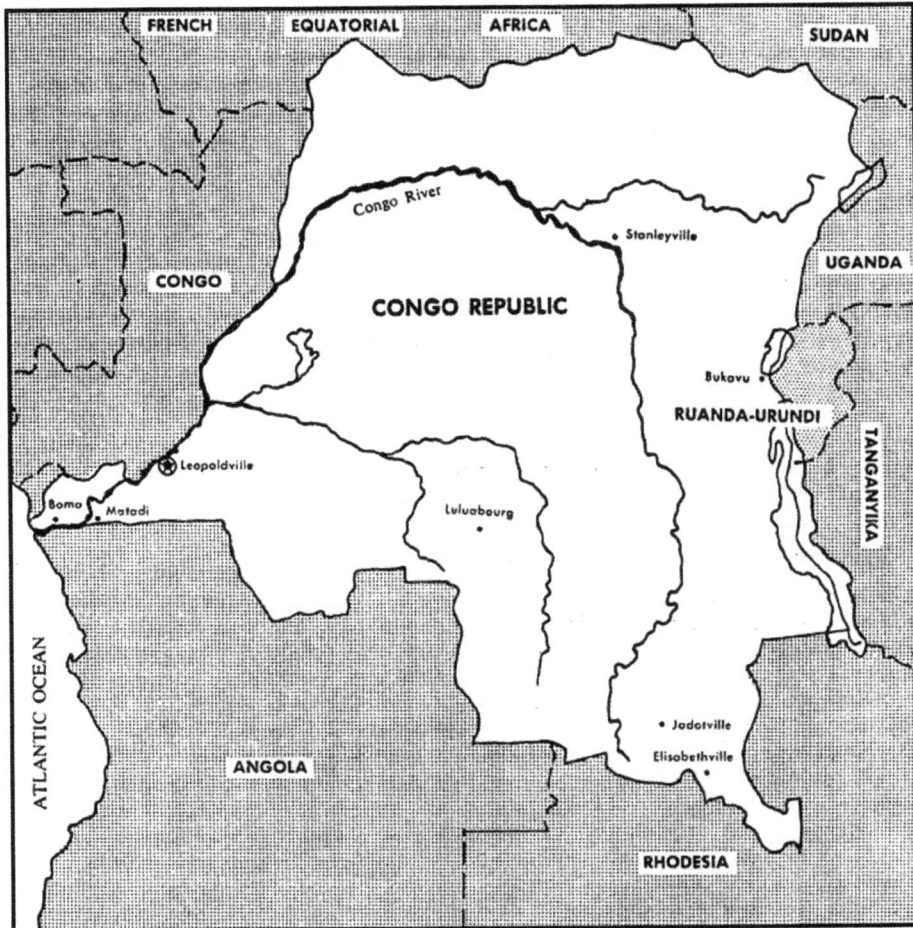

higher land, with mountain ranges in the east and southeast. Half of the land area is covered with equatorial forests, and the rest is mainly scrub and savannah. Hot and humid weather prevails in the lower western and central regions, with heavy rainfall. The higher regions in the east and southeast have a temperate climate.

The people

As of December 31, 1958, the population of the Congo Republic amounted to 13,658,185 Congolese, and 118,003 non-Congolese, 85 percent of whom were Belgians. The population density was approximately 13 persons per square mile. One-fourth of the total population are urbanized, and live in and around Léopoldville—the capital—Bukavu, Elisabethville, Jadotville, Matadi, Stanleyville, and Luluabourg. The Congolese are basically of the Bantu family, roughly divided into more than 70 ethnic groups, and subdivided into hundreds of tribes and clans. Of the more than 400 languages and dialects spoken, the principal ones are Swahili in the east and south, Tshiluba in Kasai, Kikongo in the Lower Congo, and Lingala along the Congo River.

Communications

The main outlets to the sea are the ports of Banana, Boma, and Matadi, all situated on the estuary of the Congo River. The Congo and its tributaries provide the Congo Republic also with its main channels of internal communication. Although they constitute a widespread system of communication, the many waterfalls and cataracts make frequent transshipments necessary.

Of the 87,412 miles of roads, only 20,890 miles are main roads, covered predominantly with sand, gravel, or laterite. Most of the hard-surfaced roads are in and around the major cities. As of 1957, there were 3,182 miles of railways. Some of these connect with the railways of Angola and Northern Rhodesia.[a]

Natural resources

The Congo Republic is blessed with immense mineral deposits, a rich topsoil suitable for agriculture, and timber-producing forests.

The most important mineral products are copper (in Upper Katanga Province), diamonds, tin, gold, and uranium. Others include cadmium, coal, cobalt, germanium, lead, manganese, niobium, palladium, silver, tantalum, tungsten, wolfram, and zinc. In the agricul-

---

[a] For further information see *The Worldmark Encyclopedia of the Nations* (New York: Harper & Brothers, 1960), 204–205.

tural field, the Congo Republic produces cocoa, coffee, copal, copra, rice, rubber, and sugar cane.[1]

SOCIO-ECONOMIC STRUCTURE

Economic system

The economy of the Congo Republic, prior to June 30, 1960, was based primarily on mining and agriculture.[b] Copper and diamond mining formed the industrial basis of the country; and local industries were being established to supplement the processing of mineral products. Agriculture is practiced throughout the Congo, and is geared to local requirements, although a few cash crops are raised for export.

During the Belgian administration, the economy of the Congo was expanded through a combination of private and government-owned enterprises. Mining enterprises were largely private concerns, while utilities were, as a rule, government-owned.

The main products for domestic consumption include textiles, cement, shoes, metal drums, soap, cigarettes, beer, vegetable oil, and cottonseed oil. The Congo, however, is not self-sufficient in any of these consumer-goods industries, and imports foodstuffs as well as vehicles and transportation equipment, machinery, fuels, lubricants, clothing and footwear, iron and steel products, pharmaceuticals, and chemicals.[2] Its chief exports are copper, diamonds, and uranium. Copper exports in 1960 represented 7.4 percent of the total world production, and diamond exports 15–18 percent of the total world value of diamond production. The Congo is one of the world's leading producers of uranium; however, the volume of production is not made public.

Class structure

The Congolese can be divided into two main groups: tribal and urbanized. The tribal group, comprising about three-fourths of the population, lives in rural areas, under primitive conditions, and provides the country with cheap unskilled labor. The urbanized group is beginning to form the nucleus of a modern society. It now includes a large lower class composed of semi-skilled labor, a small middle class of bureaucrats, shopkeepers, clerks, and army noncoms, and an even smaller upper class of professionals. Prior to 1960, the Belgian administrators and business people comprised the Congo's social-economic

---

[b] The civil unrest which erupted after June 30, 1960, and still persists, makes it difficult to describe the economy. If Katanga became independent, then the whole economic structure of the Congo Republic would be drastically altered. Much of the economic wealth is concentrated there.

and political elite, with an average per capita income of $2,973, compared to that of $44 for the Congolese. The present Congolese elite, or *Évolués* (those who have progressed), who replaced the departing Belgians, is composed of a small number of high school and college graduates, the skilled craftsmen of the middle class, and the army noncoms some of whom are now officers. These *Évolués* are, for the most part, missionary-trained, and form, as a whole, an emancipated westernized middle class.

The growth of Congolese urban society in the past decade seems to indicate great social mobility. This is due mainly to the policies of the Belgian administration and the mining enterprises. The colonial administration discouraged the mass immigration of Belgian nationals, and recruited Belgians only for the upper echelons of its civil service. The mining enterprises, on the other hand, limited their hiring of non-Congolese to technicians. The Congolese were encouraged to move to the cities by offers of higher pay, houses, and plots of land. They were then trained to fill the requirements of the Belgian administration and the mining enterprises for semiskilled and skilled labor.[3]

Literacy and education

Prior to 1957, there were separate schools for Belgian nationals and for the Congolese, with a total enrollment of 1,533,314. The majority of the schools for the Congolese were, however, inferior to those for Belgians. This, more than anything else, contributed to the extremely low level of literacy. In 1958, the school system was revised, and better schools and higher education were provided for promising Congolese. The two universities, the Belgian Congo and Ruanda-Urundi Government University and Lovanium University, had, in 1958, a total enrollment of 194 Congolese.[4]

Major religions and religious institutions

Catholicism, Protestantism, and Islam are the three main religions of the Congo Republic. In 1958, there were 4,546,160 Catholics, 825,625 Protestants, and 115,500 Muslims. Catholic and Protestant missionaries began work in the Congo in the latter part of the 19th century. They took the initiative in establishing schools, medical clinics, hospitals, and welfare centers.

Today, however, there is no established church. The Republic is secular, but the influence of the Catholic Church is especially strong since most of the Congolese elite are mission-trained, and most of the political parties trace their origin back to mission school associations. The *Conscience Africaine*, a Catholic-sponsored newspaper, was the first to advocate nationalism. Some of the civil unrest which erupted after June 30, 1960, was aimed at these missionaries because of the apparent cooperation between the Belgian Administration and the missions. However, this has now ended, and the missionaries have been encouraged to resume their educational work.

GOVERNMENT AND THE RULING ELITE

Description of form of government

The provisional constitution which was adopted at the Brussels Round Table Conference of January-February 1960 and is still in effect set up the basic structure of the Congolese Government. Central and provisional institutions were to be defined by June 30, 1960.

The provisional constitution provided for a legislature of two houses, with identical legislative competence—a Senate of 84 members and a House of Representatives of 137. The members of the Senate were chosen by provisional assemblies on the basis of 14 for each province, including at least three tribal chiefs. Members of the House of Representatives were elected by universal male suffrage on the basis of one representative for every 100,000 inhabitants.

The executive branch of the government was composed of the Council of Ministers, headed by a Prime Minister, and was responsible to the Parliament in all policy matters. The first Prime Minister, Patrice Lumumba, was appointed by King Baudouin of Belgium.

The President of the Republic, Joseph Kasavubu, was elected by a joint session of the two houses of Parliament. His functions are meant to be mainly honorary, and include the convening of Parliament.

The Cabinet assumed all powers relating to foreign affairs, army and national police, national finance, currency, customs, taxation, education, transportation, communication, postal services, national security, and settlement of legislative-judiciary conflicts.

The Parliament convened as a constitutent assembly to draw up a new constitution which would define the central and provincial institutions, the relationship between executive-legislative and judiciary, the legal means of amending the constitution, and establish a judiciary and judicial system. However, because of disagreement between federalists and centralists, it adjourned without achieving its purpose.

Description of political process

In the May 1960 general elections, three major political parties emerged, headed by Patrice Lumumba, Joseph Kasavubu, and Albert

Kolondji. None of them, however, achieved the majority which would enable them to form a government. These elections clearly indicated the need for a coalition government.

To prepare for independence and preserve the unity of the Congo, all the political parties represented in the Congolese Parliament joined to elect a coalition government, with Lumumba as Prime Minister and Kasavubu as head of state. This coalition was maintained until September 5, 1960.

On that day, President Kasavubu dismissed Premier Lumumba, and nominated Joseph Ileo instead. Lumumba reciprocated by dismissing Kasavubu as President. The Parliament refused to consider these dismissals, and reinstated both men, but the coalition had, in effect, ceased to exist. The political parties polarized into two major groupings—the centralists, supporting the government, and the federalists, making up the opposition.

The centralists were composed of the following parties: the National Congolese Movement (MNC) headed by Lumumba, with support throughout the Congo; the African National Unity Party (PUNA) headed by Jean Bolikango, with strength in the provinces of Equateur and Léopoldville; and National Progress Party (PNP) headed by Paul Bolya, and centered in the provinces of Equateur and Léopoldville; the Balubas of the BaKatanga (BALUBAKAT) headed by Jason Sendwe, and composed of a cartel of three tribally-based parties in Katanga Province; and the People's Party (PP), a Marxist party led by Alphonse Nguvulu. These parties, as a whole, supported the theory of a strong centralized state, which would rise above tribal loyalties.

The federalist opposition comprised two parties; the Association of the Bakongo (ABAKO), headed by Kasavubu, and centered in Léopoldville Province; and the Confederation of the Association of the Tribes of Katanga (CONAKAT), headed by Moise Tshombe, and located in Katanga Province. The federalists were also supported by prominent Congolese politicians, such as Joseph Ileo, President of the Senate, Thomas Kanza, and Justin Bomboka. These parties and politicians advocated a federal state composed of the five provinces, along tribal lines.[5]

The opposition parties were handicapped because news media were virtually closed to them. Some of the parties, both pro-government and opposition, issued newssheets, but these, as a whole, were not effective because of the high rate of illiteracy, and because distribution failed to extend beyond the major cities. Radio broadcasting was centralized under the control of the government. Most of the political leaders relied on word-of-mouth as a means of rallying their supporters.

Legal procedure for amending constitution or changing government institutions

The provisional Constitution of 1960 contained only one clause relating to its amendment. This dealt with secession. Secession by any province, or of part of the national territory, could be achieved legally only if it received a two-thirds majority in Parliament. Legal procedures for amending the Constitution were to be included in a new constitution, to be enacted at a later date.

Relationship to foreign powers

At the Brussels Round Table Conference of 1960, agreements were concluded between the Belgian Government and the Congolese leaders concerning the future relations of their respective countries. It was decided that, on June 30, 1960, the Belgian civil servants in the Congo would come under the authority of the Congo Government. Similarly, the judicial system of the Belgian Administration was to continue to function until such time as the Congo Parliament enacted a new system. General treaties of friendship, assistance, and cooperation were provided for; and a Belgian technical mission was to be sent to the Congo after June 30, 1960. However, the mass exodus of Belgian personnel from the Congo in the wake of the July 8–11 riots, and the arrival of Belgian paratroopers on July 11, 1960, made the ratification and implementation of these agreements impossible.

The Lumumba government then sought economic and technical assistance from neutral nations, and accepted the proffered help of the Communist bloc. The Western bloc, with the exception of Belgium, refused to take unilateral action vis-a-vis the Congo outside the United Nations. Katanga Province seceded on July 11, 1960, and declared its independence. Since then it has strengthened its ties with Northern Rhodesia, and the *Union Minière du Haut-Katanga* (Mining Union of Upper Katanga) has provided Tshombe, the secessionist leader, with technical personnel, Belgian officers for his army, and weapons. The independence of Katanga Province has not, however, been recognized by any state.

The role of military and police powers

The *Force Publique* (Home Guard) was established by the Belgians in 1891 to defend the frontiers of the Congo against tribal raids and to maintain internal order, prevent tribal wars, and put down rebellious tribes. Composed of volunteers, it was "a native army, almost entirely illiterate, poorly paid, and officered entirely by Europeans."[6] Mutinies were not uncommon.

On June 30, 1960, the Congolese Government changed the name of this 25,000-man force to "National Army." It maintained the above-mentioned purposes for its new army, and kept the white officers. On July 8, 1960, the central government, in an effort to pacify the mutiny, dismissed the white officers and replaced them with Congolese non-coms, who were promoted to officer rank.

Because of the unruliness of the *Force Publique* the Belgian Administration set up the *gendarmerie* (constabulary) during World War II as a parallel organization. Smaller in number, more effective and manageable, the *gendarmerie* was given the task of maintaining the internal security in and around the major cities, while the *Force Publique* was relegated to the barracks and to the maintenance of order in remote rural areas. The Congolese central government maintained this force, and used it effectively to quell the mutinous *Force Publique* in July 1960. The *gendarmerie* is recruited from the elite of the *Force Publique*, and its units represent the most reliable force at the disposal of the central government.

WEAKNESSES OF THE SOCIO-ECONOMIC-POLITICAL STRUCTURE OF THE PREREVOLUTIONARY REGIME

History of revolutions or governmental instabilities

Luinumba's government, which took over on June 30, 1960, inherited grave political, social, and economic problems. These became increasingly acute, because of the government's lack of trained administrative personnel to replace the Belgians and because of its inherent weakness.

The July 8–11 mutiny left the Lumumba government without an administration, and without an effective law-enforcing agency. The Belgian civil servants, who were supposed to train the Congolese had fled or were in the process of fleeing, and could not be replaced immediately. The national army lost its effectiveness when it mutinied against the central government, and subsequently broke up into factions of varying allegiances.

The Lumumba government was, essentially, a coalition government. The centralist-federalist compromise which had brought him to power limited the efficacy of his government. The effective powers which would have enabled him to meet the immediate problems were denied him by the federalists, because they would have enabled him to establish the strong central government which he advocated. Nor could he circumvent this parliamentary opposition and appeal directly to the masses, because his policies also struck at the tribal structure of the Congolese society.

The masses, per se, were politically unconscious and inarticulate. They were still in a tribal state of organization, and gave their allegiance to their tribal leaders. These tribal leaders opposed a strong central government because it threatened to relegate them to inferior status.

Economic weaknesses

The Lumumba government faced two immediate economic problems which it could not solve. It had come to power with practically no liquid assets in the Central Bank, and in a protracted state of chaos the subsistence-type agriculture of the Congo could not support the masses.

The 1959–60 uprisings had resulted in a flight of capital which the Central Bank has not checked, and the withholding of private international capital investments.[7] The Lumumba government found itself in the unenviable position of not being able to pay its employees and its army, thereby losing control over both.

With no visible means of paying for its imports of foodstuffs, the government found itself equally unable to supplement the food requirements of some of its tribes, thereby losing their support.

Social tensions

To establish the stability required for instituting reforms, the Belgian Administration had adopted two social measures: the creation of a Congolese middle class of *Évolués,* and the formation of tribal associations in the rural areas which would carry over when tribesmen moved to the urban centers. It was hoped that both groups would become the allies of the Belgian Administration. Both these measures, however, failed.[8]

The *Évolués* soon found themselves the subjects of Belgian paternalism. They were encouraged to rise above tribal loyalties, and, at the same time, they were denied real power or significant roles. The segregationist attitudes of the Belgian nationals further enraged them, and, in their frustration, they turned against the Belgian Administration.[9]

The formation of tribal associations tended to disrupt the amalgamation of the Congolese into a homogeneous society. It reemphasized the tribal structure of the Congolese society, resulting in the rekindling of tribal jealousies, enmities, and wars.[10]

With independence, the *Évolués*, who had risen above tribal loyalties, found themselves shunned and suspected by the tribes, and unable to lead.

The large number of tribesmen who had recently moved to the cities in search of food swelled the ranks of the unemployed, adding to the uneasiness of the situation.

Government recognition of and reaction to weaknesses

Patrice Lumumba recognized the weaknesses of his regime. By calling for the intervention of the United Nations, he had hoped to eject the Belgians, maintain security, retrain and reorganize the national army, train a civil service, and restore the authority of the central government over Kasai and Katanga Provinces. His followers succeeded in retaking Kasai, but the campaign destroyed his coalition government. In dealing with Katanga, he had hoped that the United Nations would either compel it to submit to the central government, or allow him to reconquer it while upholding him as Premier in spite of the breakup of his coalition government. The United Nations refused to accede to his demands, and suggested negotiation with Tshombe. He refused to negotiate on the grounds that negotiations could only lead to a federal solution, and that a federal solution would strengthen the hand of the federalists.[11]

Instead, he turned against the United Nations, accepted the economic and technical aid offered by the Communist bloc, and threatened to ask for the direct intervention of the Soviet Union.[12]

FORM AND CHARACTERISTICS OF REVOLUTION

ACTORS IN THE REVOLUTION

The revolutionary leadership

The leader of the coup d'etat was Col. Joseph Mobutu, Chief of Staff of the Congolese National Army. He had first enlisted in the *Force Publique* in 1949, and had been discharged 7 years later as a sergeant major. In Léopoldville, he took up writing and journalism, eventually becoming editor of *L'Avenir*, and, later on, editor in chief of the MNC's newspaper, *Actualités Africaines*.

In January 1960, he attended the Brussels Round Table Conference as a delegate of the MNC. He was appointed State Secretary for Defense by Lumumba after the Congo had been granted its independence. Shortly after the mutiny of the national army, on July 8, 1960, he was appointed Chief of Staff with the rank of colonel.

Colonel Mobutu began his political career as a centralist, closely associated with Lumumba's MNC. In the 1959–60 events, he remained in the background, and was a relatively unknown figure in Congolese politics. He seems to have enjoyed the complete confidence of Lumumba, and the split appears to have been the result of Mobutu's own personal conflicts. He is a devout Catholic, is anti-Communist (he ejected all the Communist diplomats and technicians), and is not anti-European. He seems to have no political ambitions, as is attested by the fact that he immediately appointed a Council of Commissioners, and promptly stepped down as strong man when President Kasavubu named a provisional government in February 1961 composed of members of the suspended Parliament.[13]

The revolutionary following

The military coup d'etat was carried out with the support of that part of the national army stationed in Léopoldville. The population, as a whole, was apathetic. The federalists, however, who were not displaced by this military exercise of power tended to back Mobutu.

ORGANIZATION OF REVOLUTIONARY EFFORT

Internal organization

The military coup d'etat seems to have resulted from a spontaneous agreement between Colonel Mobutu and the officers of the elements of the national army stationed in Léopoldville in the afternoon of September 14, 1960.

External organization

There is no evidence that Colonel Mobutu had the support of foreign countries. However, in the days preceding the coup d'etat, he visited the American Embassy frequently,[14] a fact which leaves the door open to speculation. Implied support was granted by the United Nations after the coup d'etat, when the U.N. command in the Congo accepted the cooperation of Colonel Mobutu in discharging its mission.

GOALS OF THE REVOLUTION

In removing Lumumba, Colonel Mobutu appears to have been prompted by a desire to end the Kasavubu-Lumumba power struggle which was furthering the constitutional disintegration of the Congo.[15] The restoration of Kasai Province by Lumumba had resulted in the massacre of a large number of Baluba tribesmen. This campaign had destroyed national unity and weakened the effectiveness of the government. Subsequent attempts by Ghanaian troops of the U.N. command to disarm the national army, and Communist machinations in the Congo, were interpreted by Colonel Mobutu as direct interference by Ghana and the Communist bloc in the internal affairs of the Congo in support of a Lumumba dictatorship. The continued absence of an effective government, in such a situation, would have played into the hands of Lumumba and his supporters.[16]

REVOLUTIONARY TECHNIQUES AND GOVERNMENT COUNTERMEASURES

The Lumumba-Kasavubu power struggle had resulted in a chaotic stalemate. The protracted state of chaos was indicative of their respective weaknesses. Both leaders lacked the support of the masses, and neither had control over all the elements of the national army. It would have been impossible for either man to gain the upper hand without outside assistance. Therefore when Colonel Mobutu seized control of the government, the Lumumba regime had already been critically weakened. The elements of the national army stationed in Léopoldville proved to be stronger than either the centralists or the federalists, and were thus able to dominate.

MANNER IN WHICH CONTROL OF GOVERNMENT WAS TRANSFERRED TO REVOLUTIONARIES

On September 15, 1960, elements of the national army in Léopoldville, loyal to Colonel Mobutu, seized control of Radio Congo, the government radio. Colonel Mobutu then broadcast an announcement that the national army was taking over until January 1961, but that his action did not constitute a coup d'etat. Parliament, however, was to be suspended until further notice. Nothing was said about the fates and future roles of Lumumba and Kasavubu. When some members of Parliament refused to obey Colonel Mobutu's orders, soldiers surrounded the Parliament building and barred entry to everyone.

Lumumba reacted first by trying to discredit Colonel Mobutu on Radio Congo. However, U.N. troops, who had taken over control of Radio Congo from Mobutu's forces, refused him entrance. Lumumba then proceeded to Camp Léopold in an effort to sway the rebellious soldiers. He was immediately seized and placed under house arrest by the soldiers. These soldiers were Baluba tribesmen, and bore a grudge against Lumumba for the Kasai campaign which had resulted in the death of a large number of their fellow tribesmen. Subsequently, Lumumba escaped to the protection of the U.N. troops and the Ghanaian embassy.

THE EFFECTS OF THE REVOLUTION

CHANGES IN THE PERSONNEL AND INSTITUTIONS OF GOVERNMENT

The military coup d'etat by Colonel Mobutu did not change any of the governmental institutions. Parliament was suspended, but it was not dissolved. The replacement of the Lumumba Cabinet and the dismissal of the Premier proved to be the only changes in personnel. A Council of Commissioners-General was appointed by Mobutu. It was a nonpolitical cabinet composed of university professors, graduates, and students. Their task was to restore effective government.

MAJOR POLICY CHANGES

Colonel Mobutu's internal and foreign policies reflected major changes. His internal policy was one of moderation aimed at restoring national unity through negotiations, and the reconciliation of the political rivals.[17] This is borne out by the fact that Colonel Mobutu made very limited attempts to extend his control beyond the provinces of Léopoldville and Equateur, to the Lumumbist stronghold of Orientale. Nor did he, initially, attempt to incarcerate Lumumba or any of his followers. He finally imprisoned Lumumba when the latter intrigued to regain his former status.

The foreign policy of Colonel Mobutu aimed at isolating the Congo from foreign interference until the internal situation could be normalized. He immediately ordered the Communist bloc diplomats and technicians out of the Congo, and threatened to demand the withdrawal of Ghanaian and Guinean troops of the U.N. command on the grounds that they were interfering in the affairs of the Congo. He resumed cooperation with the U.N. command, in the hope that the United Nations would prevent the outbreak of violence, and thereby provide the Congo with internal stability which would be conducive to negotiations.

LONG RANGE SOCIAL AND ECONOMIC EFFECTS

It would be impossible, at this moment, to assess the long range economic and social effects of Colonel Mobutu's coup d'etat. The economic and social situation of the Congo was in a state of flux when he took over, and no efforts were made, during his tenure, to change it. Colonel Mobutu concentrated on the political situation, and left the immediate and long range economic and social problems to the United Nations.

OTHER EFFECTS

When Lumumba was removed, his second in command in the Cabinet, Antoine Gizenga, retired to the Province of Orientale and declared his Cabinet to be the legal Cabinet of the Congo Republic. He was recognized as the legal Premier of the Congo by the Communist bloc, a number of neutrals and a number of independent African states.

The coup d'etat failed to reconcile the secessionist provinces of Katanga and Orientale with the Congolese central government. On February 9, 1961, President Kasavubu announced the end of the military regime of Colonel Mobutu. President Kasavubu stated that Colonel Mobutu had restored public order, and Colonel Mobutu said he had accomplished his political role, and would henceforth devote his time to the national army.[18]

NOTES

1. Roger de Meyer, *Introducing the Belgian Congo* (Brussels: Office de Publicite S.A., 1958), pp. 47–79.
2. *The Worldmark Encyclopedia of the Nations* (New York: Harper & Brothers, 1960), 211–213.
3. Gwendolen M. Carter, *Independence for Africa* (New York: Frederick A. Praeger, 1960), pp. 84–85.
4. *Belgian Congo* (Brussels: The Belgian Congo and Ruanda-Urundi Information and Public Relations Office, 1959), I, p. 445.
5. Colin Legum, *Congo Disaster* (Baltimore: Penguin Books, 1961), pp. 94–105.
6. Ibid., p. 112.
7. Ibid., pp. 91–92.
8. Ibid., pp. 49–51.

9. Maurice N. Hennessey, *The Congo* (New York: Frederick A. Praeger, 1961), p. 55.

10. Legum, *Congo Disaster,* p. 51.

11. Ibid., pp. 141–168.

12. Rolf Italiaander, *The New Leaders of Africa* (Englewood Cliffs, New Jersey: Prentice-Hall, Inc., 1961), p. 164.

13. Ibid., p. 166.

14. *Time,* (26 September, 1960), 31.

15. Hennessey, *The Congo,* p. 91.

16. Legum, *Congo Disaster,* p. 151.

17. "Journalist Strongman," *The New York Times,* 16 September, 1960, p. 4.

18. *The New York Times,* 10 February, 1961, pp. 1–2.

RECOMMENDED READING

BOOKS:

Carter, Gwendolen M., *Independence for Africa.* New York: Frederick A. Praeger, 1960. Chapter 7 discusses the political development in the Congo from 1957 through 1960, and the resulting tensions.

Hennessey, Maurice N., *The Congo.* New York: Frederick A. Praeger, 1961. A historical survey of the Congo from 1885 to the present.

Italiaander, Rolf, *The New Leaders of Africa.* Englewood Cliffs, New Jersey: Prentice-Hall, Inc., 1961. Chapter 14 discusses in particular the role of Joseph Kasavubu and Patrice Lumumba in Congolese politics.

Legum, Colin, *Congo Disaster.* Baltimore: Penguin Books, 1961. An excellent study of the Congolese crisis. Includes detailed discussions on the roles of the Belgian administration, the political elite of the Congo, and the United Nations.

PERIODICALS:

Holmes, John, "The United Nations in the Congo," *International Journal,* XVI, 1 (1960–61), 1–16. Discusses the role of Mr. Hammarskjold in the Congo, and the reactions of the Communist bloc, neutrals, and the West, to his policies.

SECTION V

....

MIDDLE EAST

TABLE OF CONTENTS

GENERAL DISCUSSION OF AREA AND REVOLUTIONARY DEVELOPMENTS

THE IRAQI COUP OF 1936

THE EGYPTIAN COUP OF 1952

THE IRANIAN COUP OF 1953

TABLE OF CONTENTS

THE IRAQI COUP OF 1958

THE SUDAN COUP OF 1958

TABLE OF CONTENTS

THE MIDDLE EAST

GENERAL DISCUSSION OF AREA AND REVOLUTIONARY DEVELOPMENTS

DESCRIPTION OF AREA

GEOGRAPHY

The Middle East, as defined in this discussion, extends from Egypt in northeastern Africa to Iran in southwestern Asia and from Turkey on the fringe of southern Europe to the Sudan in East Africa. The countries included in this region are: Egypt, the Sudan, Syria, Jordan, Lebanon, Iraq, Iran, Turkey, Afghanistan, Yemen, Saudi Arabia, and Israel. There are also several semiautonomous oil sheikhdoms, prominent among which is Kuwait. It extends over portions of two continents—Africa and Asia—and its common denominator is the predominance of the Islamic religion and the Arabic language and culture; there is also a similarity in the economic, social, and political patterns in the political divisions of the region. The so-called "Northern Tier" of the Middle East, comprising Turkey, Iran, and Afghanistan, share the Islamic religion and a considerable body of social and cultural influences derived from the Arabs, but these countries have preserved their non-Arabic languages and historical traditions. The Arabian peninsula, the so-called "Fertile Crescent" (comprising Syria, Lebanon, Jordan, and Iraq), and the two Nile Valley countries, Egypt and the Sudan, represent the Arab core of the Middle East.

Geographically, the Middle East includes the state of Israel. However, politically and culturally her problems are unique and do not fall into the general pattern as do those of the Muslim states. For this reason, Israel has not been included in this discussion, except insofar as her existence and the events of the Palestine War have contributed to the revolutionary ferment of the area.

SOCIO-ECONOMIC FACTORS

The Middle East, although it contains more than half the world's petroleum deposits, is not otherwise heavily endowed with either mineral resources or arable land. It serves as a communications hub between Europe, Africa, Southeast Asia, and the Soviet bloc. Walter Laqueur, an authority on the Middle East, citing the findings of a U.N.

research unit, summed up the socio-economic situation in the region in 1955 as characterized by:

> a rapidly growing population, which in some countries is pressing heavily on the means of subsistence; . . . high mortality, widespread disease, and low literacy rates; a paucity of mineral resources [other than] oil; a marked concentration on agriculture; a marked shortage of capital.[1]

As a result, the countries depend on foreign sources for investment. Technological development is primitive, productivity and national output are low, and wealth and income are unequally distributed. In the 1950's over 100 million people were living in the Middle East as here defined. A small number for so vast a territory, the population was actually quite dense in the few parts of the region which were arable through natural or manmade irrigation.

RELIGION

The role that Islam has played in the political, social, and cultural life of the Middle East is obviously of great importance. There are many different points of view and opinions on the specific nature of Islam's role. Abdul Aziz Said, a Middle Eastern scholar, considers Islam an important factor in the authoritarian tradition of Muslim states, in the reluctance of the Muslim to take any interest or any active role in his government, and in his indifference to corruption and mismanagement.

> Doctrinaire Islam conceives of man as being created solely for the service of God, and . . . this has produced in the Moslem mind the implication that social welfare is not the concern of the government. The modern concept of the state is alien to traditional Moslem political theory.[2]

The attempts of reformers to bring Islamic doctrine into line with modern political, economic, and social theory, he believes, have been hampered by ignorance both of Western civilization and of their own heritage and a characteristic incapacity for self-criticism.

HISTORICAL DEVELOPMENTS

THE IMPACT OF THE WEST

The modern period in the political life of the Middle East began in the early 19th century when the Ottoman (Turkish) Empire, which had exercised control over the Arab world for several hundred years, began to disintegrate. Simultaneously, the leading states of Western Europe, then in the midst of a technological revolution and an unprecedented outpouring of energy, began to extend their cultural, economic, and finally their political influence in the area. European manufactured goods competed with the products of native handicraft; Western education weakened Islamic traditions and values; and British, French, and Russian political control replaced Turkish and Persian political hegemony in the Middle East. In the 19th century the Arab world experienced non-Islamic rule for the first time in its modern history, for while political ascendency had long since passed from Arab to Turkish hands, the Turks were not considered alien in the sense that European "Unbelievers" were alien. The psychological shock caused by the decline of the Muslim power in the contest with the Christian West, and the theological implications of this decline for the Muslim Believer, were to have a definite impact on Middle Eastern attitudes in the 20th century. "Power belongs to God, and to His Apostle, and to the Believers," the Koran had promised. Yet here were European "Unbelievers" who clearly had access to greater power than the Muslim "Faithful." This undeniable fact threatened to undermine the Muslim's traditional conceptualization of the universe. Asians, Africans, and Latin Americans were psychologically and culturally better prepared for their encounter with the West than were the Muslims of the Middle East.

By the end of the 19th century, England had outdistanced both her French and Russian rivals in the area to become the dominant Western power in the Middle East; however, France continued to exert strong cultural influences throughout the Middle East, especially in Lebanon and Syria, and in Egypt, where nationalistic Egyptians frequently sent their children to French schools as a means of counteracting British influence. British strength in the Middle East was consolidated between the time of the occupation of Egypt in 1882 and the conclusion of World War I in 1918, and during the period between World War I and World War II, the British continued to control the area through an interlocking system of economic and commercial ties, military bases, and influence over the native ruling elite. During the interwar period France controlled Lebanon and Syria through having been given a League of Nations Mandate over these countries. France ruled her mandate territories virtually as colonial possessions; as a result, there was considerable nationalist agitation and violence during the interwar period, particularly in Syria.

Critics of imperialism have observed that the European presence strengthened the hand of certain landowning elements of Middle Eastern society, at the expense of both the mass of the population who were landless and the traditional ruling elite. The introduction of the Western concept of private property and the establishment of parliamentary institutions—both top priorities in British colonial policy— had the effect of creating a native landed aristocracy who were more closely tied to the European authorities than to their traditional rulers,[3] and were armed with legal and political rights incommensurate with those of the peasant population dependent upon them for their livelihood. Most critics agree that the Middle East's encounter with European imperialism was too late and too brief for the area to have derived the benefits generally associated with colonialism. As Bernard Lewis of the University of London has expressed it:

> There is a case to be made for and against imperial rule as a stage in political evolution. . . . But there is little that can be said in defense of the half-hearted, pussy-footing imperialism encountered by most of the peoples of the Middle East—an imperialism of interference without responsibility, which would neither create nor permit stable and orderly government.[4]

A pro-British expert of the area has stressed, on the other hand, the "positive community of interests between England the Middle East":[5] He credits the English with bringing security, domestic and external; investment capital and technical skill to irrigate the desert and exploit its petroleum resources; administrative competence and integrity; schools, courts, commerce. "For the dominant groups," he sums up, "—landowners, tribal leaders, merchants, religious leaders—the presence of England was usually of positive advantage, and rarely harmful."

World War II shattered the delicate machinery of British control in the area. It is true that, as Hourani writes, "England had enough strength to beat back the enemy from outside, and enough support to be able to use the resources and communications"[6] of that area in spite of a rising tide of nationalist sentiment.

As a result, however, England made enemies, particularly in Egypt and Iraq, who could cause more trouble in the postwar period than the shattered might of the British Empire could contain. British occupation of Vichy France's outposts, in the French mandates of Lebanon and Syria, eliminated the last vestiges of French control in the Middle East and brought nationalists to power in those two countries.

The dominant fact in the international relations of the Middle East in the postwar period has been the decline in British influence and the rise of Arab nationalism as a major international force. British impotence was dramatically demonstrated to the Arabs, whose cause the British had undertaken to champion before the world organization, when Great Britain suffered defeat in 1948 before the United Nations on the Palestine question, at the hands of a temporary American-Russian coalition. No longer could England pose as the final arbiter of the Arab's fate. There were also social and political forces which made cooperation with England less advantageous and which gave power to people who could obstruct the pro-British policies of the indigenous elite.

Egypt took the lead in the post-World War II revolt against Western domination of the Arab-Islamic world. The emerging social and political forces which, in the case of Egypt, were successful in removing British influences from their country early in the postwar period were not so successful in such countries as Jordan and Iraq. Here England retained her acquired prestige and many of the symbols of imperial power. "The rulers of Jordan and Iraq would still ask British advice on major matters, and still give special weight to her wishes, even when she had lost the means to enforce them . . . Thus, England remained for 10 years the strongest power in the Middle East, but by default and on borrowed strength."[7] The partial and uneven withdrawal of British influence from the area had the serious effect of creating antagonisms between ultranationalist regimes born of postwar conditions, such as the Nasser regime in Egypt, and the older pro-Western nationalist regimes, such as the Nuri as-Said regime in Iraq. Too often in the postwar period British influence in the area was sufficient only to antagonize people rather than to guide the course of events.

THE IMPACT OF ISRAEL

The defeat of the Arab states in Palestine in 1948 at the hand of Jewish irregular forces, the establishment of the state of Israel in 1948, and the Anglo-Franco-Israeli expedition against Egypt in October 1956, have had disruptive and constructive repercussions that still reverberate in the Arab states of the Middle East. The intensity of these repercussions, however, has varied from country to country.

The immediate effect of the Palestinian War of 1948 was marked by a wave of revolutionary ferment in the Arab world, and contributed to successive coups d'etat (1949–51) in Syria, the coup d'etat in Egypt in 1952, and led directly to the assassination of King Abdullah of Jordan in 1951.

The Palestinian War, moreover, led to three major and related developments: it strengthened Arab nationalism, it contributed to the further decline of Western influence, and it changed the political outlook of the Arab states, away from the West and toward closer relations with the Communist bloc.

Arab nationalism generated a political force in the Middle East which transcended political boundaries, and which forced the Arab governments into closer relationships and a community of action. Nationalism generated a hatred for Israel and the West, and impatience with the evolutionary methods which characterize Western civilization, advocating revolutionary socialism as a means of solving the socio-economic problems of the Arab world.

The Arabs blamed the West for their defeat, and held the West responsible for the establishment of the state of Israel. England, especially because of the Balfour Declaration, was made to bear the brunt of Arab wrath—a factor which hastened her complete withdrawal from Iraq and Egypt. Arab hatred for the West precipitated the eventual removal of most Arab politicians traditionally associated with the West. This factor contributed to the political reorientation of the Arabs, in that it removed the vital link between democratic evolution and the political development of the Arab states.

The Arab defeat in Palestine, and the consequent hatred and fear of Israel which followed, culminated in the adoption of two major goals: social and economic reform, and military armament, with the latter taking precedence.

In order to preserve the *status quo* and prevent the resurgence of violence in the Middle East, the West had placed an embargo on arms shipments to either the Arab states or Israel. Since Israel could manufacture most of its military requirements, and the Arabs could not, the Arabs turned to other sources for military requirements. Early in 1956, Egypt announced an arms deal with the Soviet Union, and promptly proceeded to build up its armies. The Anglo-Franco-Israeli attack on Egypt in October 1956 crystallized this hatred and fear of Israel; and similar arms deals were concluded in 1957 between the Soviet Union and Syria, and in 1958 between the Soviet Union and Iraq. Communist bloc economic and technical pacts followed in all three countries.

POLITICAL DEVELOPMENTS

The political environment in which the events described in this section occurred has been likened to "a vast cloud of popular political feeling compounded of national self-assertion, religious solidarity, social idealism, and a consciousness of weakness and dependence."[8] The major ideological force in the Middle East today is nationalism, and every Middle Eastern political leader professes to be an ardent and sincere nationalist. Some, whose nationalism is expressed in antiforeign terms, may be referred to as ultranationalists. Iran's Mosaddegh, Egypt's Nasser, and Iraq's Kassem have rivaled one another at various times as the leading exponent of militant ultranationalism in the area.

Laqueur, a Middle East scholar previously quoted, concludes that this new nationalism must be faced and should be studied and understood. Its most outstanding features, he writes, are chauvinism and xenophobia. "The fact that xenophobia appears now as anti-Westernism, not anti-Easternism, has of course to be explained by . . . the circumstance that the West has been 'in' for many years and the Soviet bloc has been 'out' . . . ."[9] Laqueur explains the xenophobic element, which is so prominent in Arab nationalism, more on the basis of the impact of foreign rule of the Arab nations than on the basis of Islam's alleged antiforeign bias. Xenophobia seems to be less pronounced in the non-Arab countries of Turkey, Iran, and Afghanistan, which have never experienced long periods of alien rule.

The emphasis on religious solidarity and social and economic reforms, invariably associated with Middle Eastern nationalism, appears to be closely related to the political aspirations of the elite. Political unification of the Islamic community has been a historic ideal in the Middle East, especially among the Arabs. The foreign policy implications of Pan-Islamism, like Pan-Arabism, for a nationalist leader such as Nasser are obvious.

The Western-educated and secular-oriented new Middle Eastern elite who head the nationalist movements find in Islam one of the most effective means of communicating with the broad masses of their people. Koranic invectives are poured out against the Unbelievers, whom the nationalists identify as the foreign imperialists and their pro-Western "stooges." But the influence of Islamic customs and social attitudes works both ways, so that the nationalist elite is constantly obliged to accommodate its secularizing policies and attitudes with the traditions adhered to by the broad masses of the "Faithful."

To the urban middle class, the nationalist elite appeals less through religious than through nationalistic symbols, such as progress. Public works projects, such as steel mills, hydroelectric dams, and irrigation systems, serve the dual function of raising the living standard and the national confidence of the middle class, who throughout the Middle East is sensitive to limited industrialization of the area. National pride in technological and material achievements is an important ingredient in the nationalism of the middle-class Middle Easterner.

Arab nationalism has been confronted with an organizational dilemma: whether to organize into national units within borders inherited from the colonial past or to organize into one Pan-Arabic union embracing the entire Arab world. "By what may seem a paradox," remarks Hourani, "the area of Arab unity is itself a cause of disunity."[10] Not only does each state think of itself as the logical nucleus around which the rest should coalesce, but the concepts of union differ widely in different sections. In northern Africa, including Egypt, the existing states have a long history and a strong sense of national identity. In most of Asia, however, nationalism is a relatively new concept and the sense of a single Arab identity is much stronger. Pan-Arabist agitation was a factor in the Syrian Coups of 1949, the Iraqi Coup of 1958, and the Sudanese Coup of 1958.

All revolutions in the Middle East have been carried out by elite groups or by those close to the elite. The underlying cause of the Middle East's elitist revolutionary tradition, and indeed the cause of elitism itself, has been sought in a variety of historical, cultural, social, and economic factors. Throughout the history of the region governments have usually been alien, hostile, and divorced from the mass of the governed. The people have traditionally regarded their government as the tax collector, conqueror, and the dispenser of privileges to its favorites; they have had little experience with the positive side of government. Recently, however, there have been certain positive aspects of government in the Middle East; but the benefits resulting have not been distributed equally and consequently have served in some states such as Iraq[a] to increase rather than reduce existing social tensions.

Already backed by historical precedent and substantially aided by religious attitudes, elitism in Middle Eastern political activity received strong support in the 20th century from the cultural, economic, and political impact of the West on the area. The rapid progress of Westernization, secularization, and urbanization which resulted from this contact created a fragmented and antagonistic society in an area which was already socially malintegrated. Since the ruling elite was most affected by contact with the West, the traditional alienation and hostility between the rulers and the ruled was greatly intensified. As

---

[a] See Discussion of the Iraqi Coup of 1958, p. 485.

the traditional hold of Islam over the mass of the population weakened, the legitimacy of the traditional elite was undermined and their place was taken by Western-educated nationalists who attempted to reconcile their policies of modernization with the traditional structure of Islamic society. The nearest approach to a common ideology between this new nationalist elite and the masses was their common antagonism toward European domination.

The nearest thing in the Middle East to a mass political party occurred in Egypt in the 1920's, when the nationalists formed the *Wafd* Party which enjoyed real mass support for many years; however, this bold experiment came to an end in the 1940's when collaboration with the British, and failure to incorporate "new blood" into the party from the rising urban lower-middle class, contributed to its downfall. It was replaced by the rule of an elitist military-socialist junta. The *Ba'th* (Arab Resurrection) Party, with a nationalistic-socialistic bent, and the ultranationalist and fundamentalist politico-religious *Ikhwan al-Muslimim* (Muslim Brotherhood) are also examples of parties whose ideology appealed to the masses. The brotherhood was doomed to failure, however, by its inability to come to terms with the realities of life in the type of modernized society which the overwhelming majority of Middle Easterners have come to desire. The *Ba'thists* still remain a significant political force in Syria and, to a lesser extent, in Jordan.

REVOLUTIONS AND THE EMERGENCE OF NATION STATES

KEMALIST REVOLUTIONS

The genesis of the modern revolutionary tradition in the Arab world occurred in non-Arab Turkey in 1922, when Kemal Ataturk, an officer in the Turkish Army, led a revolution against the Ottoman Sultan. An enlightened dictator, Kemal introduced drastic reforms into the country's social, economic, political, and administrative structure designed to transform Turkey into a modern, secular, and "Westernized" state. The influence of the Kemalist approach to the social and economic problems of the region was keenly felt throughout the Middle East; "Kemalist" revolutions occurred in two other countries of the Northern Tier—in Iran in 1923 and in Afghanistan in 1928–29. The Arab countries were controlled by the British and French, who administered the former provinces of the Ottoman Turkish Empire either as protectorates or as mandates under the League of Nations. Consequently, the Kemalist type of revolution was confined by historical conditions to the independent non-Arab states of the Northern Tier and to the decade of the 1920's. The spirit of the Kemalist revolution, however, was not so narrowly confined, as events were to prove in later years.

MILITARY-SOCIALIST REVOLUTIONS

A highly significant revolution occurred in Iraq in 1936, when some of Iraq's Turkish- and British-trained army officers, influenced by events in Turkey and Iran, allied with the left-of-center political opposition to stage a successful coup against the only independent Arab government in the Middle East at that time. The Iraqi Coup of 1936 represents a bridge between the Kemalist revolutions of the 1920's and the Nasser type of revolutionary situation that has prevailed in the postwar period throughout the Arab world.[b]

Thus was established a pattern of revolutionary coalitions, referred to by some students as "military socialism," which has been present in most of the revolutionary situations which have occurred in the Middle East since World War II. Military-socialist revolutions occurred in Syria in 1949 and 1951, in Egypt in 1952, in Iraq in 1958, and in Turkey in 1959. Like the Kemalists, these revolutionaries looked to changes in the socio-economic structures of their countries and to better exploitation of their natural resources and production capacities. In this respect, the underlying motivation for reform and modernization in the Middle East has remained unchanged since the Turkish Empire first began sending its army officers to Western military schools in the last century. Today's Middle Eastern revolutionary elites have merely broadened the base of this policy to include and involve the emerging masses in the process of modernization. Military-socialist revolutions often contain an extreme anti-Western theme that was not a dominant characteristic of the Kemalist revolution of the 1920's.

POST-INDEPENDENCE REVOLUTIONS

The Iraqi Coup of 1936 is also an early example of a revolution directed against the independent government set up after the withdrawal of direct foreign control. An important factor "prompting the rash of Near Eastern military coups in the last two decades has been the feeling of frustration caused by the inefficiency, weakness, disorganization and corruption of civilian regimes—a feeling accentuated by the high hopes which were earlier attached to independence as a

---

[b] See the summary of the Iraqi Coup of 1936, p. 421, for details of this revolution.

universal panacea."[11] The Syrian coups in 1949, the Egyptian coup in 1952, and the Iraqi, Sudanese, and Pakistani coups in 1958 all occurred within 2 to 10 years after the withdrawal of foreign troops from the country. The Syrian and Egyptian coups, and the near-revolutionary situations which developed in Lebanon in 1952 and 1958, may also be described in terms of their relationship to the feeling of general frustration that pervaded the Arab countries following their unsuccessful military encounter with Israel in 1948–49.

The Iranian Coup of 1953 is the only important counterrevolution in the Middle East in recent decades. The Iranian revolutionaries restored the delicately balanced oligarchical system composed of the Shah, the military elite, and the parliamentary clique of politician-landlords which had been upset by a dissident elite politician, Mosaddegh. That a countercoup of this nature was possible was probably due to the fact that Iran was, and to some extent remains today, more tradition-oriented, less urbanized, and less anti-Western (partly as a result of its historical enmity towards Russia) than the Arab countries in the area.[c]

The five revolutions summarized here are representative of the more than 20 military coups d'etat which have taken place in the area since 1920. In selecting specific revolutions for inclusion, such factors as availability of research material, distinctiveness of revolutionary type, and estimation of military-strategic interest were taken into consideration. These revolutions as a group are analyzed below in terms of certain functional aspects: the actors, motivations, and techniques of revolution.

GENERAL DISCUSSION OF REVOLUTIONS IN THE MIDDLE EAST

ACTORS IN THE REVOLUTION

Following the tradition of Kemal Ataturk, *military men have been the prime movers in every instance of revolutionary activity in the Middle East.* Indeed, there is no lack of historical precedent for the intervention of the military in the political affairs of the region. The Mamlukes and the Janissaries before them are historical examples of a military elite who came to dominate their political overlords. The reform attempts initiated by Middle Eastern rulers in the last century were often prompted by military defeats and hence began with the army. "The

---

[c] See the summary of the Iranian Coup of 1953, p. 465, for further details.

officer corps thus came to have early and privileged access to social and political ideas and organizational techniques taken over from Europe."[12] Consequently, the military elite became the earliest spokesmen for constitutionalism, social and economic reforms, and nationalism. The prominent role played by military leaders in the political life of the area has been explained in terms of socio-economic factors and political characteristics prevailing in Middle Eastern countries. Unlike many of the Western countries, where the officer corps has traditionally been identified with the upper classes, the Middle Eastern army officer is likely to come from more humble social origins, and his political and cultural identification is often with the rising urban middle classes rather than with the traditional upper classes. The upper class has been largely discredited in the eyes of today's politically conscious Middle Easterners, being regarded, at best, as a tradition-bound and nonprogressive group, and, at worst, as a puppet subservient to foreign interests. Since the 1940's many high school teachers, lawyers, doctors, and other professional people, becoming disillusioned with their role in society as civilians, have joined the army.[13] To this group, opposition to European rule and patriotism provided an avenue to social advancement.[14] When European rule ended and indigenous rule was set up, this important avenue was closed. The only way to reopen it was through revolutionary activity directed against the native ruling elite. Hence, an additional cause for military intervention in the political process.

The political situation in Middle Eastern countries has often been characterized by a vacuum in real leadership and the absence of a sustaining link between the ruling elite and the mass of the politically conscious population. Confronted with the realities of administering the economically underdeveloped and culturally ill-prepared political units which they have come to control, many Middle Eastern political leaders rapidly became discredited in the eyes of the general public, which expected much more than it received from national independence. Middle Eastern political leaders have lacked the political party structure which in the Western world provides a necessary link between the man in the street and the politician in the government. Because of this, military leaders often felt compelled to intervene in the political process. Their political notions generally were rather vague. They usually called themselves "liberals," "reformers," or "socialists." Most military leaders were sincerely interested, at least at first, in creating an efficient governmental structure to increase the nation's material strength and political prestige in the world. At the same time, they strove for a more equitable distribution of the economic products among the population. Lacking the proper training and background, a "military-become-political" elite finds it difficult to direct its energies and implement its good intentions. When an attempt was made to transform the public and private sectors of society into military camps, while giving lip service to civilian values and attitudes, there were obvious conflicts. The resolution of these conflicts has usually come through another military coup d'etat by a different sort of military leader.

Relations between the civilian politicians and the military personnel who collaborate to carry out a revolution have varied in each country. In Egypt there was apparently no collaboration between the two groups before the event and very little after the coup; in other countries the political opposition knew of the coup prior to the event and was asked to play a role of varying cooperation in the postrevolutionary regime. In the Iraqi Coup of 1936 the military group stayed in the background after the coup and maintained a civilian opposition politician as their "front man"; however, in the Iraqi Coup of 1958, as in the Sudanese Coup of 1958, the military leader of the revolution assumed the premiership and headed a government composed of both friendly politicians and military personnel. In the case of Iran, in 1953, the military leader, General Zahedi, was also a politician; nevertheless, his government was composed of more military figures than civilian.

MOTIVATIONS: POLITICAL AND IDEOLOGICAL ORIENTATION AND GOALS

While there have been some significant variations from the general pattern of revolutionary motivation in the Middle East, there have been more similarities than dissimilarities in the revolutionary goals and political orientation of the revolutionary actors in the region. Such things as complete national independence—whether from a European power or from some other Arab country—social and economic reforms along Western lines, and a secularized Islamic community are regularly included among the principal revolutionary goals of Middle Eastern revolutionary movements.

These revolutionary goals appeared in their most extreme form in the Egyptian Coup of 1952 and in the Iraqi Coup of 1958; they had been vaguely present in the first Iraqi Coup of 1936, but they were not so well articulated then as in 1958. The Egyptian and Iraqi revolutionaries identified the Alawi (King Farouk's) and Hashemite dynasties and their political minions with British influence in the area, and in the interest of establishing what they considered the complete independence of their countries from foreign domination the revolution-

413

ary actors overthrew these native ruling elites. As a foreign policy corollary to national independence, both the Egyptian and the Iraqi postrevolutionary regimes adopted anti-Western "neutralist" foreign policies and established diplomatic ties with the Soviet bloc. At the same time social and economic reforms were initiated, and a large portion of the former privileged class, especially the largest landowners in the country, were subjected to severe political and economic sanctions. A number of Islamic institutions were either abolished or nationalized, although the revolutionary regimes scrupulously avoided any open clashes with the religious authorities and constantly sought to identify the new regime publicly with Islamic traditions.

The leaders of the Iranian and Sudanese revolutions made certain departures from the general pattern of revolutionary motivation in the region. The Iranian military coup was led by conservatives against Mosaddegh's radicalism. In the Sudan, too, the leaders were less motivated by demands for social and economic reform than by Sudanese nationalism and secularism.

TECHNIQUES OF REVOLUTION

The range of revolutionary techniques employed in the Middle East has naturally depended upon such factors as the time period of the revolution, the number and type of participants in the revolution and their control of the armed forces of the country, and the extent of foreign support of the revolution. There never has been a successful revolutionary movement in the Middle East which was organized on a mass basis and existed over a long period of time, with propaganda directed at the general public and a protracted campaign against the regime in power. Instead, there have been narrowly based and clandestinely organized conspiratorial groups, invariably drawn from among the military elite and usually including elements of the political opposition, who conspire over a short period of time to overthrow the government in power. Of the five military coups summarized in this section, only the Egyptian revolution was the work of a revolutionary organization that had been in existence for more than 5 years; the other revolutionary organizations were temporary and *ad hoc*, existing for the sole purpose of overturning an objectionable government.

Under such conditions, the only propaganda activities engaged in by the revolutionaries have been the distribution of leaflets and radio broadcasts at the time of the coup. The Egyptian revolutionary organization carried on a vigorous propaganda campaign prior to the coup, but the circulation of this propaganda was limited to members of the military and selected members of the political elite. Foreign support of Middle Eastern revolutions has usually been in the form of radio broadcasts and printed material smuggled into the country. Other foreign support has been of a clandestine nature which is difficult to ascertain and describe.

RESULTS AND OUTLOOK

HIGHLIGHTS OF THE 1950'S

In the past decade revolutionary activity was centered in Syria, Egypt, and Iraq. The process began in Syria in March 1949, when, beginning with Col. Husni al-Zaim and followed by Col. Sami al-Hinnawi and later Col. Adib Shishakli, a series of three military coups during the course of that year swept from power the older generation of nationalist politicians who had led the nationalist movement against the French in the 1930's and had guided independent Syria since 1945. Pan-Arabist agitation for the Greater Syria scheme (incorporating Lebanon, Jordan, and Iraq) and inter-Arab rivalries, combined with frustrations growing out of the Palestine War, produced a breakdown in the Syrian political system and brought to power in December 1949, a regime of younger, more nationalistic and leftist-oriented politicians and army officers. The new regime lacked personal dynamism and popular appeal, and it remained for the Egyptian Coup of 1952 to produce an Arab leader who could project himself beyond the confines of his own country to capture the imagination of Arab nationalists and exert an influence throughout the Arab world and Middle East.

The Egyptian revolution has been compared to the French Revolution for its far-reaching impact on the Middle East's political development. At the time of the revolution, Egypt was already the most politically advanced and highly industrialized of the Arab countries and was the cultural leader of the Arab world in the Middle East. By 1955 Colonel Nasser had attained something like a Napoleonic aura in the Arab world; he had identified himself with the most progressive features of the revolution, he had wrested control of the revolutionary movement away from the older and more conservative General Naguib, and he had obtained Soviet arms—a feat which, to the Arab's, dramatically proclaimed Egypt's independence from the Western powers.

The period between the nationalization of the Suez Canal in 1956 and the annexation of Syria in 1958 represents the highwater mark of what came to be known as "Nasserism." The military encounter with

British, French, and Israeli forces in the fall of 1956 cost Nasser a considerable amount of his newly purchased Soviet military hardware, but the diplomatic support he gained from the United States-Soviet alliance in the United Nations bolstered Egyptian prestige in the face of military defeat. Early in 1958 pro-Nasser elements in Syria, particularly the *Ba'thists*, agreed to merge that country with Egypt to form the United Arab Republic, and it appeared to be only a matter of time before Nasserist elements in Jordan, Lebanon, and Iraq would follow suit, bringing about an Egyptian dominated Arab union extending from the Nile to the Euphrates. But the Iraqi Coup of July 14, 1958, brought to the scene of Arab politics Nasser's first revolutionary rival, Col. Abdul Kerim Kassem, and, instead of following Syria into Nasser's realm, Iraq emerged as the focal point of opposition to Egyptian "imperialism" in the region.

Colonel Kassem has challenged Nasserism on its own terms—ultranationalism and social reforms. Present-day bitterness between Cairo and Baghdad is nevertheless remarkably similar to the antagonism that formerly prevailed between the Nasser regime and Nuri as-Said's "feudalistic" regime and stands as a reminder of the practical difficulties involved in the Pan-Arabist idea. Mitigating strongly against the demands of the Pan-Arabist ideal is the fact that Iraq is a wealthy oil-producing area, while the rest of the Fertile Crescent and certainly Egypt are "have-not" countries. Nasser's anti-Communist police measures have caused the Communist organization throughout the Middle East to rally behind Iraq's Kassem. Kassem depended heavily on the Communists against Nasserist elements during the first year of his regime, but in recent years he has become more wary of Communist "support."

PROSPECTS FOR THE 1960'S

The revolutionary movements led by Nasser and Kassem represent the most significant development in the Middle East. The moderate revolutions which have taken place in Iran, the Sudan, Turkey, and elsewhere in the region are no guarantee that these countries are permanently immune to revolutionary activity. Even such politically unsophisticated and rather backward areas as Saudi Arabia, Yemen, Kuwait, and other "oil sheikhdoms" of the Arabian desert will very likely become involved in the revolutionary ferment going on in the Fertile Crescent.

Particularly subject to revolutionary activity are such countries as Lebanon, Jordan, and Iran. Jordan has an unstable government and a heterogeneous population of traditionalist Bedouin tribesmen and urbanized, pro-Nasser refugees from Palestine. Iran has not solved the social and economic problems which were already pressing ones before the Mosaddegh era a decade ago; the Shah holds his throne by counterbalancing rural-traditionalist, military, and urban middle-class elements, and the conditions for steering this tortuous course are becoming increasingly difficult. In the summer of 1961 pro-Mosaddegh rallies were being held both in Teheran and by Iranian student groups in England and the United States, and conservatives in Iran were looking to the ultra-nationalist and even the extreme left for support against the present reformist, pro-Western regime.

Finally, Egypt, Syria, and Iraq are by no means free from continued revolutionary ferment. Nasser has already earned the active opposition of the Communist organization, and it may be only a matter of time before Kassem too will become a prime target of Communist intrigue. The importance of the Communists in the Middle East is not dependent on their numerical strength but on their organizational strength. In spite of the intraparty factions and dissention which plague Arab Communists, they are generally admitted to possess the best political organization in the region. The Communists may provide the organizational framework within which the various opponents of Nasser and Kassem will operate. The non-Communist opposition includes rightwing elements, religious traditionalists, such as the Muslim Brotherhood in Egypt; tribal elements who resent the urban-based revolutionary regimes; national minorities, such as the Kurds in Iraq and the various sects in Syria; and, most important of all, the new social forces which the Nasser and Kassem revolutionary movements have introduced into the political arena. When the social and economic aspirations of the people are not met in the ways they have been led to expect, there is likely to be another outburst of bitterness and frustration which may turn against the revolutionary regimes.

On 28 September 1961, elements of the United Arab Republic's First Army (Syrian in composition) in the Syrian Province rebelled against the government of President Gamal Abdul Nasser in conjunction with Syrian politicians. With the apparent support of the people, the rebels declared their independence from the U.A.R., thus breaking up the union of Syria and Egypt which had come into being in 1958. The economic reforms of July 1961 (see section on Egyptian revolution), and the resignation of Abdul-Hamid al-Sarraj, Vice President in charge of Internal Security and the last Syrian in the U.A.R. Cabinet, seem to have been the immediate causes which precipitated the rebellion.

The Syrian Rebellion of September 28, 1961, appears to have been carried out primarily by rightwing elements in the local government and in the armed forces, in coalition with other Syrian political elements, alienated by Nasser's drastic social and economic policies. The successful introduction of these social and economic policies preconditioned a centralized union, rather than a loose association of equal and semi-independent states: Egypt and Syria. The successful implementation of these policies demanded the subserviency of Syria's higher economic standard to that of Egypt. Both measures necessitated the elimination of Syria's bourgeoisie which had previously played a dominant role in determining the economic and political future of Syria. As a result, the role of Syria in the affairs of the U.A.R. was slowly, but assuredly, minimized: Syrian politicians were dropped from successive U.A.R. Cabinets and replaced with Egyptian officials, until the greater majority became Egyptian; and Syria's economy reflected the change in the purchasing power of the Syrian pound in relation to that of the dollar. At the time this account is being written (October 1961) the new Syrian leaders are urging the Egyptians to eliminate Nasser and establish a new government. It is too early to predict the precise effects of the Syrian coup, but it will almost certainly result in continuing unrest and possible new revolutions in the Middle East.

In conclusion, it seems safe to assume that elitism in political activity—and especially in the making of political revolution—will continue in the Middle East, until there has emerged an integrated national society in each of the Middle Eastern countries. Such a society would first have to come to terms with both its Islamic heritage and the Western content of its national life and would have to arrive at a generally accepted political consensus of goals and values. Political leaders may then be able to attract mass support and participation in a revolutionary movement on a scale comparable to that in other parts of the world. In Tunisia, for instance, also Arabic and Islamic, a genuine mass movement has functioned successfully. That country, since it achieved independence in 1956, has not been subject to elitist coups d'etat.

NOTES

1. Walter Z. Laqueur, *Communism and Nationalism in the Middle East* (New York: Frederick A. Praeger, 1956), pp. 24–25.
2. Abdul Aziz Said (Unpublished manuscript, American University, Washington, 1961).

3. K. J. Newman, "The New Monarchies of the Middle East," *Journal of International Affairs*, XXII, 2 (1959), 160–161.

4. Bernard Lewis, "Democracy in the Middle East—Its State and Prospects," *Middle Eastern Affairs*, VI (April 1955), 105.

5. A. H. Hourani, "The Middle East and the Crisis of 1956," *St. Anthony's Papers, No. 4: Middle Eastern Affairs, No. 1* (London: Chatto and Windus, 1958), p. 18.

6. Ibid., p. 17.

7. Ibid., pp. 21–22.

8. Ibid., p. 30.

9. Laqueur, *Communism and Nationalism*, p. 10.

10. Hourani, "The Middle East," p. 30.

11. Dankwart A. Rustow, "The Politics of the Near East, S.W. Asia and Northern Africa," *The Politics of the Developing Areas*, eds. Gabriel A. Almond and James S. Coleman (Princeton: Princeton University Press, 1960), p. 423.

12. Ibid., p. 422.

13. Laqueur, *Communism and Nationalism*, pp. 20–21.

14. Said, Unpublished manuscript.

RECOMMENDED READING

BOOKS:

Hourani, Albert. "The Middle East and the Crisis of 1956," in *St. Anthony's Papers, No. 4, Middle Eastern Affairs, No. 1*. London: Chatto and Windus, 1958. Pp. 9–42. An excellent survey of the historical and political background of the Middle East by an outstanding scholar and Middle East Specialist at Oxford University.

Khadduri, Majid. "The Army Officer: His Role in Middle Eastern Politics," in *Social Forces in the Middle East*. Ed. Sydney N. Fisher. Ithaca: Cornell University Press, 1955. The best monograph on the subject by a native of Iraq and leading Middle East scholar at Johns Hopkins School of Advanced International Studies.

Kirk, George E. *Contemporary Arab Politics*. New York: Frederick A. Praeger, 1961. A political history and analysis of the Arab states with particular emphasis on revolutionary developments since 1952. The author is a professor at Harvard University and has written several standard histories on the area. Good bibliography.

Laqueur, Walter Z. *Communism and Nationalism in the Middle East*. New York: Frederick A. Praeger, 1956. A comprehensive study of ideo-

419

logical tendencies and political developments in the region by a specialist on the Middle East and the Soviet bloc.

Rustow, Dankwart A. "The Politics of the Near East," in *The Politics of the Developing Areas*. Eds. Gabriel A. Almond and James S. Coleman. Princeton: Princeton University Press, 1960. An excellent study of contemporary politics in the area by a competent American political scientist.

PERIODICALS:

Fauzi M. Najjar. "Ideological Vacuum in the Middle East," *Free World Forum*, I, 4, 16–21, 27. Excellent discussion of ideological trends and possibilities in the area.

Journal of International Affairs, XIII, 2 (1959). Entire issue of this Columbia University publication devoted to Middle East. Articles by John S. Badeau and K. J. Newman are especially relevant to the study of revolution in the Middle East.

THE IRAQI COUP OF 1936

SYNOPSIS

In October 1936 a coalition between army officers, dissatisfied with corruption and incompetence in high places, and a political group of left-of-center intellectuals, staged an almost bloodless coup d'etat against the authoritarian government that had ruled Iraq. A popular general ordered the army to march on Baghdad and issued an ultimatum to the King demanding the dismissal of the existing government and the formation of a new government composed of the political group that was an accomplice to the coup. There was no resistance, and in less than 24 hours the King acceded to the demands.

BRIEF HISTORY OF EVENTS LEADING UP TO AND CULMINATING IN REVOLUTION

In October of 1936, a coalition between army officers dissatisfied with corruption and incompetence in high places and a political group of left-of-center, liberal intellectuals, called the *Ahali* group, staged an almost bloodless coup d'etat against the authoritarian government which had ruled Iraq for the past 19 months. The army wanted to set up a Kemalist dictatorship, based on nationalism and social and economic progress. The *Ahali* group also wanted to effect socio-economic reforms, but based on democratic, parliamentary procedures. Lacking the opportunity to replace the authoritarian regime in power by parliamentary methods, this political group agreed to cooperate with the army officers in their goal of overturning the government.

Utilizing the temporary absence of the loyal chief of staff, Bakr Sidqi, a popular general, ordered the army to march on Baghdad and issued an ultimatum to the King demanding the dismissal of the existing government and the formation of a new government under the *Ahali* politician, Hikmat Suleiman. There was no resistance, and in less than 24 hours the King accepted the formation of a government by the revolutionary coalition group.

The coup is an early example of a coalition between a leftist political group and the army; in this sense it resembled the Syrian coups of 1949–51 and other Socialist-military revolutionary liaisons in the Middle East since 1945. Despite its avowed aims and intentions, the new government was no less authoritarian than the one it replaced. The social and political content of the revolutionary goals was not realized.

THE ENVIRONMENT OF THE REVOLUTION

DESCRIPTION OF COUNTRY

Physical characteristics

Situated in the Middle East between the Arabian desert and the Iranian mountains, Iraq shares extensive common borders with Iran to the east, Saudi Arabia to the south, and Syria to the west. None of those are "natural" frontiers but are the results of diplomatic compromise following the fall of the Ottoman Empire at the end of the First World War.

Iraq's 171,000 square miles—the country is the size of California—can be divided topographically into three main zones: the desert, the river valley, and the highland.[1] Typical desert climate prevails over most of the country, except for the very humid region adjacent to the Persian Gulf. The summer months (May to October) are extremely hot and dry, though blessed with cool nights. Winters are mild in the

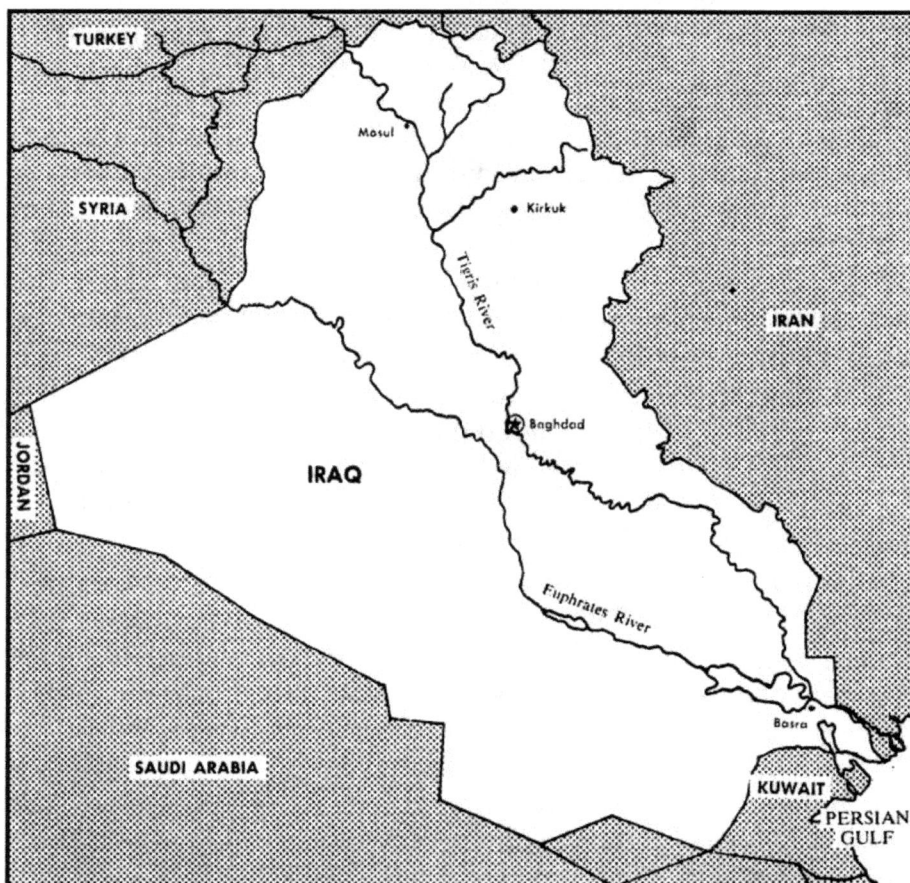

south, but may be very cold in the northern highlands. Rainfall occurs only in winter, when it averages over 12 inches in about 60 days north of Baghdad and somewhat more in the northeastern hills and considerably less in the southwestern desert.[2] Torrential and unpredictable rainfall makes the Tigris-Euphrates River system less manageable than the Nile, for instance, and alternating floods and droughts pose a major problem.[3]

The people

Iraq's first official census, taken in 1947 and of dubious reliability, showed a sedentary population of more than four and a half million persons, to which must be added at least a quarter million nomads.[4] In the 1930's, the population must have been around four million. Iraq has not experienced any serious population pressures, for with five times as much arable land as Egypt, Iraq has only a quarter of Egypt's population.[5] The areas of greatest population density are along the Tigris and Euphrates and in the northern mountain valleys. At least two-thirds of its population live in rural villages; Baghdad, Iraq's capital and largest metropolitan center, had in 1956 less than one million inhabitants.

The people of Iraq are not split along racial lines, but rather along those of religion, language, and cultural tradition. Eighty percent of the population are Arabic in language and cultural identification; the overwhelming majority of these are Muslim. The largest—and traditionally the most difficult to assimilate—minority community in Iraq are the Kurds. Comprising around 17 percent of the total population, the Kurds are concentrated in the northern and northeastern highlands along the Turkish and Iranian borders where they lead for the most part an agricultural pastoral existence, their mountain warrior days now only a part of their traditional lore. The homeland of the Kurds lies within the borders of Iran, Iraq, Turkey, and the U.S.S.R. Kurdish nationalism has been a threat to Iraq national unity, held in check both by the disposition of the Great Powers and by the absence of any articulate and unified pattern of loyalty to a Kurdish nation. Lesser minorities are the Assyrians, and the Yezidis, a much despised Muslim sect called "Devil Worshippers," who, along with Arab Christians and Kurds, are concentrated in Mosul province and speak a Kurdish dialect.

Communications

Iraq's communication routes and the centers they served were not very different in the 1930's from the caravan routes and bazaar cities of the past thousand years. Basra, located within 100 miles of the Persian Gulf on the Shatt al-Arab, formed by the confluence of the Tigris and Euphrates, served as Iraq's only seaport and as the center of the date industry; Mosul in the north, connected by rail and highway with Syria, served as a point of entry from the Mediterranean ports and as a center of the wool industry. Baghdad was connected by rail with Basra, Kirkuk, and other parts of central Iraq, but not until 1939 were rail connections with the Mediterranean coast established. Highways and river transport were important supplements to rail connections during this period. International civil airports were established in Basra and Baghdad in the early 1930's. In general, however, internal transport and communication facilities remained in an underdeveloped state.

Natural resources

Iraq's natural resources include extensive oil deposits and fertile alluvial soil, together with the all-important Tigris-Euphrates River system. However, "with the exception always of its phenomenal richness in petroleum . . . , and its agricultural potential, Iraq is a country sadly deficient in natural wealth."[6] Iraq's poverty in the 1930's contrasted sharply with the country's past prosperity and the area's natural potential. Mesopotamian agriculture depends on irrigation and elaborate manipulation of the river system for its success; however, the canal-and-dike system that had made Baghdad a major city in the medieval world was never restored after its destruction by the Mongols in the 13th century.[7]

SOCIO-ECONOMIC STRUCTURE

Economic system

By the 1930's, British and Iraqi efforts to improve the state of agriculture were taking effect in a general way, but the material benefits of their activities—building bridges, irrigation projects, railroads, and highways, fighting locusts and pests—went only indirectly to the peasants, for the tribal sheikhs and landlords received the direct benefits. In 1930, five-sixths of the arable land in Iraq was either state property or collective tribal property; after 1932, the Government began in earnest the distribution of land among the sheikhs.[8] Thus, the tribal sheikh-landlord class emerged as a powerful social and economic group.

Oil has been Iraq's leading export since the 1930's, when the petroleum industry became the country's first modern industry. Major discoveries were made in 1923 near the Iranian frontier and in 1927 in the vicinity of Kirkuk by European and American oil companies to which the Iraqi Government had entrusted the development of its oil resources. The extraction of petroleum does not require a large labor force, consequently no real working class had developed in Iraq as a significant social group before 1936. In 1928–29, the first modern trade unions were organized among such occupational groups as taxi drivers, barbers, and mechanics. When they allegedly engaged actively in politics in 1931, the government closed these unions, and until 1943, no trade union activity on a countrywide scale was permitted.[9]

Class structure

Iraqi social structure before the Second World War was typical of that found in most underdeveloped non-Western areas: (1) a vast substratum of peasants and, hardly distinguishable as a group apart from the peasantry, an inchoate working class; (2) a much smaller traditional elite composed of feudal landlords (sheikhs) and religious leaders (ulema) ; and (3) a small "middle class" loosely made up of bureaucrats, professional and business people, intellectuals and students, and others who might be considered urbanized and to some extent influenced by Western civilization. In 1936 Iraq, the social—as well as the economic and political—elite comprised about one thousand sheikhs, together with the clique of professional politicians who represented their interests in the central government. Even in 1957, "88 percent of the peasants owned no more than 6.5 percent of the land, while the rest of the land was in the hands of the state and about one thousand sheikhs—the very sheikhs who were the pillars of society in Iraq and who controlled the parliament and the government."[10]

In addition to such traditional social escalators as the army and the religious hierarchy, the panoply of representative government with its expending bureaucracy and political party system has emerged in modern Iraq as a significant avenue of upward social mobility.

Literacy and education

Literacy in Iraq in 1936 was below the current rate of 10 to 15 percent. The predominance of classical Arabic in press and radio to the exclusion of the colloquial language makes communication with the semiliterate especially difficult, since they are for the most part not familiar with the formal literary language. Within the country literacy rates have tended to be lowest in the south and in rural areas all over.[11]

Major religions and religious institutions

Islam is the dominant—and state—religion in Iraq. It is the Sunni-Shi'i schism within Islam which has been the chief religious complication in the social, economic, and political life of the nation. About half the Muslim population of Iraq are Shi'is, and in the 1930's most of these were concentrated geographically in the south and southeast (closer to Iran where they are the majority sect) and economically in the lower classes.[12] The Sunni's have dominated the government and society in general. They tend to be better educated, and the royal family have all been Sunni Muslims. It was the custom in 1930 Cabinets to include at least one Shi'i Muslim. The Shi'i Muslims claim to be in the majority and feel that they are underrepresented in the government. There were usually as many or more Kurds in the government, although that group forms only 17 percent of the population, but the Kurds are Sunni Muslims and have been more politically active than the Shi'is.

GOVERNMENT AND THE RULING ELITE

Description of form of government

From March 1925 to July 1958, Iraq was officially a constitutional parliamentary monarchy. A constitution was formulated in the 1921–24 period of the British Mandate which was never formally suspended. In 1932 the British terminated their mandate in exchange for a treaty which granted them certain military concessions. When this treaty became effective in October 1932 Iraq had achieved full independence. The Iraqi Constitution provided for a limited monarchy with a King to function as chief of state and a cabinet-type executive to serve as head of the government, a bicameral national legislature, and an independent judicial system. King Faisal I of the Hashemite dynasty came to the throne in 1921, as a result of British influence and the support of the leading families of the country, in response to a popular plebiscite. Faisal was succeeded on his death in 1933 by his son, Ghazi I, who was the ruling monarch in 1936.

The Iraqi Crown performed the usual duties of a constitutional monarch, and, in addition to these ceremonial duties, had powers of decree when Parliament was not in session and the power to confirm or reject legislation within 15 days to 3 months. The Crown prorogued and dissolved Parliament on the recommendation of the Prime Minister. In practice, only Faisal I had sufficient personal prestige to exercise fully the constitutional powers of the Crown. A major weakness in the Crown's position, in relation to the government, was its inability to dismiss a Prime Minister who refused to be responsible to the will of Parliament. The parliamentary government was circumscribed purposely by constitutional provisions calculated to give the final word to the executive branch. This was considered necessary to safeguard foreign interests and to enable the King to fulfill foreign obligations of Iraq.

The Prime Minister and his six or more ministers, all members of Parliament, were in theory responsible to the Chamber of Deputies both individually and collectively. In actual practice, however, the Prime Minister and his coterie dominated Parliament and exercised undue influence in the judicial and administrative systems. The Constitution provided for a Senate, one-fourth the size of the lower house, appointed by the Crown on the recommendation of the government. Senators were usually former holders of high political office.[13] Consequently, the Senate tended to be more independent and critical of the government than was the Chamber of Deputies, but the Senate was without power to censor the government. The Chamber of Deputies was, until 1952, elected indirectly, and it was to this body alone that the government was constitutionally responsible. In practice, however, governments were able to dominate Parliament by frequent dissolutions and manipulation of elections to such an extent that a vote of no confidence was never passed in the history of the Iraqi Constitution.

Description of political process

Political parties and power groups supporting government

The real power group in Iraq was the landlord class. Closely allied to this group and to a large extent overlapping it, were the bureaucrats and wealthy middle class, who sought respectability and economic advantage alike in absentee commercial farming. Debate over such things as the fine points of British-Iraqi relations, carried on in a highly emotional and nationalistic tenor, kept Iraqi political leaders at a safe distance from the economic and social issues around them. The eight Iraqi governments that were in power from 1932 (the beginning of full independence) to the Coup of 1936 rose and fell in response to the shifting personal alliances and cliques within a small circle of professional office holders, devoid of any substantive political content or any real popular support and participation. In the 1932–36 period, the lives of the first seven Cabinets ranged from 2 weeks to 6 months and the last Cabinet, headed by Yasin al-Hashimi, lasted for an unprecedented 19-month period. Considering that there were only five different Prime Ministers in this period, and that such Cabinet posts as Foreign Minister were held by the same person through all eight governments, there was in actuality less instability than one might think. Nuri as-Said, the venerable Iraqi political leader and perennial member of the Cabinet, made a classic comment on the

Iraqi political situation: "With a small pack of cards, you must shuffle them often."[14]

Character of opposition to government and its policies

A basic cause of Iraq's governmental instability has been, on the one hand, the acute competition among its politicians and, on the other, the absence of strong political parties. In the 1920's there had been three main parties with the same objective—termination of the British Mandate and full independence. In 1930, when the government concluded the Anglo-Iraqi Treaty, whereby Iraq would obtain full independence in 1932 in return for becoming a British Ally in time of war, rival politicians organized the *Ikha'al-Watani* Party (National Brotherhood Party) out of old political groups to oppose the treaty with Britain on the grounds that it compromised Iraqi national aspirations. In defense of the treaty, the *Ahd* Party was formed, taking its name from a pre-1920 nationalist group. After ratification of the treaty and termination of the Mandate in 1932, its *raison d'être* disappeared and the *Ahd* was dissolved.

The *Ikha* survived as a political party until 1934–35. However, *Ikha* politicians had completely compromised the ultranationalist principles on which their party was founded when for 7 months in 1933 they participated in the government under the conditions of the treaty. The party finally dissolved itself in 1935. After that there were no organized political parties through which the intense personal rivalries of the politicians and the latent feelings of social discontent among the masses might have been channeled and accommodated.

In the absence of political parties as such, the *Ahali* group performed the function of a political opposition party during the Yasin-Rashid regime (March 1935–October 1936). Formed in 1931–32 by enthusiastic young Iraqis imbued with liberal ideas, the group was characterized at first by its youthful membership—mostly recent graduates of British and American colleges and the Baghdad Law School—and by its democratic ideological orientation, ranging from advocacy of the principles of the French Revolution over to those of modern socialism. The *Ahali* newspaper, the group's organ from which it took its name, was first published in January 1932, and became at once the most prominent daily paper in Iraq. In 1934, the group adopted socialism as its first article of faith, but for practical reasons they preferred to call their ideology *Sha'biyah* (Populism). *Sha'biyah* called for sweeping social reforms to be carried out by the state, through a parliamentary system of government based on functional representation. It advocated a kind of collectivism, but did not admit the existence of a class struggle in society or revolutionary procedures in social change. Also, in contrast to orthodox Marxism, *Sha'biyah* recognized the family and religion. The *Ahali* manifesto repudiated nationalism but recognized patriotism.[15]

Opposition from Arab nationalist elements and conservative politicians in the government forced the closing of the group's Baghdad Club, a social circle for the propagation of *Sha'biyah*, and compelled further tactical retreats from the principles of socialism. In 1934–35, with the addition to the group of a number of the more liberal elder politicians, who had for various reasons broken with the regime in power, definite plans were made to work towards achieving power. An executive committee was set up and secret meetings were instituted. "But *Sha'biyah* was no longer preached . . . . , and only the demand for reforms, in a general way had become [the *Ahali* group's] chief slogan."[16]

Within the group, however, there was a leftwing who cherished more radical ideas; of these, only Abd al-Qadir was a Communist, and the rest were Populists. There was also a split between the leftwing which wanted to make the group into a mass political party and others who opposed this type of organization. The politicians, who had come to dominate the group, had little faith in legal channels in the pseudo-parliamentary system they knew so well. Consequently, the idea of a publicly organized political party lost out, and the *Ahali* group continued as a conspiratorial clique.

The main form of overt opposition to the regime carried on by the *Ahali* group was through their press, until government censorship silenced that form completely in April 1936. The group petitioned the King several times in May 1936, in protest over the government's ruthless suppression of tribal disorders and curtailment of civil liberties, and these petitions were published in Syrian and Lebanese newspapers and smuggled into Baghdad for distribution.

Legal procedure for amending constitution or changing government institutions

The Iraqi Constitution of 1925 could be amended by a two-thirds vote in both houses of the legislature, provided that the Chamber of Deputies should have been dissolved after its approval of the amendment and a newly-elected Chamber returned for a second and final vote on the amendment. Before 1936 this procedure was only used once, in 1925. From 1936 to 1941, the normal functioning of this Constitution was interrupted by the intervention of the military in the selection of the executive.

Relationship to foreign powers

The bedrock of Iraqi foreign policy was the Anglo-Iraqi Treaty of 1930 which bound Iraq to have "full and frank constitution" with Britain in foreign policy matters, to be Britain's Ally in time of war, and to permit British use of transportation and communication during war. This alliance was regarded as the guarantee of Iraqi independence by the moderates and as just the opposite by the ultranationalists. British commercial interests were much in evidence, and this may have been the cause of some anti-British sentiment. The real difficulty in Anglo-Iraqi relations, however, arose out of the frustration felt by Pan-Arabists over Britain's failure to support a united Arabic state. British support of Jewish immigration into Palestine and French reluctance to give up Mandate control over Syria and Lebanon caused Pan-Arabists to look to the Fascist bloc for support of their cause. Iraqi nationalists were torn between the attractions of Pan-Arabism—a kind of horizontal and cultural loyalty—and the attraction to "Mesopotamianism"—a kind of vertical and historical loyalty. The Government of Yasin al-Hashimi, which the Coup of 1936 overthrew, was characterized by a definite Pan-Arabist foreign policy orientation. Most *Ikha* politicians, being ultranationalists, were both Pan-Arabist and anti-British in foreign policy orientation.

The role of military and police powers

The Yasin-Rashid regime was the most authoritarian government Iraq had ever experienced. That it remained in power for the unprecedented span of 19 months was due primarily to the effective control the government exercised over the army and the police, and when this control broke down in 1936, the end of the regime was at hand. The Iraqi Army was established in 1921 by British Mandate authorities. After full independence in 1932, the army distinguished itself by dealing promptly and effectively with the frequent uprisings among dissident tribes and minority groups that plagued every government in power. Officially under the command of the King and the Minister of Defense, the army was under the operational control of the Chief of Staff. During the Yasin-Rashid regime this post was held by Gen. Taha al-Hashimi, a brother of the Prime Minister who was loyal to the government. The police was under the direct control of Rashid Ali, the other strong man in the regime, who held the coveted post of Minister of the Interior. By using the army against the dissident tribesmen and the police against their rivals in political circles, Yasin al-Hashimi and Rashid Ali were able to prolong their rule long after the disappearance of whatever popular support accompanied their entry to power.

WEAKNESSES OF THE SOCIO-ECONOMIC-POLITICAL STRUCTURE OF THE PREREVOLUTIONARY REGIME

Governmental instabilities and socio-economic tensions

A history of governmental instability combined with political irresponsibility in a society beset by such malintegrative factors as the Sunni-Shi'i conflict, the unequal distribution of wealth, and the general lack of social and economic opportunity preceded the Coup of 1936. It was thought that full independence would permit Iraqi politics to begin to turn on genuine economic and social issues instead of nationalistic slogans, but such was not the case. The Yasin-Rashid regime came to power in March of 1935, through somewhat less than legal means. Out of power for over a year and feeling frustrated in normal political channels, the *Ikha* politicians turned to the exploitation of Shi'i tribal discontent in the middle Euphrates provinces. The resulting disorders convinced the King that only this group of politicians could maintain order.

Iraq in 1936 was an economically underdeveloped, predominantly agrarian, semifeudal society. It was not, however, a static or dormant society, such as its neighbor Saudi Arabia, but was in the initial throes of economic development, with its concomitant urbanization, secularization, and Westernization of society. The peasant-worker masses were living in miserable social conditions, but were politically unconscious and inarticulate; their discontent was expressed only indirectly—through tribal revolts led by feudal sheikhs ostensibly for religious reasons, and through the malaise of the middle classes. The educated classes in the urban areas, and especially the younger generation, were to various degrees discontented with Iraq's general lag behind the economic and cultural level of Europe and Turkey. The example of the Kemalist reforms in Turkey stood in sharp contrast to the inept and ineffective policies of Iraq's pseudo-parliamentary government, and this tended to alienate the younger generation from the parliamentary system of government.

Government recognition of and reaction to weaknesses

The weaknesses recognized by the Yasin-Rashid regime included the disaffection of certain political figures, such as Hikmat Suleiman, and the perennial discontent of certain Shi'i and minority tribes. General discontent among the masses, even among the literate and politically conscious segments of the population, was not of critical importance to the government, so long as it controlled the army and the police. The Yasin Cabinet included two Shi'is, a Kurd, and a few elder politicians not connected with *Ikha*. By these moves, the Prime

431

Minister believed he had constructed a strong government in the classic Iraqi manner of reconciling all factions in a vast personal coalition. He held hopes of a reconciliation with his old colleague Hikmat Suleiman, who had refused the post of Finance Minister when he was passed over for Interior Minister, until the very day of the coup. As for the tribal sheikhs in the middle Euphrates area who continued to riot as they had done under previous governments, Yasin was in a position through his control of the army to put them down with dispatch and efficiency. The gradual alienation of key persons in the military hierarchy from the government was completely unknown to the Prime Minister. When in the spring of 1936, the *Ahali* press began criticizing the government, the regime reacted by suppressing these organs and arresting and deporting certain suspected Communists. Sensing their loss of popular support, the Yasin government stepped up Pan-Arabist propaganda both at home and abroad, including cultural and diplomatic exchanges among Iraq, Egypt, and Syria.[17] In addition, all political parties were abolished by Yasin, closing the door for parliamentary change and leaving forceful change as the only alternative.

FORM AND CHARACTERISTICS OF REVOLUTION

ACTORS IN THE REVOLUTION

The revolutionary leadership

The leadership of the coup was provided by Hikmat Suleiman of the *Ahali* political opposition group and Gen. Bakr Sidqi, a national hero and a favorite among the younger officers as a result of his leadership of the army against the Assyrians in 1933 and in subsequent tribal disorders. Hikmat broke with the regime when he was refused the Interior Ministry on account of his ties with the *Ahali* movement, which to nationalists and conservatives like Yasin al-Hashimi and Rashid Ali was nothing less than communism under another name. In point of fact, Hikmat was not even a Socialist but declared himself a Kemal-type "reformist," which brought him into agreement with enough of the *Ahali* ideology (in the less radical form it had assumed by 1935) to permit his joining the movement. He was an admirer of Mustapha Kemal and his brother had participated in the Young Turks revolution of 1908.

Hikmat's counterpart in the army, Gen. Bakr Sidqi, shared his Kemalist sentiments, although Bakr's chief concern was a military dictatorship which would carry out these Kemalist reforms. Both Hikmat and Bakr enjoyed wide popularity among the younger generation;

432

both called themselves Kemalists, and both of them were politically ambitious and opportunistic. Bakr's political ambitions were not, however, known to the Prime Minister, who trusted him completely, as attested by the fact that Bakr was made Acting Chief of Staff a few weeks before the coup. The real goal of Hikmat Suleiman and Bakr Sidqi, the instigators and organizers of the revolution, was nothing more than personal aggrandizement: Hikmat wanted to be Prime Minister and Bakr wanted to be Chief of Staff, and the al-Hashimi brothers stood in their way.

The revolutionary following

The October coup enjoyed considerable popular support among both the younger army officers and the urban intellectuals of the *Ahali* movement. This mass support was, however, not organized, and its effect on the revolution was passive rather than active. There was no active participation in the coup, except by the rank and file of the Armed Forces who merely obeyed the military orders of their officers.

ORGANIZATION OF REVOLUTIONARY EFFORT

There were two dissimilar opposition groups ranged against the Yasin-Rashid regime: the *Ahali* political opposition group and the army officers' group. The only contact between these two centers of opposition was a clandestine personal relationship between the *Ahali* politician, Hikmat Suleiman, and Gen. Bakr Sidqi of the Iraqi Army. There were no joint meetings and no overlapping of personnel between the two groups. In the authoritarian environment of 1936 Iraq, no overt opposition to the government through the press and public communications media was feasible; consequently, opposition must have taken the form of secret "cell" meetings. Overt opposition was possible only through the Syrian and foreign presses, which might be smuggled into the country and distributed clandestinely; however, such operations were incidental to the success of the revolution. In fact, there was a minimum of organizational structure and activity, as is to be expected in the case of a coup d'etat. The nucleus of the revolutionary organization was nothing more than a personal conspiracy between Bakr Sidqi and Hikmat Suleiman. In the summer of 1936, these two leaders decided to seize the first opportunity for a military coup and, in the meantime, to prepare their followers for the roles they were to play in the revolution. When Chief of Staff Taha al-Hashimi left for a visit to Turkey in late October, leaving Bakr Sidqi in charge of the army, the latter decided that the opportunity had arrived. He first confided in Gen. Abd al-Latif Nuri, probably promising him the post of Defense Minister in the revolutionary Cabinet. He then contacted a number of other officers in whom he had confidence and secured their cooperation. With the army firmly behind him, Bakr informed Hikmat of the impending coup and asked the *Ahali* group to draw up a petition to the King calling for the dismissal of the Yasin Cabinet and the installation of another government under the leadership of Hikmat Suleiman. Hikmat informed the *Ahali* leaders that the army had resolved to stage a coup and wanted them to form a new government. After some hesitation, the *Ahali* leaders drew up a proclamation to the King, which was then signed by General Bakr as "Chief of the National Reform Force" and given to Hikmat to deliver to the King on the day of the coup. Bakr contacted Gen. Abd al-Latif Nuri on the 23rd and the coup took place on the 29th of October; thus, the entire organization of the revolutionary effort appears to have transpired in less than a week. The *Ahali* leaders were informed of the conspiracy in which they were to figure so prominently only a few days before the coup.

GOALS OF THE REVOLUTION

The concrete political aims of the instigators and organizers of the revolution, and of many of their followers, went no further than the attainment of high positions in the revolutionary government. Beyond the immediate goal of governmental seizure, the leaders of the coup ascribed only to the vague slogan of "reforms," which presumably would involve a rise in living standards among the masses, enlarged opportunities for the urbanized middle class, and less corruption in high places. No specific system of government or institutional changes were proposed for the implementation of these reforms, but it was generally assumed that the role of the military hierarchy in the revolutionary government would be more active than in previous governments. On this question, there was an area of major disagreement between the *Ahali* group and the military hierarchy. Pan-Arabism did not figure prominently in the avowed aspirations of the revolution; the *Ahali* leaders were primarily concerned with domestic reforms and Gen. Bakr Sidqi was more interested in a military dictatorship in Iraq than in Pan-Arabist foreign ventures. The revolutionary following in both the army and the *Ahali* movement were more susceptible to the appeals of Pan-Arabism than was the leadership of the revolution.

REVOLUTIONARY TECHNIQUES AND GOVERNMENT COUNTERMEASURES

The seizure of the machinery of government was achieved in a virtually bloodless coup d'etat in the space of a few hours, after only a week's prior planning for the operation. The government was taken completely by surprise. On the morning of the 29th at 7:30, Bakr's Second Division started the march on Baghdad. At 8:30, five planes flew over the city dropping copies of the Proclamation, which called on the King "to dismiss the Cabinet and form a new one composed of sincere men under the leadership of Hikmat Suleiman." This was the only specific example of the use of propaganda as a method of weakening the existing government's authority. A more significant communication, which gained the revolutionaries the support of the Crown, was a telegram from the two rebellious generals which assured the King of the army's loyalty to him personally. A letter from the generals, delivered by Hikmat Suleiman to Zuhur Palace just after 8:30, gave the King until 11:30 a.m. to form a new government.

The King, the Cabinet, and the British Ambassador (who remained neutral in the crisis) met at once. Yasin wanted to resist, but the King refused to authorize any government action to counter the coup. Apparently, King Ghazi was not an accomplice to the coup, although when it came he felt no compulsion to risk civil war for the sake of Yasin-Rashid government,[a] with which he had often had personal clashes. The King's qualified support together with the British Ambassador's passive support of the coup, was of decisive importance. Yasin's only alternative was to resign, which he did at that time, and the King then called on Hikmat to form a new government.

At this point, the crisis might have been concluded without the use of violence, but Hikmat refused to form his government until the King had drawn up a formal request in writing; also, the Minister of Defense in the Yasin government, Ja'far al-Askari, insisted on going out to meet the army personally to persuade it not to enter the city, since the Yasin government had "voluntarily resigned." This action was the only countermeasure taken by the legitimate government during the coup and this was more personal than governmental. Ja'far's action reflected the general willingness on the part of the King, the British, and moderate elements of society to accept the Hikmat government, but without open intervention of the military. Ja'far was nonpartisan, despite his being in the deposed Cabinet, and had great prestige in army circles as one of its founders. Bakr feared he would

---

[a] There have been unconfirmed reports that the King was an accomplice to the coup and knew about it in advance.[18]

turn the army against the coup and so had the Minister assassinated en route. This was a political error on the part of Bakr, for not only was it unnecessary for the success of the coup, but it alienated friends and relatives of a renowned popular figure in Iraqi political and social circles. The only other violence occurred at the expiration of the 3-hour ultimatum, when four bombs were dropped over public buildings in Baghdad, injuring about seven persons.

MANNER IN WHICH CONTROL OF GOVERNMENT WAS TRANSFERRED TO REVOLUTIONARIES

At 5:00 p.m., the army entered Baghdad, with Bakr at the head of his victorious National Reform Force, amidst the enthusiastic applause of the populace. Proceeding to the Ministry of the Interior, General Bakr met with Hikmat and the *Ahali* politicians to form the new Cabinet. The Cabinet formed the next morning included Hikmat as Prime Minister and Bakr Sidqi as the Chief of Staff. Nuri as-Said and Ja'far al-Askari would undoubtedly have been in the Cabinet, but for Bakr's assassination of the latter, a relative and friend of Nuri as-Said. Thus, an important segment of Iraq's social and political elite were implacably alienated at the outset of the revolutionary regime.

THE EFFECTS OF THE REVOLUTION

The vague slogan of "National Reform" under which the unlikely coalition of liberal intellectuals, frustrated professional politicians, and authoritarian military leaders had united for the seizure of governmental power in October, was not to be translated into political results in the months that followed. The *Ahali* land reform program alienated the tribal landlords and was never put into effect. Under pressure from both tribal leaders and the military, Hikmat Suleiman dissolved the *Ahali* and exiled some of its leaders. Gen. Bakr Sidqi became increasingly unpopular because of the ruthlessness with which he crushed tribal disorders and the arrogance with which he injected the army into governmental affairs. In August 1937 a band of dissident officers assassinated him and overthrew the government. As the historians Longrigg and Stoakes note: "Although the government had adduced the despotism of its predecessor as justification for its own coup d'etat, it had itself respected civil liberties no more, had intervened no less in elections and administrative appointments, and had quelled tribal unrest with equal severity."[19] Its only accomplishments were the launching of some irrigation and communications projects, and the conclusion of a treaty of alliance with Saudi Arabia, a mutual defense pact with Persia, Turkey, and Afghanistan, and an agreement with Persia settling an ancient waterways dispute.

In summary, beyond changes in governmental personnel, there were no significant changes either in foreign policy orientation—the negotiation leading up to the treaties mentioned above had been initiated under previous governments—or in domestic socio-economic policies. The only immediate effect of the October coup was in the area of political procedures: the army had discovered its power. From August 1937 to April 1941 there were six changes of government; in each case, the army was either directly or indirectly responsible for the changes. "The old pattern of personal rivalries and intrigue was restored, with the difference, that the army, having once tasted power, constantly interfered with political developments."[20]

In 1941 the army, under pro-German influences, became embroiled in military hostilities with British forces. The defeat of the Iraqi Army by the British dealt a serious blow to the army's morale, and since 1941, it has not been the prime mover in Iraqi politics that it was from 1936 to 1941.[21]

The positive effects of the October 1936 Coup in Iraqi politics may be summed up as follows: (1) The army entered into a new role in which its leaders regarded themselves as the supreme arbiters of the political fate of the nation; (2) Iraq concluded international treaties with its Arab and non-Arab neighbors for the first time since achieving full independence in 1932. Negative effects of the coup were: (1) the complete failure of the revolutionary government, either by dictatorial or parliamentary methods, to effect any noticeable economic and social reforms, and (2) the failure of the *Ahali* group to become a permanent organized political force in Iraqi politics.

NOTES

1. George L. Harris, *Iraq: Its People, Its Society, Its Culture* (New Haven: HRAF Press, 1958), pp. 28–30.
2. Stephen Hemsley Longrigg & Frank Stoakes, *Iraq* (London: Ernest Benn, Ltd., 1958), pp. 15–16.
3. Harris, *Iraq*, pp. 30–32.
4. Ibid., p. 32.
5. Walter Z. Laqueur, *Communism and Nationalism in the Middle East* (New York: Frederick A. Praeger, 1956), p. 173.
6. Longrigg and Stoakes, *Iraq*, p. 29.
7. Ibid., pp. 64, 110–111.

8. Laqueur, *Communism and Nationalism*, p. 178.

9. Ibid., p. 178.

10. Ibid., p. 173.

11. Harris, *Iraq*, p. 227.

12. Ibid., pp. 51–52.

13. Ibid., p. 121.

14. Majid Khadduri, *Independent Iraq: A Study in Iraqi Politics Since 1932*, (London: Oxford University Press, 1951), p. 29.

15. Ibid., pp. 71–73; see also *Mutal'at fi ash-Shabiyah* (Reflections of Populism), (Ahali Series No. 3, Baghdad: Ahali Press, 1935).

16. Ibid., p. 75.

17. Ibid., pp. 56–70, passim.

18. Interview with Abdul Aziz al-Said at American University, 14 April 1961.

19. Longrigg and Stoakes, *Iraq*, pp. 99–100.

20. George Lenczowski, *The Middle East in World Affairs* (Ithaca: Cornell University Press, 1956), p. 246.

21. Harris, *Iraq*, p. 95.

RECOMMENDED READING

BOOKS:

Harris, George L. *Iraq: Its People, Its Society, Its Culture*. New Haven: HRAF Press, 1958. A comprehensive survey of economic, social, cultural, and political factors in Iraq as of 1958.

Khadduri, Majid. *Independent Iraq: A Study in Iraqi Politics Since 1932*. London: Oxford University Press, 1951. A detailed study of Iraqi politics, with particular emphasis on events surrounding the Coup of 1936. Author is a native of Iraq and is Professor of Middle East Studies in the School of Advanced International Studies, Johns Hopkins University, Washington, D.C.

Laqueur, Walter Z. *Communism and Nationalism in the Middle East*. New York: Frederick A. Praeger, 1956. Chapters 13–15 discuss ideological developments in Iraq from 1914 to 1956. Author is an authority on the Middle East and ideological developments in the Arab world.

Lenczowski, George. *The Middle East in World Affairs*. Ithaca: Cornell University Press, 1956. A standard work on the history of the area. Contains a section on Iraq which surveys the events of the 1936 coup briefly.

Longrigg, Stephen Hemsley and Frank Stoakes. *Iraq*. London: Ernest Benn, Ltd., 1958. A comprehensive history of the country from Mesopotamian days through the most recent revolution of July 1958.

PERIODICALS:

Khadduri, Majid. "The Coup D'Etat of 1936: A Study in Iraqi Policies," *Middle East Journal*, II, 3 (July, 1948), 270–292. Best brief account of the Coup; later expanded in his book *Independent Iraq: A Study in Iraqi Politics Since 1932*.

THE EGYPTIAN COUP OF 1952

SYNOPSIS

The Egyptian Coup of 1952 was in the Middle Eastern tradition of the Turkish revolution of Kemal Ataturk. It represented a Socialist-military liaison and was characterized by a strong emphasis on nationalism, social and political reform, and economic development. Unlike similar military coups in the area following the defeat of the Arab states in the Palestine War of 1948–49, the Egyptian coup d'etat resulted in important changes in the economic and social structure. To a far greater extent than other military juntas, the Society of Free Officers, who engineered the coup d'etat, have been able to institutionalize their political control over Egypt.

BRIEF HISTORY OF EVENTS LEADING UP TO AND CULMINATING IN REVOLUTION

In July of 1952 a secretly organized group of junior officers in the Egyptian Army seized control of the army, overthrew the government, and ordered the abdication and exile of King Farouk. The conspiratorial group, called the Society of Free Officers, had been in existence since World War II, but had been reorganized after the Palestine War for the specific purpose of intervening in the political processes of the country. The group acted alone and without active collaboration with other political and subversive groups; however, much of the work of the opposition was done by political elements operating parallel to but not in liaison with the Free Officers' organization. The political and ideological orientation of the revolutionary group was ultranationalist and left-of-center. The revolutionists' avowed aims included the elimination of foreign influences from Egypt and a social and economic revolution directed primarily against the landowning class.

The military phase of the revolution was accomplished with speed and efficiency; the entire operation was carried out in a few hours during the night of July 22–23 with almost no bloodshed and in complete secrecy. A popular independent politician was elected on the 23d to replace the Prime Minister and the King was expelled on the 26th leaving the military junta, called the Revolutionary Command Council, in full power. General Naguib, not himself an active member of the conspiratorial group but sympathetic to its aims and popular in the country, was selected as a figurehead, but real power was in the hands of Col. Gamal Abdul Nasser, President of the Free Officers Society and one of its charter members.

THE ENVIRONMENT OF THE REVOLUTION

DESCRIPTION OF COUNTRY

Physical characteristics

Egypt is situated in the northeastern corner of Africa and in the Middle Eastern desert belt. It is about the size of Texas and New Mexico combined, and its desert areas are similar in climate and appearance to desert regions of those states. The country owes its very existence to the Nile River, which flows out of Central and East Africa and into the Mediterranean by way of Egypt. The 3 percent of Egypt's territory which is habitable today is located along the Nile in a narrow strip, in the Nile Delta (155 miles wide at its Mediterranean base and 100 miles long), and in isolated cases along the Mediterranean littoral. The Nile's seasonal flood occurs between July and December, when

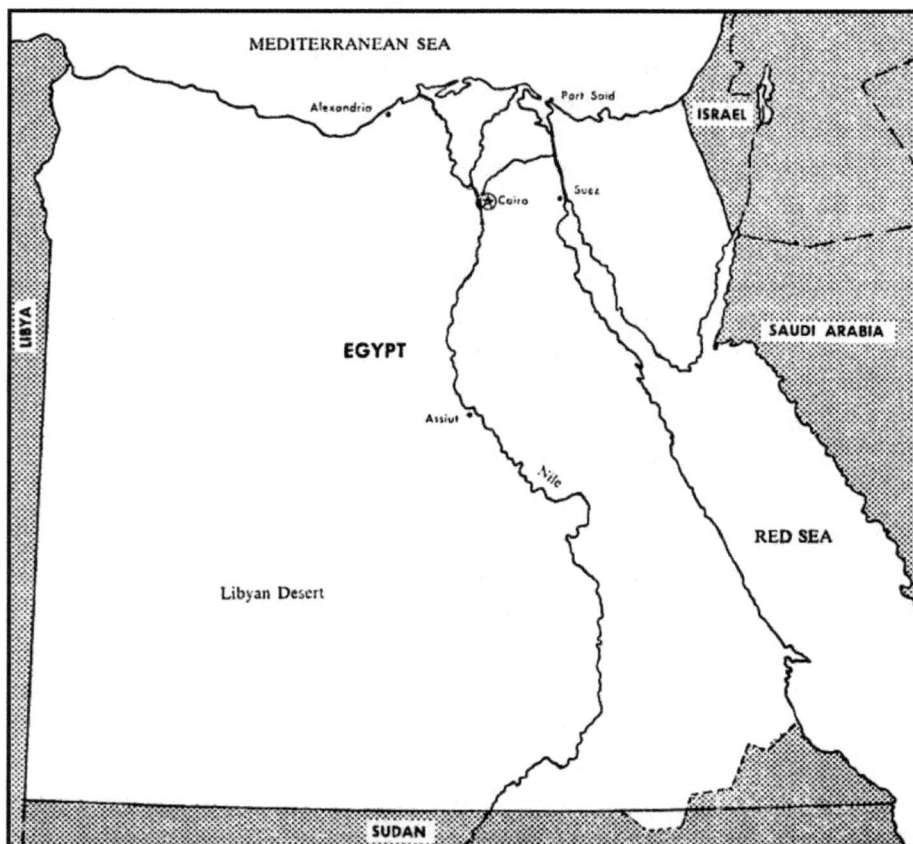

water for irrigation and fertile alluvial soil is brought down from the Nile's African headwaters.

The people

Modern Egyptians are not racially different from the population of ancient Pharaonic and early Christian Egypt. "The bulk of the population, with some Negroid admixture in the southern part of the Nile Valley are of the same brunet-white physical type which is found from northern India to Gibraltar and on both sides of the Mediterranean."[1] There has been no racial problem in Egypt and no language problem; the official language (spoken by 99 percent of the people) has been Arabic. Indigenous minorities include the Copts, an Arabic-speaking Christian community of two to three million, a cultural relic from Egypt's pre-Islamic days; several hundred thousand dark-skinned Nubians, and some nomadic Bedouins.

Egypt's population density is one of the highest in the world. A population of over 20 million is confined to less than 14,000 square miles of inhabitable territory in the Nile Valley and Delta and the Canal Zone. Egypt has one of the world's highest birth and death rates. Since 1939, the death rate has been declining, while the birth rate has remained high. Consequently, by 1947, the country's population had outgrown its income. In recent years, Egypt has experienced rapid urbanization; in 1947, 19 percent of the population were living in cities of more than 100,000. Major urban areas include: Cairo, the capital, with over two million; Alexandria, chief seaport, with one million; Port Said, Suez, Tanta, Mahalla al-Kubra, Al Mansura (located in the Delta and Canal Zone) and Assiut (200 miles south of Cairo), each with 100,000 or more inhabitants.

Communications

Through the Mediterranean seaport of Alexandria and the Suez Canal ports of Port Said on the Mediterranean side and Suez on the Red Sea side, Egypt has excellent facilities for communication with the outside world by water. By land the situation is less favorable: the extensive rail and highway connections built by the British to the east remain blocked by the political impasse of the Arab-Israeli conflict; the railroad south along the Nile stops at Aswan, and is linked through a desert road with the Egyptian and Sudanese rail systems; connections with Libya are via coastal highway. The best communication is in the Nile Valley, and the Delta, and the Canal Zone, where rail mileage approaches European standards. Since 1945, Cairo has become a major center of international air traffic.

Natural resources

The Nile is Egypt's most important natural resource. Besides fertilizing and irrigating the soil of the Valley and Delta, the river serves as an artery of internal communication. Little hydroelectric power is derived from the Nile, however. There is little mineral wealth in the country. Oil exists in limited quantities in the Sinai and Suez regions, but there is no coal in the country, a factor which has impeded the exploitation of the iron deposits known to exist around Aswan.

SOCIO-ECONOMIC STRUCTURE

Economic system

Egypt is predominantly an agricultural country; its industrial labor force is only about one million. The major cash crop is cotton which makes up 85–90 percent of exports. The country imports wheat and other foodstuffs. Egyptian industry is concentrated in Cairo and Alexandria, where it is centralized in a few large plants. These large enterprises, owned before the revolution by Egyptian and European capitalists, exist alongside of a plethora of small-scale, inefficient firms which stay in business as a result of a system of fixed prices and protective tariffs. Textile manufacturing is the leading industry, followed by food processing, cotton ginning, cement manufacturing, and petroleum refining. Egypt is more highly industrialized than the other Arab countries in the Middle East.

Class structure

The social elite of pre-1952 Egypt were the large landholders. Through a mutually reinforcing combination of economic affluence and political power, a group of about 10,000 landlords owned 37 percent of the arable land in Egypt. Social status traditionally depended less on occupation than on wealth in land. This group, composed largely of non-Egyptian (Turkish and Albanian) Pashas and Beys, scorned Egyptian and Arabic culture and affected a great interest in all things French or otherwise European; they were conspicuous for their lavish spending and display of wealth and were closely associated with King Farouk and the royal court.

The next most important social class in terms of the distribution of wealth, and, more importantly, political consciousness was the middle class. This group included professional people, journalists, businessmen, industrialists, intellectuals, and the clergy in its upper echelons, and in a lower echelon—the growing urban lower-middle-class element—minor civil servants, small businessmen, skilled workers, elementary schoolteachers, etc. That this group was not culturally and socially homogeneous is obvious, and the Egyptian middle class did not take an active part in the political life of the country before 1952. The lower-middle-class element was closest to the rural peasant in manners and attitudes; this element has also been the most nationalistic and anti-Western.

The overwhelming majority of Egyptians (about 70 percent) were and still are rural peasants, called *fellahin*. This group lived in rural isolation and extreme poverty. Rents and interest rates were exorbitant and wages low.[2] An urban proletariat, numerically small and composed of unskilled workers, domestic servants, and petty traders, stood between the *fellah* in the village and the lower-middle-class element of the city. Both the urban proletariat and, even to a larger extent, the rural proletariat were but vaguely conscious of the events carried out in their name during the revolution of July 23, 1952.

Social mobility in prerevolutionary Egypt was noticeable only in the upper echelons of the middle class. The army and the clergy were avenues of social advancement open to more humble individuals, but the landowners' dominance of the political scene precluded a high degree of social mobility in these directions. Nevertheless, a certain degree of mobility was inherent in the process of urbanization and industrialization which prerevolutionary Egypt was undergoing.

Literacy and education

Despite Egypt's great concern with the problem of illiteracy since 1925, as late as 1956 about 75 percent of the population was illiterate. This was largely due to the extreme poverty of the mass of the population and the difficulty of the Arabic language. The spoken or colloquial language was almost never written or used in mass communications media before the 1952 revolution. In addition to the government's secular public schools, the traditional religious schools—the village *kuttabs* and Al Azhar in Cairo—continued to exist. Al Azhar, the world center of Islamic learning, offered courses in the humanities and sciences as well as religious and Arabic studies and awarded degrees in law and theology.

Major religions and religious institutions

Over 90 percent of the Egyptians were adherents to Islam; the ancient Coptic Christian Church claimed most of the rest, but the Coptic Church had no political significance and seemed to be dying out. The Egyptian Government, both before the revolution and after, has maintained the mosques and appointed and paid the imams who preach in them, as well as other officials of the Islamic hierarchy. The imams and the ulema (teachers) and Sheikh (rector) of Al Azhar University had not been opponents of the government in power, although some tensions existed between the secularist politicians and the traditional religious leaders.

GOVERNMENT AND THE RULING ELITE

Description of form of government

General

The Constitution of 1923 under which the Egyptian Government operated before 1952 reflected in its contradictory provisions the irreconcilable conflict between Western democratic ideals and the authoritarian traditions of Egypt's governmental structure. Although the Constitution stated that all power resided in the people, the Crown was given, in addition to extensive executive authority, legislative power cojointly with the Senate and Chamber of Deputies. Moreover, the King had a strong suspensory veto. He could prorogue and adjourn Parliament for as long as a month, during which time he had powers of decree. Cabinet ministers in theory were collectively responsible to Parliament, but in fact they were only responsible to the King. The Senate was selected partially by royal appointment on the advice of the Prime Minister and partially by the body itself. The Chamber was elected through an indirect ballot on the basis of universal suffrage for a term of 5 years.

Responsibility of executive to legislature and judiciary

In theory, Cabinet ministers were responsible to Parliament for their actions and policies, but they were also appointed by and responsible to the King. In actual practice, the executive branch of the Egyptian Government was not responsible to the legislative; moreover, it intervened arbitrarily in the working of the judiciary and local administration. The Prime Minister was in fact the creature of the King. "Elections were usually controlled; and the reasons for the frequent rise and fall of cabinets had little connection with their relative strength in parliament."[3] King Farouk was influenced in his selection of Prime Ministers only by pressures from the landlords and the conservative religious leaders; periodically, as in 1942, British influence determined the choice.

Description of political process

Political parties and power groups supporting government

Prerevolutionary Egyptian political processes must be described in terms of an interplay of three forces: the British, the King, and the *Wafdist* politicians. In almost every election between 1923 and 1950 a *Wafdist* majority was returned to Parliament; more often than not, however, the King formed a government composed either of *non-Wafd* members or *Wafdists* amenable to royal influence. If a government became anti-British, the British intervened in behalf of its opponents in Parliament. Although elected by a wide majority in 1950, the *Wafd* Party was no longer representative of and responsive to the realities of the social situation. Its previous success was primarily due to general discontent with the *non-Wafdist* bloc previously in power which had been responsible for Egypt's part in the Palestine War (1948–49). Originally a broadly based, nationalist political movement, the *Wafd* Party became completely discredited in the postwar period. This was due to its history of accommodation, compromise, and finally *de facto* alliance with the British; moreover, in the postwar period new social classes had emerged with which the *Wafd*, as a bourgeoisie-oriented group, had little in common. Rustow observes that "the *Wafd* movement gradually was converted from a comprehensive-nationalist movement into a party machine for the pursuit of patronage, nepotism, and graft."[4]

Character of opposition to government and its policies

The only real competition to the *Wafd* came from the Muslim Brotherhood (al-Ikhwan al-Muslimin). Founded in 1928–29 by Hassan al-Banna, a schoolteacher in the Canal Zone city of Ismailia, the movement began as a religious lay organization and at first enjoyed the support of the ulema in Cairo. Its xenophobic, anti-Western, anti-modern, fundamentalist brand of religion soon alienated modern-thinking Muslims, but among the semiliterate and illiterate masses, recent recruits to the city from *fellahin* villages, the Brotherhood flourished. It served their social needs in a strange environment and provided an outlet for their economic and political frustrations. A fundamentalist sect, the Brotherhood was more negative than positive in its ideology; its positive content consisted only of the vague slogan: "Back to Islam as preached by Mohammed" and called for the establishment of a theocratic state and denounced Western influence. After 1945, the Brotherhood had attracted a number of educated but bitterly anti-Western Egyptians.[5] Hassan al-Banna then had around a half million followers and, "starting a prolonged campaign of terrorism against all those whom he accused of collaboration with the

447

British,"[6] he definitely aimed at achieving political power. In 1948–49 there occurred a series of terrorist acts in which the Prime Minister and other officials, as well as Hassan al-Banna, were assassinated and the Brotherhood was outlawed. It continued as a subversive organization with wide support.

The other leading subversive political group in prerevolutionary Egypt was the Communist movement. Egyptian communism—developing from Marxist study groups during World War II—has been particularly beset by factionalism. At the time of the revolution, the leading faction was called the Democratic Movement of National Liberation (MDLN in French and HADITU in Arabic). There were around 7,000 Communists in Egypt at this time, and they were better organized than either the *Wafd* Party or the Muslim Brotherhood, though by Communist standards party discipline was weak and factionalism a serious threat to the group's unity of action. Never a mass party, the Communists were subjected to frequent arrests and imprisonment by both *Wafdist* and non-*Wafdist* governments.[7]

The revolutionary organization which overthrew the government of Prime Minister al-Hilaly during the night of July 22–23, 1952, included none of the above but was a secret organization in the Egyptian Army, officially called the "Society of Free Officers" (Zubat al-Ahrar). Political opposition groups, such as the *Wafdists,* the Muslim Brotherhood, and the Egyptian Communist factions, took no active part in the events surrounding the seizure of power, but they did give their passive support to the revolutionaries. There was an especially close collaboration between individual members of the Free Officers' movement and the Muslim Brotherhood. Both movements were ultranationalistic and both favored some type of radical social reform; however, the fundamentalist ideology of the Brotherhood ruled out any real merger of the two movements. The various manifestations of overt political opposition to the government (which for the last 6 months meant *Wafdist* opposition) had no connection with the Free Officers' movement. In December 1951, the Free Officers, according to one of the revolutionaries, had made their last effort to collaborate with this political party and were rebuffed.

Legal procedure for amending constitution or changing government institutions

According to the 1923 Constitution, a two-thirds majority in each house of the legislature was necessary for a revision of the constitution, subject to approval by the Crown. As in most Middle Eastern political systems, constitutional forms did not significantly affect the actual functioning of the Egyptian political organism.

Relationship to foreign powers

Foreign relations have long been inextricably involved in Egyptian domestic affairs. As a strategic outpost of the British Empire, Egypt has been closely associated with Great Britain since 1882. After 1922, Egypt became nominally independent, but British economic interests and military forces in the country, confined only after 1947 to the Canal Zone, gave little substance to this fact. In the post-World War II period, relations with Britain worsened and in October 1951, the *Wafd* government abrogated the Anglo-Egyptian Treaty of 1936 which permitted British occupation of the Sudan and the Canal Zone. The nationalistic agitation accompanying this act and the British refusal to withdraw from the canal and the Sudan set off a series of terrorist and guerrilla activities in the Canal Zone and led to anti-Western demonstrations and riots in Cairo and Alexandria which directly affected the course of domestic politics.

The most significant factor in Egyptian foreign relations after troubles with Great Britain was the question of Pan-Arabism in general and relations with Israel in particular. Pan-Arabism came to the political scene in Egypt later than in the Fertile Crescent, but after 1945 both Parliament and Palace showed an interest in asserting Egyptian leadership of an Arab bloc. The menacing appearance of Israel, together with a sense of declining Western influence in the area and perhaps a desire to divert attention from pressing social problems within Egypt, led *Wafdist* politicians to adopt a vigorous Pan-Arabist foreign policy. The *Wafdists* bitterly attacked the government for its alleged mishandling of the Palestinian conflict in 1948–49. Intra-Arab rivalry, especially that between Jordan and Egypt, was a source of considerable embarrassment to proponents of Pan-Arabism. As a result of the deterioration of relations with Britain and being angered with the United States for its support of Israel and Britain, Egypt assumed a posture of "positive neutrality" during the Korean conflict. Egypt refused to join either NATO or the Baghdad Pact, which would have satisfied British demands for defense of the Suez Canal. Overtures to the Soviet bloc were also initiated during the prerevolutionary regime.

The role of military and police powers

After independence, Laqueur remarks, the police became a personal agency of the King and one of the most effective instruments of repression. It was strengthened, the administration was centralized, and the force took on a paramilitary character. This made the transfer of authority to the Revolutionary Command Council relatively easy.[8]

WEAKNESSES OF THE SOCIO-ECONOMIC-POLITICAL STRUCTURE OF THE PREREVOLUTIONARY REGIME

History of revolutions or governmental instabilities

On January 26, 1952, a major riot occurred in Cairo during which mobs burned approximately 700 commercial, social, and cultural establishments, mostly foreign-owned or associated with foreigners. Twenty-six persons were killed and many wounded. The *Wafdist* government failed to use the police to check the disturbances, and King Farouk seized this opportunity to dismiss the *Wafdist* government. During the next 6 months there was continual mob violence and a procession of Palace-appointed anti-*Wafdist* Prime Ministers who had to face either a hostile *Wafdist* Parliament or, after its dissolution in March, a dangerous political vacuum. King Farouk's open intervention in the selection of the government was especially obnoxious to enlightened Egyptians. A descendant of an alien and unpopular dynasty, Farouk I became unpopular in his own right on account of his personal excesses and his arbitrary political activities. By 1952 the prestige of the dynasty of Mohammed Ali had reached its nadir, so that the continued existence of the monarchy constituted a major weakness in the governmental structure.

Economic weaknesses and social tensions

The Census of 1947 revealed for the first time that Egypt's population had outrun its income—and this income was dependent on the vulnerable cotton economy. The *fellahin* masses and the urban proletariat were subsisting at a standard of living just above famine conditions but not high enough to protect the mass of the population from the effects of undernourishment and poor health.[9] Inflation was a major problem in postwar Egypt; while the cost of living increased 400 percent, wages for urban workers rose only about 100 percent in the 1945–50 period.[10]

Government recognition of and reaction to weaknesses

In addition to the usual suppression of subversive political activities by its police and military power, the *Wafdist* government, in power from 1950 to January 1952, instituted certain positive reforms. Outstanding Egyptian intellectuals were appointed to head the newly created Ministries of Social Affairs and Education. The establishment of Rural Social Centers in the villages was the most significant result of these reforms; however, far from solving the agrarian problem, these Centers probably stimulated in the young Egyptian social worker cadres (who were specifically selected for their *fellahin* social origins) a strong desire for a more revolutionary approach to the problems of the *fellahin*.[11] There were also certain reforms in labor legislation; labor unions had been allowed to form as early as 1942 by a previous *Wafd* regime, apparently in an attempt to control the industrial workers through a government-supervised union structure. Aware of growing discontent among the junior army officers, the government attempted to stem this dissatisfaction by a wave of promotions, according to one of the leading revolutionaries.[12] Finally, there was a halfhearted attempt to ally middle-class criticism of corruption in government by the appointment as Prime Minister in March 1952 of Ahmed Naguib al-Hilaly, popular former Minister of Education, who claimed "he would carry on a comprehensive 'purification' campaign."[13]

FORM AND CHARACTERISTICS OF REVOLUTION

ACTORS IN THE REVOLUTION

The revolutionary leadership

The instigators, organizers, and leaders of the Free Officers' movement came from a group of junior officers in the Egyptian Army, most of whom were below the rank of colonel. In social origins, most of them were of middle or low-middle-class backgrounds. Their generation had come to maturity in the anti-British atmosphere of the late 1930's and had seen the Allies fight for their lives in World War II. British interferences in Egyptian politics during the war and the continued presence of British troops and civilians in large numbers in Egypt after the war had wounded their national pride, and the frustration caused by Arab defeats in Palestine in 1949 had left its mark on their personalities. These young officers, most of them around 30 years of age, were Egyptian nationalists in ideological orientation; more socialistic than were the Kemal-type nationalists of the interwar period, they were nevertheless following in the Middle Eastern pattern of "military socialism" first established by Kemal Ataturk in Turkey in the 1920's. Their complex of goals and aspirations was based on a desire for what Nasser called "an Egypt free and strong."[14] To implement this simple desire they looked to the elimination of remaining British and Western influences in Egypt and, at the same time, to the elimination of "feudal" landlord economic interests.

The leading figure in the Egyptian revolution and *de facto* head of the Free Officers' group was Gamal Abdul Nasser. He was born in upper Egypt in 1918, "the son of a good middle-class family."[15] He studied in Cairo from the age of 8, and in 1937 he entered the Military College. In 1939, as a young graduate, he served in the Egyptian Army with Hakim Amer, Anwar al-Sadat, and others of the conspiratorial group, with whom he shared a similar social background and similar views on politics. Nasser has been described as a passionate nationalist, but one who was, at the same time, methodical and intelligent and possessed of an attractive personality. He was later an instructor at the Military College in Cairo, along with most of the other Free Officers, and he served in the Palestinian Campaign of 1948–49.

The president of the military junta which ruled Egypt, after the revolution, was Mohammed Naguib. His participation in the prerevolutionary developments was limited, but he played an important role immediately after the coup. Naguib was born in Khartoum in 1901, the son of a graduate of the Egyptian Military Academy and an officer in the Anglo-Egyptian military forces in the Sudan. Naguib's large family of nine brothers and sisters all became professional or middle-class people, and he rose to the rank of major general in the Egyptian Army. Naguib attended Gordon College in the Sudan and later the military academy there; in his relations with the Egyptian military hierarchy, Naguib was often frustrated, and he frankly admits that this bitterness contributed to his participation in the Free Officers' movement.[16]

The revolutionary following

As a secret conspiratorial group within a specific element of society, the following of the Free Officers' movement was necessarily limited to those officers whom the leadership of the movement came in contact with and felt they could take into their confidence. Not essentially different in social origins and political orientation from the leaders of the movement, the following numbered more than a thousand in 1947, before the reorganization that followed the Palestine War. Membership in the organization was limited to military personnel; however, the movement enjoyed considerable sympathetic support in civilian quarters.

ORGANIZATION OF REVOLUTIONARY EFFORT

Internal organization

The revolutionary organization, later known as the "Free Officers' Movement," had its origin in a group of officers who began to discuss politics together in 1938. One of these officers was Gamal Abdul Nasser. By early 1939 a secret revolutionary society was formed within the army dedicated to the task of freeing Egypt from British influence. From this small nucleus the organization expanded. Each member formed a new cell around him and the society gained adherents, primarily within the army, but also among civilians. "Soon there was a militant organization covering the entire country. Members used secret signs and passwords. Military rank was observed inside the society as it was in the Army."[17]

By 1942 the society was organized into sections, each of which was composed of several cells. Control over the organization was exercised by a Central Committee. During the war the society established informal liaison with the Muslim Brotherhood, also a secret revolutionary organization that operated with a much wider popular base. Liaison was also established with the German Army in Egypt and plans were made to utilize the opportunity of British weakness in 1942 to stage an uprising within the army in collaboration with the German forces. However, the events of World War II—the German threat to Egypt and the Suez Canal—increased the influence of the British Army and the vigilance of its intelligence system in Egypt. The attempts at collaboration with the Germans came to naught and resulted in the arrest of many members of the society who were also known to be favoring a German victory over British forces.

Toward the end of 1944 and the beginning of 1945, a reorganization of the society occurred. A military group under the leadership of Nasser was established along with a parallel civilian group under the leadership of Anwar al-Sadat. A Supreme Committee coordinated the activity of the two groups, which, otherwise, were kept completely apart. It was, however, the military which was the decisive factor in the revolutionary organization and little is known about the function or organization of the civilian or "popular" group, as it was called.[18]

By 1945, a definite organization had been established, which issued clandestine circulars and had an inchoate revolutionary administrative system. Anwar al-Sadat, one of the leaders, attributes to Gamal Abdul Nasser the institution of this system, which consisted of five subcommittees: (1) Economic Affairs, which collected and administered revenues of the group (derived in part from contributions from the members amounting to up to 2 months' salary and spent for arms and dependents' allowances); (2) Combat Personnel, which recruited, indoctrinated, and assigned new members to the appropriate cell (only Nasser and Abdul Hakim Amer knew the exact number and names of all the members); (3) Security, the elite committee which "supervised recruits, making sure of the orthodoxy of their revolutionary belief"[19] and dealt with all matters relating to secrecy; (4) Terrorist, described by al-Sadat as "largely theoretical," since the group did not generally approve of terror as a tactic; and finally (5) the Committee of Propaganda, which issued circulars and operated through word-of-mouth communication. The leaders constituted the Supreme Committee and also served on the five subcommittees.

During the Palestine War, many of the officers in the secret society were killed and activities had to be curtailed to fight the war. However, also as a result of the war, the political climate had become such that army officers were much more ready than in the past to support a revolutionary organization. In 1950, another reorganization took place "to prepare for direct action."[20] The organization now came to be called the "Society of Free Officers" and the vertical structure was tightened. The Free Officers' group control was delegated to an Executive Committee composed of 9 or 10 members; most of these had been members of the old Superior Committee, and later they were to constitute the Revolutionary Command Council. Nasser, at the age of 29 and with the rank of major, was elected president of the Executive Committee in 1950; reelected in 1951, he was president at the time of the coup, according to al-Sadat. Naguib claims in his autobiography, however, that Nasser was secretary general and that he was president. According to his account Naguib was unable to play as active a part in the planning of the coup on account of government surveillance of his activities.[21]

After 1949, the propaganda circulars issued by the group became a real force in Egyptian politics and the name of the group became widely known. The cell-type of organization was used. A cell was composed of five members, each of whom formed a fresh cell whose members remained unknown to the other members of the parent cells, thereby limiting for security purposes, the individual's knowledge of the whole organization.[22] Members were required to make monthly payments into a savings account, to build up an emergency fund. The officer in charge of the section (which before the reorganization had consisted of 20 cells) had power to expel or admit new members, though subject to review from the Executive Committee. As a clandestine conspiratorial group, the membership of which was known only to the leadership and not to the rank and file members, the Free Officers' organization was not democratically controlled. Real power resided in the Executive Committee, and in this group Nasser played the dominant role.

External organization

Although much of its organization took place in Palestine, the movement had no foreign connections except briefly during World War II.

GOALS OF THE REVOLUTION

Concrete political aims of revolutionary leaders

The immediate aim of the revolutionists was the overthrow of the monarchy and the establishment of a democratic parliamentary republican system of government. The Free Officers confidently—and, as it turned out, naively—believed that with Farouk out of the picture there would come forth on the Egyptian political scene a new type of political leadership, one revitalized and "purified" (a recurrent word in the language of the times), which would carry out the Free Officers' goals and aspirations for an Egypt free from the twin evils of foreign domination and feudal-capitalist oligarchy. Nasser himself wrote later[23] that the officers had supposed that their role would be no more than that of a "commando vanguard," and that as soon as Farouk was overthrown, a united nation would be ready to take up "the sacred advance towards the great objective." The reality, he found, was quite different.

The military junta, called the Revolutionary Command Council, apparently intended to relinquish their power to civil authorities as soon as they could find a body of political leaders whom they could trust to administer and execute the social and economic reforms the Free Officers deemed necessary to Egypt's regeneration. Their initial choice was Ali Maher, an independent politician popular with the masses for his recent price reductions (he was Prime Minister in February 1952) and known to the Free Officers both for his liberal inclinations and for his having been deposed by the British in 1942.

Social and economic goals of leadership and following

The social content of the revolutionary goals of the military junta reflected the middle-class origins of the leading participants in the movement. Lenczowski notes that the revolutionists claimed to be "animated by the unselfish desire to see Egypt emancipated from imperialism and feudalism and served by an honest government that would ensure social justice, economic progress, and dignity to all citizens."[24] The feudalistic and capitalistic oligarchy associated with King Farouk bore the brunt of the social criticism. Tangible social and economic goals of the Free Officers' movement included, first of all, agrarian reform, and then industrialization, tax reform, and wage-price adjustments.

REVOLUTIONARY TECHNIQUES AND GOVERNMENT COUNTERMEASURES

Methods for weakening existing authority and countermeasures by government

The Free Officers group carried on a vigorous anti-Government propaganda campaign through the mails "not only to military men but also to a large number of civilians known for their free and progressive tendencies and their opposition to corruption."[25] These pamphlets served to publicize the group and to galvanize public attention around the specific goals and aspirations of the opposition group within the army. During the war at least some of the members conspired with German forces in North Africa, but these attempts proved fruitless, as they were uncovered by British intelligence agents. After the war, the group became well known as a secret organization in opposition to the regime, but the individuals who belonged to the movement were almost never found out. Other than this, the Free Officers' organization carried on no overt activity against the government. It engaged in no terrorist or guerrilla activity, although it did apparently encourage the attacks on the British in the 1951–52 period, especially in the Canal Zone. Operating parallel to the Free Officers' movement in opposition to the government were such major terrorist groups as Hussein's "Socialists" and the Muslim Brotherhood; also, the Communist organization and the *Wafd* (in opposition to the government after January 1952) carried on an active campaign, both legally and illegally, to discredit the government. In such an atmosphere of terror and intrigue against an already crumbling governmental system, the Free Officers had only to conspire among themselves and decide on the timing of the revolutionary act. They exercised effective control over the army, although the High Command and Chief of Staff were loyal to the government. Their greatest asset lay in the secrecy which surrounded their organization. Many of the leaders were trusted by the regime and enjoyed the confidence of high government officials.

Methods for gaining support and countermeasures taken by government

During the earlier phases of the organization, the members relied almost completely on word-of-mouth means of influencing fellow officers or civilians. It was decided that each member should go into the streets, cafes, mosques, and other public places to meet people and talk politics. But there was never an attempt to rival the more popular opposition movements such as the terroristic Muslim Brotherhood, with which some liaison was established. The tactic adopted seems to have been to let these organizations spread discontent with the *status quo* among the masses and thereby prepare the people for revolutionary developments. The real power lay in the army, and the Free Officers group did not contemplate a violent revolution with mass support. If they could gain control over the army they could successfully dispose of the government. The propaganda circulars and manifestos that were issued were designed to gain support for the ideas of the Free Officers. They were not meant as appeals for action.

The only attempt to collaborate with an existing political group occurred in December 1951, when the leftwing of the *Wafd* was contacted and informed that the Free Officers would support the *Wafdist* government then in power in its dispute with Farouk. Had the *Wafdist* politicians agreed to this, the Free Officers had planned to act in January 1952, when Farouk dismissed the *Wafdist* government. The King would have been expelled by the military junta and the *Wafdist* government might have remained under a republican system of government.

MANNER IN WHICH CONTROL OF GOVERNMENT WAS TRANSFERRED TO REVOLUTIONARIES

The original plan was to stage a coup in March 1952; this was changed when a key figure hesitated to go through with his part in the move. Action was postponed until the night of July 22–23. The immediate cause of the revolution was the King's decision, on the 15th of July, to dissolve the Military Club's executive committee on which the Free Officers held a majority. In the Club's annual election, General Naguib, openly supported by the secret organization of Free Officers against the King's choice, had been elected President. This allied Naguib with the movement, although he was not an active conspirator and did not know more than five members of the group before the coup, and it served as a trial of strength which enabled the Free Officers to demonstrate publicly their opposition to Farouk. After July 15, the Free Officers believed that the government was about to arrest Naguib and that their own safety was in jeopardy. A Cabinet crisis on the day before the coup had no influence on the course of events. The die had been cast between the 16th and 20th of July.

The chief conspirators were all in Cairo and they had the support of key figures in the provinces. With the troops under their command, the Free Officers marched on the General Headquarters, seized the High Command meeting there in emergency session, and then took possession of government buildings, telephone exchanges, radio stations, and other strategic points throughout the country. Notes were sent to foreign envoys, especially the British, to warn them against interfering in what was "a purely internal affair"; armored and infantry units were stationed along the road to Suez as an additional precaution. The coup was accomplished swiftly and almost bloodlessly, while most of Cairo slept and the streets were silent, as one of the participants recalls.[26] In the morning, General Naguib, who had remained at home during the coup, was informed that he was to be Commander in Chief of the Armed Forces; Nasser, in control of the Revolutionary Command Council, was the real power in the revolutionary regime about to be set up. The leaders of the coup described three distinct stages in their revolutionary process: seizure of the army, which they achieved around midnight; overthrow of the government, which they achieved during the morning of the 23d; and expulsion of the King, which was not accomplished until the 26th. Before his abdication, Farouk served to provide a tinge of legality to the revolutionary regime, in that Ali Maher, the choice of the military junta for Prime Minister, had been duly invested by the King as Prime Minister.

THE EFFECTS OF THE REVOLUTION

CHANGES IN THE PERSONNEL AND INSTITUTIONS OF GOVERNMENT

A series of drastic changes in both the personalities and the institutions of Egyptian Government resulted from the 1952 coup. Most significant among these were: the removal of King Farouk and members of al-Hilaly's government, along with numerous palace and government appointees; the abolition of the 1923 Constitution in December 1952; the abolition of political parties in January 1953; and in 1955–56, the abolition of the Sharia and non-Muslim religious courts. Purges took place in the armed forces and in government-controlled economic enterprises, as well as in local administration of the government. The dominant power group in Egypt since the July coup has been the Revolutionary Command Council, composed of around 10 former members of the Free Officers' movement, who at the outset administered the revolutionary regime's program collectively. In 1954–56 an intragroup power struggle took place in which Colonel Nasser, the younger and more radical member, supplanted the elder and more conservative figurehead, General Naguib. The Egyptian Constitution of 1956, and that of the Egyptian-Syrian union in 1958, reflected the dominant political role of the military junta and the mass organization that they brought into being. Nasser was elected President, first of Egypt and later of the larger political union, the United Arab Republic.[27]

MAJOR POLICY CHANGES

Revolutionary Egypt has undertaken a vigorous foreign policy, which is not so different in general orientation from that of pre-1952 Egypt as it is different with respect to the intensity and degree of its application. The revolutionary regime early distinguished itself for its ultranationalist and anti-Western stands; it aspired to—and to a large extent achieved—the goal of leadership of the Arab world; beyond this, Nasser looked to the Islamic community throughout the world and to sub-Sahara Africa as an outlet for Egyptian energies.[28] The regime became closely aligned economically with the Soviet bloc in its first few years, and there has been a large degree of foreign policy alignment with the bloc, especially on matters involving the colonial powers vis-a-vis Afro-Asian and Latin American states. The regime emerged successfully from its nationalization of the Suez Canal and the ensuing military encounter with the British, French, and Israelis.

LONG RANGE SOCIAL AND ECONOMIC EFFECTS

The most significant impact the coup of 1952 had on Egyptian society and economic relationships was its attack on the landlord-Pasha class. This group was effectively removed from power, and the social elements immediately below came to the fore. The poorest of the peasants have not benefited from the much-publicized agrarian reform of September 8, 1952, as have the "middle" and "upper" peasantry. Less publicized but of more immediate benefit to more people have been the improvements in social services which the regime has been able to achieve.[29] In general, the lower-middle and middle-class elements have benefited from the policies of the revolutionary government; there is today considerably more social mobility than was the case before 1952 in Egypt.

The effect of the revolution on religious institutions is also interesting to note. One authority notes that Nasser maintains the tradition of Friday worship but that this appears to be political expedience rather than an expression of genuine piety. "Realizing that the mass of the population has little understanding of . . . politics, Egypt's leaders are seeking support for government policy from the pulpit by appeals couched in the language of the Islamic faith."[30] In June 1961, the revolutionary government nationalized Al Azhar University, the leading religious academic institution in the Islamic world.

On the ninth anniversary of the revolution (July 1961) sweeping economic and political reforms were introduced which transformed Egypt into a Socialist state and served to concentrate all effective political power within the United Arab Republic in the hands of a centralized government. All banks and insurance companies were nationalized; all major companies and business firms were taken over by the state; maximum landholding was limited to 100 instead of 200 acres; and maximum individual income was limited to 15,000 Egyptian pounds (about $40,000) a year. On the political plane, separation between federal and regional governments in the United Arab Republic was eliminated and a new central government with 37 executive ministers (of whom 16 were Syrian) was established, and Cairo was made the official seat of government, though a provision was made to have it convene for 4 months every year in Damascus, Syria.

OTHER EFFECTS

In addition to the drastic changes in governmental personnel and institutional structure, the vigorous participation in foreign affairs, and the effective beginnings of a social revolution which the coup of 1952 brought short, the Egyptian revolution contained a charismatic quality that distinguishes it from the ephemeral military coups and ruling juntas that have long been associated with the Middle East. The flavor of this revolutionary charisma or mystique is readily discernible in the writing and speeches of Nasser, al-Sadat, and other leaders of the movement. Its appeal is strong throughout the Middle East and Islamic community, so that the Egyptian revolution is not solely a domestic development. John Badeau, a Middle East specialist and the current U.S. Ambassador to the United Arab Republic, has likened it to the French Revolution, which also "had its self-seeking leaders, its power cliques, its political nationalism, but [which] let loose forces that finally changed the pattern of social life in most of Europe."[31] He believes the Egyptian revolution is doing the same in the Middle East.

In view of the recent events during which Syria severed her connection with Nasser and the U.A.R., this statement seems to attribute more force to the Egyptian revolution than is warranted. The Syrian rebellion was basically a conservative move to save that country from the radical reforms which Nasser is attempting to introduce.

NOTES

1. George L. Harris, *Egypt* (New Haven: HRAF Press, 1957), p. 44.

2. Beatrice McCown Mattison, "Rural Social Centers in Egypt," *Middle East Journal*, V, 4 (Autumn 1951), 461–480; see also Afif I. Tannous, "Land Reform: Key to the Development and Stability of the Arab World," *Middle East Journal*, V, 1 (Winter 1951), 9.

3. Harris, *Egypt*, p. 74.

4. Dankwart A. Rustow, "The Politics of the Near East, S.W. Asia and Northern Africa," *The Politics of the Developing Areas*, eds. Gabriel A. Almond and James S. Coleman (Princeton: Princeton University Press, 1960), p. 410.

5. Harris, *Egypt*, p. 334.

6. George Lenczowski, *The Middle East in World Affairs* (Ithaca: Cornell University Press, 1956), p. 408.

7. Walter Z. Laqueur, *Communism and Nationalism in the Middle East* (New York: Frederick A. Praeger, 1956), pp. 42–62.

8. Ibid., p. 97.

9. Ibid., p. 275.

10. Anwar al-Sadat, *Revolt on the Nile* (New York: John Day, 1957), p. 117.

11. Mattison, "Social Centers," passim.

12. al-Sadat, *Revolt*, p. 101.

13. Dr. Rashed el-Barawy, *The Military Coup in Egypt: An Analytic Study* (Cairo: Renaissance, 1952), pp. 180–181.

14. Gamal Abdul Nasser, *Egypt's Liberation: The Philosophy of the Revolution* (Washington: Public Affairs Press, 1955), p. 49.

15. Ibid., 115.

16. Mohamed Naguib, *Egypt's Destiny* (Garden City, New York: Doubleday, 1955), pp. 13–128.

17. al-Sadat, *Revolt*, p. 13.

18. Ibid., p. 64.

19. Ibid., p. 84.

20. Ibid., p. 113.

21. Naguib, *Egypt's Destiny*, pp. 32–33.

22. al-Sadat, *Revolt*, p. 113.

23. Nasser, *Egypt's Liberation*, pp. 32–33.

24. Lenczowski, *The Middle East*, pp. 419–420.

25. el-Barawy, *Military Coup*, p. 195.

26. al-Sadat, *Revolt*, p. 141.

27. Don Peretz, "Democracy and the Revolution in Egypt," *Middle East Journal*, XIII, 1 (Winter 1959), 26–40; see also Curtis Jones' study, "The New Egyptian Constitution," *Middle East Journal*, X, 3 (Summer 1956), 300–306.

28. Nasser, *Egypt's Liberation*, pp. 79–114.

29. Charles F. Gallagher, "The United Arab Republic Today; Part I: The Liberation of Egypt, *American Universities Field Staff Reports, Northeast Africa Series* , VII, 1 (January 8, 1960).

30. Harris, *Egypt*, pp. 332–335.

31. John S. Badeau, "The Middle East: Conflict in Priorities," *Foreign Affairs*, XXXVI, 2 (January 1958), 240.

RECOMMENDED READING

BOOKS:

Harris, George L. *Egypt*. New Haven: HRAF Press, 1957. 370 pp. A comprehensive study of contemporary Egypt.

Lenczowski, George. *The Middle East in World Affairs*. Ithaca: Cornell University Press, 1956. 576 pp. A standard text in the history of the Middle East since 1900. Contains a survey of recent Egyptian history.

Naguib, Mohammed. *Egypt's Destiny* . Garden City, New York: Doubleday, 1955. 256 pp. Autobiography of General Naguib. Contains detailed account of the revolution, more factual and objective than Nasser's book.

Nassar, Gamal Abdul. *Egypt's Liberation: The Philosophy of the Revolution*. Washington: Public Affairs Press, 1955. 119 pp. Subjective and highly personal, impressionistic account of the events of the revolution.

al-Sadat, Anwar. *Revolt on the Nile*. New York: John Day Co., 1957. 159 pp. Autobiographical account of Free Officers' movement and the revolution by one of the leaders of the movement. Detailed and factual as well as impressionistic account.

PERIODICALS:

Gallagher, Charles F. "The United Arab Republic Today; Part I: The Liberation of Egypt," *American Universities Field Staff Reports, Northeast Africa Series*, VII, 1 (January 8, 1960). An objective study of the social, economic, and political effects of the Egyptian revolution.

Hourani, Albert. "The Anglo-Egyptian Agreement: Some Causes and Implications," *Middle East Journal*, IX, 3 (Summer 1955), 246–247. A study of Anglo-Egyptian relations, with particular emphasis on the Egyptian side of the question.

Mattison, Beatrice McCown. "Rural Social Centers in Egypt," *Middle East Journal*, V, 4 (Autumn 1951), 461–480. A detailed study of the prerevolutionary Egyptian Government's most important reformist undertaking.

Peretz, Don. "Democracy and the Revolution in Egypt," *Middle East Journal*, XIII, 1 (Winter 1959), 26–40. Surveys political results of the 1952 coup.

THE IRANIAN COUP OF 1953

SYNOPSIS

The Iranian coup of August 1953 was counterrevolutionary and conservative in political orientation. The Shah had dismissed Prime Minister Mosaddegh and appointed General Zahedi in his place several days before the coup d'etat took place. Mosaddegh had refused to give up the reins of government and turned against the Shah who fled the country. General Zahedi, the legal head of the government, went into hiding, and Mosaddegh attempted to stay in power with the support of the Communist-front *Tudeh* Party, ultranationalist groups, and the police and army. The Iranian Army, however, remained loyal to the monarchy, turned against Mosaddegh and staged a coup to return political conditions to pre-Mosaddegh normalcy. Although left-of-center political support was also present in the anti-Mosaddegh coalition, it was essentially a conservative and pro-Western counterrevolution.

BRIEF HISTORY OF EVENTS LEADING UP TO AND CULMINATING IN REVOLUTION

In August of 1953 the government of Prime Minister Mohammed Mosaddegh was overthrown by units of the Iranian Army loyal to the Shah and Gen. Fazlullah Zahedi. Mosaddegh had come to power legally in 1951, but during the summer of 1953 he set up a personal dictatorship based primarily on the support of the lower-middle class and unemployed elements of Teheran, organized and directed by members of the Communist-front *Tudeh* Party and certain ultranationalist groups. The Shah, Iran's constitutional monarch, dismissed Mosaddegh and appointed General Zahedi Prime Minister in his place on the 13th of August. Mosaddegh defied the Shah, and the monarch fled the country. General Zahedi went into hiding, and the *Tudeh* organized a series of riots and street demonstrations in support of Mosaddegh and in denunciation of the monarchy; these lasted until the 18th, when the regime became alarmed over their excesses and called out the army. Elements of the army and the civilian population speedily quelled the anti-royalist demonstrations and then turned on the Mosaddegh regime. Little is known about the prior planning and organization of the coup, and the element of apparent spontaneity was a significant characteristic of this coup d'etat.

The arrest of Mosaddegh on the 20th of August marked the end of the most nationalistic and anti-Western government Iran had ever experienced. The two-and-a-half-year period of Mosaddegh's rule has often been described as one of internal disorder and political demagoguery, and, in foreign relations, a period of isolation and hostility growing out of Iran's sudden nationalization of foreign oil company holdings in 1951 and the ensuing Anglo-Iranian oil dispute.

THE ENVIRONMENT OF THE REVOLUTION

DESCRIPTION OF COUNTRY

Physical characteristics

A strategic land bridge between India, Central Asia, and the Mediterranean world, Iran borders the U.S.S.R. on the north, Afghanistan and Pakistan on the east, and Iraq and Turkey on the west. In area, the country is slightly smaller than Mexico and about one-fifth the size of the United States. The terrain is mostly that of a mountainous plateau, from 3,000 to 5,000 feet above sea level. Only about one-tenth

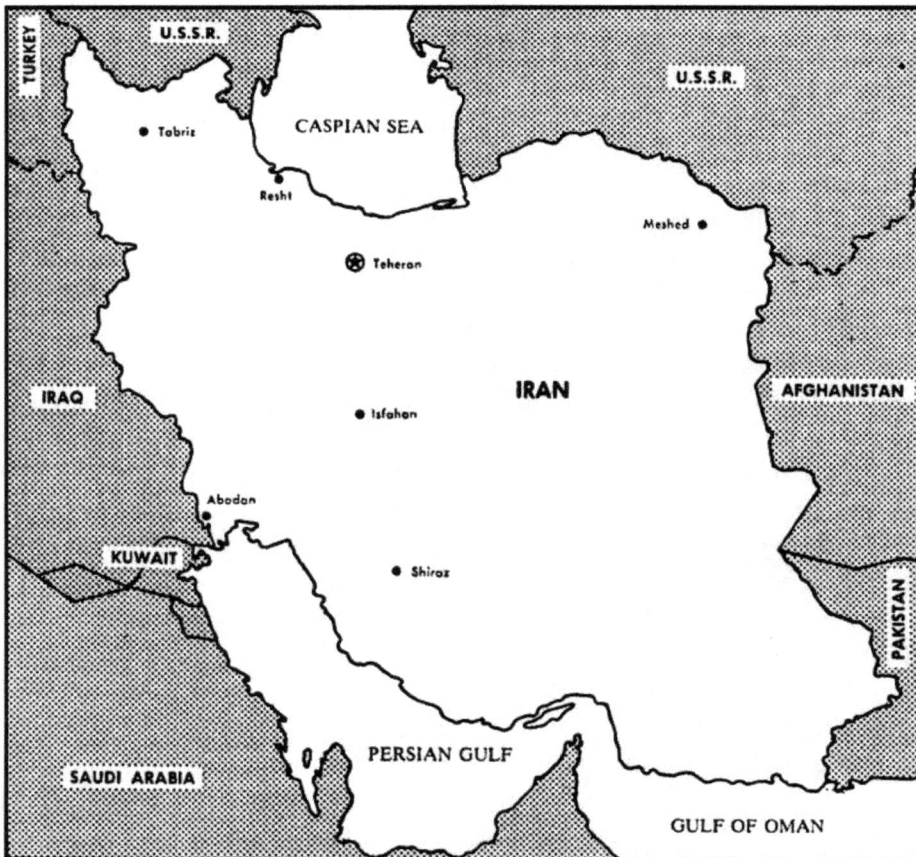

of the land is arable; two-tenths is in forest and grazing lands, and the remaining seven-tenths is in mountains and deserts.[1]

The people

The population of Iran, according to U.N. estimates in 1954, was slightly above 20 million. Most of these people live in only 30 percent of the country, population density being highest in the valleys of the northwestern Elburz Mountains and in the Caspian lowlands. The average density of population in inhabited areas was approximately 106 persons per square mile. Teheran, the capital of Iran, had a population of over 600,000 in 1950 and was a rapidly growing urban center; there were seven smaller cities of 100,000 to 200,000 inhabitants. About two million Iranians were nomadic or semisettled; around three million lived in cities of more than 20,000 and could be considered urban dwellers. Iran's birth rate is high (about 45 per thousand); however, there has been no official concern over population increases.[2]

The dominant ethnic group in Iran is the Persian, who together with closely related subgroups make up more than two-thirds of the population. Persian language, culture, and historical traditions give the Iranian national image its characteristic stamp.

Around the rim of the central plateau, which is the historical homeland of the Persians, live numerous tribal and nomadic peoples. In the western Zagros Mountains live the Kurds, Bakhtiari, and Lurs, semisettled, seminomadic pastoral tribes, who are closely related to the Persians and tend to identify with them. Iranian Kurds, numbering around one million, have assimilated better than the Kurds in Iraq and Turkey. Turkic and Dravidian peoples live along the northern frontier, a most important group being the 2 million Azerbaijani on the border of Soviet Azerbaijan, who have a culture of their own despite strong historical ties with the Persians. Finally, there are several hundred thousand Arabs living along the Persian Gulf, who neither identify with the Persian national image nor consider themselves an oppressed minority. Despite its heterogeneous population, Iran has remained relatively stable through the prestige and energy of the Persian core of Iranian society. Due to the interplay of power politics it has preserved its territorial integrity despite Soviet-inspired separatist movements.[3]

Communications

Persian caravan routes fell into disuse and ill-repair after the 16th century when navigation supplanted overland routes. Reza Shah, founder of the present dynasty, initiated an ambitious railroad plan

467

to link Teheran with all parts of the country. By the 1950's rail connections had been established with the Caspian Seaport of Bandar Shar in the north and the Persian Gulf ports of Bandar Shahpur and Khorramshahr in the south. Reza Shah's efforts and those of the British and Russian Armies during the Second World War produced an extensive system of highways connecting all major points in the country. There were a number of small ports on the Caspian, but for political reasons these ports were not so important in external communication as the ports on the Persian Gulf which also served the petroleum industry. Large international air terminals were located at Teheran and Abadan on the Persian Gulf. Smaller airdromes, suitable for internal air traffic, were located in all larger cities of the country. In 1953 Iran's communication with the outside world was better developed than its internal communication—a situation not at all uncommon in underdeveloped areas.[4]

Natural resources

Except for oil and natural gas deposits, for which it ranked as the world's fourth largest oil producing country, Iran had little mineral or agricultural wealth. Recent geological surveys have reported a wide range of mineral wealth in the northern mountains, but these resources had not been exploited to any significant extent. Mountainous terrain and scanty rainfall throughout most of the country precluded the exploitation of its otherwise fertile soil.[5]

SOCIO-ECONOMIC STRUCTURE

Economic system

In 1953 Iran was an underdeveloped country with only one major industry—petroleum and natural gas. Petroleum products were the country's major export and chief source of foreign capital. In 1951, the Iranian Government nationalized the industry, and British technicians left the country, causing output to decline drastically. The state already owned most of Iran's modern industrial enterprises, since Reza Shah initiated an industrialization campaign in the 1920's. In 1946, a 7-year economic plan was begun which was financed entirely by oil revenues. In addition to modern and semimodern state-owned industrial plants producing textiles, cement and bricks, ceramics and pottery, there were numerous small handicraft shops, which turned out the traditional Persian rugs and handwoven goods. These traditional industries employed more than half of the country's very small industrial labor force.

Primarily an agricultural economy, about four-fifths of Iran's population were peasants. They were, for the most part, landless tenants or at best owners of small, poor tracts of land, and they produced a variety of crops for subsistence and limited domestic distribution. Iran was not a significant exporter of any agricultural products, but there was no food shortage within Iran.

Class structure

As an economically underdeveloped agrarian society, Iran had the same general type of tradition-oriented social structure found elsewhere in the Middle East, although for historical reasons Persian class structure exhibited greater rigidity and internal stability. At the apex of the social pyramid was the small but powerful urban elite, whose members owned most of the land and water resources of the country and the paramount means of production. Through a mutually reinforcing combination of land ownership and high governmental or religious position, this group, variously estimated to comprise from 200 to 1,000 families, completely dominated Iranian society. There was a greater sense of historical continuity and perhaps more consciousness of *noblesse oblige* among these landlords than in the case of their Arab counterparts; a great number of the present elite came into power in the 1920's with Reza Shah and the Pahlavi dynasty.[7]

Of considerable social significance were the urban groups immediately below the landed gentry. Traditionally, this group, called the "Bazaar group," was made up of merchants, artisans and craftsmen, and members of the lower clergy. In modern Iran, however, this class has given rise to three new classes: the moderately well-to-do modern industrialist type, "the *nouveau riche*, who associate with each other in Western-style clubs, but who remain highly visible as social climbers and are largely avoided by the older elite";[8] the semiliterate skilled industrial workers, few in number but subject to Western—and Communist—influences and increasingly conscious of their rights and special interests; and the educated white-collar workers, many of them civil servants, who are probably most symbolic of the gradual differentiation of the new middle groups from the bazaar. These groups tended to emulate the elite, and adopted Western habits.[9] These middle groups had little group consciousness or cohesion, and they presented a concerted front only in their resistance to foreign domination and rural conservatism. They also represented the strongest force for progressive social action.

The urban proletariat often played a part in modern Iranian politics, mostly by providing the participants for the street demonstrations which bazaar agents would, for a price, organize and train to march through the streets or riot, in support of or in opposition to, almost any issue.[10]

Literacy and education

The level of general education and literacy was low among the vast majority of the population of Iran; only the traditional elite and upper strata of the new middle classes were well educated. Literacy estimates for the Teheran areas ranged up to 20 percent among adult males. Women and rural inhabitants were the least literate elements of the population. There was a modern University of Teheran and university colleges of Isfahan, Meshed, Shiraz, and Tabriz. The University of Teheran had a student body of around five thousand, 5 percent of whom were women, and most of these came from the rising middle class. "Politics [was] the life blood of the students. Most [took] jobs in government when they [graduated]. Nepotism and family connections [played] an important role in getting some of them into and through the school as well as placing them in positions."[11]

Major religions and religious institutions

"Nine out of ten Iranians were members of the Shi'i sect of Islam. For a thousand years influential Persians have favored this sect and for four and a half centuries it has been the state religion and bulwark of Persian nationalism."[12] Shi'ism is the Persian national variant of Islam. The only Sunni Muslims in Iran were the Kurds, some of the Turkic groups, and the Arabs. The *Baha'i*, a deviationist group from Islam, has been the cause of some serious civil disturbances in recent years.

Fundamentalist and antiforeign politico-religious organizations have played a major role in Iranian politics. During the Mosaddegh era, the most important of these was the *Fedayan Islam* (Devotees of Islam), associated with the Shi'i religious leader and politician, Ayatolah Sayyid Abd al-Kasim Kashani. The motto of this Muslim action group bespoke its militant anti-Western orientation: "Death to all who follow Western customs."[13] Members of this group figured prominently in political assassinations and threats and acts of violence during times of political crisis. In 1951 members of *Fedayan Islam* assassinated General Razmara, the Shah's military strong man and Prime Minister; the organization was banned by the government in 1955 after another attempted assassination. During the first part of Mosaddegh's rule, the *Fedayan Islam*, under the influence of Kashani, was one of Mosaddegh's chief sources of support. Later, when Kashani broke with Mosaddegh, the *Fedayan Islam* threatened Mosaddegh's life and made an attempt on the life of his Foreign Minister, Fatemi.

GOVERNMENT AND THE RULING ELITE

Description of form of government

General

Since 1906, when enlightened elements of the Teheran "bazaar bourgeoisie," supported by the British, forced the declining, Russian-backed Qajar dynasty to accept a Western constitution, Iran has been in theory a constitutional parliamentary monarchy. Patterned after the French Constitution of 1875, the Constitution vested executive power in the Shah, a Prime Minister, and a Cabinet responsible to Parliament; legislative power in a Parliament (*Majlis*) of two houses, called the *Senat* and the *Shora;* and judicial power in an independent judiciary system. The powers of the Shah included the usual Crown prerogatives such as command of the armed forces, power to convene and dissolve parliament, to appoint and dismiss Cabinet ministers, to introduce legislation and issue decrees, to declare war and make peace, and absolute veto power over legislation; moreover, the ruling Pahlavi dynasty has taken these constitutional powers very seriously, especially those relating to the armed forces. The lower house, elected for a term of 2 years, was composed of 136 members (prior to 1956), between the ages of 30 and 70, most of whom were elected on a geographical basis. Since the representative was not required to reside in his electoral district, about a third of the members lived in Teheran; these were usually absentee landlords, however, who did not necessarily represent urban interests but their own conservative, rural-traditionalist interests. Although called for in the 1906-07 Constitution, the upper house did not become operative until 1950. The Shah appointed half of the 60 member *Senat*, the other half being elected on the basis of universal male suffrage; half of the Senators came from Teheran and half from the 10 provinces (*ostans*).[14]

Responsibility of executive to legislature and judiciary

The Iranian executive was not responsible in the Western European sense to the legislative and judicial branches of government, both of which it often dominated and manipulated to its own ends; however, there has developed under the present Shah what one writer calls "an informal political equilibrium . . . between the executive and legislative branches."[15] The *Majlis* and the Senate on the one side, the Shah on the other, respected the "invisible and indeed elastic boundary" that protected the prerogatives of the other. When Mosaddegh and part of the legislature attempted to overthrow the Shah, the rest of the legislative branch and the army came to his defense and the balance was reestablished.

Legal procedure for changing executive and legislature

Only the Shah had the constitutional power to select the Prime Minister, and in practice he also selected the other ministers of the cabinet; however, this was not done without reference to the political climate of the *Majlis*, to which the government must look for support of its program. Legally, the Prime Minister and his government served at the pleasure of the Shah, until there was a vote of no confidence in the lower house of the *Majlis*, or until they resigned of their own accord. According to a constitutional amendment passed in 1949, the *Majlis* could be dissolved by the Shah and new elections held within 3 months.

Description of political process

Political parties and power groups supporting government

The Mosaddegh regime had, at the outset, the political support of a loose coalition called the National Front, organized around the personality of its founder and leader, Dr. Mohammed Mosaddegh. Such personal political groupings have frequently occurred in Iranian politics, which like that of many underdeveloped societies is characterized by an embryonic political party system in which personalities are more important than issues. Political parties were first permitted in 1941, and, with the exception of the well-organized Communist-front, the *Tudeh* Party, the political parties which were then formed were invariably organized around prominent members of the ruling elite who were already in the *Majlis*. Their party programs were expressed in nationalistic platitudes and were indistinguishable from one another.

A strong deterrent to the development of issues, and consequently of political parties in the Western sense, was the fact that the government controlled the elections in the urban areas, and in the villages the local landlord, himself usually a candidate, controlled the vote of the peasants. The dominant power group was the landowning elite, who controlled the *Majlis*; the urban groups attempted sporadically in the 1940's to challenge the domination of the ruling elite but ended by coalescing with it in Mosaddegh's National Front emphasizing nationalism over divisive social issues.

At the peak of his power, Mosaddegh's political supporters included ultranationalist elements, such as the Iran Party and the *Fedayan Islam*, most of the students and intellectuals, the socialists, the *Tudeh* Party, the bazaar merchants, and some nationalistic landowners. Mosaddegh's identification with the nationalization of the Anglo-Iranian Oil Company gave him the widest popular support. "Although his National Front Party had a delegation of only 8 out of 136 in Parlia-

472

ment when he became leader of the government, his following and influence was widespread in all classes."[16]

So great was Mosaddegh's popular support from extraparliamentary forces that he was able to dominate both the *Majlis* and the Shah long after his parliamentary support had collapsed. This personal coalition held together so long as Mosaddegh aimed at relatively simple and negative objectives, such as the expulsion of the British, but when economic chaos resulted from the disruption of the oil industry and when Mosaddegh became too dictatorial even by Iranian standards, his support started to decline.

In January 1953 Kashani turned against Mosaddegh, and in July a bloc of his erstwhile allies in the *Majlis* resigned in protest, so that the body no longer had a quorum. In August 1953, Mosaddegh held a referendum on the question of dissolving the *Majlis*, a power constitutionally held only by the Shah. Winning the referendum, as everyone expected, the Prime Minister dissolved Parliament and ruled the country in defiance of the Shah's order dismissing him. Only the *Tudeh* Party and extreme nationalists then remained in Mosaddegh's camp.

Mosaddegh is a member of the landowning elite and was educated in the West. He was elected to the *Majlis* in 1941, after Reza Shah's abdication, and in 1946 forged a number of splinter groups into a so-called "National Front" in opposition to Qavam. A superb orator, he played upon the latent and inflammable nationalism and xenophobia to build up a following. The Anglo-Iranian oil dispute gave him his great opportunity. As chairman of the *Majlis* oil committee, he drafted the Oil Nationalization Law of 1951. In a politically immature country, as one author notes, eloquence and the appearance of sincerity can quickly raise a man to a position of power. "To all of Iran he was a Messianic protagonist of the forces of good engaged in a self-sacrificing struggle with the forces of evil."[17]

Character of opposition to government and its policies

Mosaddegh's National Front coalition began to disintegrate in January of 1953, when Mullah Kashani and another leading figure in the National Front went over to the opposition and other defections followed. The causes of these defections were varied and included such factors as personality clashes over appointments and retirement of personnel, Mosaddegh's loss of charismatic appeal among the elite if not among the masses, and, ironically enough, the growing suspicion that Mosaddegh had become too moderate in his dealings with the British over the oil crisis. The most important organized opposition group was the Toiler's Party, formed in 1950 by Dr. Baghai, a former supporter of Mosaddegh and member of the National Front.

The party was nationalistic and left-of-center in the spectrum of Iranian politics; it was loyal to the Constitution and the institution of the Shah, however, and had wide support among the oil workers. The Toiler's Party left the Mosaddegh government in 1953 when he violated constitutional procedures. The Baghai group also disapproved of his increasing reliance on the *Tudeh* Party and objected to Mosaddegh's retention of certain Iranian officials suspected of being pro-British.

The decisive factor in the opposition, however, was the army. Initially, it supported Mosaddegh wholeheartedly for his nationalistic posture, but the officer elite turned against him when he attempted to wrest control of the army from the Shah and retired several generals. When the issue between the Shah and Mosaddegh came to a head during the summer of 1953, the army went over to the opposition, which now included all conservative elements of Iranian society.[18]

Legal procedure for amending constitution or changing government institutions

The Constitution of 1906 permitted constitutional amendments by a two-thirds vote in a specially elected constituent assembly; the Iranian Constitution was written in such general terms, however, that most of its subsequent constitutional development has come about through regular acts of Parliament. Moreover, the Shahs have not felt bound by its provisions and have often ignored it, as when Reza Shah selected a foreign princess and Sunni Muslim to be the first wife of his son and heir, although this was specifically contrary to the provisions of the Constitutional Amendment of 1925 which sanctioned the Pahlavi dynasty's overthrow of the Qajar ruling family.

Relationship to foreign powers

Iran was never a colonial dependency of any foreign power; however, it was divided into Russian and British spheres of influence in the 19th century, and the current conflict between its northern neighbor, Soviet Russia, and the Western powers has directly influenced Iranian domestic politics since the Second World War. Relations with Iran's Muslim and Arab neighbors have been reasonably friendly despite minor border disputes. These relations were governed in 1953 by the Sa'adabad Treaty of 1937 between Turkey, Afghanistan, Iraq, and Iran.

Relations with Britain were traditionally close and reasonably friendly until the oil crisis developed in 1950–51, when failure to obtain economic aid from the United States for the 7-Year Economic Development Plan initiated in 1946 spurred Iran to ask for more favorable terms. The Iranians pointed to the more generous terms given by American oil companies in Saudi Arabia and Latin America. In the ensuing quarrel, which was taken before the World Court and the Security Council of the United Nations, Iran nationalized the holdings of the Anglo-Iranian Oil Company in March 1951. British technicians left the country in October 1951, when negotiations over compensations broke down, and in October 1952 Mosaddegh's government broke off diplomatic relations with London. British influence in the world oil market prevented Iran from selling oil abroad.[19]

The United States continued to lend Iran limited aid under the technical assistance programs already in existence, lest the country collapse completely and slip into the Soviet orbit. At the same time, the United States was unwilling to ally itself with Mosaddegh by buying the country's oil or granting aid in any large-scale amounts. U.S. policies toward the Mosaddegh regime, appearing inconsistent to some observers, reflected the American dilemma over support of an uncommitted but strategically situated country against a reliable and important European ally. The Middle East historian Sidney Nettleton Fisher sees evidence that the change of administrations in 1953 brought a change in U.S. policy. In the spring of 1953, he contends, "Secretary of State Dulles changed the American tune and helped to pull out the rug from under Mosaddegh."[20] In support of this thesis he notes that on his swing through the Middle East that spring Dulles omitted Teheran from his itinerary, on the plea of insufficient time, although he visited every other capital in the area. The following month President Eisenhower informed Mosaddegh that Iran could expect no further aid until the dispute with Britain was settled, and shortly afterward Secretary Dulles openly accused Mosaddegh of tolerating and apparently cooperating with *Tudeh*.

The role of military and police powers

The Iranian Army has been a decisive force in the political life of the country since Reza Shah, a military man, came to power in the 1920's. Those who were progressive young junior officers in those days were by the 1950's conservative retired people who had merged with the traditional elite.[21] The younger officers were attracted at first to Mosaddegh on account of his vigorous nationalistic leadership, but when he attempted to wrest control of the army from the Shah, they were divided in their loyalty. The police, composed of the provincial *gendarmerie* and the municipal division, were under the control of the Minister of the Interior and were more subject to the influence of the government of the day.

WEAKNESSES OF THE SOCIO-ECONOMIC-POLITICAL STRUC-
TURE OF THE PREREVOLUTIONARY REGIME

History of revolutions or governmental instabilities

Iran had not experienced any revolutions or coups d'etat in its recent history. On the other hand, pre-Mosaddegh governments had not been particularly stable. In the past, political instability had been counteracted by the institution of the Shah and order had been maintained by his military strong men in times of crisis.

The equilibrium of Iranian politics was upset when Mosaddegh was appointed Prime Minister in May 1951. His rule was characterized by violence and civil discord. In 1952, Mosaddegh demanded and obtained power to rule by decree for a period of 6 months. When he asked for a year's extension of this power in January 1953, opposition developed in the ranks of his own National Front, and the Mosaddegh coalition began to crumble, although he obtained his immediate goal of dictatorial powers. Conflict over government appointments and retirements also undermined Mosaddegh's regime.

Toward the end of his regime, Mosaddegh was himself engaged in a revolution. His refusal to relinquish his power and recognize General Zahedi as the new Prime Minister and his attempt to abolish the institution of the Shah were, in effect, revolutionary acts against the *de jure* government. The greatest weakness of Mosaddegh's government and the reason for his fall lay in his overestimation of his own power. He defied the institution of the Shah and Iranian political traditions, relying on extraparliamentary support from the masses. At the time of crisis this support proved insufficient to sustain him in power. Mosaddegh's main line of communication with the urban masses who were his chief supporters lay through the organization of the Communist-front *Tudeh* Party; however, while the *Tudeh* had supported Mosaddegh on the nationalization question, they were not reliable on other issues. Mosaddegh found himself in the position of many other Middle Eastern extreme nationalists: definitely not a Communist himself, he nevertheless was dependent on the Communist organization for marshalling the support of the urban masses, since there was no non-Communist mass political party upon which he could base his regime when he lost the confidence of the narrow clique of politicians and officers who supported him initially.[a]

---

[a] There is considerable disagreement on the nature and extent of Mosaddegh's dependence on the *Tudeh*. Charles Issawi, a well-known Egyptian student of the Middle East who is not considered anti-Western, has described Mosaddegh as one "who passionately fought the British but refused to make much use of Soviet support or of that of the *Tudeh* party."[22]

Economic weaknesses

Iran had no pressing food shortage, and the country never experienced anything comparable to Egypt's economic troubles. By the summer of 1953, however, the stark reality of an oilless, and therefore cashless, economy posed a serious threat to domestic stability. Iranian currency declined in value as the demand for foreign currency drove the exchange rate from around 50 rials to the U.S. dollar to around 120 in the summer of 1953.[23] Lenczowski writes: "Fed for a year and a half on patriotic slogans, the populace could not live forever in a frenzy of enthusiasm while basic goods became scarce and the gap between prices and incomes increased."[24] The urban groups, rather than the largely self-sufficient peasants, were the ones most directly affected by this economic crisis. "The army was short on supplies; and the wealthy landowners who governed the country soon discovered that loss of the royalty revenues on which their corrupt governmental practices battened was forcing them to change their ways."[25]

Social tensions

Social tensions between rural-traditionalists and urban elements were a contributing factor in the downfall of the Mosaddegh regime. The more conservative bazaar businessmen and the landowning elite had become alarmed during the summer of 1953 over the state of the economy and the growing influence of the *Tudeh* among the urban masses. While the conservative elements lacked, at this time, the power required to oppose Mosaddegh outright, they were nevertheless significant as a potential base of political support for whatever effective opposition to the regime would later emerge. Like the Shah and Mossadegh's opponents in the *Majlis*, the conservatives were intimidated by mob violence and the police power of the regime, but when the army turned against Mosaddegh, these elements rallied around it. A large part of the "bazaar bourgeoisie" continued to support Mosaddegh, however, after his overthrow.[26]

Government recognition of and reaction to weaknesses

Mosaddegh turned to the urban masses of Teheran when he saw his Parliamentary support slipping away. He resorted more vigorously than before to his powers and skills as a political orator. The August referendum was supposed to be a further demonstration to the world and to his domestic opponents that the masses of the people were behind him. His fellow politicians were well aware, however, of the corruption of Iranian electoral practices. This referendum was conducted in such an obviously controlled fashion that it was discredited even in a country where elections were customarily rigged. Finally, when he became alarmed over the excesses of the *Tudeh*-led street demonstrations, Mosaddegh called in the army to support the police in putting down the demonstrations against the Shah. He trusted in his charismatic personal qualities as the champion of Iranian nationalism to insure the army's loyalty to him, but was disappointed.

FORM AND CHARACTERISTICS OF REVOLUTION

ACTORS IN THE REVOLUTION

The revolutionary leadership

The personnel of the leadership instigating and organizing the coup d'etat which overthrew Prime Minister Mosaddegh on August 18–19, 1953, is not generally known. The key figure in the movement was Gen. Fazlullah Zahedi, whom the Shah had appointed Prime Minister in place of Mosaddegh on August 13. Zahedi, an Iranian officer and politician, had been in the Senate in 1950. He became Minister of Interior in 1951 before Mosaddegh came to power and had remained in the government during the early months of Mosaddegh's premiership. In July 1953, Zahedi left the Parliament Building where he had been taking sanctuary for 10 weeks under a guarantee of personal safety. He was by this time an acknowledged opponent of Mosaddegh.[27]

The political and ideological orientation of those who combined to overthrow Mosaddegh may be regarded primarily as traditionalist and conservative. Their goal was to restore the constitutional balance between the Shah, the *Majlis*, and the army, which Mosaddegh had upset. The left-of-center Baghai group, Mullah Kashani's religious followers, and the few conservative bazaar business people and the landowners were opponents of Mosaddegh's regime who supported the countercoup.

The revolutionary following

Those who supported the Zahedi coup, either by turning physically on the antiroyalist demonstrators in the streets or by giving their loyalty and moral support to the Zahedi regime after the arrest of Mosaddegh on the 20th of August, included the leaders discussed above as political opponents of the Mosaddegh government and the conservative and moderate elements of the general public. The peasants were largely untouched by the events of the revolution. Large segments of the urban proletariat, such as the oil workers who supported the Baghai politicians, were loyal to the Shah and resented the *Tudeh* as the agent of a foreign and traditionally hostile power.

ORGANIZATION OF REVOLUTIONARY EFFORT

As indicated above, nothing is known of the precise nature of the organization that must have preceded the outbreak of hostilities between the army and the Mosaddegh authorities. It is known, however, that there was a secret conspiratorial clique among a group of anti-Mosaddegh officers associated with General Zahedi and loyal to the Shah. Liaison with political and social groups sharing their loyalty to the Shah and opposition to the Prime Minister may have existed, but nothing definite is known. There were no ostensible ties with foreign powers, although the Zahedi group could not have been unaware of the regime's widespread unpopularity in the Western world.[b]

GOALS OF THE REVOLUTION

There is no evidence that the revolutionary group had any specific social and economic goals which they hoped to achieve by their action against Mosaddegh, other than a general desire to return to a state of normalcy. To achieve this aim, they intended to remove Mosaddegh and his associates from their position in government, and to return to the institution of the Shah.

REVOLUTIONARY TECHNIQUES AND GOVERNMENT COUNTERMEASURES

The revolutionary events surrounding the Iranian Coup of August 18–20, 1953, are unlike the events in most revolutionary situations in that to a large extent the techniques customarily associated with the revolutionaries were in this instance being used by the government in

---

[b] There have been persistent reports, neither denied nor confirmed, that the U.S. Central Intelligence Agency was instrumental in the organization and planning of the Zahedi coup d'etat. The Middle East historian Sydney Nettleton Fisher mentioned that Allen Dulles, the head of the CIA, met with Princess Ashraf, the Shah's sister, in Switzerland during the Mosaddegh crisis in 1953,[28] and newsman Richard Harkness discussed the role of the CIA in the Mosaddegh overthrow in an article in the *Saturday Evening Post*. According to Harkness' account, Allen Dulles, Princess Ashraf, and LeRoy Henderson (U.S. Ambassador to Iran) met in Switzerland, ostensibly on vacations but actually to plan the overthrow of Mosaddegh. Brig. Gen. H. Norman Schwarzkopf, an American police official who had been detailed in 1942–48 to assist the Shah in reorganizing the Iranian *gendarmerie* and became a close friend of Zahedi, returned to Iran in August of 1953 supposedly "just to see old friends again." According to Harkness, Schwarzkopf was involved in the detailed planning of the operations.[29]

power. In a sense, the Mosaddegh government was itself engaged in a revolution. The Shah had dismissed Mosaddegh and appointed General Zahedi in his place, as the Constitution empowered him to do, so that after August 15 the Mosaddegh government was a *de facto* regime and the Zahedi government the *de jure* government. In such a situation, the Mosaddegh regime made use of street demonstrations, riots, and acts of terrorism directed against the opponents of the regime. The *de facto* government also used the usual police methods to silence political opposition. When these techniques appeared to be getting beyond control—and possibly because they were causing an unfavorable reaction to his regime abroad—Mosaddegh turned to the classic countermeasure available to the government in power: he sent the army against the street demonstrators, who ironically represented his strongest support in the country.[c]

This military action, which started as a countermeasure of the government in power, was transformed—to some extent spontaneously and to some extent in response to prior planning—into a revolutionary technique used by the revolutionists against the government in power. This turn of events occurred during the day in which the army moved against the *Tudeh*-led mass demonstrations in the streets of Teheran. Once the anti-Shah demonstrators had been dispersed, the officers and rank and file persons who took part in this operation continued the assault against the Mosaddegh government and delivered the reins of power into the hands of the *de jure* Prime Minister, General Zahedi, who was in hiding.

MANNER IN WHICH CONTROL OF GOVERNMENT WAS TRANSFERRED TO REVOLUTIONARIES

After 2 days of antimonarchist demonstrations, culminating in a tank battle that lasted for several hours, the army got the upper hand. According to the historian Lenczowski,[30] the identity of the officers who defied Mosaddegh and his Foreign Minister, Hussein Fatemi, who had been active in the demonstrations, is not certain, and it is not known to what extent they were in communication with General Zahedi in his hiding place. It appears clear, however, that he did not direct the actual fighting. Public opinion, meanwhile, had turned against the antiroyalists, and when the rioters were dispersed

---

[c] At this time, U.S. Ambassador LeRoy Henderson conferred with the Prime Minister and warned him that the United States was undecided whether it should continue to recognize his government or not, in view of his close dependence on the Communist elements in the country. This may have contributed to his decision to use the army against the antiroyalist demonstrators.

on August 19, Zahedi emerged from hiding and assumed the office to which the Shah had called him earlier.

THE EFFECTS OF THE REVOLUTION

CHANGES IN THE PERSONNEL AND INSTITUTIONS OF GOVERNMENT

There were no changes in governmental institutions, this having been essentially a revolution to restore constitutional government to the country. The most important change connected with the coup was the removal from power of Mosaddegh and his body of ultranationalist associates. They were given varying prison sentences and his foreign minister, Hussein Fatemi, was condemned to death. Mosaddegh himself was sentenced to a prison term of 3 years.

MAJOR POLICY CHANGES

In foreign policy, the Zahedi government moved quickly to normalize relations with Britain, entered into negotiations with the Anglo-Iranian Oil Company, obtained considerable financial aid from the United States, and won several concessions from the Soviet Union. In the winter of 1953–54, an international consortium was set up to process and distribute Iranian oil. Iran agreed to pay compensation to the Anglo-Iranian Oil Company for its property seized in 1951, and the consortium agreed to return to the Iranian Government half of its profits.

LONG RANGE SOCIAL AND ECONOMIC EFFECTS

No immediate change in social and economic relationships within the country could be perceived, although the return of normal relations with the outside world and the return of oil revenues had a positive effect on the national economy.

OTHER EFFECTS

After the Zahedi coup, Iranian politics resumed their pre-Mosaddegh pattern. General Zahedi's military dictatorship, like General Razmara's regime before Mosaddegh, was not oppressive by traditional Iranian standards, although civilian elements in the cities were less than fully satisfied with the enlarged influence of the army in the government. In the spring of 1955, after having concentrated on the restoration of internal order and security (Communist activities were sharply curtailed, as were, incidentally, the activities of the Shah's political opponents who were not Communists), Zahedi resigned and was replaced without incident by a more reform-minded political leader, the venerable Iranian statesman and close advisor to the Shah, Hussein Ala.

By 1960, Iranian politics showed signs of increasing unrest. Student demonstrations in Teheran and by Iranian students in London demanded the return of Mosaddegh; politicians from the Mosaddegh era began to be heard from; and there was a general breakdown in relations between the Shah, who dominates the government and is interested in economic and social reforms in order to stave off revolution, and the landowning elite, who dominate the *Majlis* and oppose the Shah's reforms. In 1961, Mosaddegh, who best personifies the tradition of militant opposition, remained an important rallying point for all shades of opposition to the Shah and the *status quo* in Iran, and it seemed highly possible that ultraconservative members of the elite might join forces either with leftist, nationalist and middle class opponents of the Shah under the leadership of a Mosaddegh-type politician, or with the army under a military strong man to overthrow the liberal reformist and pro-Western government installed by the Shah in May of 1961.[31]

NOTES

1. Herbert H. Vreeland (ed.), *Iran* (New Haven: HRAF Press, 1957), pp. 28–31.
2. Ibid., pp. 31–35.
3. Ibid., pp. 31–35, 39–51.
4. Donald N. Wilber, *Iran: Past and Present* (Princeton: Princeton University Press, 1955), pp. 163–172.
5. Ibid., pp. 132–151.
6. Vreeland, *Iran*, pp. 135–138, 152–155, 203–211.
7. Ibid., pp. 245–248.
8. Ibid., p. 248.
9. Ibid., p. 249.
10. Ibid.
11. Ibid., p. 273.
12. Ibid., p. 290.
13. Ibid., p. 298.

14. Ibid., pp. 56–81; also see Wilber, *Iran*, pp. 173–204.

15. Vreeland, *Iran*, p. 61.

16. Sydney Nettleton Fisher, *The Middle East* (New York: Knopf, 1959), p. 515.

17. Vreeland, *Iran*, pp. 87–88.

18. George Lenczowski, *The Middle East in World Affairs* (Ithaca, N.Y.: Cornell University Press, 1956), p. 201; also see *Ost-Probleme* (Bonn, W. Germany), VII, 39, 1496–1499.

19. Benjamin Shwadran, "The Anglo-Iranian Oil Dispute," *Middle Eastern Affairs*, V (1954), 193–231; see also Lenczowski, *The Middle East*, pp. 192–200.

20. Fisher, *The Middle East*, pp. 518–519.

21. Andrew F. Westwood, "Elections and Politics in Iran," *Middle East Journal*, XV, 2 (Spring 1961), 153–164.

22. Charles Issawi, "Middle East Dilemmas: An Outline of Problems," *Journal of International Affairs*, XIII, 2 (1959), 103.

23. *Time*, (July 13, 1953), 40–42.

24. Lenczowski, *The Middle East*, p. 201.

25. Fisher, *The Middle East*, p. 517.

26. Ibid., p. 519.

27. Ibid., p. 518.

28. Ibid., p. 519.

29. Richard and Gladys Harkness, "America's Secret Agents: The Mysterious Doings of CIA," *Saturday Evening Post*, (November 6, 1954) 66, 68.

30. Lenczowski, *The Middle East*, p. 202.

31. Kennett Love, "Iran's Three R's," *New Leader*, XLIV, 39 (December 11, 1961), 8–11.

RECOMMENDED READING

BOOKS:

Fisher, Sydney Nettleton. *The Middle East*. New York: Knopp, 1959. A standard history text. Contains a balanced account of the Mosaddegh era.

Lenczowski, George. *The Middle East in World Affairs*. Ithaca, N.Y.: Cornell University Press, 1956. A standard text in the history of the area. Contains good summary of events surrounding the Zahedi coup.

Upton, Joseph M. *The History of Modern Iran: An Interpretation.* Cambridge: Harvard University Press, 1960. The most up-to-date and complete history of Iran.

Vreeland, Herbert H. (ed.). *Iran.* New Haven: HRAF Press, 1957. A comprehensive study of social, economic, political, and cultural factors in contemporary Iran.

Wilber, Donald N. *Iran: Past and Present.* Princeton: Princeton University Press, 1955. A general survey of Iran's economic, social, and governmental structure, and a concise history of the area from ancient times to the present.

PERIODICALS:

Free World Forum. I, 4, 39–64. Special supplement on Iran, containing articles on Iranian history, economy, and foreign relations with the United States and with the Soviet Union.

Love, Kennett. "Iran's Three R's," *New Leader*, XLIV, 39 (December 11, 1961), 8–11. Excellent discussion of Prime Minister Ali Amini's reformist government and its future prospects. The three "R's" are for reform, revolution, and Russia.

Ost-Probleme (Bonn, W. Germany). VII, 39, 1496–1499. A German-language article on Iranian political parties and groups with particular emphasis on the *Tudeh* Party.

Shwadran, Benjamin. "The Anglo-Iranian Oil Dispute," *Middle Eastern Affairs*, V (1954), 193–231. A detailed analysis of British-Iranian relations from 1952 to 1954.

Westwood, Andrew F. "Elections and Politics in Iran," *Middle East Journal*, XV, 2 (Spring 1961), 153–164. Best account of Iranian political developments since 1953. Also contains useful data on social groups and their relations to political groups. Details of 1960 electioneering.

THE IRAQI COUP OF 1958

SYNOPSIS

On the morning of July 14, 1958, Brig. Gen. Abdul Kerim Kassem led a coup d'etat against the pseudo-parliamentary monarchy of Iraq. After a day of mob violence and limited military action, in which the royal family and the pro-Western "strong man" of Iraq, Nuri as-Said, were killed, a republic was proclaimed under the leadership of a military and civilian junta. The immediate cause of the revolution was the government's decision to move Iraqi troops under Kassem's command into the allied Kingdom of Jordan to protect that country from pro-Nasser elements within its borders.

BRIEF HISTORY OF EVENTS LEADING UP TO AND CULMINATING IN REVOLUTION

The long range cause of the revolution was the general discontent with the prerevolutionary regime's pro-Western, anti-Nasser foreign policy, its repressive internal policy, and its failure to identify with the rising middle-class elements and the urban masses. The groups allied against the old regime included rightwing ultranationalists, leftwing intellectuals, military personnel, pro-Nasser Pan-Arabists, and the Communists. The group which emerged on the day of the revolution and comprised the cabinet of the new government was made up of the leaders of all the above groups except the Communists. The Communists were informed of the coup just prior to the event, and their assistance in organizing mobs and street demonstrations was used by the revolutionary clique; however, no actual Communists were included in the governing junta set up after the coup.

The revolutionary goals included a strong emphasis on a neutralist foreign policy with close ties with Nasser and the Soviet bloc, and a general program of wider social and economic reforms than the prerevolutionary regime had pursued, including agrarian and fiscal reforms, anti-inflationary measures, and increased social welfare activities relative to economic development. Political party activity, which had been in abeyance in Iraq since 1954, was restored and a variety of political groups sprang up in the wake of the revolution. No centralized mass movement in support of the revolutionary regime was organized, as was the case in Egypt.

THE ENVIRONMENT OF THE REVOLUTION

DESCRIPTION OF COUNTRY

There were no significant differences in the physical characteristics, racial and linguistic structure, population density and urbanization, communications, and the natural resources of Iraq between the time of the 1936 revolution and the one in 1958. Iraq was still underpopulated in 1958, with a total population of around six million. (See map on p. 422.)

SOCIO-ECONOMIC STRUCTURE

Much of the socio-economic structure of Iraq in 1958 was still what it had been in 1936 (see preceding discussion on the Iraqi Coup of 1936). There were, however, significant changes within the ruling elite as well as among the lower classes. The position of the tribal chiefs was somewhat weakened. The older generation of tribal chiefs or large landholders had become closely identified with the 1920 rebellion against the British and with the subsequent achievement of independence. For this reason, many of them had been almost national heroes. The 1958 generation did not command the same respect. Moreover, many of the tribal chiefs had begun to move to the cities together with their families, thus becoming absentee landlords. This trend to the city had begun before World War II, but accelerated during and after the war.

Another important change took place as a result of the government's economic development plan in the emergence of small but important privately-owned industry. Tens of thousands of rural workers migrated to the cities. They received higher pay, but their expenses too were much higher. Separated from the traditional social controls they had experienced while they were working the land, they became easy recruits for various opposition movements. The government concentrated on long range economic programs, but did not meet the problems created by this new social group.

The middle classes too experienced important changes. Between 1936 and 1958 the number of educated people increased tremendously. There were not enough jobs for intellectuals, especially not in the traditional government service. Thus intellectuals also became easy recruits for opposition groups In addition, many of the civil servants had become demoralized because they had changed from a fairly well paid satisfied group to an underpaid dissatisfied group.

This trend started during the war and continued. Bribery and competition became even more accepted than they had been previously.

Another noticeable difference between Iraqi society in the 1930's and that of the 1950's is the fact that after World War II, and especially as a result of the Palestinian conflict in 1948–49, the Jewish and foreign minorities were allowed to emigrate and lost some of their erstwhile positions of dominance in the commercial sector of the Iraqi economy, their places being taken by both Sunnis and Iraqi Shi'is. Thus, the socio-economic position of this long-suppressed element in Iraq's society has been on the rise in the postwar period.

GOVERNMENT AND THE RULING ELITE

Description of form of government

Until July 14, 1958, Iraq's form of government was that established by the 1925 Constitution, a constitutional parliamentary monarchy. See the summary referred to above for a discussion of the form and working of this government system.

A major change in the Iraqi Constitution occurred in 1943, when an amendment was passed empowering the Crown to dismiss the government. It had been found that in practice the Prime Minister was not responsible to the Chamber of Deputies; hence, this amendment to provide for the removal of an unpopular and irresponsible Prime Minister and Cabinet. Although never used by the Crown, this royal prerogative existed as a shadowy threat.[1] Another significant constitutional development took place in 1952, when the indirect electoral system was replaced by direct elections.[2] The Crown was not as popular in 1958 as it had been during the 1930's. Indeed, the relationship of the Hashemite dynasty to the people of Iraq was never comparable to that between modern European monarchs and their subjects. The 23-year-old King, Feisal II, was not an unpopular figure, but neither was he a national hero among the masses, and his uncle, Crown Prince Abdul Illah, was the object of considerable hatred. Abdul Illah served as Regent during Feisal's minority from 1939 to 1953, and it was thought that he continued to exert influence over the King and the government after Feisal came of age in 1953. The Regent had called on the British to support him against the anti-British regime of Rashid Ali in 1941, and Iraqi nationalists have never forgiven Abdul Illah for this. Consequently, the Crown was discredited in the eyes of most Iraqis and offered no hope of becoming a popular national institution.

Description of political process

The working of the political process was conditioned by the social situation in the country. As in the 1930's, the dominant interest group in postwar Iraq was the landlord class. Closely allied to this group and to a large extent overlapping it, were the government bureaucrats and wealthy middle class, who sought respectability and economic advantage alike in absentee commercial farming. The changes in government that occurred during this period were more often the result of shifting personal alliances and cliques within a narrow circle of professional officeholders than the result of a real political disagreement. Popular support of the government and participation in the governing process was limited to the ruling elite, said to include around four hundred wealthy landlord families. In such circumstances, political parties in the usual sense could not be expected to function, although from 1946 until 1954 the organization of political parties was permitted by the government. Throughout the postwar period, Nuri as-Said, a venerable statesman and astute politician, and an enlightened conservative and friend of the Western bloc, governed the country either directly as Prime Minister or indirectly through his influence over the members of the government.

Political parties and power groups supporting government

The "Government" Party was organized in 1949 by Nuri as-Said and was called the Constitutional Union Party (*Ittihad al-Dostur*). Into this grouping were gathered the dominant conservative landowning elements, especially the tribal sheikhs of the Middle Euphrates provinces. In 1954, when the ruling elite felt themselves threatened by the rising tide of popular discontent expressed as Nasserism and communism, Nuri decided to abolish all political party organizations, his own "Government" Party included. The group continued as an informal organization held together by the personality of Nuri as-Said.

Character of opposition to government and its policies

In 1954, a crucial year in Iraq's political history, opposition to the government was composed of the following groups: (1) a "loyal" opposition, made up of the Socialist Nation Party, headed by a conservative Shi'i dissident from Nuri's group, and the loosely organized United Popular Front, led by a respected elder politician; (2) a militant opposition, which included the *Istiqlal* (Independence) Party, an ultranationalist group, and the National Democratic Party, a leftist group containing a number of former members of the *Ahali* movement; and finally, (3) the Communist Party, a subversive organization. After 1954, the *Istiqlal* group and the National Democratic Party went underground along with the Communist movement, and an active collaboration among these three groups ensued.[3] In 1956 they were joined by the *Ba'th* Party, a Pan-Arabist "Socialist" movement based in Syria but with branches throughout the Fertile Crescent. *Ba'th* elements in Iraq opposed both the Nuri regime and the Communists and favored union with Syria and Egypt under Nasser.[4] These subversive opposition forces drew popular support from middle-class urban intellectuals and semi-intellectuals who objected to the authoritarianism and nepotism of the Nuri regime, as well as the regime's pro-Western foreign policy; from the urban and rural proletariat's economic discontent in the face of inflationary pressures caused by oil wealth and industrialization programs of the government's Development Board; and from the Shi'is and Kurds' traditional resentment of the Baghdad government. Criticism of the regime in the press and other domestic news media was severely limited, and the electoral system was thoroughly rigged by the government so that the opposition forces turned increasingly to covert means of opposing the regime.

Legal procedure for amending constitution or changing government institutions

There had been no change in amending procedures under the Iraqi Constitution of 1925. (See preceding discussion on the Coup of 1936, p. 421.)

Relationship to foreign powers

The Conservative clique who ruled Iraq in the postwar period feared the military might of the Soviet Union, assisted as it was both by geography and by the presence of an organized Communist underground within the country; thus, the clique looked to the Western bloc, along with Turkey and Iran, for its defense. The Baghdad Pact of 1955 replaced the Anglo-Iraqi treaty of 1930, which had met with the opposition of involving Iraq's non-Arab, but Islamic neighbors, Turkey and Iran as ostensible equals with Britain and Iraq. The fallacy of this equality was pointedly revealed in 1956 when Britain proceeded to attack Egypt without prior consultation with Iraq and other Baghdad Pact partners. Ultranationalists, Pan-Arabists, and pro-Nasser elements in Iraq and throughout the Middle East attacked the Nuri regime after 1956 with renewed vigor as an accomplice and stooge of Western imperialism. The identification of the Iraqi Government with the Western bloc and the acrimonious attack on Nuri as-Said personally which the propaganda organs of the Egyptian, Syrian, and Saudi Arabian Governments assisted by the Soviet bloc, conducted in the 1956–58 period contributed significantly to the atmosphere of violence in which the July 14 coup took place.

Iraqi attitudes toward Egypt—and vice versa—reflected the traditional hostility between the Mesopotamian and Egyptian centers of civilization and power. However, in recent times, it has been primarily a struggle for leadership of the Arab world and Arab nationalism. Iraq lost face when Egypt attracted Syria away from its sphere of influence. The Arab Union between Iraq and its "poor relation," Jordan, effected in the last months of the Nuri regime, was an effort to offset the prestige gained by Nasser through the merger of Egypt and Syria.

Relations between Iraq and the West had never undergone the difficulties experienced in Egypt's relations with the West. This was largely due to the fact that the British had only a small contingent at Habaninah Airfield and not the sizeable force they had maintained in Egypt. The continued dependence of Iraq's oil economy on Western European markets also contrasted with the declining Egyptian cotton trade with Western Europe.[5] The participation of France and Israel in the British attack against Egypt made the position of pro-Western Iraqis extremely difficult.

The role of military and police powers

In the 1950's as in the 1930's, the life of the regime in power depended ultimately on the army and police. On the surface both of these forces appeared to be loyal to the ruling elite until the day of the coup. They had long been used to suppress the political foes of the regime, and the army was being used to strengthen the Jordanian regime at the time of the revolt.[6]

WEAKNESSES OF THE SOCIO-ECONOMIC-POLITICAL STRUCTURE OF THE PREREVOLUTIONARY REGIME

History of revolutions or governmental instabilities

After the chaotic period before the Second World War (1936–41), when Iraq experienced a series of seven military coups d'etat, there was a long period of what appeared on the surface to be political stability in the country. Nuri as-Said held together a personal coalition of the leading Iraqi families and politicians that lasted from 1941 down to July 14, 1958. There were frequent riots and demonstrations, especially during the Palestinian trouble, and in 1952 after the Egyptian revolution, and again in 1956 during the Suez crisis, but the government kept these in check by its use of police and military power. The period of 1952–54 was of crucial importance to the political development of Iraq; after some tentative "reforms from above" a return to repressive policies occurred in 1954. After that, subversive activity against the regime flourished. In the absence of institutional criteria for qualification to political office, nepotism and personal favor became the deciding factors. This led to intense jealousies even among the most favored political appointees. Serious frictions in the inner circle around Nuri arose often on personal grounds and contributed to the weakening of his authority. Others objected vehemently to the repeated appointment of certain individuals to cabinet posts on the basis of purely personal favor.

Economic weaknesses

Iraq had a comparatively sound economy. With no population pressures and plenty of fertile soil, Iraq had a ready supply of capital. Its oil revenues were sufficient to finance both the current needs of the state and the government's long range program of agricultural and industrial development. The economic weakness of the system must be sought in the distribution of this national wealth. The government's Development Board, "a well-considered and systematic utilization of 70 percent of Iraq's oil revenues for development of the national economy,"[7] functioned competently and led to some improvement in the real wages of urban workers despite inflationary pressures. The government achieved substantial improvements in public health and education among the rural peasants. But these achievements were not able to overcome the traditional alienation of government from the masses, and the mercantile and entrepreneurial middle class in the cities, who reaped the immediate economic benefits of the government's Development Board activities, were outside the ruling elite and so felt no identification with the regime.[8] Nuri's solution to the problem of land tenure, worse in the 1950's than in the 1930's, was to change and enlarge the area of arable land by irrigation projects. This was a slow process, but it satisfied the landowning elite on whom Nuri depended for his political support. It definitely did not satisfy the rising middle class nor the majority of the peasants.[9]

Social tensions

Social tensions present in the 1930's were greatly intensified by 1950. Differences between Iraqi social conditions in 1958 and in 1936 were more a matter of degree than of substance. Yet the increased social tensions had an important bearing on the 1958 revolution. Iraqi society in 1958 was torn by all those social antagonisms it had inherited from the past—the Sunni-Shi'i rivalry, Arab-Kurd hostility, and urban-rural antagonism. To these were added the social conflicts engendered by urbanization and economic development: an urban proletariat, hard pressed by economic conditions (housing and living costs were inflated) and socially uprooted from a rural background; a lower-middle class of semi-intellectuals who suffered the economic and social plight of the working class but with the difference that they were politically conscious and articulate (especially in ultranationalistic groups); and an upper-middle class of wealthy contractors, merchants, entrepreneurs (who shared in the economy but not the governing of the country), and intellectuals, and, finally, at almost all social levels, the younger generation of Iraqis. The social group of the ruling elite was largely closed to these elements, being based on kinship and possession of land and influence in the regime.[10] The ruling elite, itself, was torn by personal jealousies and opportunistic tactics. Even some of the old politicians whose interests rested on the maintenance of the regime indulged occasionally in bitter attacks on Nuri as-Said either to ingratiate themselves with the opposition or to force Nuri to make concessions to their personal demands.

Government recognition of and reaction to weaknesses

The Nuri regime certainly recognized the fact that it lacked popular support, either among the masses or the middle-class elements, but it depended first upon the use of force and second on its manipulation of key figures within the ruling elite—the tribal-based landowning families—and finally on its gradual development of the national economy which would mollify both the middle class and the working class by providing improved working conditions and even great wealth to some among the middle class. Nuri looked to Western economics, not to Western politics, to secure his power in Iraq. One observer of the prerevolutionary regime's methods of control writes:

> . . . It bound tribal chiefs, business men, and religious leaders collectively with ties of economic interests and social prestige. . . . The army was purged and wooed, mob action anticipated by police techniques. . . . The apparatus of state . . . expanded with the injection of oil revenues.[11]

Meanwhile a facade of political democracy was maintained through rigged elections, and labor and agrarian arrest was countered with palliatives rather than with genuine reform.

Shortly before the revolution Nuri embarked on an extensive purging of the administration. Hundreds of officials were dismissed from their jobs by the decision of a special committee, including a large number of police commissioners and officers. Although most of them probably deserved the treatment, personal feelings and predilections played a role in the treatment of some. Moreover, some other officials, equally deserving of dismissal, were not touched because they were protected by dominant political influence. This disparity created a large body of disgruntled persons who were eager for vengeance. The feeling of grievance was enhanced because of the social stigma attached to the loss of a government position. Hence a large number of this group joined the mob which applauded the revolution at the outset.

FORM AND CHARACTERISTICS OF REVOLUTION

ACTORS IN THE REVOLUTION

The revolutionary leadership

The leadership of the revolution was provided by an informal organization of army officers and opposition political leaders. It included: Brig. Gen. Abdul Kerim Kassem, Commander of the 19th Brigade in the Iraqi Army; Col. Abd al-Salam Arif, Commander of the 20th Brigade; Mohammed Mahdi Kubba, head of the subversive ultranationalist and rightist *Istiqlal* (Independence) Party, the scion of a prominent Shi'i family in Baghdad; Gen. Nijib al-Rubai, conservative conspirator whom Nuri as-Said had "exiled" by appointing him Ambassador to Saudi Arabia; Khalid Naqshababandi, a Kurd and former governor of the northern province of Erbil; and an indeterminate number of other leaders of the underground elements involved in the coup. There was no definite agreement among these leaders as to the political and ideological orientation of the revolutionary movement, other than that it stood for the overthrow of the Nuri regime and the Hashemite monarchy and the establishment of a regime composed of themselves and oriented away from the Western bloc, which would pursue a vigorous program of social and economic reforms Opposition to the *Status quo* was sufficient as the group's *raison d'être* and this alone served as its political and ideological touchstone. The goals and aspirations, never announced prior to their coming to power, included the usual nationalistic and neutralist slogans on foreign relations and the usual reformist and "socialistic" slogans on domestic matters.

The revolutionary following

The revolutionary following included the members of the *Istiqlal*, the *Ba'th*, the Communist, and the National Democratic Parties; the latter group apparently had not possessed advance knowledge of the coup but welcomed it. The National Democrats, heirs to the *Ahali* group of the 1930's[a] and the only non-Communist leftist party, supplied several cabinet members in the revolutionary regime. The *Istiqlalists* were rightwing ultranationalists, in the *Ikha* tradition of the 1930's, and they attracted the support of enthusiastic students and professional people. The *Ba'th* attracted support from pro-Nasser elements; it was militantly anti-Communist. No estimate of the number of people in these groups is available, although according to a leading student of communism in the Middle East, the Communists claimed to have as many as 17,000 in 1950.[12] The revolution also enjoyed wide popular unorganized support, for the prerevolutionary regime was genuinely unpopular with the masses and the middle class, as well as support from the majority of the army officers and from some of the Kurdish tribesmen.

ORGANIZATION OF REVOLUTIONARY EFFORT

Internal organization

Type of organization

The conspiratorial group was an informal organization among the army officers and the leaders of the political opposition groups. General Kassem has said that he first got the idea of forming an officers' group to stage the coup d'etat from Syrian officers whom he met in Jordan at the time of Suez. He organized the group a few months before the coup. Colonel Arif apparently formed a separate officers' group, but the two collaborated in the July coup.[13] The political opposition groups with which Kassem and Arif conspired[b]—the *Istiqlalists, Ba'thists*, and Communists—were all three subversive organizations by virtue of the 1954 decree forbidding political party activity in the country.

Relationship between leadership and following

In a conspiracy such as this, the number of persons actively involved before the outbreak of the revolution—and therefore qualifying as members of the leadership—must have been limited to few more than those named above. The rank and file members of the political groups and the other officers in the armed forces provided the following and became engaged in the revolution during the morning of the 14th and thereafter. The leadership, and certainly the following,

---

[a] See discussion of the Iraqi 1936 Coup, p. 421.

[b] Although the presence of a conspiracy between Kassem's military clique and the opposition *Istiqlal* and *Ba'th* parties prior to the revolution, on July 14, 1958, cannot be documented in this study, the formation of a cabinet on that same day, comprised of members of the above parties, indicates strongly the presence of some form of understanding between the military and the civilian groups.

of this revolutionary coalition movement was not homogeneous as to final social and economic goals, and especially were they divided on foreign policy. The *Ba'thists*, and Colonel Arif, as was later disclosed, favored Iraq's annexation to or federation with Nasser's United Arab Republic. The Communists, aware of Nasser's anti-Communist and dictatorial regime, were stout defenders of Iraqi independence. The *Istiqlalists* were opposed to Nuri's pro-Western regime, but they were not interested in socialistic reforms.[14] Some Kurds were hesitant about joining the revolution, although they did not like the old regime's union with Arab Jordan.[15]

External organization

The only revolutionary group actually organized outside the country was the *Ba'th*, a Pan-Arabist and pro-Nasser movement based in Syria with branches in Lebanon, Jordan, and Iraq. The Communist organization was indigenous to Iraq, but, like its rival, the Syrian *Ba'th*, it had substantial support from outside the country. The Nuri regime was the object of violent propaganda attacks emanating from both the Soviet bloc and its Arab neighbors (Saudi Arabia, Egypt, and Syria). Press and radio were used in the attack on the regime, the attacks becoming especially vehement after 1956. According to one source, one effect of the Nasser-*Ba'th* and Communist propaganda attacks on Nuri was to strengthen his government in the eyes of Iraqi nationalists who were opposed to Nasser.[16] Going beyond mere propaganda, the Egyptian and Syrian military attachés in Baghdad in 1955–56 conspired with Iraqi Army officers to arrange a pro-Nasser coup d'etat; in 1956, after the Suez crisis, Nasser secured the formal alliance of Rashid Ali al-Gailani, an important Iraqi politician in exile in Saudi Arabia since his unsuccessful anti-British coup in 1941; thereafter Egyptian liaison with Rashid Ali's *Istiqlalist* friends in Iraq improved. King Saud of Saudi Arabia hired Gen. Najib al-Rubai to assassinate Nuri in 1957 and the Syrians attempted to apply economic pressure in 1956 by sabotaging the Iraq Petroleum Company's main pipeline.[17] These propaganda, terrorist, and conspiratorial activities by foreign powers were not the decisive factors in the July coup but they did contribute to the violent emotionalism that accompanied the revolution.

GOALS OF THE REVOLUTION

The leadership on the day of the revolution agreed on the following concrete political aims: (1) the overthrow of the Hashemite Monarchy and the Nuri regime and that clique of politicians, (2) the establishment of a republican form of government in which they

would hold the key positions, and (3) the realignment of foreign policies along neutralist lines and away from the pro-Western orientation
of the prerevolutionary regime. In addition, the leadership and their
followers looked to the revolution for the attainment of the following
social and economic goals: (1) a rise in industrial and agrarian productivity, together with fairer distribution of the social product; (2)
an expansion of educational, medical, and housing facilities; and (3)
agrarian and fiscal reforms. The rightist elements of the revolutionary
coalition, the *Istiqlalists*, were less interested in these longrange goals
than they were in the immediate political aims of the movement.

REVOLUTIONARY TECHNIQUES AND GOVERNMENT COUNTERMEASURES

In the months immediately preceding the coup d'etat there was a
noticeable increase in antigovernment propaganda and *Istiqlalist* activities. Since 1956, such activity had been normal in Iraq, but events in
Lebanon and Jordan following Nasser's annexation of Syria made the
situation in the Fertile Crescent more tense than at any time since Suez.
When Egypt and Syria formed the United Arab Republic (U.A.R.) in
February 1958, 6,000 Iraqis descended on the post office in Baghdad to send Nasser congratulatory messages and when the authorities
refused them, the mob delivered them to the Egyptian Ambassador in
person.[18] In April, Nuri attempted to strengthen his regime by joining
with the Hashemite Kingdom of Jordan to form the Arab Federation
and to act as a counterweight to Nasser's U.A.R. In June, Nuri urged
the West to send troops to Lebanon and Jordan to prevent their being
carried by the *Ba'thists* into the U.A.R. Hashim Jawad, Iraq's U.N. representative, dealt a blow to the prestige of the government when he
voted against his instructions and with the Nasser bloc on the question of Cypriot Greek self-determination. Against all signs of political
opposition to this government, Nuri used the usual, tried and true
methods—police arrests and censorship—while he called upon the
West to intervene in the area militarily. These final actions did not
serve to endear the Prime Minister to Iraqi nationalists.

Trusting the army to the end, Nuri ordered General Kassem,
Colonel Arif, and another commander to take their brigades (the
19th, 20th, and 8th) to Jordan to shore up that Kingdom's defenses.
When these troops, supposedly on the way to Jordan, reached Baghdad before dawn on July 14th, they surrounded the King's Palace,
the home of Nuri as-Said, and the radio station. The King and royal
family were assassinated at once, probably to prevent their escaping

and becoming a rallying point for opposition to the coup, as had happened in 1941.[19]

Nuri as-Said escaped but was hunted down and assassinated on the 16th. Other members of the old regime disappeared from public life. The British Embassy was sacked and partially burned, and violence continued throughout the day. Apparently, the revolutionary leadership intended violence to be directed only against the royal family and possibly against Nuri as-Said; however, the mobs, organized and directed by the Communists, went further than originally intended.

The coup had apparently been planned some months earlier, and the first opportunity to carry out these plans came on July 14. The movement of Iraq troops to Jordan was rumored to mean that the Nuri regime planned to attack the U.A.R. in support of the Western powers. This may have contributed to the fact that none of the troops opposed the coup. The 12,000 Iraqi troops already in Jordan returned on the 15th, not to restore the old regime and monarchy, as reported by the Jordanian radio, but to support the Kassem regime. The Communists were undoubtedly informed of the impending event a short time before, for early on the 14th "they were busy organizing crowds, painting slogans, distributing leaflets and interpreting the revolution in their own terms."[20] Though not among the leaders who actually instigated and organized the coup, the Communists were apparently used by the revolutionary group to organize the urban masses in support of the revolution.

MANNER IN WHICH CONTROL OF GOVERNMENT WAS TRANSFERRED TO REVOLUTIONARIES

The transfer of power was accomplished by violence. There was no attempt to maintain the appearance of a legal and orderly transition from the Hashemite Monarchy to the revolutionary republic. The Baghdad radio simply announced during the morning of the 14th that the Monarchy and government had been overthrown and a republic established by a military and civilian junta. Free elections and the restoration of constitutional government were promised.

THE EFFECTS OF THE REVOLUTION

CHANGES IN THE PERSONNEL AND INSTITUTIONS OF GOVERNMENT

In place of the Crown, the revolutionary group set up a three-man Council of Sovereignty, composed of Gen. Najib al-Rubai, Mohammed Mahdi Kubba, and Khalid Naqshababandi, and representatives of the three major national and religious subgroups of the country (Sunni, Shi'i, and Kurd). Real power was vested in the Council of Ministers, where General Kassem became Prime Minister, Minister of Defense and Interior, and Commander in Chief of the Army, with Colonel Arif second in command as Deputy Prime Minister. Other Cabinet posts went to members of the *Istiqlal, Ba'th,* and National Democratic Parties; the Communists, though active in organizing street demonstrations and mob violence in support of the revolution, were not offered an official place in the revolutionary government. Institutional changes that followed in the wake of the coup included the withdrawal of Iraq from the Arab Federation with Jordan, the abolition of tribal judicial systems, and the coordination of the Development Board under closer Cabinet control.

MAJOR POLICY CHANGES

Foreign policy under the new regime differed from that of the old regime in that it was no longer pro-Western and at least temporarily not anti-Nasser. One of the surprises of the revolution, however, was the moderation and the "embarrassingly proper"[21] attitude towards Iraq's prerevolutionary commitments to the West which characterized the Kassem regime in its first few days. Later, the Baghdad Pact was declared ineffective though not immediately renounced. Good relations with all nations, including the Western states, based on mutual interest and not on any form of dependence, was the avowed policy—indicated in an interview by Kassem on July 22—of the new Government. Jordan was assured of Iraq's military assistance in the event of attack from Israel, although Nuri's federal union between Jordan and Iraq, established a few months earlier, was dissolved. Diplomatic relations with Red China, the U.S.S.R., and other Soviet bloc states were entered into, although not with East Germany out of deference to the Bonn Government. Relations with the U.A.R. remained cordial for the first few months, although in August Kassem was observing that any union with the U.A.R. would necessarily be slow.[22] In the winter of 1958–59, the old antagonism between Egypt and Iraq returned and the air waves were once more full of bitter denunciations. The Communists in Iraq took the lead in denouncing Nasser, his *Ba'thist* sympathizers, and Colonel Arif; Arif and *Ba'th* had disappeared from the revolutionary coalition by the end of the year.

LONG RANGE SOCIAL AND ECONOMIC EFFECTS

In the early days of the revolutionary regime, rents were reduced by 10 to 20 percent, and in September limitations on land tenure were enacted. These were not as drastic as the Egyptian limitations, but Iraq has more land relative to population; in any event, extensive lands had come into the possession of the new regime through confiscation of royal and royalist holdings and this was available for distribution to the landless *fellahin*. Social services have been extended (including a system of marriage loans) and less money was spent on long range development schemes than before the revolution. Early in the new regime assurance was given to the Iraq Petroleum Company that nationalizations were not intended, despite goadings from both Nasser and the Communists. Negotiations have gone on, but no serious threat of nationalization has yet arisen.

OTHER EFFECTS

The Kassem regime, unlike the Nasser-Naguib government, has made no attempt to institutionalize into a progovernment party, and thereby control, the revolutionary enthusiasm created by the revolution. This may be due to Kassem's distrust of political party activity, or to the inability of the civilian politicians who make up the majority of the new regime's Cabinet to agree on permanent political goals. The result has been only a loosely organized National Liberation Front, involving the leaders of the various parties but not the mass following. In the first year of the new regime, this decentralization of the revolutionary movement, and the freedom of political organization under the regime, was exploited by *Istiqlalists* and *Ba'thists*, and also by the Communists who gained control over mobs of peasants and urban workers and threatened the government. The *Ba'th* was put down first and later the *Istiqlal*, and by the first anniversary of the coup the Communists, at least for the time being, had retired from the field. In 1960, after a year of self-imposed dissolution, the National Democratic Party returned as an organized group in support of the Kassem regime.[23]

NOTES

1. Majid Khadduri, *Independent Iraq: A Study in Iraqi Politics Since 1932* (London: Oxford University Press, 1951), pp. 215–216.

2. George L. Harris, *Iraq: Its People, Its Society, Its Culture* (New Haven: HRAF Press, 1958), p. 121.

3. George Grassmuck, "The Electoral Process In Iraq, 1952–1958," *Middle East Journal*, XIV, 4 (Autumn 1960), 397–415.

4. F. R. C. Bagley, "Iraq's Revolution," *International Affairs* (Toronto), XIV, 4, 284.

5. *Economist* (London), July 17, 1958, p. 184.

6. Benjamin Shwadran, "The Power Struggle in Iraq," *Middle East Affairs*, XI, 5 (May 1960), 150–161.

7. Fahim I. Qubain, *The Reconstruction of Iraq: 1950–1957* (New York: Praeger, 1958), VIII, pp. 245–262.

8. Ray Alan, "Iraq after the Coup," *Commentary*, XXVI, 1 (July 1958), 198–199.

9. Edward Atiyah, "Die Mittelost Krise in Sommer 1958," *Europa Archiv*, XIII (20 Nov.–5 Dec.), 11193.

10. Alan, "Iraq after the Coup," pp. 195–196, 198–199.

11. "A Year of Republican Iraq," *The World Today*, XV, 7 (July 1959), 287.

12. Walter Z. Laqueur, *Communism and Nationalism in the Middle East* (New York: Praeger, 1956), p. 336.

13. Bagley, "Iraq's Revolution," p. 288.

14. Ibid., pp. 283–285.

15. Robert F. Zeidner, "Kurdish Nationalism and the New Iraq Government," *Middle East Affairs*, X, 1 (January 1959), 24–31; see also C. J. Edmonds, "The Kurds and the Revolution in Iraq," *Middle East Journal*, XIII, 1 (Winter 1959), 1–10.

16. Atiyah, "Die Mittelost Krise," p. 11191.

17. Alan, "Iraq after the Coup," p. 194.

18. Atiyah, "Die Mittelost Krise," p. 11191.

19. Alan, "Iraq after the Coup," p. 195.

20. "A Year of Republican Iraq," *World Today*, p. 292.

21. Ibid., p. 428.

22. "Chronology, July 1–September 15, 1958," *Middle East Journal*, XII, 4 (Autumn 1958), 426.

23. "A Year of Republican Iraq," *World Today*, pp. 288–290; see also Walter Z. Laqueur, "The Iraqi Cockpit," *The New Leader*, XXXIII, 5 (February 1, 1960), 7–8.

RECOMMENDED READING

BOOKS:

Harris, George L. *Iraq; Its People, Its Society, Its Culture.* New Haven: HRAF Press, 1958. A comprehensive and detailed study of contemporary Iraq.

Khadduri, Majid. *Independent Iraq; A Study in Iraqi Politics Since 1932.* London: Oxford University Press, 1951. A standard work on Iraq's political history down to 1950.

Laqueur, Walter Z. *Communism and Nationalism in the Middle East.* New York: Praeger, 1956. A comprehensive and scholarly work dealing with ideological forces in that area. Contains several chapters specifically devoted to Iraqi developments down to 1955. Laquer is an American specialist on Soviet relations with the Middle East and Africa.

Qubain, Fahim I. *The Reconstruction of Iraq: 1950–1957.* New York: Praeger, 1958. A detailed analysis of the prerevolutionary regime's program of economic development written on the eve of the revolution.

PERIODICALS:

Alan, Ray. "Iraq After the Coup," *Commentary,* XXVI, 1 (July 1958). A detailed discussion of conditions and events leading up to the revolution, written soon after the coup by a political commentator and Middle East specialist for the American Jewish Congress' organ *Commentary.*

Bagley, F. R. C. "Iraq's Revolution," *International Affairs,* XIV, 4. A survey of events and developments during the revolutionary regime's first year in power, by a British professor of Middle Eastern Affairs at Durham University.

Edmonds, C. J. "The Kurds and the Revolution in Iraq," *Middle East Journal,* XIII, 1 (Winter 1959). An analysis of Kurdish attitudes towards the Iraqi revolution.

Grassmuck, George. "The Electoral Process in Iraq, 1952–1958," *Middle East Journal,* XIV, 4 (Autumn 1960). A detailed study of the prerevolutionary regime's limited experiments with political freedom during this period.

Habermann, Stanley J. "The Iraq Development Board: Administration and Program," *Middle East Journal,* IX, 2 (Spring 1955). A general survey of the prerevolutionary regime's economic development program.

Laqueur, Walter Z., "The Iraqi Cockpit," *The New Leader*, XXXIII, 5 (February 1, 1960). A brief discussion of relations between the revolutionary regime and its Communist and non-Communist supporters since the coup.

Shwadran, Benjamin. "The Power Struggle in Iraq," *Middle East Affairs*, XI, 5 (May 1960). A discussion of the revolutionary regime's economic reforms and its attitudes toward Iraqi Communists.

Zeidner, Robert F. "Kurdism Nationalism and the New Iraq Government," *Middle East Affairs*, X, 1 (January 1959). A discussion of relations between the revolutionary government and Iraq's largest national minority groups.

THE SUDAN COUP OF 1958

SYNOPSIS

On 17 November 1958, a military junta composed of senior officers and headed by the Commander in Chief of the Army, Gen. Ibrahim Abboud, overthrew the Sudan's parliamentary government of Prime Minister Abdullah Khalil. It was a nonviolent revolution, in which the members of the prerevolutionary regime rather willingly acquiesced, out of a desire to head off a more radical solution to the social, economic, and political impasse in which the country found itself. It was a conservative move designed to prevent a radical, Nasser-type revolution; it was directed against a British-modeled parliamentary system of government which had broken down under the social tensions and economic difficulties confronting the country. The Sudanese revolution has been called a "sham" revolution by some Western observers, and by some Sudanese dissatisfied with the military regime's conservative orientation; however, the coup is important as a specific type of conservative reaction to a revolutionary situation.

BRIEF HISTORY OF EVENTS LEADING UP TO AND CULMINATING IN REVOLUTION

Between the 1820's and the 1880's the Sudan, then little more than a vague geographic concept, was administered by the Egyptians. When, in 1882, the British Empire engulfed Egypt, the Sudan passed under British control, although to accommodate Egyptian claims to the country the area was administered as a "condominium" and was called the Anglo-Egyptian Sudan. Egyptian cultural influences remained strong in the Sudan; however, opposing these influences was a strong sentiment of local Sudanese patriotism, which found expression in the various mystical sects of the Islamic religion which flourished throughout the area. It was an account of a violent Sudanese uprising against the Egyptians, led by one of these religious mystics, called the *Mahdi*, that the British came into the Sudan in the 1880's. After violently suppressing the *Mahdist* rebellion, the British supported the followers of the *Mahdi*, whose Ansar sect has ever since been regarded as pro-British, against its rival sects which tended more towards assimilation with Egypt. The history of the Sudan has thus revolved around the rivalry of religious sects identified with British and Egyptian influences.

The British, who dominated the condominium, advanced the interests of the Sudanese natives at the expense of Egyptian interests. Britain sent some of its best talent in colonial administration to the Sudan, and it has generally been acknowledged that the Sudan was one of the best-administered colonial territories in the British Empire. Against these positive features in the Sudan's colonial background there are such negative factors as the striking lack of cultural homogeneity and the general socio-economic primitiveness on the part of most of its population. The Sudan emerged in 1952–53 from the old Anglo-Egyptian condominium as a semisovereign state with local autonomy, and in 1956 it gained full sovereignty and complete independence from both Britain and Egypt. There was little national consciousness to provide the necessary substance to the form of political independence. Ironically, the relative ease with which independence had been achieved, and the well-administered paternalistic system which the country had lived under, may have been contributing factors to the difficulties which the Sudan encountered in its postindependence period. For instance, there was in the Sudan, unlike most newly independent states, no nationalist political organization and no nationally-known leader of the independence movement.

THE ENVIRONMENT OF THE REVOLUTION

DESCRIPTION OF COUNTRY

Physical characteristics

Situated in northeastern Africa, the Sudan is a bridge between the Middle East and Africa. It is the world's ninth largest country in area and is nearly one-third the size of the continental United States. The most important geographical feature in the Sudan is the Nile River, which flows through the Sudan for more than two thousand miles of its course; the life of the country is concentrated along the banks of the Nile's two branches. To the north and west are sparsely populated desert areas; in the south are dense forests and tropical swamplands typical of central Africa, and along the Red Sea coast is a rugged mountainous region.

The people

According to the official census of 1956, the population of the Sudan was then above 10 million. Around half of the population were Arabic-speaking Muslims who lived in the northern and central regions of the country. Ethnically akin to the Egyptians and racially a

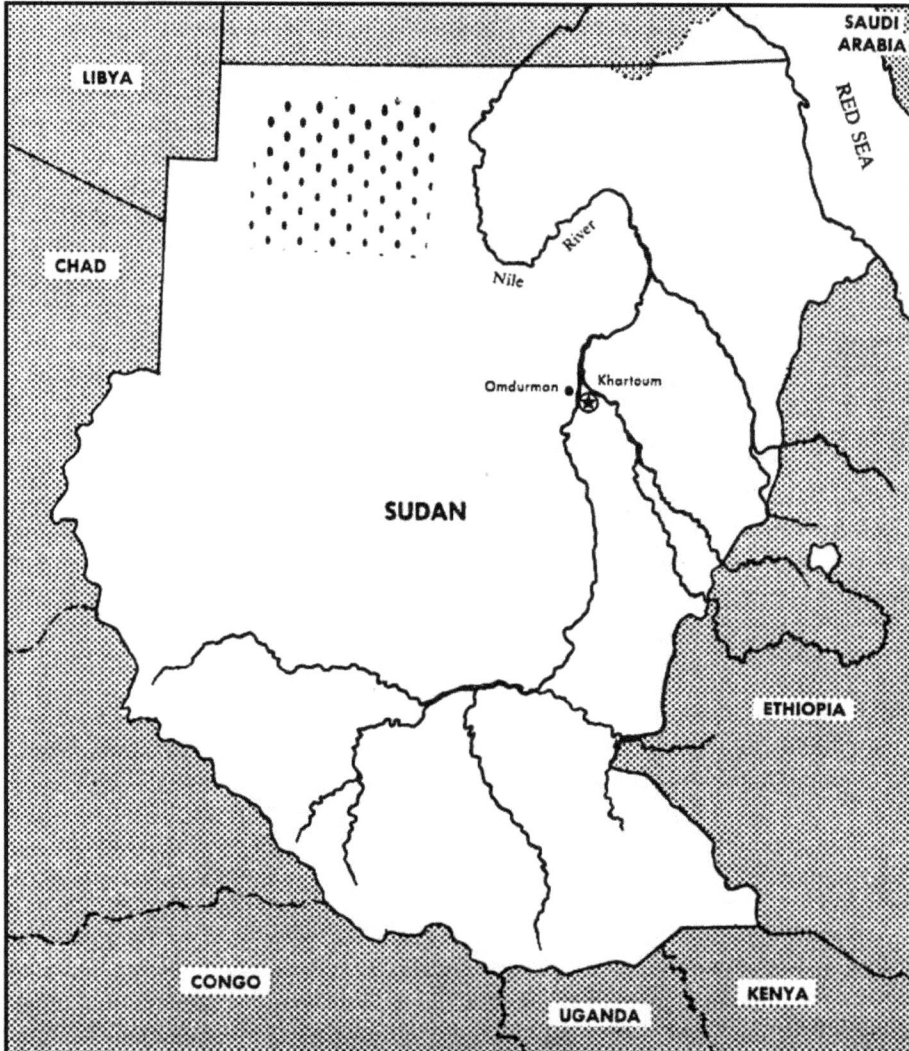

mixture of Mediterranean and Negroid types, these Arabic-speaking Sudanese are still the dominant group in the Sudan. Moving from north to south in the Sudan the population becomes more Negroid and less Mediterranean. The remaining half of the population are seminomadic tribesmen living either in the southern region or in the mountains along the Red Sea. Racially akin to the Negroid tribes of central Africa, the southern tribes are in various stages of primitive social development and speak hundreds of separate dialects. Their participation in Sudanese developments has been minimal.

The capital city of Khartoum, situated at the confluence of the White and Blue Niles, is the political, cultural, and commercial hub of the Sudan. In 1956 over a quarter million Sudanese were living in the urban complex of Khartoum, Khartoum North, and Omdurman, the populous suburb on the west bank of the Nile. The vast majority of the population were either living in rural villages or in a seminomadic state.[1]

Communications

In 1958 the Sudan had almost three thousand miles of railroad, located for the most part along the Nile and the Blue Nile; this line is connected with the Red Sea port of Port Sudan. Port Sudan, the country's only major seaport, has been described as a "modern harbor with excellent berthing, fueling, and handling facilities."[2] River transport is naturally a vital component of the country's internal communication system. In 1958 highways supplied the most numerous lines of communication; in most cases these roads are only graded earth tracks. Both international and domestic air transport had been developed in the Sudan, and there was in 1958 "a domestic state-owned air line which [provided] reasonable coverage for passengers, mails, and freight for the principal centres in the country. The postal and telegraph network [was] widespread and efficient."[3]

Natural resources

The Sudan's foremost natural resource is the country's fertile soil along the valley of the Nile and its branches and tributaries. No important mineral deposits have been found and the country is without a fuel supply adequate for the needs of modern industry. Sudan's waterways reveal very little hydroelectric potential, although they have considerable potential for irrigation purposes and have already been put to good use in that connection.[4]

SOCIO-ECONOMIC STRUCTURE

Economic system

The Sudanese economy is based almost entirely on cotton. Cotton, in 1958, accounted for about 70 percent of the country's foreign exchange earnings and supplied the government with more than half its revenues. A State Department survey on the Sudan reported in 1958:

> Owing largely to its administrative heritage from the condominium, the Sudanese Government dominates the economy. It is the largest employer of labor in the country. It owns and operates some of the principal enterprises such as the railroads, communications systems, and other public utilities . . . about half the output [of cotton] is produced by Government-controlled

cooperatives, and the Government markets about 75% of the crop. The Sudan is self-sufficient in most basic food-stuffs.[5]

Industrial development in the Sudan has been extremely limited and consists principally of industries processing agricultural products—cotton weaving, tanneries, soap manufacture, and vegetable oil presses.

The most striking economic enterprise in the Sudan is the government-controlled agricultural cooperative located in the triangle of land between the White and Blue Niles. The project was started around the beginning of the century as a joint enterprise among the British Government, two private British firms, and native tenant farmers. In 1950, the private firms were replaced by an official Sudanese Board and, in 1956, the British administration by the Sudanese Government. The government provided land and specified its most economic and efficient uses. The Board administered the project and promoted social services among the tenant farmers. All grain and fodder grown according to the government's specifications belonged exclusively to the tenant farmer. The Sudan derived around 40 percent of its total export revenue from this project.[6]

Class structure

The Sudan represents a prime example of a poorly integrated society, characterized by differences between the Arab Muslims in the north and Negroid pagans of the southern region, conflicts between the Islamic sects within the Arab core, and clan loyalties within these religious sects.

The Negroid tribesmen of the southern region live in a panoply of separate societies, and even members of the dominant Arab core of the population are often more conscious of status within their local clans and religious sects than on the national Sudanese level. Thus, the country's social elite consists of both the various tribal and sectarian elite and the new Westernized administrative elite who held positions in the government and the army. There is no large landowning elite in the Sudan as had generally been the case in the Arab world. Small landowners are numerous, and the tenants on the government-controlled lands of the Gezira project enjoy "the highest standard of living of any peasant group in Africa or Asia."[7] There were also the beginnings of an urban proletariat, especially around Khartoum.

Social mobility in modern Sudan has been associated with the rapid development of the country's cotton-dominated agricultural economy under the British after World War I and with the "Sudanization" of the civil service and the army after 1924, when the British dismissed large numbers of Egyptian military and civil administrators. Sudanization at the expense of the Egyptians continued down to the 1950's, after which time many positions held by the British were made available to qualified Sudanese applicants. Thus, for a variety of reasons, a relatively high degree of social mobility prevailed throughout the British colonial period.[8] Apparently, this condition continued into the period under consideration in this study.

Literacy and education

It has been estimated that a mere 6 percent of the entire population were literate at the time of the revolution;[9] however, literacy standards were higher in the Khartoum province, where, according to the 1956 census, half of the male population claimed to have attended school.[10] The independent Sudanese Government embarked on an ambitious program to expand greatly the educational facilities established under the British. In 1957 the government nationalized British and American missionary schools in the southern region in a move designed more to strengthen national unity than to discriminate against the Christian religious training provided by these institutions.[11] The Sudan's leading institution of higher education is the University of Khartoum, developed in 1956 from the old Gordon Memorial College. The educated, elite cadres are almost all graduates of this institution, which is held in high esteem by both the government and the opposition. The first Sudanese political party was fittingly named the "Graduates Congress."

Major religions and religious institutions

Islam is the dominant religion in the Sudan, although the three million Negroid tribesmen of the southern provinces are predominantly pagan. There are only a few Christians and Muslims in the entire southern region.[12] The Islamic community is divided into many rival factions or sects. The two leading sects, each with a following of more than three million, are the Ansar sect and the Khatmiya sect.

GOVERNMENT AND THE RULING ELITE

Description of form of government

The government before the coup d'etat operated under a transitional constitution adopted in 1956. This instrument provided for a parliamentary system of government, with a prime minister and cabinet responsible to the popularly elected lower house of the national legislature. There was a Supreme Commission, composed of five persons selected by Parliament, which served as nominal head of state, with supreme command of the armed forces and many of the constitutional functions of the British Crown. The Supreme Commission was also responsible for the conduct of national elections, and it appointed 20 of the 50 members of the upper house. Legislation required the approval of both legislative bodies. The Prime Minister's Cabinet consisted of from 10 to 15 ministers and of these at least 2 were required by the constitution to be from the southern region. The judiciary comprised two divisions, the civil and the *Sharia* court systems, and was, according to the Sudanese Constitution, a separate and third branch of government. The legal procedure for changing the executive and legislative branches followed closely the British pattern.[13]

Description of political process

Political parties and power groups supporting government

The prerevolutionary government of Prime Minister Abdullah Khalil was a coalition government headed by the dominant *Umma* Party, which had won 63 of the 173 seats in the national legislature in the February 1958 elections, and participated in by the minority People's Democratic Party (PDP), and an unstable bloc of Southern Liberals representing the non-Muslim southern constituencies. The term "liberal" in this case carries no ideological significance. The distinguishing characteristic of the Sudanese Southern Liberals is their advocacy of a decentralized federal system of government which would allow the non-Muslim southern region greater political autonomy. The *Umma* Party, founded in 1945 to campaign for complete Sudanese independence without any links with Egypt, was associated with the nationalistic Ansar sect and its venerable religious leader, Sayed Abe-al-Rahman al-Mahdi. Because of the *Umma's* emphasis on political gradualism and its pro-British rather than pro-Egyptian foreign policy orientation, the *Umma* Party appeared conservative in the Sudanese political spectrum.

The People's Democratic Party (PDP) was formed in July 1956 by elements of the rival Khatmiya sect who broke away from the National Unionist Party (NUP). The Khatmiyas, more urbanized than the Ansars and heavily concentrated in the east of the Sudan, were pro-Egyptian; they opposed the Ansars both on religious grounds and out of traditional rivalry. They favored a multiple, commission-type, instead of a single, head of state, since they feared that the position might be captured by the leader of the numerically superior Ansar sect. The PDP, led by Ali Abdul Rahman, also favored closer ties with Egypt and a "neutralist," anti-Western foreign policy. The unifying bond between the *Umma* and the PDP, both political manifestations of traditionally antagonistic religious sects, was their common fear of the secular-oriented and urban-based National Unionist Party. The Southern Liberal bloc was loosely organized and was never a dependable supporter of the Khalil regime; they favored a federal system in which the southern region would enjoy separate and equal political status with the Muslim north.

Character of opposition to government and its policies

Opposition to the Khalil government came from three general sources: dissident elements of the PDP within the governing coalition; the parliamentary opposition politicians of the National Unionist Party (NUP); and extraparliamentary subversive elements such as the Nasserists and the Communists. Of these three general areas of discontent, only the first—the PDP—played a decisive role in the November coup. The Khatmiya-dominated PDP feared a coalition between the Ansar-dominated *Umma* and the parliamentary opposition group, the urban-based and secular-oriented NUP. Such a coalition was indeed being considered in some quarters of the *Umma* governing party as a means of broadening the base of the regime in the face of rising opposition from urban, secular-minded elements among whom the Communists and Nasserists were increasingly active in the summer of 1958. This was the year of Iraq's Nasserist revolution and Western intervention in the Lebanon crisis; public opinion throughout the Middle East was running strongly in favor of Nasserism and violently against identification with the Western powers.

The other two sources of opposition which, taking advantage of prevailing popular discontent, might have joined forces to stage a Nasserist revolution, were the NUP and the Communists. The Communist Party was officially banned; however, it had always operated more through front organizations—particularly through the Sudan Workers' Trade Union Federation—than through a regular political party structure. Nasserist influence was strongest in the NUP, and it remains unclear why this group, the only real political opposition group in the country, acquiesced in the conservative revolution staged by the PDP and elements of the *Umma*. Apparently, the NUP decided its influence would be greater in a military regime than it had been in a parliamentary regime; at least one of the army officers prominent in the coup represented NUP political sentiments. The NUP may have believed that subsequent military coups would increase their influence over the regime.

Sudan had inherited from its colonial days a relatively free and articulate press. In the months preceding the November coup the

Sudanese press accurately mirrored the sharp decline in the regime's popularity. The independent Khartoum daily *al-Ayam* expressed the general tenor of public opinion among politically conscious Sudanese when it wrote, a few days before the coup: "Nobody will be sorry to see the present government go. It has been the worst government the Sudan has ever witnessed. It has given the country instability, misrule, disunity, and economic disaster . . . But what is the alternative? . . . . Some people are in favor of a coalition between *Umma* and NUP, others support a coalition of PDP, NUP, and Southern Liberals. But none of these will give the Sudan the stability it aspires for. . . ."[14]

Legal procedure for amending constitution or changing government institutions

The Sudan's transitional constitution contained no amending provisions, since a special committee was in the process of drafting a permanent organ of government for the Sudan. The activities of this constitutional committee were suspended, along with the parliamentary system of government, by the military junta which came to power in November 1958.

Relationship to foreign powers

The Khalil government of the Sudan was pro-Western in foreign policy orientation, although the Sudan was not formally aligned with any Western power by treaty obligations or political ties of any kind. The Sudan was a member of the Arab League, and the Khalil government gave lip service to the neutralist slogans currently popular in the Middle East. In June 1958, the regime agreed to accept a U.S. loan of $30 million, but only after a long and acrimonious debate in which Khalil was accused by members of his own coalition of delivering the Sudan back into the hands of the colonialists. Radio Cario repeated these accusations and even went so far as to compare Khalil to Iraq's pro-Western Nuri as-Said, the most hated man in the Arab nationalists' vocabulary. Opponents of the Khalil government looked to Egypt as the leading exponent of the type of vociferous anti-Western foreign policy orientation which was noticeably absent from the Sudanese Government's foreign policy. Relations between the Khalil government and Egypt grew worse during the events of 1958. Sudanese and Egyptian national interests conflicted over distribution of the waters of the Nile, marketing of cotton and livestock, and disputed territories along the Red Sea; however, the Egyptian High Dam at Aswan, as Cairo announced in October 1958, would back water over parts of the Sudan. Egypt has traditionally looked upon the Sudan as an extension of Upper Egypt and has regarded Sudanese nationalism as a British creation.[15]

The role of military and police powers

The Sudanese Army evolved from the Sudan Defense Force, a colonial constabulary which the British organized in 1925 from purely Sudanese units of the Egyptian Army evacuated that year from the condominium as a result of political conflicts between the British and the Egyptians. Prior to the November coup, the Sudanese Army was regarded as a thoroughly nonpolitical force, though it was generally recognized that the ties of individual officers and men under their command to one of the two dominant religious sects could bring the army into a political contest involving sectarian interests. A contributing factor in the army's intervention in the political affairs of the country was the high degree of public pride in the army and the high morale among Sudanese military men. In February of 1958 the army had demonstrated its usefulness to the country, when Sudanese troops were dispatched to the disputed Halaib area on the Egyptian border. It has also been noted that the Sudanese troops, who fought well with the British during World War II, have never been on a losing side.[16] The police were not involved in politics and played no significant role aside from normal police duties.

WEAKNESSES OF THE SOCIO-ECONOMIC-POLITICAL STRUCTURE OF THE PREREVOLUTIONARY REGIME

History of revolutions or governmental instabilities

A brief but violent mutiny occurred in 1955, on the eve of the Sudan's complete independence, when noncommissioned officers of the Latuka tribe rebelled against their Northern Muslim officers. The mutiny quickly spread to other southern Negroid contingents of the army, which at that time was organized along tribal lines and was under northern Muslim officers. After the government had put down the mutiny, the army was reorganized on a nationally integrated basis and some southerners were commissioned as officers; however, antagonisms and distrusts continued to separate northern and southern Sudanese.

During the summer and fall of 1958 Sudanese political parties became increasingly irresponsible. The parliamentary opposition group, led by ex-Premier Ismail al-Azhari as leader of the Nationalist Unionist Party (NUP), disagreed with the government on virtually every issue, making constructive legislation an impossibility. "We are the opposition," declared Al-Azhari, "and the Opposition must oppose everything the government does."[17] Members of the PDP, itself a participant in the governing coalition, joined the NUP in opposing almost every government measure. Opposition to Khalil's government centered primarily around his identification with the West. Paralyzed by a three-way split between the uncompromising forces of the *Umma*, the PDP, and the NUP, the parliamentary system had broken down completely and become publicly discredited by November of 1958.

Economic weaknesses

The Sudan was experiencing severe economic difficulties by the fall of 1958. As a result of the overpricing of Sudanese cotton during the Suez Crisis of 1956, at a time when the U.S. Department of Agriculture was making available on the world market large quantities of American cotton, much of Sudan's cotton crop for the years 1956–58 remained unsold. Egypt's unofficial boycott of Sudanese livestock further aggravated the economic crisis facing the Sudan. Sudanese gold and foreign currency reserves had shrunk from $178 million in 1956 to $80 million in May of 1958. This shortage of reserves compelled the Khalil government to seek foreign loans.[18]

Social tensions

In the face of economic difficulties which were beginning to be felt by the urban population and with the ruling political elite bitterly divided among itself, the Sudan's social tensions had developed to the point where a violent Nasserist revolution might have occurred. There was growing antagonism between urban-middle and lower-middle-class elements and the rural-based traditional elite who monopolized Sudanese politics. The election to Parliament of 45 members of the urban-based and secularly oriented National Unionist Party demonstrated the emerging political strength of the detribalized nonsectarian urban population of the country. But the NUP still did not have enough parliamentary strength vis-a-vis the sectarian bloc to exercise an influence in the political affairs of the country commensurate with the political aspirations of the urban population, which was becoming increasingly alienated from the governing sectarian bloc. The Sudanese labor force was more politically conscious and better organized than that of most other Middle Eastern countries; both Communist and Nasserist influences were reported to be widespread among the Sudanese working class.[19] Added to the social tensions in the cities were the traditional hostilities between Arab Muslims of the north and Negroid tribes in the south, and the antagonisms between Ansar and Khatmiya Islamic sects.

Government recognition of the reaction to weaknesses

The Khalil government was under no illusions as to the impact of the Sudan's economic difficulties and social tensions on the country's political system. One of the advantages of a relatively free press is, as British experience has shown, that the government is better able to assess its public support. The ruling political elite realized that the parliamentary system did not command sufficient respect among politically conscious Sudanese to survive a continuation of the type of political stalemate and interparty feuding which had gone on since the election in the spring of 1958. They also realized that the most likely outcome in such a situation was a radical, Nasserist revolution, led by some popular junior officer in the army and supported by the detribalized and nonsectarian urban elements of the population.

Correctly or not, the Sudan's ruling elite considered the social and political situation in the country to be ripe for revolution. To the ruling group, it was only a matter of choice between a radical revolution, which would cost most of them their lives and political fortunes, and a mild revolution, which would permit the ruling elite to continue its former influence over the government under a different guise and would incorporate some of the revolutionary demands of the discontented urban elements. Thus, there were in reality two revolutionary sources in the Sudan: one characterized as radical and Nasserist which never developed into an overt revolutionary movement, and the other, a conservative revolution designed to ward off the radical revolution before it occurred by replacing a vulnerable parliamentary regime with a strong military regime, which would neutralize revolutionary ferment and preserve the *status quo.*

The leaders of the two rival religious sects and their political arms, the *Umma* and the PDP, reacted by staging a sham revolution against their own creation, the Khalil government. One knowledgeable observer of Sudanese politics has written, "When confronted with the alternative of cooperating in a military truce or continuing the roller coaster ride toward an unknown and possibly radical end, both the 'ins' and the 'outs' acquiesced in the plan to set parliamentary government aside in the interests of establishing order and insulating the country against a non-Sudanese solution of its coming-of-age problems."[20]

514

FORM AND CHARACTERISTICS OF REVOLUTION

ACTORS IN THE REVOLUTION

The revolutionary leadership

It has been generally concluded that the behind-the-scenes insti-gators and organizers of the Sudanese coup were the two politico-religious leaders who headed the Ansar and Khatmiya Islamic sects (Sayed Abe al-Rahman al-Mahdi and Sayd Ali Mirghani) and certain *Umma* and PDP political leaders associated with these sects. Their political and ideological orientation was conservative and traditional-ist, and their revolutionary goal was to anticipate and ward off a radi-cal social upheaval in the country. The overt leaders of the revolution were Gen. Ibrahim Abboud, Commander in Chief of the Sudanese Army, Maj. Gen. Ahmad Abd al-Wahab, Deputy Chief of Staff of the Army, Gen. Hassan Beshir Nasr, and other generals of the Sudanese Army High Command.

The key figure in the coup, General Abboud, was a member of the Khatmiya sect and, though not active politically, was inclined toward the PDP. Born in 1900 in Suakin, a small town on the Red Sea, Abboud studied engineering at the Gordon Memorial College and was graduated from what later became the Sudanese Military College in Khartoum. During World War II, he served with distinction in the Ethiopian-Eritrean and Libyan campaigns; after the war he rose rap-idly to become a full general and Commander in Chief of the Sudan's Army by 1956. In 1955, Abboud headed an arms-purchasing mission to Britain and, in 1957, toured Europe and Asia on a similar mission. General Abboud has been described as having a "pleasant, gregari-ous manner, but also a reputation for toughness in a military situa-tion that had made him a respected but not a beloved general."[21] *The New York Times* assessed Abboud's political philosophy soon after the November coup as "likely to be west of neutral and right of center."[22]

Other members of what has sometimes been called "the revolu-tionary triumvirate" were Generals Wahab and Nasr. Wahab, born in 1915 and graduated from the Sudanese Military College in 1935 at the top of his class, had an impressive army career, serving in World War II campaigns and taking charge of the suppression of a mutiny in the southern region in 1955. Wahab was a son-in-law of Prime Minis-ter Khalil, a member of the Ansar sect, and was associated, though not actively, with the *Umma* party. General Nasr, reportedly sympathetic to the NUP, was the triumvirate's most outspoken advocate of closer relations with Egypt. Thus, the personnel of the military junta which served as the overt leadership of the revolution represented the three contending factions in the country, but with the greater emphasis on conservative and traditionalist interests.[23]

The revolutionary following

There was never a mass revolutionary following, either before or after the coup d'etat. The population remained passive throughout the transfer of power. It may be assumed that those whose interests were closely identified with the conservative, traditionalist leaders responsible for the coup were its most enthusiastic supporters; however, there were no public manifestations either supporting or opposing the coup.

ORGANIZATION OF REVOLUTIONARY EFFORT

Internal organization

The smooth conduct of the revolutionary effort bespoke effective prior planning; the organization of the revolution was accomplished through informal meetings among the participants prior to November 17. Nothing is generally known of the details of these conspiratorial meetings.

External organization

There has never been any suggestion of foreign involvement in the November coup d'etat. Foreign influences were very definitely present in the general revolutionary environment in which the November coup occurred. In the months preceding the coup the Egyptian radio denounced the Khalil regime as a "Western stooge" and a "traitor to Arab nationalism." In October the leader of the PDP had met in Cairo with the leader of the NUP, former Prime Minister Ismail al-Azhari, who was then returning from a visit to the revolutionary regime in Baghdad. This meeting between the two most pro-Egyptian Sudanese political groups was viewed with considerable alarm among the *Umma* politicians and hastened their decision to replace the PDP with the NUP as the junior partner in their governing coalition.[24]

GOALS OF THE REVOLUTION

Concrete political aims of revolutionary leaders

The primary political aim of the revolution was the suspension of political activity and the parliamentary system of government and their replacement by a nonpolitical, military-dominated junta.

Social and economic goals of leadership and following

There was no significant socio-economic content in the revolutionary goals of the Sudanese military junta, despite General Abboud's vague references to national prosperity and welfare in his first proclamation to the nation. Abboud declared: "In changing the prevailing state of affairs, we are not after personal gain nor are we motivated by any hatred or malice towards anyone. Our aim is the establishment of stability, prosperity, and welfare of this country and its people."[25]

He listed as faults of the previous regime, which the army intended to correct, such things as "bitter political strife, instability, failure to safeguard independence, and misuse of the national resources."[26] Essentially the goals were conservative and aimed at the preservation of the *status quo.*

REVOLUTIONARY TECHNIQUES AND GOVERNMENT COUNTERMEASURES

Methods for weakening existing authority and countermeasures by government

Since this was a coup staged by one part of the ruling elite against another part of itself, there were no government countermeasures, and the only revolutionary technique employed was the use of the army to make what was actually a sham revolution appear to be a genuine use of force to overthrow the legal government.

Methods for gaining support and countermeasures taken by government

There were no appeals for mass support, and the principals among the ruling political elite were in basic agreement concerning the military coup. The nature and extent of Prime Minister Khalil's participation in the planning of the revolution is not clear. He may not have had advance notice of the specific details of the coup, for example; however, he had recently expressed a desire to retire from the political scene, and it appears likely that the instigators of the revolution informed Khalil that provision would be made for his retirement with a pension after the military came to power. If such was the case, this constituted the revolution's only technique to gain support.

MANNER IN WHICH CONTROL OF GOVERNMENT WAS TRANSFERRED TO REVOLUTIONARIES

On November 17, 1958, in a swift and bloodless action before dawn, the army took control of all public buildings and communication centers in Khartoum. General Abboud broadcast a proclamation to the nation announcing that a military junta headed by himself had seized control of the Government, suspending the Constitution, dissolving Parliament, and dismissing the Khalil government. The parliamentary system came to an end without a shot being fired; there were no street demonstrations.

THE EFFECTS OF THE REVOLUTION

CHANGES IN THE PERSONNEL AND INSTITUTIONS OF GOVERNMENT

The 13-man military junta formed itself into a Supreme Military Council, headed by Abboud as president of the council, which governed the country through a Council of Ministers, headed by Abboud as Prime Minister and composed of senior officers in the junta and in Sudanese political parties. The only member of the Khalil cabinet who remained in the government after the revolution was the Minister of Education and Justice, Ziyadeh 'Uthman Arbab, who held the same post in the postrevolutionary Council of Ministers. He was an active *Umma* politician and member of Parliament.[27] While parliamentary politicians were removed from the reins of government and were replaced largely by the military, civil servants and nonpolitical administrators continued to run the country as before.

MAJOR POLICY CHANGES

The revolutionary regime continued Sudan's "west of neutral" policy orientation; however, it had to withdraw from the exposed and too openly pro-Western position taken up by the previous regime. In April 1959, at the Accra conference in Ghana, the military regime's Foreign Minister enunciated the broad outlines of the Sudan's foreign policy: neutrality in the East-West conflict of interests; nonalignment with any of the Arab blocs; support of African independence movements; avoidance of military pacts except for defense of the Sudan against overt aggression; and acceptance of foreign economic assistance which does not compromise Sudanese independence and sovereignty.[28] The military regime accepted a longstanding Soviet offer to barter timber, cement, and light machinery for Sudanese cotton and other agricultural goods, valued at about $7 million in 1959 and about $8 million in 1960.[29]

Sudan's relations with Egypt improved considerably after the military coup. General Abboud was no more pro-Egyptian than Prime Minister Khalil had been; however, while Khalil had needed the dispute with Egypt to maintain his shaky political coalition, Abboud was free to negotiate with Cairo without worrying about parliamentary support. Nasser enjoyed a similar freedom. In the year following the coup most of the major disagreements with Egypt were settled through negotiations.

LONG RANGE SOCIAL AND ECONOMIC EFFECTS

While the Sudanese revolution changed the faces of officeholders from civilian politicians to army generals, it did not challenge the basic internal or external policies of the previous regime, and it did not, at least directly, change the political and social fabric of the country, which had been described as one "in which family and sectarian loyalties tend to carry more weight than party, ideological or national considerations."[30] It is likely, however, that the longrun effect of the military coup will be to weaken the power of traditionalists and conservatives. Military personnel in the Middle East have invariably been less tradition-oriented and socially conservative than the political leaders of the region.

OTHER EFFECTS

A significant result of the Sudanese coup d'etat has been the military's increasingly strong hold over the reins of political power in the country. Having come to power more or less at the behest of two of the three rival political factions, the military regime has been reluctant to "return to the barracks" and restore parliamentary government as the politicians intended them to do after a time. The Abboud regime put down four attempted coups during 1959, led by dissident Army officers, most of them junior officers, who accused the regime of being too similar to the previous Khalil regime; the Supreme Military Council dealt with these disturbances by reorganizing itself to exclude General Wahab, Khalil's *Umma*-Ansar son-in-law, and to include some of the dissident officers.

By 1961, there was increasing opposition from the ousted political leaders, who now were demanding a return to parliamentary government. In June 1961, a railway workers' strike coincided with the politicians' appeal for the reconvening of Parliament. The Abboud regime replied by arresting the Sudan's only two ex-Premiers, NUP leader Ismail al-Azhari and *Umma* leader Abdullah Khalil, on July 11. The only remaining center of opposition to the regime is Sayed Abe-al-Rahman al-Mahdi, politico-religious leader of the Ansar sect, whom the Abboud government has not dared to arrest for fear of the effect this would have on the many Ansar adherents in the army. "The crucial political question still unanswered," Sudan expert Helen Kitchen writes, "is whether the many Ansars in the army would remain loyal if the regime were to come into open conflict with *Mahdi*."[31]

Having once experienced the wider participation in government that the parliamentary system permitted, politically conscious Sudanese are not likely to remain content with their present military dictatorship, even though it may provide more efficient and more socially progressive government than did the previous regime. The military regime has recently taken steps to accommodate these political aspirations with the goals of the regime by establishing institutions of local government on a limited scale.

NOTES

1. Helen Kitchen, "This is the Sudan," *Africa Special Report*, IV, 1 (January 1959), 2, 15; see also "The Sudan," *The Middle East*, 1958 (6th Edition), 324.
2. "The Sudan," p. 320.
3. Ibid.
4. Ibid.
5. U.S. Department of State, *The Sudan: Middle East Bridge to Africa*, (Publication 6572, Near and Middle Eastern Series 28; released January, 1958), p. 13.
6. Ibid., p. 14.
7. Helen Kitchen, "The Gezira Scheme—A Man-Made Miracle," *Africa Special Report*, IV, 1 (January 1959), 13.
8. Leo Silberman, "Democracy in the Sudan," *Parliamentary Affairs*, XII, 3 and 4 (Summer and Autumn 1959), 363–364.
9. Kitchen, "This is the Sudan," p. 15.

10. Karol J. Krotki, *Twenty-One Facts About the Sudanese* (Khartoum: Republic of Sudan, Ministry for Social Affairs, Population Census Office, 1958), p. 29.

11. Helen Kitchen, "Uniform Education: Key to Unity?" *Africa Special Report*, IV, 1 (January 1959), 8.

12. Kitchen, "This is the Sudan," p. 15.

13. "The Sudan," p. 327.

14. Helen Kitchen, "The Government of General Abboud," *Africa Special Report*, IV, 1 (January 1959), 3.

15. John S. Badeau, "The Sudan: Reasons for Military Coup," *Foreign Policy Bulletin*, XXXVIII (March 15, 1959), 97.

16. "The Sudan Today," *The Islamic Review*, (January 1960), 19–25, passim; see also Helen Kitchen, "The Army," *African Special Report*, IV, 1 (January 1959), 5.

17. Kitchen, "The Government of General Abboud," p. 3.

18. Helen Kitchen, "Crisis in the Economy," *Africa Special Report*, IV, 1 (January 1959), 9.

19. Peter Kilner, "A Year of Army Rule in the Sudan," *World Today*, XV, 11 (November 1959), 436.

20. Helen Kitchen, "The Sudan in Transition," *Current History*, XXXVII (July 1959), 25.

21. Helen Kitchen, "Sudanese Personalities," *Africa Special Report*, IV, 1 (January 1959), 11.

22. Ibid.

23. Ibid.

24. Peter Kilner, "A Year of Army Rule," p. 433.

25. Ibid., p. 431.

26. Ibid.

27. Kitchen, "Sudanese Personalities," p. 11.

28. Helen Kitchen, "Foreign Policy of the Sudan," *Africa Special Report*, IV, 1 (January 1959), 4.

29. "Economic Notes," *Africa Special Report*, V, 6 (June 1960), 11.

30. Kitchen, "The Sudan in Transition," p. 35.

31. "Opposition Leaders Arrested in Sudan," *Africa Special Report*, VI, 8 (August 1961), 10.

RECOMMENDED READING

BOOKS:

Arkell, A. J. *A History of the Sudan*. London: Athlane Press, 1955. A reliable and well-written history of the Sudan.

Fisher, Sydney Nettleton. *The Middle East*. New York: Knopf, 1959. A standard work on the history of the area. Contains a brief historical survey of political developments in the Sudan immediately prior to the revolution, pp. 638–642.

Laqueur, Walter Z. *Communism and Nationalism in the Middle East*. New York: Praeger, 1956. Contains a short account of Communist activity in the Sudan prior to 1955, pp. 63–69.

Shibeika, Mekki. *The Independent Sudan*. New York: Robert Speller & Sons, 1959. The most recent history of the Sudan. Primary emphasis on the colonial background of independent Sudan.

PERIODICALS:

Kitchen, Helen (ed.). *Africa Special Report*, IV, 1 (January 1959). Entire issue devoted to the Sudan and the events surrounding the November coup d'etat. Most of material written by the editor, Mrs. Helen Kitchen, who is an expert on the Sudan.

Badeau, John S. "The Sudan: Reasons for Military Coup," *Foreign Policy Bulletin*, XXXVIII (March 15, 1959). An analytical discussion of some of the factors leading up to the revolution by a leading Middle East scholar.

Kilner, Peter. "A Year of Army Rule in the Sudan," *World Today*, XV, 11 (November 1959), 430–441. *A* review of Sudanese developments in 1959 and some discussion of factors leading to the revolution.

Rawlings, E. H. "The Sudan Today," *Contemporary Review* (March 1960), 159–162. A survey of postrevolutionary developments in the Sudan with special emphasis on political opposition to the military regime.

Silberman, Leo. "Democracy in the Sudan," *Parliamentary Affairs*, (Summer and Autumn 1959), 349–376. A discussion of the 1958 election in terms of the subsequent breakdown of the Sudanese parliamentary system.

SECTION VI

························

FAR EAST

General Discussion of Area and Revolutionary Developments
The Korean Revolution of 1960
The Chinese Communist Revolution: 1927–1949

TABLE OF CONTENTS

GENERAL DISCUSSION OF AREA AND REVOLUTIONARY DEVELOPMENTS

THE KOREAN REVOLUTION OF 1960

THE CHINESE COMMUNIST REVOLUTION: 1927–1949

TABLE OF CONTENTS

FAR EAST

GENERAL DISCUSSION OF AREA AND REVOLUTIONARY DEVELOPMENTS

GEOGRAPHICAL DEFINITION

Generally speaking, the Far East is the east central portion of the Asian continent. As used here, the Far East consists of: Mainland China, including Manchuria, Inner Mongolia, and Sinkiang; Tibet; Outer Mongolia (Mongolian People's Republic); Formosa (Taiwan); Hong Kong and Macao the Korean peninsula; and Japan. Close to one billion people occupy the coastal areas and alluvial plains on the Far East. China, the largest country, claims control over 650 million of these people, and its population continues to grow at approximately 15 million a year. Japan's population has also multiplied rapidly, but it has a very limited land area to absorb the growth: here there are over 93 million people occupying an area smaller than California. Birth control has recently been effectively advocated by the government.

Much of the Far East is composed of vast mountain ranges and barren desert plains, such as the Himalayas, the Greater Khingans, the Altais, and the Gobi and Takla Makan deserts. The climate varies from north to south and from mountains to plains. Tropical monsoons in South China result in much rainfall and moderate temperatures. North China and Manchuria have less rainfall and a greater temperature range.

HISTORICAL BACKGROUND

CHINA

From the second half of the 17th century until the second decade of the 20th, China was ruled by the Manchus, who established the Ch'ing dynasty. Under Manchu rule, China experienced almost 150 years of peace and economic prosperity. By the beginning of the 19th century, however, the Ch'ing dynasty had lost its vigor and China showed signs of decay in every field. One important factor in the growing economic difficulties was a phenomenal increase in population. Toward the end of the 17th century there were about 100 million Chinese; by the middle of the 19th century the population exceeded

400 million. Landholdings decreased in size to an average of less than half an acre per capita and unemployment became widespread.

Simultaneously with the decline of the Ch'ing dynasty, Western powers expanded their influence. The Chinese, at first, rejected Western overtures for trade and economic concessions because they considered the foreigners uninvited guests and ignorant barbarians. Europeans, on the other hand, regarded the Chinese as backward and decadent. The Western powers, particularly Great Britain, were able to impose economic treaties and obtain concessions from the Chinese by virtue of their superior technology and military strength. A major war broke out in 1839 between Great Britain and China after the Chinese Government imposed severe restrictions on the opium trade conducted by Great Britain. The Chinese were defeated and forced to acquiesce to British demands. "The pattern by which foreign powers kept China subjugated had been set: each time the Chinese stirred to resist foreign pressure, their inevitable defeat by overwhelming military force was used by the foreigners to demand more concessions."[1] The "treaty ports" along the coast became more and more dominated by the West and China ceased to be completely sovereign.

In addition, the countries on the fringe of China which had acknowledged Chinese suzerainty were taken over by European powers. Russia advanced into Mongolia, Manchuria, and Turkestan; France, after a short war with China, established her rule in Indochina; Great Britain took over Burma; and Japan took possession of Korea and Taiwan, also defeating China in war.

China was impoverished from within and humiliated from without; the Manchu government was impotent to solve her problems. Unrest, riots, banditry, and rebellion were the natural consequence. In 1911, the Ch'ing dynasty was overthrown and replaced by a republic, and China entered into a transitional stage when factional power struggles and revolutionary efforts went side by side to formulate a new political pattern. Thus, 2,000 years of monarchial rule came to an end. After the establishment of the Republic, Outer Mongolia became "independent" through Russian manipulations. Although Chinese suzerainty was still recognized by Russia and Outer Mongolia, the latter became, in effect, a Russian protectorate during the remainder of the tsarist period. Similarly, Tibet became a British protectorate, while China, nominally, retained her suzerain status.

JAPAN

The history of Japan during the same period is in sharp contrast. While China desperately but unsuccessfully clung to her old ways, Japan decided to adopt Western methods and Western technology. Within a short time Japan became the leading Asian nation and a contender for world power. Having defeated China in 1895, Japan challenged Russia in Manchuria in the beginning of the 20th century. War ensued in 1904–5, and Japan emerged victorious. Her gain in prestige was enormous and far outweighed the territorial concessions she secured. As a result of World War I, Japan was able to consolidate her position further by taking over Germany's former spheres of influence, over Chinese protests, despite the fact that China too had declared war on Germany.

In the postwar period, Japan expanded into Manchuria, where she established the puppet state of Manchukuo in 1932. After several lesser "incidents," a full-scale war with China began in 1937 and did not end until 1945. Japan succeeded in occupying the entire coastal area and taking over 95 percent of Chinese industries. Meanwhile, Japan had challenged the Western powers also, and had made clear her intention to evict them from China. The outbreak of World War II prevented Japan from defeating and occupying China completely, but her conquests in Indochina, the Philippines, Burma, and Indonesia effectively isolated China from her Western allies.

* * * * * * *

Two Western powers have retained a precarious hold on port cities in southeast China; Great Britain in the Crown Colony of Hong Kong and Portugal in Macao. Both places were occupied by Japan during World War II, but were returned to British and Portuguese authorities after the War.

REVOLUTIONARY DEVELOPMENTS

CHINA

According to the "theory of the Heavenly Mandate," a Chinese ruler may be faced with a rebellious force which could end his rule if he fails to follow the heavenly patent drawn for him by the Chinese deity. By the same token, the failure of an attempt to end his rule by a rebellious force is proof that his rule is approved by heaven. Thus,

China's long history of rebellions, some successful but most unsuccessful, is justified by this "unwritten Chinese constitution."[2a]

According to one scholar of Chinese history, China's traditional rebellions fall into two classes: "great peasant uprisings, often associated with religious movements, and the insurrections of powerful generals . . ."[3] Most of the peasant uprisings were defeated,[b] but "In each case the weakened dynasty a few years later succumbed to some military adventurer who had risen either in the ranks of rebellion or in the armies raised to suppress the rebels."[4]

There was widespread revolutionary ferment in China throughout the Manchu rule. The factors contributing to it were many. The Manchus were foreigners who ruled over a large and populous country. They found in Neo-Confucianism a philosophy which preached the "right of the superior to rule and the duty of the subjects to obey and be content . . ."[5] Civil service examinations favored the Manchus and official appointments discriminated against the men from southern China. Intellectual life became stagnant. Thus, unemployment and discontent were not confined to peasants and poorer people. "Men of ability, finding the way to government service blocked, became, in resentment, the leaders of the popular movement against the dynasty."[6] An ideological justification for a revolutionary spirit was found in the older form of Confucianism. The three main causes of the unrest that prevailed in the 19th century and the revolution that overthrew the dynasty in 1911 appear to have been: (1) widespread unemployment and poverty, caused by a tremendous increase in the population and unrelieved by governmental measures; (2) the dissatisfaction of many scholars and intellectuals with Manchu rule; and (3) resentment and hatred of the Westerners who had humiliated China and did not recognize the right of China to be master in its own house.

The disintegration of the Ch'ing Empire was accompanied by insurrections which, beginning early in the 19th century, broke out from Taiwan to Turkestan. Secret societies gained ground, and "embattled farmers, driven by social inequities to take up arms, formed organizations animated by nationalism and religious ideologies. By the middle of the century, anti-Manchu insurrections reached a floodtide . . . ."[7]

Only one rebellion, the Taiping Rebellion between 1850 and 1864, reached really large proportions. It was based on an ill-assorted blend

---

 One study concludes that the "basic justification for the ruler's power manifests itself only through the acceptance of a ruler by his people; if the people kill or depose him it is clear that he has lost heaven's support."[3]

 Some peasant uprisings were successful. The most striking examples are those which led to the establishment of the Han dynasty in the third century B.C. and the Ming dynasty in the 14th century A.D.

of misunderstood Christianity and native Chinese beliefs. The revolutionary force was composed of thousands of Chinese peasants and supported by Protestant missionaries. Its short-lived success was due to the social reforms which the revolutionaries instituted; these gave way to widespread terror as the leaders lost their reforming zeal. The revolutionary leadership had no unified command and the revolt was crushed mainly by the Han-Chinese forces of the Hunnanese army and the Huai Valley army, with some support from British and American officers.

Two other rebellions in the same period resulted from discriminatory practices against Chinese Muslims in western China. The Yunnan uprising lasted for 18 years, from 1855 to 1873; its defeat is attributed to the large Muslim faction which sold its services to the Manchus. The Sinkiang revolt beginning in 1862 was successful in establishing a Muslim state and conquering areas in Kansu, Shensi, and Hopei provinces. It was defeated by a strong Chinese Army which drove the Muslim tribes out of China proper and regained all of Sinkiang except Ili Province, a large portion of which was returned to China by Russian occupation troops in 1881.[c]

The Boxer Movement, concentrated in northern China, led to the Boxer Rebellion of 1899–1901;[d] it was an expression of popular hostility toward the Western powers and the Christian religion. The movement was strengthened by resentment over foreign demands for more economic concessions and the military and moral defeat China suffered at the hands of the Japanese in 1895. Secret societies were organized and supported by reactionary Chinese officials. These engaged in widespread terrorists acts against Chinese Christians and foreigners alike. The British intervened by sending naval detachments against the Boxers. In retaliation, the Chinese attacked the foreign legations in Peking, and the Empress Dowager declared war on all foreigners in China. An Allied expeditionary force brought Boxerism to an end.

By the beginning of the 20th century, more and more Chinese had come to the conclusion that the only hope of improving condi-

---

[c] However, the defeat of the Muslims in Sinkiang did not bring an end to Muslim unrest in China. A similar revolt took place in 1932 when Muslim elements in Sinkiang again nearly succeeded in breaking away from China. Fearing that Muslim independence in Sinkiang would have serious repercussions in Soviet Central Asia, Soviet troops suppressed the revolt and established a Soviet protectorate but left the administration in Chinese hands. This state of affairs lasted until 1942, when Soviet troops were withdrawn and China reestablished her authority.[8]

[d] The Boxer Movement was not really revolutionary. Its violence was directed against foreigners rather than against the regime. Boxer hostility to aggressive foreign powers was encouraged by the antiforeign Manchu clique, at court, which hoped to use them as auxiliaries to expel all foreigners from China. The Boxer Rebellion is an example of the unrest that prevailed during this time.

tions lay in outright revolution and elimination of the Ch'ing dynasty. The leader of the revolutionary movement was Sun Yat-sen, who later organized the Kuomintang, China's first political party. The revolutionaries were greatly aided by the unrest that prevailed throughout the country. The nationalist Kuomintang had succeeded in infiltrating the army and in obtaining strong support in south and central China and particularly among Chinese overseas who had absorbed Western ideas. A general uprising, led by the military, was touched off by a rather insignificant bombing incident in 1911 and dealt the Ch'ing dynasty its coup de grace. During the final days of the dynasty, the government called on Yuan Shih-k'ai to negotiate with the revolutionaries, who had set up a republican government with Sun Yat-sen as President. As a result of these negotiations, the Ch'ing dynasty abdicated, and, to maintain national unity, Sun resigned the Presidency in favor of Yuan. Yuan's own ambitions made him a virtual dictator thereafter until his death in 1916.

The overthrow of the dynasty did not lead to the establishment of orderly government. Monarchists attempted to reestablish the Manchu dynasty, constitutionalists wanted to create a constitutional empire, republicans tried to set up parliamentary government, and self-styled war lords struggled to retain control over their own little empires in the various provinces.

This again was in accord with tradition: in the past, fallen dynasties had not been replaced immediately by new dynasties. In another particular history repeated itself. The founders of enduring regimes had been careful to relieve some of the peasant distress and to enlist the backing of the scholar class. The Manchus fell because they had alienated the scholars and the peasants. Nationalist rule was to fail eventually for the same reason.[9]

During World War I and immediately afterward, Western ideas continued to penetrate China. With them came some industrialization, European liberalism, and Marxism. Sun Yat-sen again became the dominant personality. His idea was to rebuild China with Western help, but the Western powers seemed interested only in keeping their colonial advantages and regaining their economic concessions. Sun thereupon turned to Russia. The new Soviet Union, having just experienced a revolution and a civil war, was unable to help materially, but was able and willing to treat the Chinese as equals and to send them advisers. Thus, communism began to develop in China shortly after World War I while the national revolution was still in progress. A bitter struggle began between the Communist-oriented forces and the other Nationalist forces which kept China in a constant state of turmoil until the victory of the Communists on the mainland in 1949.[c]

OTHER AREAS

Tibet

Attention has been drawn to Tibet again recently by increased resistance among the local hierarchy and some sections of the population to Chinese domination. Armed uprisings were reported throughout 1956, 1957, and 1958, but all were suppressed by the Chinese Communists. The Chinese retreated temporarily in 1957 after a number of reversals, but returned 6 months later in an attempt to establish absolute control. The offensive provoked a general Tibetan uprising in March 1959. Red Chinese military superiority quickly crushed the rebellious Tibetans. An estimated 10,000 rebels withdrew to the remote valleys and mountains in east and southeast Tibet.

The revolutionary force in the mountains has split into several factions. The Khambas, a Tibetan ethnic mountain tribe, are not interested in political intrigues to the same extent as the Tibetans, and only demand some representation in a new government should the Chinese Communists be defeated.[10] The Khambas themselves are split into two major groups, a fact which undoubtedly weakens their position.

Korea

The principle of self-determination popularized during World War I aroused enthusiasm for democracy in Korea and stimulated a serious revolt against Japanese domination in 1919. Nonviolent mass demonstrations throughout Korea caught the Japanese by surprise. They were quelled by rather brutal police methods, but some concessions to self-rule resulted. A "Provisional Government of Korea" was established in Shanghai; it appealed unsuccessfully to the powers at the Paris Peace Conference.

Japanese authority over Korea was terminated by the Second World War. In accordance with Allied agreements, northern Korea was occupied by Russian troops and southern Korea by American troops pending the establishment of a national government. The division was perpetuated, however, by inability to agree on the makeup of that government, and the 38th parallel became the fronter between rival regimes. In South Korea a pro-Western government was headed by Syngman Rhee; in North Korea a Communist regime established

---

[c] For details of the Chinese Communist Revolution see p. 571.

itself under the tutelage of Communist China. In 1950 North Korea forces struck southward in an attempt to take over the entire country; driven back by U.N. troops—chiefly American—they were reinforced by powerful Chinese armies. The truce signed in 1953 left the border much the same as before.

In 1960 a revolution toppled the Rhee government and replaced it with a new government which brought power to different people of the same traditional ruling elite.[f] A military coup d'etat followed in May 1961. The military junta which has ruled since the coup is definitely rightwing and while governing by decree, has promised to restore democratic government sometime in the future.

Japan

Japan's expansionist program in the 20th century increased greatly the importance of a military elite which was taking an active part in the political process. In February 1936 a military Fascist organization formed by a group of young officers joined with other extremist elements in the army in an attempt to overthrow the government and abolish the party system. Before dawn on February 26, groups of 30 soldiers each quietly moved into Tokyo public buildings, assassinated political dignitaries and took over the buildings. The next day, the government declared martial law and brought in troops of unimpeachable loyalty to suppress the rebellion. Three days later the insurgents capitulated. The coup d'etat technically had failed. However, the army's prestige, rather than decreasing, increased to a point where extremist military leaders became dominant in Japan. A military totalitarianism was established and brought Japan into war against the Western Allies in 1941.

RESULTS AND OUTLOOK

The outlook for political stability in the Far East and Southeast Asia will depend largely on developments within Communist China and its attitude toward the rest of the world. Moreover, the situation will continue to be affected by the conflicts between the Nationalists and the Communists. The Nationalists are threatening to renew the civil war by invading the mainland, simultaneously with a hoped-for internal revolution against the Communists, while the Communists have announced their intention of incorporating Taiwan into the Chinese People's Republic. Each of the adversaries is being restrained by its major ally: the Communists by the Soviet Union and the Nation-

---

[f] See summary of Korean Coup of 1960 on p. 551.

alists by the United States. Because of the Communist-Nationalist rivalry and the impact of this rivalry on other countries, Communist China and the Nationalists on Taiwan will be discussed rather fully in the following section. Should Red China succeed in eliminating the Nationalists, other Far East countries would have great difficulty in withstanding the internal and external Communist pressure.

CHINA

International policies

As a bloc country

The "monolithic structure" of the Communist world has been threatened by recent ideological differences between China and the Soviet Union. The differences arise mainly from the varying interpretations of the revolutionary rules laid down by Marx and Lenin. China's special problems have prompted Communist leaders to promote a more "Trotskyite" concept of "permanent revolution" and to interpret it as constituting a mandate to work actively for the "liberation movements" in Afro-Asian countries. China believes that Communist countries should not only furnish the ideas behind revolutions, but should instigate revolutions and furnish the arms and supplies with which to conduct them. The Soviet Union is equally interested in promoting national liberation movements, but advocates a "gradualist" strategy of revolution and holds to the belief that communism will inevitably triumph without risking the high cost of a thermonuclear war. Contradictions between the Soviet Union and Communist China as of 1961 are probably more in the nature of an intraparty doctrinal dispute than an international power struggle. The gradualist strategy appears to have been adopted in some areas,[11] and in these, the Soviet Union may maintain its leading position as an exporter of revolutionary ideology.

Sino-Soviet differences on revolutionary ideology may continue for a long time; but mutual recognition that China and the Soviet Union can differ only so far without harming bloc relations may tend to preserve their close ties. There are four factors which tend to preserve the alliances: (1) both countries share a belief in the historical inevitability of their success; (2) they share similar aspirations and enemies; (3) they are striving to reconstruct their societies along similar patterns; and (4) they share similar operational concepts and organizational devices.[12]

As a model for revolution

Communist China has presented itself with some success to Afro-Asian and Latin American countries as a model for revolution. Although the basic social, political, and economic systems of the undeveloped areas do not necessarily duplicate conditions that previously existed in China, the Chinese Communists are exerting a strong influence on many of these countries.

As an Asian country

Communist China's status as an important power became evident during the U.N. negotiations to stop the Korean War. Small Asian countries cannot afford to ignore Communist China's position as a power in Asia. To some Asian countries China appears not only as a Communist state but also as a leader of Asian progress. Since 1954, China has announced its intention of following a "good neighbor" policy and has persuaded some countries of its sincerity. This is shown, for example, by India's response to some of China's actions. Chinese conquest of Tibet was accepted by India; and during the Tibetan revolution in 1959, India was determined not to intervene in the "internal affairs" of China, although Communist officials in Peking have made continued accusations that India was intervening.[13]

China concluded boundary treaties and treaties of friendship with Burma and Nepal to implement the theme of peaceful coexistence. Peking hoped that these friendly gestures would convince India of its sincerity in seeking a settlement of all differences. However, Chinese claims to border areas between China and India and Chinese occupation of some of the territory have, in 1960–61, given rise to heated disputes between the two governments. China's attitude toward other Asian countries varies considerably. China fears Japan's alliance with the United States and has constantly pressed the Japanese Government to take a neutralist position and increase its trade with China. The Japanese leftist opposition to the alliance, which has caused disturbances in Japan, is strongly supported by China. China has friendship and mutual assistance agreements with the government in North Korea and the Ho Chih Minh government in North Vietnam which are identical with treaties signed with the Soviet Union. There appears to be a tendency on the part of North Vietnam, however, to side with the Soviet Union in the "intraparty struggle."

The existence of a rival Chinese government on Taiwan (Formosa) presents Red China with a problem that has been difficult to resolve. The main issue is that the Nationalist Government on Taiwan and the Communist regime on the mainland both claim to be the legal government in China, and both agree that the offshore islands, as well as Taiwan, are an integral part of China. Taiwan's position has been strongly supported by the United States and a dwindling number of other nations. A number of newly emerging African nations have sided with neutralist and Communist bloc countries and have attacked Nationalist China's right to continued membership in the United Nations and a seat on the Security Council. Mao Tse-tung's ultimate goal is to seize the offshore islands, the Pescadores chain, and Taiwan.[14]

The Nationalist Government on Taiwan bases its hope of returning to power on the mainland on the belief that the Chinese people will eventually rise against the Communist regime in Peking and, with the help of Chiang Kai-shek's forces, will overthrow the so-called "People's Republic" and reestablish nationalist rule. Such a general uprising, to be successful, would require support by the peasants, leadership from the intellectuals, and massive defections of Communist military forces to the Nationalist cause. The question therefore arises whether Communist domestic policies are likely to alienate these three classes to such an extent that they would attempt to overthrow the regime.

The peasant class suffered its worst blow from the Communist government in 1958, when the "Great Leap Forward" was announced. This was a program of intensive industrialization which was to change the appearance of China within 3 years. The "Leap Forward" led to the introduction of communes—a system representing the most extreme application of Communist revolutionary theory in the rural areas of China.[g] Approximately 98 percent of the peasants were organized in these communes, which consolidated all local controls under a tightly-knit, semi-military authority.[16] The commune system on the whole destroyed the last vestiges of Chinese family traditions. Ancestral graves were plowed up; children were moved into government nurseries; old people were moved into government homes; men and women lived in barracks, ate in mess halls, and owned nothing; even the clothes on their backs were not their own. Peasants were forced to work long hours to increase productivity.[17]

Coupled with the natural calamities which caused crop failures in 1959 and 1960, communication had such a disastrous effect on agriculture that the regime had to halt and even reverse parts of the program. Since then there have been considerable variations in the implementation of the agricultural program in various areas. Yet even during the worst months of this crisis, in the winter and spring of 1961, there was no general revolt. Rumors circulating in Hong Kong

---

[g] From the beginning, plans were also made for urban communal units, although their establishment has been much slower and their scope narrower.[15]

of peasant riots and operations by guerrilla units in Kiangsi cannot be confirmed. Some uprisings do appear to have occurred since the imposition of the People's Communes, but only on a small scale. "Peasants have thrashed or killed their foremen on many occasions; they have attacked commune and party offices and sacked food stores. Workmen have refused to work and cadres have ignored government directives."[18] However, another source concludes, "All evidence so far indicates that the instruments of control—party, army, bureaucracy—are firmly in the hands of the regime."[19]

The regime has also made cautious overtures to China's intellectuals. In May 1956 the policy of the "100 flowers" was announced, permitting and even inviting criticism of Communist Party policies. Effective response by the intellectuals did not begin until after Mao's February 1957 speech "On the Correct Handling of Contradictions Among the People." Less than 4 months after the speech was made the policy was reversed, and all those who had ventured to criticize the party became "right deviationists" and enemies of the party. The Hungarian and Polish episodes in 1956 served to intensify the regime's advocacy of the "100 flowers" policy as a means to alleviate any unrest similar to that which took militant forms in those countries.[20] But when the antagonism of the intellectuals gained momentum, Peking decided to suppress the movement and launch a rectification drive.

In 1961 a decision to "encourage academies to go their several ways in research" was reported in the party's theoretical journal, *Red Flag*.[21] The regime appears desirous of giving the impression that intellectuals can have academic freedom even if they do not adhere to Marxist-Leninist theories, so long as they remain politically loyal. Qualitative increases in education are being emphasized and wide educational reforms are being talked about.

Thus, the Communist regime has attempted to make its authority more palatable by modifying some policies. The party has offered a wider range of incentives to the peasants, in the hope of increasing their initiative and, thereby, agricultural production. It has also resumed its attempt to woo the intellectuals by appearing to offer more academic freedom.

The strongest and perhaps the most organized opposition to the regime comes from the ethnic minorities. Tibetan opposition has been obvious, but there also has been some hostility shown by Uigurs, Muslims, and Mongols. This opposition, however, is traditional and is as much anti-Chinese as it is anti-Communist. "Go slow" policies have been adopted in certain areas of China to win over the minority groups. Minority opposition may be expensive for the Communist regime and may affect Chinese relations with other Asian countries, but the opposition does not pose a basic threat to the regime.[22]

If the regime maintains firm control through its police powers and remains willing to modify policies that provoke too great disaffection, then, in the words of one Asian expert, "any successful revolt against the Peking regime will probably come from its own ranks."[23] Factional splits have been reported between moderates and extremists in the Central Committee, but the Chinese Communist leadership has nonetheless been remarkably stable and secure. There is no immediate sign that Mao's death will bring on a power struggle.

TAIWAN

In 1949, Chiang Kai-shek transferred the seat of his government to Taiwan. Since that time, the Nationalists have been protected by the United States against any invasion attempt by the Communists. Chiang Kai-shek, supported by the United States, continues to claim that his is the legal government of China. There seems to be little if any chance that the Nationalists will realize their dream of once again ruling over all China. Yet this hope is necessary for their existence: it provides them with a *raison d'etre* and a basis for claiming international recognition as the legal government.

Even within Taiwan, however, the Nationalists are not secure. They are threatened from without by the Communists, and from within by dissatisfied Taiwanese elements. The fact that their army, once composed primarily of mainland Chinese, is forced to rely more and more on Taiwanese replacements also weakens their position. Chiang Kai-shek, too, has reached an age that makes him unlikely to remain a Nationalist symbol for many years longer.

Taiwan is an island inhabited by some 10 million people, most of them descendants of migrants from the coastal provinces of southeast China. Two million are Nationalist refugees from Communist China; there are a few indigenous Taiwanese.[h] Taiwan is essentially an agricultural country which produces a surplus of rice and sugar. Some industries were developed under Japanese rule and expanded by the Nationalists. At one time 60 percent of the industry was owned by the Nationalist Government, but the government turned over stock interests to landowners in return for their land, so that it could be

---

[h] The term "Taiwanese" refers to the people, Chinese and descendants of aborigines, who lived on Taiwan before the mass influx of the Nationalists. The term "Nationalists" as used here refers to those Chinese who have left the mainland as the result of the Communists' victory.

distributed to tenant and other farmers. Political life is controlled almost completely by the Chinese Nationalists, although indigenous elements have been granted some participation in the government. Taiwan has a constitution, but the "emergency power" provisions enable the Nationalists to rule legally in an authoritarian fashion. The Kuomintang controls the largest number of seats in the assembly and has organized a network of cells throughout the island with which it controls political activities. The party has no mass support. Two other smaller parties play only an insignificant role in political life.

Only a few Taiwanese have gained important posts with the government. The dominant position of the Nationalists has resulted in some dissatisfaction and frustration on the part of the Taiwanese elite groups, who feel that they have no proper outlet for their abilities and training.[24] Moreover, according to one Taiwanese intellectual, ordinary citizens "can be arrested at night by a squad of secret police, tried by a military court martial and sentenced, with little opportunity for appeal."[25] Thus, there is some feeling in Taiwan that the island is being ruled by unwanted intruders. However, any potential resistance is kept in check by the police. All revolutionary activities appear to be centered in Peking, Tokyo, and the United States. The Communist organization in Peking probably has little or no support on Taiwan itself.

There is no doubt that some Taiwanese feel that they should be independent of both Chinas. The Taiwanese Democratic Independence Party is the official organization that supports the spirit of resistance against the Nationalists. The Nationalists belittle its significance, but the party claims a membership of 3,000 in Tokyo alone.[26] Moreover, Taiwan is not without revolutionary traditions. The Taiwanese on several occasions have rebelled against both Chinese and Japanese rule. In the 18th and 19th centuries there were numerous rebellions against the Manchu Empire. During the 50 years of Japanese rule 15 uprisings occurred. In 1947, an uprising was staged against the Nationalist governor, who was accused of "maladministration and plunder." The uprising was put down ruthlessly after reinforcements arrived from the mainland, but Chiang changed the provincial governors in an attempt to appease the population.[27] In 1950, and apparently again in 1958, some army officers attempted coups d'etat against Chiang.[28] These attempts probably resulted from struggles for power among the Nationalists, and were not properly supported movements.

The factors which have made the Nationalists more palatable to the Taiwanese are mostly economic. A land reform has been instituted, irrigation and other improvements in food production have increased crops, credits have been made available to farmers, and the living standard in general has been improved. Much of this has been achieved through extensive U.S. economic aid.

Not all problems have been solved, however. The rate of unemployment is large, the island has to support a sizable bureaucracy and an army of 400,000 men, and the refugees from the mainland constitute a burden on the government. The Nationalists will continue to be dependent on the United States to defend themselves against threatened Communist attacks and to keep the economy and the military establishment running smoothly.

Soon after the Nationalists had established themselves on Taiwan, they began planning to return to the mainland. The hope was that an attack on the coastal province of Fukien would be followed by a general uprising against the Communist regime. However, a 1954 U.S. Government official report stated that "it would be unrealistic for even the most ardent supporters of the Chiang government to expect any successful all-out attack upon the mainland, at least in the near future."[29] Later the Nationalist Government altered its invasion plans and made them contingent upon a general uprising within China.[30] This plan presupposes that internal unrest and dissatisfaction on the mainland are approaching a point where a mass uprising is imminent. As discussed above, however, such an uprising in the near future does not appear very likely.

JAPAN

Japanese politics has been predominantly the politics of the powerful Liberal-Democratic Party supported by business and government officials. The intelligentsia and the masses of the working class form the opposition. Although the greater part of the opposition is left-oriented, it does not support the Japanese Communist Party. Nevertheless, the opposition sees in government policies a trend toward fascism, which it feels was clearly displayed, for instance, when the Mutual Security Pact with the United States, strongly supported by the government and strongly opposed by the leftist groups, was ratified in Parliament after lengthy debate. Despite evidence to the contrary the opposition claimed that the bill was "steamrollered" through Parliament. Although there is a cultural gap between the elite and the masses, it appears to be no cause for alarm. However, the political gap between the intelligentsia and the government is profound, and this is cause for unrest.

Tokyo was the scene of riots in May and June 1960 which symbolized a continuing struggle between rightists, and leftists,[i] who favor a neutralist position and demilitarization. Japanese officials declared the 1960 demonstrations the handiwork of international communism, and Communist participation in a united front council did make this organization appear Communist. However, it is still a matter of dispute as to what extent the demonstrations were actually Kremlin-directed.[31] Government officials, the ruling Liberal-Democratic Party, and business and industrial circles comprise the rightist group. The opposition is composed of trade unions, student associations, and other leftist organizations.

Mass organization began in Japan when the American occupation authorities abolished legal restrictions on labor unions. Labor unions grew rapidly, and as the rightwing organizations degenerated into company unions, the leftwing factions increased in strength. From 1951 on, they have openly opposed any Japanese security pact with the United States.

Also significant was the organization in 1947 of a teacher's union whose activities were strictly controlled by the government. Leftist tendencies in the union greatly increased when the government required that textbooks comply with government standards defined by the "Basic Law on Education." This union also opposed the security pact and influenced large numbers of postwar primary and junior high school students, who have now reached college age. Some of these students played an important role in the 1960 demonstrations.

* * * * * * *

For some time to come, events in the Far East can be expected to be largely influenced by the relative positions of power of Communist China and the Nationalists. Barring a major uprising on the mainland, for which there is little hope, the position of the Nationalists is bound to become more vulnerable. They still represent China in the United Nations, but it is doubtful how much longer their claim to that seat will be recognized. There is no doubt that the Communists exercise *de facto* control on the mainland. As long as the United States considers Taiwan a vital link in its defense system, a Communist invasion appears unlikely, since it would lead to war with the United States. Developments not only in Taiwan, but in other countries—Japan, Korea, and Southeast Asia—will depend on the attitudes and actions of the United States and Communist China. The mere exis-

---

[i] Two major sources of differences between right and left groups at that time were the fluctuating Japanese economy, and an international security system centered on a United States-Japanese pact.

tence of Taiwan as a Chinese anti-Communist base probably has considerable influence over the feelings and attitudes of the more than 650 million Chinese on the mainland. Thus, the United States is interested in supporting Taiwan not only for military reasons, but for psychological and political reasons as well. Revolutionary sentiment in the non-Communist countries of Southeast Asia is considerable[j] and the greater Communist China's success and prestige on the international scene, the more likely it is that other countries will succumb to Communist propaganda and subversion. There does not appear to be much pro-Communist revolutionary sentiment in Japan and South Korea, but in both countries a revolution that would upset the present equilibrium is not impossible. If this should happen, the Communists would be certain to gain.

NOTES

1. Hsiao Hsia (ed.), *China, Its People, Its Society, Its Culture* (New Haven: HRAF Press, 1960), pp. 25–26.

2. C. P. Fitzgerald, *Revolution In China* (London: The Cresset Press, 1952), p. 12.

3. Edwin O. Reischauer and John K. Fairbank, *East Asia: The Great Tradition* (Boston: Houghton Mifflin Company, 1960), p. 81.

4. Ibid., pp. 12–18.

5. Fitzgerald, *Revolution*, p. 12.

6. *A General Handbook of China* (New Haven: HRAF Subcontractor's Monograph, 1956), I, p. 49.

7. Ibid., pp. 49–50.

8. Ibid., p. 50.

9. G. F. Hudson, "The Nationalities of China," in *St. Anthony's Papers No. 7, Far Eastern Affairs: No. 2*, ed. G. F. Hudson. (London: Chatto and Windus, 1960), p. 58.

10. Fitzgerald, *Revolution*, pp. 14–15.

11. George N. Patterson, "The Situation in Tibet," *The China Quarterly*, 6 (April–June 1961), 86.

12. A. Kashin, "Behind the Scenes in Laos," *Bulletin*, VIII, 6 (June 1961), 27–35.

13. Zbigniew Brzezinski, "Pattern and Limits of the Sino-Soviet Dispute," *The New Leader*, XLIII (September 19, 1960), 8.

14. Hugh Seton-Watson, *Neither War Nor Peace* (New York: Frederick A. Praeger, 1960), p. 390.

---

[j] See section on Southeast Asia, p. 11.

15. Bruno Shaw, "The Quemoy Controversy," *The New Leader*, XLIII (October 31, 1960), 16–17.

16. *People's Communes in China* (Peking Foreign Languages Press, 1958), passim.

17. Central Intelligence Agency, *The Economy of Communist China: 1958–1962* (January 1960), p. 29.

18. *Hearings Before the Committee on Foreign Relations*, United States Senate, Eighty-Sixth Congress, First Session on S. 1451, Part I (Washington: United States Government Printing Office, 1959), p. 375.

19. Robert S. Elegant, "A Visitor in China," *The New Leader*, XLIV, 36 (October 30, 1961), 5.

20. H. F. Schurmann, "The Dragon Treads Lightly: Peking's New Line," *The Reporter*, XXV, 6 (October 12, 1961), 34.

21. Roderick MacFarquhar, *The Hundred Flowers Campaign and the Chinese Intellectuals* (New York: Frederick A. Praeger, 1960), pp. 169, 267–268.

22. "Quarterly Chronicle and Documentation," *The China Quarterly*, 6 (April–June 1961), 189.

23. Committee on Foreign Relations, *United States Foreign Policy, Asia* (Washington: Studies prepared by Conlon Associates, United States Government Printing Office, 1959), no. 5, p. 131.

24. Elegant, "Visitor," p. 5.

25. Committee on Foreign Relations, *Asia*, pp. 140–141.

26. Li Thian-hok, "The China Impasse: A Formosan View," *Foreign Affairs*, XXXVI, 3 (April 1958), 441.

27. Werner Levi, "Formosa and the 'China Issue'," *Current History*, XLI, 244 (December 1961), 323–324.

28. Li, "The China Impasse," pp. 445–446.

29. Ibid., p. 443.

30. Report of Senator H. Alexander Smith, Chairman, Subcommittee on the Far East, *The Far East and South Asia* (Washington: United States Government Printing Office, 1954), p. 8.

31. Interview of T. F. Tsiang, Permanent Representative of the Republic of China to the U.N., *Meet the Press*, N.B.C., V, 39, (October 8, 1961) (Washington: Merkle Press, Inc., 1961).

32. I. I. Morris, et al, "Japan Today," *The New Leader*, XLIII (November 28, 1960), Section 2 et passim.

RECOMMENDED READING

BOOKS:

Borton, Hugh. *Japan's Modern Century*. New York: The Ronald Press Company, 1955. A general review of modern Japan is imperative for a study of modern Asia. Mr. Borton's is a good study of the rise and fall of the Japanese empire.

Chanakya, Seni (ed.). *Tibet Disappears*. New York: Asia Publishing House, 1960. This is a documentary account of the Tibetan rebellion and its consequences.

Fitzgerald, C. P. *Revolution in China*. London: The Cresset Press, 1952. This volume is one of the better analyses of the Chinese revolution.

Reischauer, Edwin O., and John K. Fairbank. *East Asia: The Great Tradition*. Boston: Houghton Mifflin Company, 1960. This volume is the first of a proposed set of two, and does not treat the contemporary history of the area in question. However, the patterns of tradition carried into contemporary history are found here—administration, education, religion, etc.—as well as the environmental setting.

Seton-Watson, Hugh. *Neither War Nor Peace*. New York: Frederick A. Praeger, 1960. Although this volume covers the present conflicts of the world in general, the section on Asia merits close examination.

Tang, Peter S. H. *Communist China Today*. New York: Frederick A. Praeger, 1957. The author presents a very comprehensive study of the organizational structure of the government and party of Communist China. A revised edition will be available in the near future.

Thorp, Willard L. (ed.). *The United States and the Far East*. New York: The American Assembly Graduate School of Business, Columbia University, 1956. This book is a collection of background papers written for the major problems confronting the United States in the Far East.

Vinacke, Harold M. *A History of the Far East in Modern Times*. New York: Appleton-Century Crofts, Inc., 1959. This is one of the best general overviews of modern East Asia, well presented and well analyzed.

THE KOREAN REVOLUTION OF 1960

SYNOPSIS

On March 15, 1960, a presidential election was held in South Korea which resulted in the reelection of Syngman Rhee as President. Widespread charges of irregularities and resentment against the administration's treatment of the opposition during and before the elections led to popular demonstrations, mostly started by students in some provincial cities. Mass protests against Rhee's regime spread to Seoul, where, between April 16 and 26, up to 100,000 demonstrators demanded Rhee's resignation. The administration tried to quell the riots, first through police action, later by declaring martial law and bringing in troops. However, on April 27, Rhee resigned and turned the Government over to a caretaker Premier pending new elections.

BRIEF HISTORY OF EVENTS LEADING UP TO AND CULMINATING IN REVOLUTION

In 1910 Korea was annexed by the Japanese; it remained under Japanese domination until the end of World War II. At that time, as a result of previous agreements between the Allies, Korea was divided into two zones of occupation. Russian forces accepted the surrender of Japanese troops north of the 38th parallel and took control over that part of the country, while U.S. forces assumed responsibility over the southern half of Korea. As the relationship between the Soviet Union and the United States deteriorated, Korea became a battleground in the cold war.

A United Nations Commission was created in 1947 to supervise general elections in North and South Korea. The Commission arrived in January 1948 but found itself unable to carry out its task in the area under Soviet domination. However, elections were held south of the 38th parallel and resulted in the establishment of an independent Republic of [South] Korea (R.O.K.). Syngman Rhee became its first President and the United States turned over all administrative functions to the new Republic.

Syngman Rhee had been active for Korean independence for most of his life. While Korea was under Japanese occupation, he became President of a provisional government, set up in Shanghai in 1917.

After World War II most Koreans thought that their struggle for an independent country had come to an end, but instead the country was divided into two hostile camps. In June 1950, a Soviet-trained North Korean Army invaded South Korea. South Korea, had no force strong enough to repel the invaders, but the United Nations, benefited from the temporary absence of the Soviet delegate, intervened and initiated a "police action" under the command of the United States. Three years of war resulted in an armistice which acknowledged the military stalemate that developed after Communist China had intervened on the side of North Korea. The armistice simply reaffirmed the *status quo ante bellum*, except that the United States, acting for the United Nations, continued to maintain troops in South Korea.

The invasion from the north pointed up the danger of communism to South Korea. It was obvious that the Communists had received some support from within South Korea and Rhee was anxious to create a strong and unified anti-Communist country. To achieve this objective and stop Communist subversion, he resorted to tactics which were anything but "democratic." During the Korean War, democratic procedures had been set aside. However, after the war Rhee continued to rule in authoritarian fashion. The instrument of his control was the Liberal Party; he allowed some opposition to exist, but if the opposition dissented too strongly, its members were subjected to strongarm attacks, beatings, and even murder. Rhee was reelected in March 1960 for a fourth term, but the election was generally believed to have been corrupt and fraudulent. Moreover, in the months preceding the election, members of the opposition had not been allowed to campaign freely. The constitutional rights of freedom of assembly and of a free press were constantly violated by the administration. On April 27 Rhee was forced to resign in the face of tremendous popular pressure exerted by up to 100,000 demonstrators in Seoul alone.

THE ENVIRONMENT OF THE REVOLUTION

DESCRIPTION OF COUNTRY

Physical characteristics

Korea is a mountainous peninsula extending southward from Manchuria and Siberia for almost 600 miles.[1] Following the Korean War, the peninsula was divided at the 38th parallel, leaving to the Republic of Korea (R.O.K.) the southern 44 percent of the land area—approximately the size of Indiana.[2]

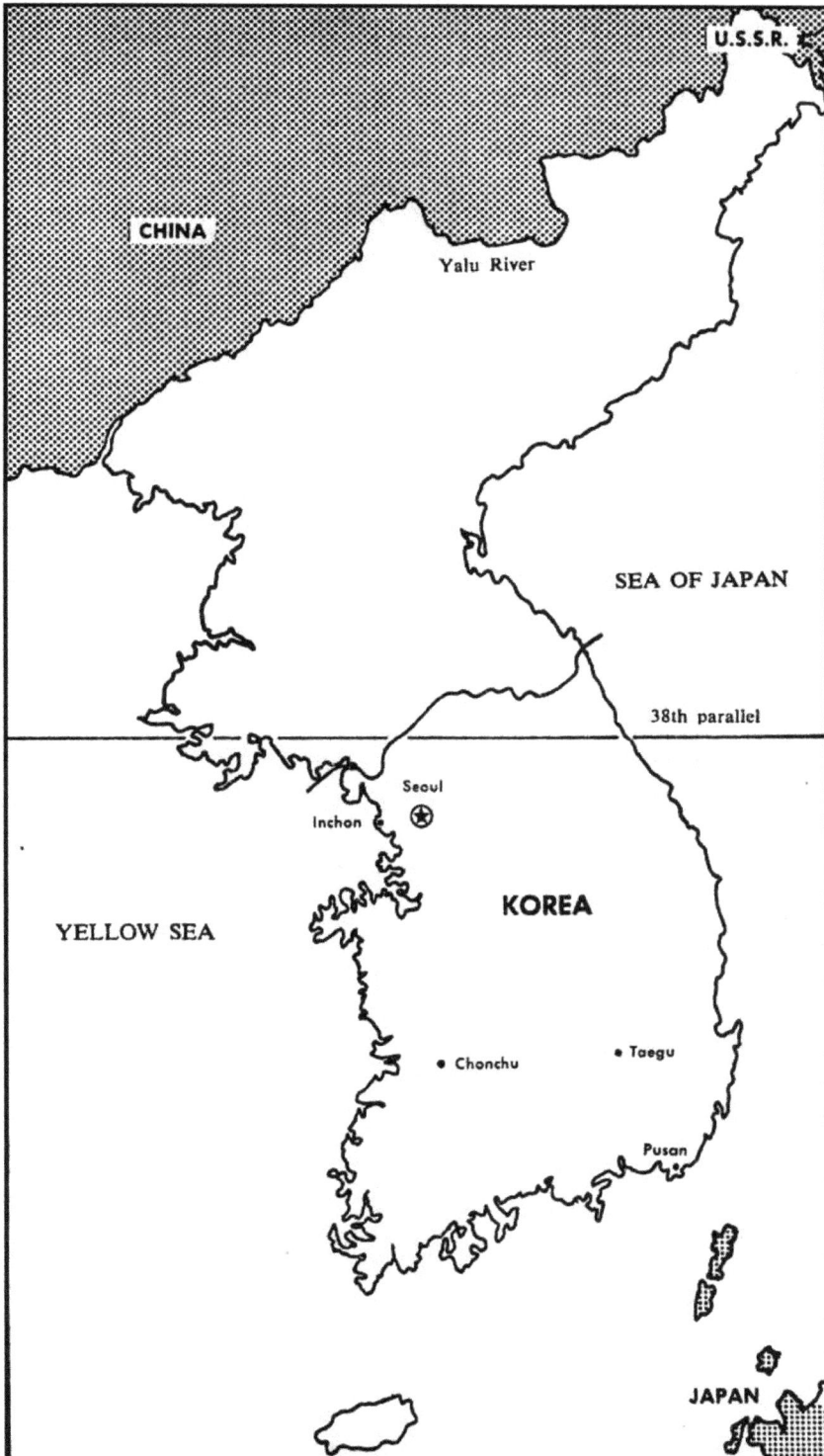

The people

The Koreans are a Mongolian people distinct from the Chinese and Japanese. The official language is Korean; there are a number of minor local dialects. Knowledge of English has become widespread. The population in 1959 was estimated at almost 23 million. Seoul, the capital city, had about 1.8 million; Pusan, Taegu, Chonchu, and Inchon are other large cities. More than half the population lived on farms.[3]

Communications

Although travel through certain areas may be treacherous, most of the Republic is accessible by some means of communication. There are more than 9,000 miles of roads, but less than 1,000 miles are paved.[4] The railway network totals about 3,500 miles and radiates from Seoul.[5] The largest port facilities are at Pusan on the southeast tip and at Inchon near Seoul on the west coast. There are 7 other large ports, 60 secondary ports, and numerous fishing ports. The larger cities have airports and there are regular flights to Japan, Hong Kong, and Taipei. Seoul's Kimpo Airport provides international service.

Natural resources

In South Korea approximately one quarter of the total land is cultivable. Along the coast is some of the best fishing water in the world. North and South Korea together have considerable quantities of rare minerals. In 1939 Korea ranked as the world's fifth gold producer. It was a leading producer of anthracite and bituminous coal, tungsten, and graphite. However, most of the mineral wealth is concentrated in the north, and South Korea lost its major supply sources as a result of the division.

SOCIO-ECONOMIC STRUCTURE

Economic system

The Republic of South Korea is still basically an agricultural country. Rice is the principal crop and the chief export. Fishing is the second most important occupation. About 85 percent of the animal protein in the national diet is derived from sea products. Korea is the sixth largest fish exporter in the world. Manufacturing has been on the increase in the last decade. Foreign aid, from the United States and the United Nations, has resulted in expanded industrial and mineral output and an increase in electric power facilities. The total labor force was estimated in 1960 at 9 million, of whom 73 percent were engaged in agriculture. The per capita income in 1958 was $105,[6] considerably more than it had been a few years earlier.

Manufacturing was confined to production for domestic consumption; a substantial proportion of the output was produced in homes and small factories. Only 7 industrial establishments out of a total of 8,810 employed 1,000 or more workers in 1955.[7] Most industrial establishments were formerly owned by Japanese enterprises, but eventually were sold by the government to private interests. South Korea received its sovereignty in 1948.

Class structure

The Korean class structure traditionally was composed of the upper class—government officials and large landowners—and the commoners—farmers and agricultural laborers. A small middle class has recently developed in the towns and cities, composed of small businessmen, and skilled and unskilled workers. Formerly, education was the primary prerequisite for high position and influence; but under the leadership of President Syngman Rhee, wealth became a more important factor. With some exceptions, the well-to-do element supported Rhee and many owed their wealth to the administration.

The Rhee government instituted land reforms in the late 1940's and the majority of the peasants now own the small landholdings on which they formerly were tenants. But landownership lost its importance as a symbol of social status under Rhee, and the economic, social, and political position of the peasants improved little in spite of the land reforms. The peasants still employed traditional methods of agriculture and the majority of them lived in the L-shaped thatched huts which had evolved many centuries ago. With the cost of living high and the proceeds from their products low, they were constantly in debt, and lived at a bare subsistence level.

The industrial labor force has had a relatively unimportant role in the Korean class structure. Lack of capital and of technical and managerial skills has retarded the rural to urban shift. Factory legislation exists but the failure to enforce it has kept the average wage at a subsistence level.

Literacy and education

Owing to several factors, including poverty, the lack of educational facilities, and Japanese suppression of the Korean language, 77.8 percent of the people were illiterate in 1945. By 1959 the percentage had been reduced to 4.1. When the Japanese capitulated, Korea found its educational facilities in a deplorable condition:

. . . there was not a single textbook in the Korean language, there were only a handful of Korean schoolteachers, and school buildings existed for less than one-tenth of the school age population. By 1958 textbooks had been written,[a] printed, and distributed for every elementary grade, for all high school subjects, and for about half the subjects taught in the universities. Some 96 percent of all children aged 6 to 11 were in school, and over 85,000 students were enrolled in 15 universities and 10 professional colleges . . . ,

which included Seoul National University and Ewha Women's University.[8] The school system was centrally controlled by the National Ministry of Education.

Major religions and religious institutions

Korean Buddhism maintains an organization entirely separate and distinct from that of other countries. In 1958 the Buddhists in South Korea numbered almost 4 million. Confucianism is also strongly embedded in the Korean culture and has many adherents. The sect of the Heavenly Way, which is uniquely Korean, numbered 1,500,000 followers in 1957. A 1958 census showed an equal number of Christians.[9]

GOVERNMENT AND THE RULING ELITE

Description of form of government

Under the Constitution of 1948, the Korean Republic established a representative government, and an amendment adopted in July 1952 provided for direct popular election of the President by secret ballot. Suffrage was extended to all who were 21 years or older.

The President was the official head of state and was elected for a 4-year term with a maximum service of two terms. However, when Syngman Rhee's second term was nearly over, an amendment was passed that eliminated the clause which limited the maximum service to two terms. The National Assembly was composed of an upper house, the House of Councilors, and the House of Representatives, proportionally representing 14 provinces (9 of them in South Korea) and the city of Seoul. The term of office in the House of Representatives was 4 years. The upper house was not yet functioning. The cabinet of the Republic, the State Council, consisted of not fewer than

[a] Many of these textbooks were translations from the Japanese, rather than especially prepared texts.

8 and not more than 15 members, appointed by the President and confirmed by the Assembly. The chief duties of the Council were to approve policies (by majority vote), assist in the supervision of executive agencies, pass on drafts of treaties, approve the budget before it was submitted to the legislature, and appoint the prosecutor general, ambassadors, and ranking military officers. Treaties, declarations of war, and the appointment of the Chief Justice required approval of a majority of the National Assembly.

Description of political process

Political parties and power groups supporting government

In 1951 Syngman Rhee founded the Liberal Party, and from that time on his regime relied upon it. This party was developed into an effective instrument for securing Rhee's reelection. The Liberal Party received many of its votes from rural areas rather than the cities. Until 1955 no political group was strong enough to challenge Rhee in the Assembly or at the polls. In that year two leaders of the old Democratic Nationalist Party (DNP), P. H. Shinicky and Cho Byong-ok, formed the Democratic Party, which attracted a large number of new members, including independents, whose spokesman was John M. Chang. The establishment of a second strong party, which constituted a conservative opposition, did not create an effective two-party system, however, since anyone connected with the opposition was faced with harsh discriminatory measures.

Rhee controlled the Assembly by a combination of support from the legislature and the use of a strong-arm means. In May and June 1952, in November of 1954, and again in 1958, fist fights broke out in the Assembly over Rhee's efforts to pass legislation which would have curtailed anti-government activities. In 1954 a constitutional amendment to limit the power of the Assembly was defeated by one vote. Dissatisfied with the result, Rhee had the speaker announce several days later that reconsideration indicated the passage of the bill by one vote.

The prevalence of guilds and other organizations indicated that Koreans were "joiners." Although not avowed political organizations, many of these groups became action arms of the Liberal Party and supported the regime. One such group was Rhee's National Society, earlier known as the Rapid Realization of Independence; another was the Anti-Communist Yough Corps, which was apparently employed in an unofficial capacity as Rhee's "hoodlum" squad and in turn profited from his patronage. Other organizations appealed to various social groups: the Korean Federation of Trade Unions, the Korean Women's Association, and the League of Korean Laborers.

Character of opposition to government and its policies

"After Rhee's near-destruction of his opposition in 1952, almost four years lapsed before anti-administration forces were able to join together in any kind of cohesive group. Rhee was not entirely without opposition in the meantime, for individual attacks on the administration within the Assembly became less rare once Rhee was reelected and martial law raised in Pusan. Not until 1956, however, were Rhee and the Liberal Party confronted with an opposition party which commanded significant popular support."[10] This was the Democratic Party, which was able to bring together a large number of opposition groups under one organization, despite problems of factionalism.

Shinicky was nominated as the Democratic Party candidate for President in the 1956 elections and John Chang was nominated for Vice President, thus balancing the ticket between the two main wings of the party. Shinicky campaigned vigorously, charging the Liberal Party with corruption and a "devil-may-care" attitude toward the lot of the people. Shinicky died of a heart attack on May 5, just before the May 15 elections. Nevertheless, the elections were held. Out of 8.7 million votes, nearly 1.8 million were invalidated, primarily because they had been protest votes for Shinicky and against Rhee. Though Rhee was reelected, Chang was elected Vice President, carrying not only the urban areas but also the stronghold of the Liberal Party, the rural areas. However, in his position as Vice President, Chang was not able to participate actively in ruling the country.

A Progressive Party had also played a role in the 1956 elections, receiving almost two million votes. However, the party was dissolved in 1958, following the trial and conviction of its leader, Cho Bong-am, as a Communist.

There was a political lull from 1956 until the Assembly elections in 1958. Then the Democratic Party scored a notable gain, increasing its representation in the Assembly from 47 to 77 seats out of a total of 233. The trend presented a serious threat to Rhee's position and his prospects for the 1960 presidential elections. The *Kynghyang Shimmun* of Seoul, the press organ of the Democratic Party, with the second largest circulation in Korea, was closed in May 1959 by governmental orders.

The Liberal Party also began to introduce new legislation designed to make its victory in the 1960 elections more certain. The Democrats reacted by staging a 6-day sit-down strike, which was broken up by the Liberal Party. The Democrats chose Pyong-ok Chough as their Presidential nominee; Chang was renominated for Vice President. Pyong-ok Choung died in February while undergoing brain surgery in the

United States. The elections were held on March 15 despite the death of the opposition candidate and the Democrats were again defeated by Rhee's Liberal Party. Ki-Poong Lee, speaker of the House of Representatives, replaced Chang as Vice President, and Rhee remained President. The Democrats protested the elections in the National Assembly, claiming corruption. They began to boycott Assembly activities, and sought an injunction to invalidate the election.

Legal procedure for amending constitution or changing government institutions

In order to change the Constitution, a two-thirds vote of the National Assembly was required. The same margin was needed to override a Presidential veto. All other legislation required a simple majority. The Cabinet was obliged to resign if the House of Representatives passed a resolution of "nonconfidence" on any proposal which the government considered so vital to its policy that it made it a "matter of confidence."

The Constitution also stipulated that if the President, in the exercise of his official duties, violated the Constitution, or other laws, the Assembly had the right to impeach him provided 30 members of the House of Representatives signed a motion of impeachment and the motion was approved by the majority of each House. Concurrence of an impeachment court was needed for a conviction.

Relationship to foreign powers

Between 1945 and 1959 the United States had given the Korean Republic $2.5 billion in nonmilitary aid and between 1950 and 1959 $1.3 billion in military aid.[11] Extensive aid removed the threat of possible starvation and greatly bolstered the country's currency. The relationship with the United States had not been entirely amicable, however. A 1955 dispute between the United States and Korean Governments culminated in the withdrawal of U.S. Ambassador W.S.B. Lacy, and Washington indicated its displeasure by withholding a new appointment for 6 months. A mutual security pact had been signed by the two nations in 1953, and South Korea's dependence on the U.S. South Pacific defense system precluded the possibility of an end to U.S.–R.O.K. military relationship. Technically, South Korea is still being protected by the United Nations.

Relations with Korea's normal major export market, Japan, had been restrained by the Rhee administration. The historical background of the two countries and more recent events (repatriation of Koreans from Japan to North Korea, confiscation of Japanese property by the R.O.K., and seizure of Japanese fishing vessels by the R.O.K.) had kept alive antagonism and distrust. There was some import-export trade between the two countries, but they did not enjoy a flourishing postwar trade.

The role of military and police powers

The security problem presented by the division of Korea requires South Korea to maintain an extremely large and costly army. This situation has increased the importance of military leaders. Military leaders were rated lower than government civil officials until recent years, but they are now "assuming a greater social and political role."[12]

The army, the police[b] and the courts became arms of the administration and were openly used to suppress opposition activity, often violently. Rhee also used the army and police as a personal bodyguard for himself and his supporters. Demonstrations were fought by the police, and time and again police units threatened and intimidated the opposition. The military appeared loyal to Rhee until the uprising in the spring of 1960; at that time they remained inactive during the demonstrations.

WEAKNESSES OF THE SOCIO-ECONOMIC-POLITICAL STRUCTURE OF THE PREREVOLUTIONARY REGIME

History of revolutions or governmental instabilities

During the long period of Japanese occupation many Koreans deeply resented their loss of independence. There was one important incident in 1919, when mass demonstrations, though peaceful in nature, were brutally suppressed by the Japanese. However, the unrest of this period resulted in some concessions toward self-government. Most of the anti-Japanese activities after that were carried out by Koreans in China and the United States.

During the rule of Syngman Rhee, corruption within the administration and between the government and its supporting business industries was a matter of general knowledge, although the record did not become public until after the revolution. Government policies, which denied the right of dissent, became one of the fundamental causes of discontent. Anyone connected with the opposition risked harsh countermeasures. Some persons were arrested for crimes allegedly committed years earlier. Although the press was allowed con-

---

[b] The R.O.K. National Police Force is about 40,000 strong, and is a counter-intelligence, internal security, and civil defense force as well as a conventional police force. It has a sizable detective force limited to approximately 15 percent of the total police personnel.

siderable freedom, newspapers overstepping the bounds set by the government were either sacked by Liberal Party hoodlums, as was the *Taegu Mail* when it criticized the constitutional amendments proposed by the Government, or closed, as was the Seoul daily *Donga Ilbo* after publishing a "typographical error" which substituted "puppet" for "president" in a reference to Rhee.[13] Violence and murder were widespread. The public was never satisfied that justice was being done. Kim Koo, one of Rhee's greatest rivals in the formative years of postwar Korea, was killed by a second lieutenant. The assassin's life sentence was suspended after he had served a short term in prison, and he returned to active duty, eventually retiring as a colonel.

Another of Rhee's foes who met a violent death was Kim Sung-ju. Charged with having violated the National Security Law, he was tried, convicted, and sentenced to a prison term. A new trial was ordered and he was sentenced to die. However, records show that he was dead 3 days before he was condemned.[14] There was a "Christmas Eve Incident" when the administration announced its intention of amending the National Security Law to grant the police wider powers and provide harsher penalties for subversives. The Democrats, trying to block this move by parliamentary measures, were beaten, kicked, and forcibly removed from the Assembly Hall by special security officers. The bill was passed and no one was punished for the incident.

Economic weaknesses

South Korea's greatest economic weakness is its military and political division from the north. The Japanese authorities developed Korean agriculture in the south and Korean industry in the north. The south is overpopulated and the north is underpopulated. The division greatly impedes a self-sustaining economy in the south.

The Korean economy therefore had to depend almost entirely on foreign aid, chiefly from the United States. This aid made possible the imports needed to maintain industrial productivity. The system of exchange rates under Rhee's administration impeded the chance of improving the overall living standard, since it favored special interest groups while reducing the real income of farmers and nonfavored business enterprises.[15] A long list of inequities characteristic of the Korean economy can be compiled: (a) special favors were granted to businessmen supporting the administration; (b) favored enterprises obtained raw materials at low prices and overexpanded their facilities to a point where the products were turned out faster than the market could absorb them; (c) other products of equal necessity had to be imported at higher prices; (d) this increased the costs of production in some not-too-favored industries and in agriculture ; (e) nonetheless, the farmer received low prices for his rice; (f) the price of wheat flour was artificially depressed; and (g) the system made it possible for clever importers to amass fortunes and discouraged the growth of export industries. The contradictions of the Korean economy were evident. New buildings and factories were being constructed, while unemployment and prices continued to rise. Wealth was accumulating in the hands of some, while many lived in dire poverty. New housing developments were being constructed, while the indigent built shacks from waste materials on public lands and drainage ditches. Farmers found living costs extremely high, but received little for what they sold.[16]

Social tensions

World War II and the 1950–53 conflict with the Communists brought drastic changes to the Korean social structure. Educated English-speaking Koreans who were willing to work with Americans and the United Nations qualified for positions of power and influence. This group later developed active interests in the nation's economy and jealously guarded their privileged position. The mass of the people found that, on the whole, their economic, political, and social position had improved but little since the Japanese occupation.

Government recognition of and reaction to weaknesses

Many of the weaknesses of South Korea were beyond the control of the administration. The country was divided, most industrial resources were in North Korea, and the economy was largely dependent on foreign aid. Democracy in Korea was not easily implemented and there was definitely a threat from communism. Rhee's solution to the problems facing Korea was to rule in an autocratic fashion and attempt to prevent his political opposition from gaining popular support. Before the election in March 1960 the administration realized that the Democratic Party was a definite threat to the reelection of Rhee and particularly to the election of his vice-presidential candidate, Ki-Poong Lee. The methods employed to assure political stability and a Liberal Party victory were those of threats and coercion. Even the Assembly was browbeaten into following Rhee's dictates. However, instead of forcing the Korean people to acknowledge Rhee's leadership, these actions increased the resentment of the people to such a point that a revolution overturned the administration which, after the election of 1960, had, it was widely alleged, remained in power only by manipulating votes.

FORM AND CHARACTERISTICS OF REVOLUTION

ACTORS IN THE REVOLUTION

The revolutionary leadership

It is difficult to ascertain exactly who the revolutionary leaders were. Unplanned and spontaneous as the revolution appeared, the chief figures were the leaders of the student demonstrations, followed later by others who took the responsibility for seeing that the change of government was orderly. John M. Chang, for example, who emerged as one of the most important figures of the revolution, had no apparent role as an instigator or organizer. Chang was a scholarly Catholic layman who served Rhee as Acting Premier during the war and who seemed to have been a popular choice for President within the Assembly in 1952. Onetime Ambassador to Washington, Chang became one of the leaders of the new Democratic Party when it was formed in 1955 and brought together the fragmented opposition to the Rhee government. He was elected Vice President in 1956, but had no duties to speak of except to receive his salary. He ran for reelection in 1960, but was defeated by Rhee's candidate, Ki-Poong Lee, speaker of the House of Representatives.

The revolutionary following

The Korean revolution seemed to have been conducted by students from colleges and universities in Masan, Seoul, and other cities, supported by university professors and other elements of the Korean people. Workers did join the demonstrators willingly and enthusiastically, but there were no reports that worker or peasant groups led demonstrations. The number of demonstrators who supported demands for Rhee's resignation varied widely from time to time and in different places. Several thousands demonstrated in Masan, and in Seoul up to 100,000 people supported several thousand students in protest meetings during the final days of the Rhee regime.[17]

ORGANIZATION OF REVOLUTIONARY EFFORT

Internal organization

The apparently spontaneous character of the revolution precludes the possibility of identifying a definite revolutionary organization. Among the student demonstrators, *ad hoc* decisions had to be made as to precisely where and when a demonstration was to take place. It can therefore be assumed that a small nucleus of students became the decision-makers and formed the organization of the revolutionary effort. This nucleus materialized on April 26, when a small representation of the demonstrators was chosen, apparently *ad hoc*, to present a list of revolutionary demands to Rhee. The previous day 189 professors from 12 universities led students to the National Assembly to voice their demands.

External organization

There is no evidence that any organizational support existed outside of South Korea. However, from the time of the election on March 15, 1960, the United States applied pressure on Rhee to alter his policies toward the opposition and in the final days strongly suggested that he resign. The sequence of steps taken by the U.S. Department of State will be discussed below. (See p. 566.)

GOALS OF THE REVOLUTION

Concrete political aims of revolutionary leaders

Chief among the announced political goals were: the resignation of Rhee, the resignation of all members of the National Assembly, the resignation of the Attorney General, release of all students arrested during the demonstrations, a new election for President and Vice President, deprivation of the Vice President-elect, Ki-Poong Lee, of all public offices, and establishment of a cabinet system of government.

Social and economic goals of leadership and following

In the day-to-day events of the revolution, the political goals of the leadership and following were quite explicit. Not so the social and economic goals. However, Rhee's opposition felt that the social and economic weaknesses of South Korea had been aggravated as the result of administration policy. The object of the revolution was first to overthrow the administration and to establish a more equitable government, and then embark upon reforms, including new monetary exchange rates, that would provide employment for the vast numbers of unemployed. The Koreans definitely desired a better standard of living.

REVOLUTIONARY TECHNIQUES AND GOVERNMENT COUNTERMEASURES

There is no specific date which clearly marks the outbreak of the revolution. It appears to have been truly a spontaneous reaction of the people in the larger cities against Rhee in general and against the fraudulent election of March 15 and the rather brutal actions of the police against student demonstrators in particular. The first signs of major unrest occurred during the election campaign, when small groups of students demonstrated in Taegu on February 8. On election day itself, more serious riots and demonstrations occurred in Masan, where thousands of people joined students in protest demonstration against Rhee's suppression of political opposition. Several other demonstrations occurred during the first part of April, in Masan, Taegu, and Pusan. The police countered these demonstrations with many arrests and in several instances students were shot down. Widespread charges of police brutality were the result. The police were accused of hiding the number of demonstrators killed by disposing of their bodies. The discovery of one student's body in the sea led to further unrest in Masan. The atmosphere in the whole country was very tense, but, as yet, there was no sequence of events that could clearly be characterized as a revolution.

By April 19 mass demonstrations, which had previously occurred only in provincial cities, spread to Seoul. Again students were the main instigators. Students in Seoul were enraged by the treatment that had been accorded students in other cities. Between 2,000 and 5,000 of them were joined by large numbers of other people until about 100,000 people milled in front of government buildings demanding Rhee's resignation, new elections, and freedom for political prisoners. Some violence occurred; the targets for attacks were mostly police installations. Large crowds in Seoul tried to reach the National Assembly, and the presidential and vice presidential mansions. In no case, however, did the demonstrators act as an organized force. Their actions rather resembled mob violence. Stones, fists, and sticks were the weapons with which the crowds defended themselves against the police, and overturned streetcars became barricades.

Until the outbreak of large-scale demonstrations in Seoul, the police had attempted to disperse unarmed demonstrators by using tear gas, fire hoses, and rifles. As a result numerous students and other people were killed and hundreds were arrested. Accurate figures on casualties are not available, but the manner in which the police acted increased the resentment of the people. At the height of the disorders in Seoul, on April 19, the government decided to declare martial law. On that day 10,000 troops marched into the capital with tanks and kept the crowds at bay. The government also decreed a curfew from 7 p.m. until 5 a.m. As a result of these steps, the capital remained relatively quiet on the 20th, yet it was clear that the public could not remain passive indefinitely.

Syngman Rhee realized that further steps were needed to appease the people. On April 21 the resignation of the Cabinet was announced, and Rhee promised reforms and punishment for those who had been accused of torturing political prisoners. Huh Chung, an independent, was charged with the formation of a new government. On April 23 martial law was relaxed. The next day Rhee promised that he would disassociate himself from the Liberal Party, and that he would remain only as a symbolic figure as the "Father of Korean Independence," while real power would revert to the office of the Premier, which Rhee had abolished previously. Despite all these concessions, the people were not satisfied, and continued demands for Rhee's resignation were heard. When this resignation failed to come, events reached a climax.

On April 25 a large number of professors from many universities led several thousand students to the National Assembly to read a manifesto in which the professors apologized for not having supported the students earlier. The manifesto contained demands for political reforms, new and fair elections, the resignation of Rhee and the Vice President, and the reestablishment of democratic government. Many people had joined the demonstration and at least 30,000 remained in the streets in the evening despite the curfew. Army loudspeakers appealed to them to disperse, and professed sympathy with "their just demands." However, only the professors were persuaded; the demonstrators refused to disperse even after the army used tear gas. In some places the troops joined the demonstrators, who in turn shouted, "Long live our soldiers."

General Song Yo Chan met with Rhee several times, warning him that the Korean Army would refuse to fire on students if they continued rioting. The unrest did not subside and on April 26 crowds attacked the police headquarters in Seoul and kept up large-scale demonstrations in front of the National Assembly and the presidential mansion. Rhee made a conditional offer to resign "if this is the will of the people."

In the meantime, the American Government had put considerable pressure on Rhee to take note of the demands of the people. Immediately after the elections of March 15 Secretary of State Christian Herter conveyed to the South Korean Ambassador his concern over "the many reports of violence and irregularity."[18] On April 19 the U.S. State Department again urged the Korean Government to take all necessary steps to protect "democratic rights of freedom of speech, of assembly, and of the press."[19] The reply of the South Korean Ambassador that the demonstrations were instigated by Communists and anti-American elements was brushed aside with a statement that the State Department had received no reports that would substantiate his allegations.[20] On the crucial day of April 26, after Rhee's conditional offer of resignation, the American Ambassador in Seoul called on Rhee to discover his real intentions. Later that day, the American Embassy issued a statement to the effect that there was an "obligation on the part of authority to understand the sentiment of the people . . ." and that "this [was] no time for temporizing."[21]

Five representatives of the demonstrators met with Rhee on the 26th, presenting him with a declaration which demanded his definite resignation. The following day Rhee gave in to the pressure and resigned. Huh Chung became Acting President pending new elections.

MANNER IN WHICH CONTROL OF GOVERNMENT WAS TRANSFERRED TO REVOLUTIONARIES

Syngman Rhee forwarded his resignation to the National Assembly on April 27 and appointed Huh Chung Acting President to arrange for a new election. The National Assembly remained intact; and although nearly two-thirds of the members were nominally members of the administration party, the motion which demanded Rhee's immediate resignation was passed without debate. Even before Rhee's resignation many Liberal delegates had defected and sided with the Democrats.

THE EFFECTS OF THE REVOLUTION

CHANGES IN THE PERSONNEL AND INSTITUTION OF GOVERNMENT

Syngman Rhee and Ki-Poong Lee were replaced by the interim government of Huh Chung, which immediately began efforts to restore law and order. Lee and his family committed suicide, while Rhee left the country. Many other high-ranking officials of the previous administration were arrested and tried for various crimes. On June 15 the old National Assembly, reorganized under Democratic Party leadership, amended the Constitution, and replaced the presidential form of government with a parliamentary form. Elections were held on July 29 and the Democratic Party won a large majority. On August 19 John Chang was accepted as Prime Minister and on September 12 he announced his Cabinet. A struggle between the "old guard" and the "new faction" of the Democratic Party, however, made compromise necessary for general accord. A law passed in the new Assembly provided for the elections of provincial governors and assemblies, mayors, town and village heads, and municipal and local councils.

MAJOR POLICY CHANGES

Some effort was made by the Chang government to effect major policy changes, but a military coup in May 1961 brought an end to parliamentary government and substituted a military junta. The 1960 revolution had been necessary to reintroduce democratic methods and safeguard civil liberties. The policies of the Chang government were directed to correct abuses of power, reduce corruption, and eliminate weaknesses and deficiencies in the politics of the Republic. It made some attempts to raise the living standard by new public works programs to provide employment, and also made attempts at a rapprochement with Japan to increase import-export trade.

LONG RANGE SOCIAL AND ECONOMIC EFFECTS

It is unlikely that the 1960 revolution has had any long range effects. The new regime was short-lived and the events of 1960 and 1961 indicated that South Korea's prospects were as uncertain as ever. The economic problem appeared to be by far the greatest. The population toward the end of 1961 was estimated at 25 million. Nearly 75 percent of the people depended on the land for a livelihood, and 40 percent of the rural families had to subsist on one acre or less. Unemployment was responsible for similar conditions in the cities. The per capita income in 1961 was under $100. The aid program had not raised the standard of living, since its positive effects did not benefit the masses.[22]

OTHER EFFECTS

The events of 1960 and 1961 led one author to sum up that the military coup d'etat in May 1961, like the April 1960 revolution, represented "little more than a shift of political power from one segment of the traditional ruling elite to another." Both General Chang and General Pak, the leaders of the 1961 coup, "are products of essentially the same social mold in which were cast John M. Chang and Syngman Rhee," and there is little hope that they will commit themselves to the social revolution that is needed in Korea. The new leaders have shown many of the personal qualities needed for a general rehabilitation of Korea, and they have promised to restore constitutional government by mid-1963. However, "they believe that the road to 'true democracy' lies through a period of police state tutelage. This fallacy is both naive and perilous."[23]

NOTES

1. Kyung Cho Chung, *Korea Tomorrow, Land of the Morning Calm* (New York: Macmillan, 1956), p. 6.

2. *The Worldmark Encyclopedia of the Nations* (New York: Worldmark Press, Inc., Harper & Brothers, 1960), p. 569.

3. Ibid.

4. Ibid., p. 570.

5. Oh Chae Ktung, *A Handbook of Korea* (New York: Pageant Press, 1957), p. 268.

6. *The Worldmark Encyclopedia,* p. 572.

7. Ibid., pp. 572–573.

8. Ibid., p. 575.

9. Ibid., p. 570.

10. Allen, Richard C., *Korea's Syngman Rhee: An Unauthorized Portrait* (Rutland, Vermont: Charles E. Tuttle Co., 1960), p. 203.

11. U.S. Department of Commerce, *Statistical Abstract of the United States, 1960* (Washington, D.C.: 1960), pp. 873, 881.

12. *Republic of Korea,* The Military Assistance Institute, The Department of Defense, June 25, 1959, p. 3.

13. Allen, *Rhee,* p. 204.

14. Earl, David M., "Korea: The Meaning of the Second Republic," *Far Eastern Survey,* XXIX (November 1960), 169–175.

15. *Report of the United Nations Commission for the Unification and Rehabilitation of Korea,* A/4466 and Add. 1, (GOAR, 15th Sess., suppl. No. 13), New York, October 1960.

16. Ibid; see also Earl, "Korea," pp, 169–175.

17. "The April Revolution," *Korean Report,* I, 1 (April 1961), 3–6.

18. *Keesing Contemporary Archives,* XII (1960), 17399.

19. Ibid., 17400.

20. Ibid.

21. Ibid.

22. Edward W. Wagner, "Failure in Korea," *Foreign Affairs,* XL, 4 (October 1961), 130–132.

23. Ibid., p. 134.

RECOMMENDED READING

BOOKS:

Allen, Richard C. *Korea's Syngman Rhee: An Unauthorized Portrait*. Rutland, Vermont: Charles E. Tuttle, Co., 1960. Allen presents a very biased critical analysis of Rhee.

Oliver, Robert T. *Syngman Rhee*. New York: Dodd, Mead & Co., 1955. An account that presents Rhee in rather favorable light.

PERIODICALS:

Berr, John M., "The Second Republic of Korea," *Far Eastern Survey*, XXIX, 19 (September 1960), 127–132.

Earl, David M., "Korea: The Meaning of the Second Republic," *Far Eastern Survey*, XXIX, (September 1960), 169–175. This article contains information on "economic distress," "official corruption," and "lack of freedom and dissent."

Sanders, Sol, "Korea: Symbol of the 'Herter Doctrine'," *The New Leader*, XLIII, 21 (May 30, 1960), 7–8. United States-Korean relations and effects of the revolution.

"The April Revolution: The Beginning of the End," *Korean Report*, I, 1 (April 1961). A datelined account of the revolution issued by the Korean Ministry of Foreign Affairs in Seoul.

Werner, Dennis, "Korea Without Rhee," *Reporter*, XXIII (September 15, 1960), 23–26. A journalistic account of postrevolutionary activities of the new government.

OTHER:

Republic of Korea, The Military Assistance Institute, The Department of Defense, June 25, 1959. Washington, D.C. A general very brief discussion of R.O.K.'s social, political, and economic institutions, including some background material.

THE CHINESE COMMUNIST REVOLUTION:
1927–1949

SYNOPSIS

From 1927 to 1949 the Nationalist Government of China and the Chinese Communist Party were engaged in a struggle for control of Mainland China which ultimately resulted in the formal establishment of the (Communist) Chinese People's Republic. Mao Tse-tung aimed at imposing his Communist revolutionary ideology on the minds of the Chinese people and defeating the enemy forces by employing both unconventional and conventional methods of warfare.

BRIEF HISTORY OF EVENTS LEADING UP TO AND CULMINATING IN REVOLUTION

For 200 years the dynastic rule of the Manchus had preserved the traditions of China, adhering to ancient practices and refusing to consider experiments or innovations. The emperor ruled the semi-autonomous provinces, with governors and viceroys under him. By 1850, however, China had become a prime target for the economic expansion of Western powers. A series of unequal treaties shattered the ancient structure beyond repair. The impact of the West increased the corruption of the ruling elite, and, toward the end of the 19th century, invasion of the mainland by a modern Japan contributed to the decline of the Ch'ing dynasty.

Reform movements, from the 1860's through the 1890's and again from 1902 to 1911, failed to alleviate social and economic problems, and by October 1911, after many revolutionary attempts by Sun Yat-sen and other revolutionary leaders, an uncoordinated general revolt, breaking out in various parts of China, ended the dynastic rule. The fall of the dynasty stirred the national consciousness and the desire for change in only a small number of Chinese, and there was little agreement as to what should replace the old order. Revolutionary and counterrevolutionary elements competed for domestic and foreign support:

Although the Ch'ing dynasty was overthrown in 1911, the emperor's abdication in favor of a republican government was not announced until 1912. Sun Yat-sen was elected Provisional President, but he stepped down in favor of a former Manchu official, Yuan Shih-k'ai. Yuan ruled in dictatorial fashion until his death in 1916. Under Yuan's successors the central government further weakened, and the nation disintegrated into a number of local regimes ruled by war lords. A faction of the defunct republican Parliament, led by Sun Yat-sen, moved to Canton to form a separatist government.

Sun Yat-sen moved into the forefront among the competing revolutionary and counterrevolutionary forces by consolidating his support, formulating a comprehensive ideology which became the gospel of the Nationalist movement, and beginning the organization of a strong party and a strong army. Disregarded by the Western powers to whom he had appealed, and intrigued with the success of the Russian revolution, Sun Yat-sen in January 1923 accepted offers of assistance from the Soviet Government. An agreement between Sun and a Russian emissary resulted in the establishment of a Nationalist Government (Kuomintang), strongly supported by individual members of the Chinese Communist Party (CCP), and headed by a leftwing group. In 1924 the first Kuomintang (KMT) Congress in Canton accepted the organization established by Sun and Mikhail Borodin, his Russian adviser. A constitution and a party manifesto were adopted. The Whampoa Military Academy was established with Soviet assistance in May 1924 to train officers for the new National Army and was placed under the command of Chiang Kai-shek. Sun Yat-sen was elected lifetime President.

Following the death of Sun Yat-sen in 1925, a struggle for party leadership took place between the right- and leftwing groups. A successful coup d'etat in March 1926 resulted in Chiang Kai-shek's assuming leadership of the Kuomintang. Determined to unite a divided China, Chiang led his National Revolutionary Army on an expedition to the North. During his absence, Borodin formed an opposition group to oust Chiang from power. The Nationalist Government was moved from Canton to Hankow, and a leftwing Kuomintang-Communist coalition expelled Chiang and his rightwing associates from the party on April 17, 1927. He entered into an alliance with wealthy merchants and bankers and established his own government in Nanking on the following day. Meanwhile, the Chinese Communists overreached themselves in their attempt to seize power and were purged from the Nationalist Party in 1927. A slight incident in the Russian Embassy in Peking (Peiping) disclosed certain evidence that the Soviet Government was intervening in the internal affairs of China and resulted in the expulsion of Russians from China. The struggle between Kuomintang factions continued, but by the spring of 1928 Chiang had regained command of the Kuomintang, and established the "Second Nanking Government."

The Communist Party activities in Shanghai, directed by Li Li-san, then by Ch'en Shao-yu and the "returned students' group" (Chinese

Bolsheviks recently returned from Moscow), continued to promote "revolutionary seizures of power in the cities and to challenge the KMT forces directly . . . " At this time, Mao Tse-tung was assigned the task of organizing the peasants into unions. The party failed in its attempt to secure mass support through the urban proletariat. However, Mao, acting independently after the failure of the Autumn Harvest Uprising in September 1927, from which he was forced to flee, established a soviet, based on peasant support, in the mountain region between Hunan and Kiansi Provinces.[a] The Communist Party leadership, after being driven out of Shanghai by the KMT, was transferred to Juichin, which became the capital of Mao's "Chinese Soviet Republic." Party leadership was exchanged several times, and it was not until January 1935 that Mao received undisputed leadership of the Communists in China. He eventually gained acceptance for his ideas.

Meanwhile, the Nanking government prepared its army, under Chiang's command, to execute a series of "extermination drives" in an effort to wipe out all Communist elements in China. The Nationalists chased the Communists all over the countryside but failed to hit Mao's vital bases in the first four campaigns. However, the "Fifth Anti-Communist Campaign" was conducted more effectively and the Communists were forced to abandon Kiangsi. In October 1934 the Red Army began its historical "Long March" to Yenan in northern Shensi, a retreat which took it over 8,000 miles of mountains and steppes.[2]

While the Nationalists were fighting the Communists, Japan was advancing into the Chinese mainland. By 1931 it appeared evident that Japan had chosen China as the main target of her economic expansion program. Staging a bombing incident which caused minor damage in Manchuria in September 1931, Japanese troops occupied Mukden. They drove on to complete the occupation of Manchuria by February 1932. They met virtually no resistance; the Manchurian war lord, Chang Hsueh-liang, under Chiang's command, withdrew his small force without attempting a stand. An "independence movement" was set up by the Japanese, and the Japanese puppet state of Manchukuo was formally established in March 1932 to consolidate Japan's military gains. Military operations against China continued until May 1933, when a Sino-Japanese truce was signed which formalized the state of Manchukuo and established demilitarized buffer zones around it.

---

[a] Around the time when Mao established this mountain base there was a similar move, namely, the establishment of the Hailufeng Soviet—the first Communist regime in China—on November 1, 1927.

Following the events in Manchuria the Communist leadership called for an anti-Japanese National United Front, and appealed to all Chinese, and to Chiang especially, for cooperation against the invading Japanese forces. The truce proposed by the Communists was accepted by Chiang and the Central Executive Committee of the KMT in September 1937. It was preceded by a series of intrigues which included the abduction of Chiang (referred to as the "Sian Incident"). The military organization of the Communist Party was reorganized under Chiang's supreme command in the new alliance.

Japanese resident troops in Peking claimed in June 1937 that Chinese troops had made an unprovoked attack on them. Using this as a pretext, the Japanese occupied Suiyuan Province. Other incidents followed, and Shanghai, Nanking, and Hankow fell into the hands of the Japanese as they pressed on along the Yangtze River.

By 1940 a military stalemate had been reached. The Japanese occupied all the northern ports, a large number of ports to the south, and a corridor up the Yangtze River as far as Ichang. Chinese guerrilla units operating behind the Japanese lines confined Japanese control to the cities and major communication routes. A Chinese victory in 1940 regained Kwangsi Province. During all this action there had been no declared war on Japan and the other Axis Powers.

Although the Kuomintang and the Communists had formed an anti-Japanese coalition, their military forces were engaged primarily against each other rather than the Japanese. Much of the military aid sent to Chiang to fight the Japanese was used against the Communists, to the dismay of the Allied Command.

Following the Japanese capitulation in 1945, KMT and CCP forces competed to take over the areas that had been held by the Japanese. U.S. emissaries attempted to reconcile the two parties. Not yet strong enough to attempt a complete takeover of China, the Chinese Communist Party agreed to a political conference, which resulted in a cease-fire agreement in January 1946. By June, however, the negotiations had collapsed, and the CCP called on the people to overthrow the Nationalist regime and establish a "people's democracy."

The new Constitution drawn up by the Kuomintang National Assembly in November 1946 offered little to entice the Communists, and it appeared that the rightwing group had no intention of liberalizing the government. Chiang wanted to incorporate all of China under one central Kuomintang government, with only token representation for the CCP.

In January 1947 the U.S. representatives abandoned their peace efforts. By early 1948 the Kuomintang seemed to have lost the initiative, and by the end of that year, the Communists held virtually all the territory north of the Yangtze. The Communist drive to occupy the cities resulted in one defeat after another for Chiang. In January 1949 the Generalissimo retired from active leadership, leaving the national forces under his Vice President, Li Tsung-jen, and in December of that year he retreated, along with the remaining Nationalist troops, to the island of Formosa (Taiwan). Mao Tse-tung and the Chinese Communist Party established the Chinese People's Republic.

THE ENVIRONMENT OF THE REVOLUTION

DESCRIPTION OF COUNTRY

Physical characteristics

Virtually isolated, China is bordered on the east by the Yellow Sea and the China Sea; on the south and west by Vietnam, Laos, Burma, and 5,000 miles of the "world's most gigantic mountain rampart"— the Himalayas, Karakorum, Pamirs, and T'ien Shan; and to the north by two enormous deserts—the Takla Makan in Turkestan, and the Gobi in Mongolia. Central mountain ranges and the Yangtze River divide China into two parts "which differ greatly in climate, in the general appearance of the land, and in the character of the people." The north is virtually treeless, except for the Manchurian forests, and dry except for a midsummer rainy season. The south is semitropical and has a heavy rainfall during the summer monsoon season.[3] In latitude, China stretches over a span equivalent to that from the middle of Canada to southern Mexico.

The people

There are more than 650 million people in China, and the human diversity is as great as that of the land. Thirty percent of the population live in eastern China, on less than 7 percent of the total land area.[4] In 1950, 20 percent of the population lived in cities. Some of these, the port cities in particular, were Westernized, but others had changed relatively little since ancient times. Peking was restored as the capital city in 1949, and had nearly three million people at that time.[5]

Communications

In the middle of the 20th century, "China had barely one mile of road (130,000 total miles) for each 6,000 people, as compared with the United States, which had one mile for each 45 people."[6] There were

KOREA

CHINA SEA

YELLOW SEA

Shanghai

Nanking

Formosa (Taiwan)

Hong Kong

Peiping

Yellow River

Hankow

Juichin

Canton

Ichang

Yangtze River

Yenan

Chungking

VIETNAM

LAOS

MONGOLIA

BURMA

TURKESTAN

TIBET

CHINA

INDIA

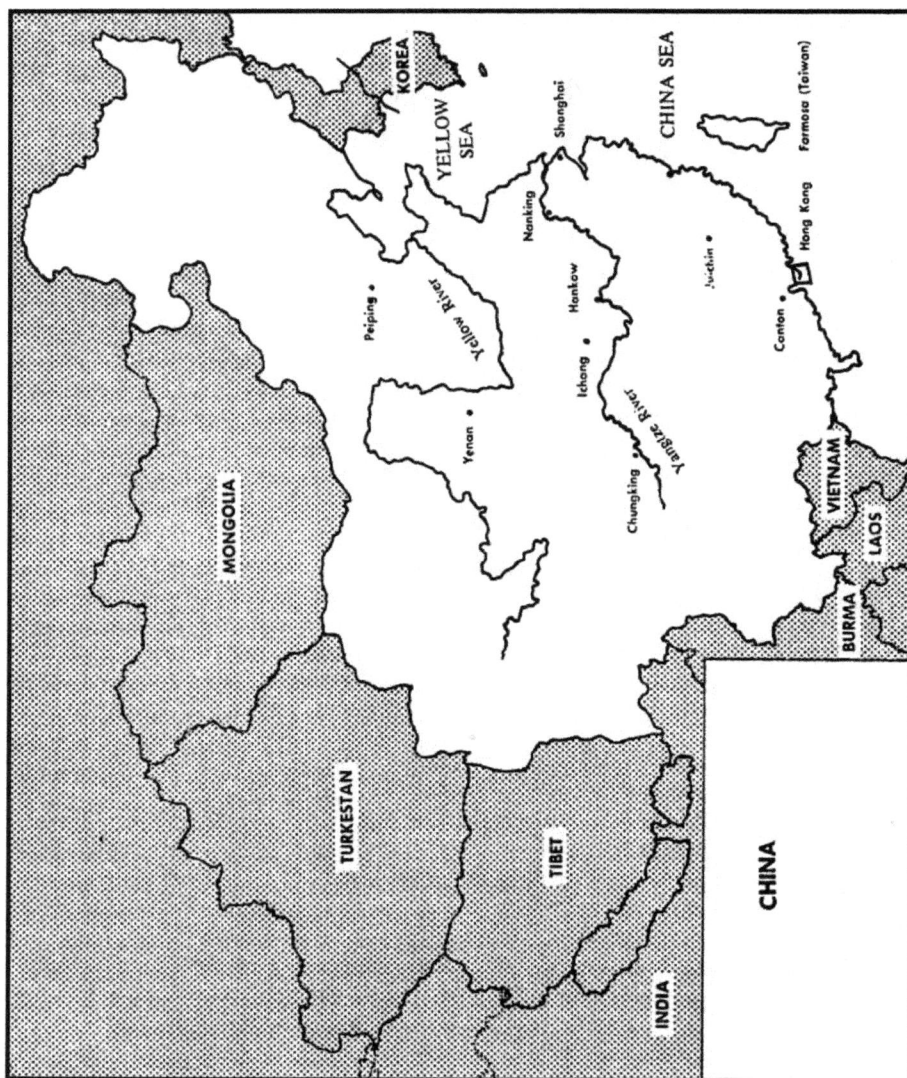

fewer than 20,000 miles of railroad, far from adequate. Until recently there was not one railroad line traversing China from east to west, and there were only two lines linking the north with the south. Most of the mileage in 1950 was concentrated in North China and Manchuria.

Rivers, canals, and other waterways provide a slower means of communications. The artificial Grand Canal is the most important; it extends 1,200 miles from Hangchow to Tientsin and connects the Yangtze, Yellow, and Peiho Rivers. Airway development began in 1929 and airlines now connect all major cities. Transportation by wheelbarrows, the backs and shoulders of men and women, and human-propelled or -pulled junks was still widespread in 1949.

Natural resources

China is primarily an agricultural country and its greatest resource has been its population. However, existing mineral resources, even if not as large as those of major industrial powers, are sufficient to allow for the development and expansion of modern industries. Coal is found in large quantities in the north in the regions of Shansi and Shensi and the supply of iron ore in Manchuria, though of poor quality, is believed adequate to meet the country's industrial needs for many years to come. Bauxite and petroleum have recently been discovered and the potential supply is thought sufficiently large for internal needs. Tungsten and antimony have long been produced in significant quantities. In general, coal, ferrous metals, and petroleum deposits are located in the northern parts, while nonferrous metals are found in the southwest. Some uranium has been reported to exist in Tibet and Chinese Turkestan. Tung oil, used in the production of fast drying varnishes, has represented a major fraction of exports. Water power for the production of electricity is abundant, especially in the mountainous plateaus of the southwest, but had not been highly developed by 1949. China's natural resources are more abundant than has been believed in the past. By 1949, they had not been fully exploited or explored. The greatest progress before World War II had been made by the Japanese in Manchuria.

SOCIO-ECONOMIC STRUCTURE

Economic system

The pre-World War II Chinese economy was overwhelmingly agricultural, with grain production predominant. China produced more than one-third of the total world output,[7] and was self-sufficient in food.

More than 75 percent of the population were employed in agriculture. In order to maintain a subsistence or slightly above subsistence level, many people cultivated relatively small areas of land. Natural disasters periodically caused famines affecting millions.[8] Land was very unevenly distributed and there existed a widespread tenancy system. The peasants were forever in debt to private money lenders.

A modern sector of the economy was devoted to the conduct of trade and the manufacture of textiles and other light industrial products in the coastal cities. The heavy industry in Manchuria was created by the Japanese after 1931, but was isolated from the Chinese economy until the Communists came to power. Silk, tea, coal, and iron ore were exported.

Class structure

The impact of the West contributed to the weakening of the Confucian way of life and its traditional class structure, and aided in establishing a more complicated social system. At the base of the structure was the vast majority of the population—the peasants—who worked small tracts of land in family units. The average peasant was concerned mostly with rent and taxes, interest rates on his debts, and the share of his production which he could keep for his own.

A small section of the rural population, the landlords, were well-to-do. This class also was dependent on good harvests, but was able to resist the pressures of the political administration. These included survivors of the "gentry" class under the Confucian system, augmented by peasants who had succeeded in raising their living standards through effort and chance. Usurers and owners of pawnshops also were considered members of this class. The "gentry" had no political philosophy except holding their economic position.

Another social group was the military. The military, as a power group, date only from 1911. Graduates of military academies filled leading positions in the central government after being military commanders of the regional armies. The common factor of this class was its military background.

Around the large treaty-ports, which were exempt from Chinese territorial jurisdiction, a new Chinese middle class emerged before and during the republican period. This class was virtually free from official interference. The establishments of the members could not be taxed arbitrarily and the cities were safe from official exactions. This class was divided into two groups: the lower, consisting of storekeepers and retail traders with a low social status and standard of living; and the upper, consisting of bankers, industrialists, and owners of large trading firms, with a higher status and living standard.

The industrial workers were of minor political importance. Politically vocal, however, were the intellectuals. They continued the tradition of the elite, deciding the political course of the nation. Under the influence of Western thought, they were hostile to tradition. The Kuomintang officials were a mixture of military leaders, former business leaders, and intellectuals.[9]

Literacy and education

Officials in China traditionally had been recruited from an intellectual elite whose eligibility to administer at a prefectual, provincial, or metropolitan level was determined by an "examination system." Thus the wisdom and knowledge of the classics, the ideas and sayings of the ancients, were important in the Chinese way of life. The emphasis on learning, however, did not lead to the establishment of a popular educational system. The system of education began to change in the 17th century, although scientific activity did not begin until the first years of the Republic. The "examination system" was abolished in 1905.

The Nationalist movement brought with it a new intellectual and cultural atmosphere, and a renaissance took place, outside the political sphere, in the late 1920's. Nevertheless, under the Kuomintang the mass of the population remained illiterate. The Nationalist Government attempted to broaden and redefine education from 1927 to 1945 and did manage to increase the number and efficiency of schools, colleges, and universities. However, war and internal strife retarded the process and the percentage of illiterates remained about 80 percent.

The number of highly educated people "was limited to Western-educated intellectuals and a part of the bureaucracy and the urban middle class . . ." The traditional gap between a "small educated upper group and a large mass of uneducated producers still continued in spite of the greater diversification of Chinese society in republican times."[10]

Major religions and religious institutions

In the pre-Christian era the Chinese worshipped the spirits of nature—of the river, the mountains, the trees, etc.—as well as the spirits of their ancestors. The Emperor acted as mediator between these heavenly forces and men. Taoism, which taught the mythical power of nature and the insignificance of man, and a Confucian ethical system, which taught the right way of living and the family relationship, formalized the ancient beliefs and institutionalized nature and man as a harmonious whole.

Buddhism was introduced during the early Christian era. It not only assigned man his spiritual place in the natural universe, but also indulged in metaphysical speculation on life, death, existence, and consciousness. Buddhism became a bridge between Taoism and Confucianism. The three philosophies were greatly modified to harmonize with Chinese tradition and blended with each other in such a way as to eliminate exclusiveness. Because of this it becomes impossible to break down the Chinese population on the basis of religion.

A pre-World War II Buddhist revival led to the founding of several Buddhist colleges and institutes to add to the existing monasteries, youth organizations, hospitals, and orphanages. Taoism followed and imitated Buddhism in this respect. However, Taoism was loosely organized and was on the verge of decay prior to the revolutionary period. Confucianism survived the crash of the empire, but existed only within the narrow confines of the intellectual world.

Roman Catholics in post-World War II China numbered over three million,[11] and had a number of institutions, including 3 universities and 288 hospitals. Protestant churches maintained 13 colleges and 216 hospitals, and had approximately one million baptized followers.[12]

GOVERNMENT AND THE RULING ELITE

Description of form of government

After the fall of the Ch'ing dynasty, China was never completely unified under one centralized authority. Traditionally the country had been divided into semiautonomous provinces, generally administered by governors or viceroys. They were responsible to the Emperor, who was the supreme lawmaker. This system preserved internal variations rather than creating national unity. After the collapse of the Manchu Empire, attempts were made to consolidate all the provinces under a centralized government. The Nationalist Party under Chiang Kai-shek secured recognition from minor war lords who accepted the authority of the Nationalist Government in the late 1920's, but it would be inaccurate to conclude that at that time the government in Nanking had completed the work of national unification, since some areas were still controlled by war lords.

The constitutional theory embodied in an outline prepared by Sun Yatsen divided the work of "national reconstruction" into three periods: (1) the period of "military government," in which the Nationalist forces would crush resistance to unification; (2) the "tutelage" period, in which the government would train the people in their four rights of election, recall, referendum, and initiative; and (3) the "constitutional period," which would be the period of constitutional government. By the end of 1928, the government was entering the second period, the "tutelage" period.

Under the new Nationalist Government Organic Law, enacted by Chiang Kai-shek's Nanking Government in 1928, the head of state was the Chairman, who was also ex-officio commander in chief of the Armed Forces. There were 12 to 16 state councilors most of whom were concurrently the presidents and vice presidents of the Executive Yuan (Committee), Legislative Yuan, Examination Yuan, and the Control Yuan. The Chairman and the presidents and vice presidents of the Yuans were appointed by the Executive Committee of the KMT, establishing a direct personal link between party and government. The Chairman appointed all other state officials—diplomats, envoys, mayors of special cities, etc.—with the concurrence of the Central Political Committee of the Nationalist Party. Chiang Kai-shek was the Chairman of the Nationalist Government. "This time it [the government] was a more permanent one but it was overshadowed by his [Chiang's] own dictatorial powers."[13] Chiang promulgated laws and issued decrees without the countersignature of the Yuan presidents or ministers concerned. He was raised in the Confucian tradition and believed that the professional ruling class should be drawn from an educated elite.

The Organie Law continued in force until a provisional constitution was adopted in May 1931. The same scheme of government was continued under the new law, except that it substantially increased the powers of the Chairman of the State Council and provided for the establishment of local governments.

The legal procedure for changing the executive and the legislature in the Nationalist Government, according to the Organic Laws, was based on the people's four "political rights" of election, recall, referendum, and initiative. However, the changes which took place within the government from 1929 resulted largely from the political pressures of party factions. Only two National Congresses were convened between 1924 and 1931. Thus, national direction gravitated around the Central Executive Committee of the Kuomintang. The calling of a National Congress was postponed time and again until it was finally decided that it would not be convoked until after the war.

Description of political process

Political parties and power groups supporting government

Political parties in China were different in character and origin from parties in Western democracies. Political power in China rested with the party that had the strongest military force. A party could not exist without an army. The Kuomintang was the oldest party. Initially organized by Sun Yat-sen, and collaborating with existing secret societies, the KMT was instrumental in the 1911 revolt. It was set up as an open political party following the fall of the Ch'ing dynasty in 1912. It took part in a Western-type parliament in 1913, but upon the dissolution of the parliament in the same year, the party was outlawed. It was later reorganized by Sun;[b] and in 1923 another reorganization (the alliance with CCP members) took place to admit individual members

___

[b] The KMT was reorganized five times during Sun's lifetime, each time with a change of name.

of the Communist Party and create a unified and disciplined party. Most of the southern war lords supported the party.

There were two distinct factions in the Kuomintang. The leftwing, supported strongly by the Communist members of the party, wanted continued cooperation with the Soviet Union and continued admission of Communists within the party ranks, whereas the rightwing element took a negative attitude on these two subjects. The two factions were united in their hostility toward the "imperialist" Western powers, particularly the British. The leftwing controlled the party apparatus until shortly before the Nanking government was established in 1928. The rightwing had control from 1928 on, but fought a constant struggle to preserve itself.

After the enactment of the Organic Law, the Nationalist Government entered the revolutionary stage of political "tutelage." During this stage the control of the government was to remain vested within the party, as prescribed by Sun, and directed and supervised by a Central Political Council composed of the members of the State Council (consisting of the Chairman and the State Councilors). This in effect made government and party synonymous.

A number of small parties, or "middle parties," were organized in the 1920's and 1930's, but they had no mass organization or following. The most important were the Young China Party (YCP), established in France; and the China Democratic Socialist Party (CDSP), established on a merged Chinese and Western philosophical foundation with socialist orientation. After World War II, when the Nationalist-Communist alliance was dissolved, both the YCP and CDSP sided with the Nationalist Government, whereas the other "middle parties" went to the side of the Communists. In the period preceding the final breakup of the alliance, a coalition of the above two "middle parties," the Third Party (also a "middle party"), and three national groups formed the China Democratic League. The League, officially an organization of the "middle parties," aimed at bringing about a constitutional development by prompting negotiations between the Nationalists and the Communists."[14] It was officially dissolved in 1947, but actually continued to exist and sided with the Communists.

Character of opposition to government and its policies

Many of the war lords who had established themselves as rulers of various provinces after the 1911 revolution refused to unify their area under the Kuomintang. The Nationalist Army made a northern expedition to break down their resistance and bring them under control.

However, the government's greatest opposition came from the Chinese Communist Party. From the time the CCP was formally established, in Shanghai on July 1, 1921, it had an official connection with the Comintern (Communist International) and became subordinate to it in 1922.[15] At the Second Party Congress in 1922 the Russians decided it would be best for the "national bourgeois-democratic revolution" in China if the CCP and Sun Yat-sen's Kuomintang entered into an alliance, or united front. The two parties arrived at a compromise (CCP members could individually become KMT members), and in October 1923, Mikhail Borodin, a personal adviser to Sun, representing the Soviet Government, arrived in Canton.

The CCP and KMT leftwing suffered several reversals in the period that followed. The Soviet Union was undergoing similar reversals in its domestic policies (the Trotsky-Bukharin disputes), and these greatly affected the international policies of the Comintern. By 1927 it became obvious to the KMT leadership that the Chinese Communist Party was challenging the KMT in some areas of eastern China. Chiang's army marched against the Communists first in Shanghai, then in Hangchow, Nanking, and Canton. By February 1928 the remaining CCP members of the KMT had been expelled and the Russian advisers evicted from Chinese soil.

Overt opposition also came from other organizations which were under CCP control. Mao Tse-tung's All-China Peasant Union, for example, attacked the KMT for favoring the landlords and other members of the "exploiting" classes. The press was regulated by the Kuomintang, but no stringent controls were established.

Legal procedure for amending constitution or changing government institutions

In 1933 the Nationalist Party decided to establish a People's Political Council composed of both elected and appointed members. This was finally set up in 1938. The Council had the power to make recommendations to the government, organize investigating committees, and pass on foreign and domestic measures of the government in resolutions to be enacted into law. However, the Council's decisions required ratification by the Executive Director of the National Party and the President of the Supreme Council of National Defense, Generalissimo Chiang Kai-shek, and he could also issue emergency decrees and orders having the force of law. ". . . Chiang's authority was practically unlimited."[16] He had the authority to change the constitution and make institutional changes.

Relationship to foreign powers

After the Russians had been expelled in 1928 and China seemed nominally under control of the Nationalist Government, the Western powers were willing to negotiate with Nanking. By 1929 unequal treaties which had favored the Western powers were revised in part, giving China almost complete tariff autonomy in import-export shipping.[c] Further revision of treaties to terminate extraterritorial rights was suspended in 1932 when attention was directed to Communist and Japanese problems. By 1940 Japan held vast areas in China and created the so-called "Reorganized National Government of China."

After 1939, most of the nations of the West were too preoccupied with fighting Hitler to be concerned with China's problems. However, the United States was sending aid to China and was training a Chinese Army first in India and later in China under the command of General Stilwell. In December 1941 China joined the Allies by formally declaring war on the Axis Powers.

Throughout World War II the U.S. Government maintained a close political and military relationship with the Nationalist Government. Maj. Gen. Patrick J. Hurley was sent to China in 1944 as the President's personal envoy to prevent Chiang and the Nationalist Government from collapsing, and to work toward closer KMT–CCP relationship for a greater war effort. After the war, in 1945, the United States sent Gen. George C. Marshall to help Chiang arrive at some compromise solution to the KMT–CCP conflict. In the course of the preliminary negotiations between the KMT and the CCP, General Marshall made suggestions to effect a cease-fire. The negotiations broke down, however, and in 1947 the Marshall mission was withdrawn. The U.S. Government became more aware that the position of the Nationalist Government was deteriorating, and it continued to press Chiang to undertake social, economic, and political reforms as the only hope to turn the tide. Some reforms were carried out by Chiang, but, because of the slow pace at which they were introduced and the manner in which they were implemented, they had little effect.

U.S. grants and credits to Nationalist China between 1945 and 1949 totaled just over $2 billion. In 1947, after the Nationalist Government had achieved some military successes and territorial gains, Congress authorized almost $1½ billion in aid to Chiang (more than half of it military), not including munitions and supplies given under surplus property arrangements.[17]

---

[c] Unequal treaties were not completely abrogated until 1943, when Sino-American and Sino-British treaties were signed on the basis of equality. Unequal treaties with many other countries were nullified by the war.

Soviet policy toward the Nationalist Government of China was ambiguous and appeared to have been maintained at two levels: the official level, where the Russians continued formal diplomatic relations with the KMT, "the sworn enemy of the Communists in China;" and the operational level, where the Soviet attitude was cynical and incapable of being understood.

> As late as 1949, the Soviet ambassador in China continued to negotiate with the National Government over important economic and transportation concessions in Sinkiang, apparently with a view to obtain for Russia a special position in the Province during the very months when the Chinese Communist armies were driving rapidly into the northwest.[18]

The role of military and police powers

The Department of Police Administration was under the jurisdiction of the Ministry of Interior within the Executive Board of the Nationalist Government. The police had various functions, of which the most important were: the establishment and distribution of police stations; suppression in banditry; suppression of dangerous articles; punishment of police offenders; and organization of special police and militia. The maintenance of power was always a problem for the KMT "and the police played an important role in this objective." After 1928, the Blue Shirts, a terroristic secret police organization numbering over 300,000, had a police network reaching to all parts of China, and had the task of eliminating KMT opposition and suppressing radical deviates.[19] Most of the provinces had their own local police organizations.

The Nationalist Army, which was under the tutorship of the chief Russian military adviser, General Blücher (Galen), and later advised by some members of the German general staff, had been organized primarily to force the northern war lords to accept the authority of the Nationalist Government. At the beginning the army, under the control of Chiang, was very successful, largely because an extensive propaganda campaign was conducted among the peasants in advance of military movements. The propaganda campaigns won peasant recruits for the Nationalist Army and caused defections from the ranks of the war lords. These tactics bore a striking resemblance to those used later by Mao Tse-tung against Chiang.

Chiang never controlled all the troops representing the Nationalist Army. When war lords consented to join forces with Chiang and the Nationalist Government, they usually retained control of their armies. Theoretically these were under Chiang's command, but his orders to the war lords were not always obeyed and uncoordinated military operations resulted. The Nationalist Army was also used to combat the "subversive" activities of "heretic" armies. Thus, in November 1930, Chiang's Nationalist Government at Nanking began a series of "extermination drives" against Mao's Central Soviet Area. In July 1931, during the third campaign, there were "300,000 troops under the personal supervision of Chiang," and in the fourth campaign in June 1932, Chiang had a reported total of 500,000 troops.[20]

WEAKNESSES OF THE SOCIO-ECONOMIC POLITICAL STRUCTURE OF THE PREREVOLUTIONARY REGIME

History of revolutions or governmental instabilities

China's history records many rebellions and insurrections. To the Chinese, such occurrences indicated corrupt and inefficient rule. It seemed natural to them, therefore, that corruption and inefficiency be met with rebellion. The Manchu Empire, in its declining years, was faced with the Taiping Rebellion from 1850 to 1864; the Nien Fei Rebellion from 1853 to 1868; the Miao Rebellion in Kweichow from 1855 to 1881; the Yunnan Rebellion from 1855 to 1873; the Mahometan rebellions in Shansi from 1868 to 1870, Kansu from 1862 to 1873, and Sinkiang from 1862 to 1878; and the revolution of 1911 which led to the eventual establishment of the Nationalist Government.[d]

Under the new republican government after 1912, China experienced some political stability, but shortly afterwards was faced with personal power struggles and the rule of war lords in some of the provinces. Conditions under Sun Yat-sen changed little, and the situation under Chiang led one author to write:

> There was never a time in the history of the Nationalist Government when a situation prevailed which might be termed loosely as "normal." The country, under the Nationalist Government was continually in a state of emergency or in the process of development and change. There was never a time when the government organization was stabilized and functioning under a well-defined system. The brief period following the establishment of the government in 1927 and prior to the outbreak of hostilities with Japan in 1937 was a period of many problems and crises. By 1931 the Chinese government was concerned with the undeclared

---

[d] See General Discussion of Far East, pp. 531–549.

war with Japan over the invasion and conquest of Manchuria. Following this there was trouble with the Communists in the south until the latter made their famous "Long March" to Yenan. By 1937 China was at war with Japan. Following the end of World War II, the Nationalist Government was again occupied with the Communist problem, a matter which had only simmered during a long and uneasy truce. During this period there necessarily was little stability. Government organs were under continual revision and change due to changing circumstances.[21]

Communist elements were not the sole target of the rightwing Kuomintang after 1928. Chiang continually had to protect his leadership from war lords in Szechwan and Yunnan who were interested only in maintaining personal kingdoms, and opposing groups who attempted to set up rival regimes. For a few years in the 1930's, the Nationalist Government showed considerable vigor. Many Communist-controlled soviets were taken and several rebellious "People's Governments" lasted less than 2 months.

Chiang also put great effort into suppressing liberal elements that advocated social and political reform, especially after 1945. Arrests of labor leaders, university students, and other intellectuals who were suspected of not adhering to the KMT party line spread fear and bad feeling throughout China. The scholar class was driven further toward radicalism, or at least weakened. Instead of trying to unite the groups that had supported the national revolution of Sun Yat-sen and the anti-Japanese United Front, KMT through its repressive measures alienated the groups and parties of the left and center and drove them into irrevocable opposition.

Economic weaknesses

The traditional restrictions on land concentration had been swept away, primarily by the impact of Western ideas and the resultant disintegration of the Confucian system. Western concepts of private property, individualism, and machine-made merchandise, as well as the extinction of the village industries, weakened the economic position of the Chinese peasantry considerably. Landlordism expanded when agrarian profits were used to buy up more land. According to some estimates, 5 percent of the population owned 50 percent of the cultivated land prior to World War II. Government credit under the Nationalists was seldom available to the small-scale cultivator, and when he was in need of emergency loans, pawnshop interest rates soared and land rentals were increased. ". . . his very misery became the basis for enrichment of others who had cash incomes and money to invest, notably of rent collectors, merchants, money lenders, and officials."[23] The poverty of the peasants was widespread; and under the Nationalist Government there was no significant reform of land tenure, no alleviation of taxation, no reduction of rent, and no curbing of usury.

Little was done in economic reform by the Nationalist Government after 1937. The KMT failed to take advantage of the patriotic surge of the early war period to carry out reforms. At that time it might have been able to overcome the opposition of groups which had a vested interest in maintaining the *status quo*. Japan by that time had gained control of Manchuria and many of the northern ports. The Kuomintang tried to make its wartime national economy self-sufficient to meet the needs of the people in the interior, at the same time attempting to sustain the war effort with some outside aid.

In 1945 China found itself in a deplorable economic situation. Not only were 70 percent of the people living at a bare subsistence level, but the war had reduced the weak middle classes to a poverty almost indistinguishable from that of the peasants. Wartime production failed to keep up with consumer demands, and the resulting scarcity of goods, coupled with the overissue of currency, fed a spiraling inflation. The high cost of food and other available essentials caused great hardship to millions of city and white collar workers. Savings were wiped out.[24]

Social tensions

Western concepts were also instrumental in changing the whole pattern of the traditional Chinese social order. The Western impact was all the greater because it came at a time, late in the Ch'ing dynasty, when the Chinese society was already well on the decline. It helped to destroy Chinese self-confidence and contentment with the traditional way of doing things. The ancient social groups—family, clan, and villages—which helped to fill the needs of individuals, changed in character along with the central authority that the dynasty represented. All these social groups which were symbols of individual security were weakened by the 1930's and nothing had developed to take their place.

The Kuomintang was tied to its own history and to a social system that was both ancient and in the process of decay. It attempted to fix the revolution and to "stabilize society" without first laying out the proper foundation.

> The attempt to modernize China without interfering with the land system, the endeavor to fit some rags of Confucian doctrine to a party dictatorship, which itself was supposed to be temporary, to deny the practice of democracy and still pretend to be preparing the people for it, to proclaim and teach nationalism, and yield to the national enemy, this medley of contradictions could not form a coherent policy which would win mass support.[25]

The peasants, who might have hoped in 1911 that the revolution would bring a change for the better, were frustrated in their efforts to improve their economic and social status. There was no strong national government to institute local reform. And the local gentry, on whom Chiang relied for support, prompted by their private interests, used their influence in the Nationalist Government to prevent renovation. Thus, there was no local order which could meet the emerging aspirations of the peasants and provide an opportunity for satisfying their needs. Looking for the security lost in the reshuffling of the ancient social system, the peasants were attracted by Communist appeals and were turned against the Kuomintang.

Peasant animosity toward the new social classes which supported the Nationalist Government grew to a point where revolution was a constant threat. Landlords and money lenders failed to soften their demands. When the war with Japan broke out, military conscription brought large segments of the peasant population into direct contact with Nationalist Army officers who were often cruel and inconsiderate. These officers tended to support the propertied landlord-merchant groups. In some instances troops were required to work on projects unconnected with the war effort for the benefit of officers and landlords.

Traditionally, the military had never been held in high esteem. Tensions were generated because of methods of recruitment which forced the poorer peasants to provide recruits for military units without orderly or just procedures. Also, the attitude of military men toward the local population created resentment. Military units were often arbitrary in requisitioning supplies and quarters. Looting was not uncommon. One of the appeals of communism was that the Communist officers made a special effort to create a different image of their own armed forces. Their troops were better disciplined and attempted to avoid any act that might antagonize peasants or villagers.

Historically, the intellectuals tended to offer allegiance to the existing order. If this order alienated the intellectuals, and was challenged and overthrown, as was the Ch'ing dynasty in 1911, the intellectuals would transfer their allegiance to the new rulers. However, after 1911, the new rulers failed to utilize adequately the mental resources of the intellectual class. Throughout most of the revolution the intellectuals were torn between foreign ideas that concentrated on commerce and industrialization, and a discredited Chinese way of life against which they rebelled. The Communists continued to direct appeals toward the intellectuals, and shortly before the Communist takeover, the intellectuals were assured that they would have a place in the new regime.

Government recognition of and reaction to weaknesses

The Nationalist Government fully recognized the weaknesses it had inherited from the old order. Its basic policy, therefore, was directed toward synthesizing the traditional way of life and the new Western ideas. The national and provincial governments were modernized; however, little was done to renovate the lower administrative levels. The constant threat of communism and the continuous pressure from Japan encouraged the Nationalist Government to stress administration from above; local problems were regarded with indifference, and local initiative was not mobilized to support Nationalist objectives. The Kuomintang thus relied primarily on the support of the new social classes which controlled the nation's wealth rather than on the lower classes which made up the bulk of the population.

In the late 1930's some elements of the KMT recognized that Communist appeals to form a united front against the Japanese should not remain unheeded. Divided, the Chinese were defenseless. However, during World War II, Chiang continued to ignore the demands of the liberals and the advice of American Ambassador Clarence E. Gauss and Gen. Joseph W. Stilwell, who was accused by KMT officials of intrigue and undermining Chinese sovereignty. Chiang insisted that any compromise with the CCP would have impeded or ended effective national unity, which was necessary for substantial political and economic reform. Compromise, Chiang felt, would not only have meant abandoning the revolutionary principles of national unification laid down by Sun Yat-sen, but would have opened a potential channel for Russian encroachment.

Chiang alienated large segments of China's educated classes. The loyalty of his own armies became questionable after World War II. Voices of criticism became louder from within the Kuomintang, but they went unheeded. By the end of the war Chiang had lost the support of his own people, and the Chinese Communists proved able to "exploit decisively the relative military and political weakness of the post-1945 Nationalist structure."[26]

FORM AND CHARACTERISTICS OF REVOLUTION

ACTORS IN THE REVOLUTION

The Revolutionary leadership

The revolutionary elite were a well-disciplined group of Communists with a "respectable" educational background. Of the top 13, only two had less than a middle or normal school education. Three had formal military training. All of the 13 came from densely populated rural areas having revolutionary traditions and relatively high cultural levels. Eight of them are known to have received some form of training in the U.S.S.R.[27]

The leaders were not of poor backgrounds by Chinese standards. They did not become revolutionaries to improve their own lots; instead they knowingly submitted themselves to hardship, hazard, and uncertainty. They became revolutionaries to direct China on a path they felt would be more desirable. Compared to the Western powers, China was politically, militarily, economically, and socially backward; and the revolutionary leaders sought solutions to these problems.[28]

The characteristics and backgrounds of the top four revolutionary leaders were varied. Mao Tse-tung, the son of a peasant, was an assistant librarian at Peking National University in 1918 when he was introduced to Marxism. His interest in politics increased, and as he became more radical, he took a direct role in political agitation and organization. He attended the foundation meeting of the Chinese Communist Party and received important posts in party work. While recovering from illness in Hunan in 1925–26, he formed peasant unions, which he later described as the "nucleus of the great peasant movement."[29] His ideas concerning the importance of the peasants in the Chinese revolution were at first rejected by the Central Executive Committee of the Communist Party.[e] Mao nevertheless organized "peasant-worker revolutionary armies" and incited the peasants to uprisings. He was dismissed from the Politburo and placed on probation for his independent actions. Between 1927 and 1934, he gradually rose within the party leadership and, at 42, in January 1935, he was elected Chairman of the Central Committee. He has retained the party leadership to the present.

---

[e] Lindsay (Michael Lindsay, "Mao Tse-tung," *The Atlantic*, December 1959, 56–59) states that Mao did not attempt to modify Marxist theory to fit his practice. Mao had a firm conviction that the peasants could realize their true interests only by "serving the working class and becoming part of it." The peasants were a means to serve the cause of revolution.

Chou En-lai, whose father was a mandarin and a Manchu official, received his education at Nankai Middle School and Nankai University. He helped organize the Chinese Communist Party branch in Paris while traveling and studying abroad. When he returned to China in 1924, as a well-known revolutionary leader (he had participated in several rebellions prior to his travels), he was made Chief of the Political Department at the Whampoa Academy, then directed by Chiang Kai-Shek. Chou became a fugitive in 1927, and in 1931, at the age of 33, he received an appointment as Political Commissar to the Commander in Chief of the Red Army, Chu Teh. After the Communist force arrived in Shensi following the "Long March," he was put in charge of foreign affairs; and when the Chinese People's Republic was established, he became Premier and Minister of Foreign Affairs.

Chu Teh, also son of a peasant, became a militarist and accumulated wealth. He studied in a military academy in Yunnan. While studying military science in Berlin in 1922, he abandoned his riches and joined the Communist Party. He was expelled from school in Germany for taking part in a student demonstration. After his return to China he was appointed Chief of the Bureau of Public Safety in Nanchang, where he also instructed a regiment of cadets. Chu joined Mao in the Chingkang mountains in April 1928, and became a Red Army commander, with Mao serving as his party commissar.

Lin Shao-ch'i, the son of a wealthy peasant, was a member of the so-called "returned students group" of young Chinese Bolsheviks who had studied in Moscow. Although less in the limelight than Mao, Chou, or Chu before the early 1940's, Lin played an important role in trade union activities and underground work. He was the author of the 1945 party constitution and has since become one of the leading theoreticians.

The revolutionary following

Mao insisted that the peasants, and not the practically non-existent "proletariat," were the key to the Chinese revolution. Between 1927 and 1937, the Communist movement developed from a workers' party into an agrarian revolutionary party, incorporating the support of the embittered peasant population. The rallying cry was land reform.

After 1937 the CCP recruited many of the liberal elements, including large numbers of students, workers, and intellectuals. Many doctors and lawyers who had supported the KMT went over to the Communist side. These intellectuals were fearful of the Communists, but they wanted a regime that would offer them democratic values and the KMT had not done so. Communist promises won them over, or, at least, kept them from active opposition.[30]

ORGANIZATION OF REVOLUTIONARY EFFORT

Internal organization

The organization of the revolutionary effort was that of the Chinese Communist Party. During its rise to power, the party's organizational structure changed considerably to fit the prevailing conditions. In 1927, the party was outlawed, and Communist bases were threatened by KMT attacks. In 1934, the Communist military force was defeated and lost its base in Kiangsi. In 1937 the Communist organization formed a united front with the KMT, while gradually extending its rural operational bases, where it executed its land policy and carried out nationwide propaganda against the KMT. The large rural areas which were under Communist control by 1945 afforded the party strong bases of operation.

Another point to consider is the theoretical change which resulted from Mao Tse-tung's "revelations" in 1925–26. From that point, to Mao at least, the revolution was no longer to be an urban revolution of the workers, but rather a rural revolution of the peasants. Mao's peasant movement had been regarded by the Comintern and the Chinese Communist leaders "only as a peasant war supporting urban revolution."[31f] But communism in the cities waned under KMT attack, and by 1931 the strategy had proved a failure. Mao then became the dominant force and his peasants the protagonists of the revolution.

The period of the Kiangsi Soviet or the Chinese Soviet Republic, from 1931 to 1934, could therefore be regarded as the first successful attempt to establish a form for the exercise of political authority by the Communist Party in order to implement its major revolutionary strategies. The Chinese Soviet Republic proclaimed in November 1931 "was officially designated a 'democratic dictatorship of the proletariat and peasantry,' according to Lenin's formula of 1905."[32] There was, however, no semblance of democracy, and lacking the workers, it was hardly a coalition of workers and peasants.

The 1931 Constitution of the [Chinese] Soviet Republic vested power in the All-China Congress of Soviets of Workers', Peasants', and Soldiers' Deputies. The Central Executive Committee appointed a

---

[f] The development of the role of the peasantry is one of the basic principles of the Communist revolutionary strategy in China. This can also be checked in Peter S. H. Tang's, "Stalin's Role in the Communist Victory in China," *American Slavic and East European Review*, XIII, (October 1954), 375–388.

Council of People's Commissars to conduct governmental affairs, and retained supreme power between congresses. Radical land reforms were passed by the First All-China Soviet Congress in 1931.

Within the stabilized areas of Communist occupation, after the "Long March," a hierarchy of "soviets" was created.[33] The village soviet was the smallest unit; above it were the district soviets, the county soviets, the provincial and central soviets. Each soviet elected officials to the soviet at the next higher level.

Under the soviet system, various committees were appointed to perform specific duties. The most important committee was the revolutionary committee, usually elected at a mass meeting. This committee governed an area after it was occupied by the Red Army[g] and worked in close cooperation with the Communist Party. There were also committees for education, political training, military training, land, public health and welfare, partisan training, enlargement of the Red Army, and many other functions.

The Central Committee of the KMT adopted a truce proposal made by the Communists in February 1937, designed to combine the KMT forces and the Communist forces in the war against the Japanese. Chiang accepted the Communist proposal in September 1937. To redesign the revolutionary organization under the United Front alliance, the Communists undertook to abolish the party's particular soviet government and to suspend land distribution and confiscation. The party also abolished the Red Army and placed it under the supreme strategic direction of the Nationalist Government. The party and the army remained intact under a "national defense government" incorporating pro-KMT officials, pro-Communist officials, and neutral officials. During the "Yenan" period, as it is called, the struggle for power between the KMT and the Communists continued, but the appearance of a unified effort against the Japanese was maintained.

In 1945 a new constitution of the Chinese Communist Party established various principal organs of its hierarchy—a Congress of the Party, a Party Committee, and a Conference of Party Representatives at the national, provincial, border region, local, municipal, urban, and rural levels. Factories, mines, villages, enterprises, army companies, public organizations, and schools had Plenary Party Meetings, Party Cell Committees, and Conferences of Party Cell Representatives. The Central Committee, which was the highest authority, elected the members of the Central Secretariat to attend to daily administrative duties. The Secretariat also directed and supervised the work of the

---

[g] Intensified propaganda in Nationalist-held territory generally preceded the Red Army occupation, after which the revolutionary committee was established.

various specialized agencies under the Central Committee. The Secretariat had its counterparts in the lower party organizations at all levels. Orders flowed in a direct line from the Central Committee to the cells below. The new constitution increased centralization of power, gave greater weight to rural areas, and put more emphasis on "intra-party democracy."[34]

The Red Army, referred to as the military arm of the party, built its strength on peasant recruits and captured arms and ammunition. The close relationship between the army and the Communist Party hierarchy made the army "politically conscious." Commanders successfully fused the art of civil administration and political propaganda with conventional military arts.

Supreme command of the army was vested in the Communist Military Council, consisting of a Chairman, who was also Chairman of the CCP (Mao Tse-tung), a Vice Chairman (Chu Teh), a second Vice Chairman, a Chief of Staff, the Chiefs of the General Staff, the Inspector General and his two deputies. Political commissars represented the party at all levels. All commanders and leaders were trained to fight and act independently. Thus, decentralization became possible when necessary, while the machinery for centralized command remained intact.

The Red Army consisted of irregular and regular forces. The irregular forces included the Self-Preservation Corps and the Militia, and were collectively called the People's Armed Forces. These specialized in the production and use of hand grenades and land mines, transportation and communication, and reconnaissance. Both forces were organized into battalions, companies, platoons, and squads. Squads and platoons were formed in the villages, the lowest organizational bases. Squads consisted of 5 to 15 men, and two or more squads formed a platoon. Two or more platoons formed a company, and the company was based on the township. The battalion was based on a rural district and consisted of two or more companies. The battalion was the highest operational unit of the People's Armed Forces. Irregular units were also organized according to age and sex. Some women, for example, were organized into the "Women's Vanguard," and some children under 16 were organized into "Anti-Japanese Youth Corps" and "Mending and Cleaning Units." By 1945 the irregular troops numbered two million.[35]

The regular forces were organized in similar units, except that the division was the highest operational unit, and the divisions were grouped into war areas. The activities of the People's Armed Forces were coordinated with those of the regular forces through a chain of command which ran from the country to a group of 10 to 12 countries, and then to the provincial headquarters and headquarters of the army. The political commissars played an important role in the coordination of the forces.

Mao started out with 1,000 men. By 1928 the number had increased to 11,000. In 1934, going into its fifth major battle against the Nationalists, the Red Army consisted of 180,000 troops. Of the 100,000 that started on the "Long March," fewer than 20,000 reached Shensi.[36] By 1945 the Red Army had increased to 1 million regular troops.[37] "Communist strength, built mainly on KMT surrenders, had risen to over 1,600,000" by early 1949.[38]

External organization

The organization of the revolutionary effort remained within the frontiers of China. However, the Chinese Communists did receive some support from the Soviet Union.

According to A. Doak Barnett, the "relationships which developed between the Chinese Communists and Moscow during the 1930's and 1940's up to the end of World War II are still obscure in many respects. However, there is no definite evidence that Moscow seriously attempted to intervene in the internal politics of the Chinese Communists, who were then isolated in remote rural areas." The Communist expansion in northwest China during the war had "no significant outside material support." Mao developed his own revolutionary strength.[39] During the period from 1945 to 1949, the Communist forces received "indirect" military aid in captured Japanese war material from the Soviet Union in Manchuria, which had "an important and direct effect on the situation in China and helped the Communists substantially in building up their strength." However, the Soviet Union continued diplomatic relations with the Nationalist Government during this period, "and even though the Communist victory in the Chinese civil war was clearly imminent, they negotiated with the Nationalists throughout 1949 to secure special Soviet economic rights in Sinkiang."[40] Stalin's ambition was more Russian than Communist.

The extent of the support received from Chinese nationals on foreign soil is also obscure. "Although both the Chinese Nationalists and the Communists have competed actively for control of the Overseas Chinese, neither side has won a total victory. The Overseas Chinese are highly opportunistic and have shifted their political loyalties several times in response to changing situations and pressures." Overseas Chinese sentiment for the Communists is "pro-Peking" rather than "pro-Communist," since ideology is frequently "not basic in determining their political orientation."[41]

GOALS OF THE REVOLUTION

Concrete political aims of revolutionary leaders

Initially, the revolutionary leaders' basic political aim, which was made clear prior to and during the period of the Kiangsi Soviet, was to establish a socialist society by expanding Communist control over all areas of China, forming soviets in the controlled areas, and organizing all soviets under one central authority through a Congress of Soviets with its executive committee, "supported and defended by a unified Red Army."[42]

During the "United Front," Mao's basic aims did not change. "The CCP will never forego its socialist and Communist ideals," said Mao, "which will be realized by the transition from the bourgeois-democratic stage of the revolution."[43] The Communists, according to Mao, were leading an armed Chinese people in a fight against feudalism and dictatorship toward independence and democracy.

Social and economic goals of leadership and following

The basic social and economic goals included independence from illegitimate foreign interests, increase of production at all economic levels, and increase in consumption for the millions of Chinese. The Communist regime also wanted to free the Chinese from their obligations to landlords, usurers, and tax collectors, which would give the people more time for creative activity and would raise the cultural level of the whole population.

The revolutionary leadership's specific social and economic goals were publicized in a number of documents. According to the 1931 Constitution of the Soviet Republic, the regime wanted "to improve thoroughly the living conditions of the working class, to pass labor legislation, to introduce an 8-hour working day, to fix a minimum wage, and to institute social insurance and state assistance to the unemployed as well as to grant the workers the right to supervise production." For the peasant the regime from the outset wanted to confiscate the land of all landlords and redistribute it "among the poor and middle peasants, with a view to ultimate nationalization of land." During the Yenan period, however, the regime suspended land confiscation, but insisted on rent reduction in its areas. Burdensome taxes were to be abolished and a progressive tax was to replace miscellaneous levies. And above all the regime wanted to free itself from "capitalist" and "imperialist" domination.[44]

REVOLUTIONARY TECHNIQUES AND GOVERNMENT COUNTERMEASURES

Methods for weakening existing authority and countermeasures by government

Mao's military philosophy was drawn from several sources but primarily from the Chinese military philosopher Sun Tzu (500 B.C.). Other sources were his practical experience, both inside and outside the Kuomintang, and his conclusion that the key to revolutionary success lay with the masses.[45] The primary concern of this study is Mao's 1927–45 guerrilla warfare, and the conventional warfare of the 1946–49 period. The Kremlin-directed strategy of the 1921–26 period did not contain Mao's philosophy of revolution as did that of 1946–49, and is, therefore, excluded from this discussion.

Mao Tse-tung already had an established organization in Hunan Province when the Kuomintang outlawed the Communist Party in 1927. He created his first armed guerrilla force out of a handful of Nationalist fugitives and trained peasants from the Peasant Unions he had established prior to 1927. Based in the mountain villages, the small bands first conducted propaganda campaigns to win the support of the local peasants. Mao felt confident of victory, for he knew the "needs and the hopes of the masses." The attractive offer of "land to those who till it" enticed large numbers of peasants and Nationalist peasant soldiers to the Communist camp.

In this period Mao drew heavily upon the enemy's manpower and weapons, strengthening his forces while weakening the Nationalist forces. As the Nationalist Army penetrated Communist territory, the Red Army would concentrate its forces to attack isolated and weaker units of the Nationalists, destroying them whenever possible. They avoided giving battle to stronger Nationalist formations and succeeded in preserving their own strength. Thus, in 1927, Mao's units not only surprised and defeated a division of Nationalist troops, but also captured men, recruited many of them into the Red Army, and captured weapons and supplies. This strategy has been referred to by military writers as "parasitic cannibalism." Red forces defended and strengthened themselves by defeating and sapping the strength of the enemy forces.[46] By 1931, Mao had succeeded in expanding and connecting the areas (soviets) under Communist control, and formally established the first Chinese Soviet Republic.

In countering the Red Army, the National Army, under Chiang's command, stepped up its anti-Communist campaign with a series of "extermination drives," which did not succeed in wiping out the Communists. The Nationalist forces failed to capture Mao's vital bases, although he was forced to move them again and again. Mao defended his Republic successfully against four of Chiang's drives, which started in December 1930. However, the "Fifth Anti-Communist Campaign" in 1934 forced him to abandon Kiangsi and make a strategic retreat into northwest China.

In this campaign, Chiang mustered close to one million troops, more than half of them experienced in warfare. Gen. Hans von Seeckt of the German High Command became one of Chiang's military advisers. Adopting a new strategy, Chiang first surrounded the soviet area in an attempt to blockade and finally exhaust the Reds. The "great wall" built around the soviets gradually moved inwards, in an attempt to crush the Red Army in a "vise." The fifth campaign was nevertheless inconclusive; it failed to destroy the main body of the Red Army.

While Mao and other Communist leaders were establishing soviets in Southern China in the early 1930's, Red soviets and their armies appeared in other scattered areas of China. The largest of these was established in the central Yangtze Valley and embraced a population of over two million people. Foundations were already being laid in Shensi and other northwest provinces for what later became the new headquarters for Communist forces. The final destination of the Red Army after it broke through the encircling Nationalist Army was not then known and was not decided upon until a few months before the end of the journey.

The Red Army was divided into three major forces in Szechuan Province during the "Long March" in 1935. Chiang's pursuing troops succeeded in driving a wedge between two segments of the Red Army, and this separation was completed during the summer of 1935, when one of the rising rivers of Szechuan became impassable. Mao and Chou En-lai led one of the major forces to Shensi. The other two major forces, one under the command of Chu Teh, remained in Szechuan. The three forces were consolidated in 1936.[47]

In Shensi, the Chinese Soviet government reestablished itself and brought the area under Communist control. New radical reforms, social and economic, were instituted. These included changes in marriage and family relationships, education, civil law, agriculture, and scores of other activities where civil rights and duties needed to be defined. Administrative agencies were established not only within the party hierarchy, but also among the numerous non-Communist peasants. Both political and military training became compulsory.

Following Mao's arrival in Yenan in October 1935, a shift in strategy took place. Instead of placing sole reliance on military force, the Communists initiated what is referred to as the strategy of "psychological disintegration." Agitators aroused the masses through propaganda conducted on a "person-to-person" basis, and force was used only when necessary. The next step was to organize these aroused masses, which also included defectors from the Nationalist camp, into a well-trained peasant army. In this manner, not only did the Communists expand their control, but they proceeded to "disintegrate the enemy, dissolve his loyalties, destroy his organizations, and demoralize, confuse, and reduce him to general ineffectiveness."[48]

With their reputation preceding them as they advanced, the Communists worked within the "framework of fear." Uncertainty and insecurity made the Chinese people more receptive to propaganda. Fear was dispelled, however, by the initially decent behavior of the Reds, which caught the peasants by surprise and made them even more open to Communist influence. Local leaders, under Communist control, aided in completely mobilizing the people and establishing Communist bases. From these bases guerrilla warfare could be developed and maintained for long periods. Mao waged a psychological-ideological battle against the vulnerable points of his enemy, and the technically and militarily trained Nationalists were ill-prepared for this type of warfare.[49]

The psychological-ideological battle was also carried on within the ranks of the Communists themselves. The "Long March," for example, was described by Mao as being a march against the Japanese rather than a retreat from the Nationalist Army. This deception was strengthened by Mao's earlier declaration of war against Japan, and the Japanese attack on China became a useful tool in Mao's propaganda campaign. From Kiangsi to Shensi, the morale of the Red soldiers was kept at a peak by the belief that they were marching "to meet imperialist invaders."

In Shensi the propaganda reached down to elementary school level; young children who were learning their characters also learned to glorify the Red soldiers. For adults, theaters staged dramas depicting Nationalists, capitalists, imperialists, and Japanese as grotesque and bloodthirsty enemies of the Chinese people. These melodramas, rallies, and meetings kept the soviet citizens constantly fighting for the Communist cause in a "hate campaign" that was apparently very successful.

During the anti-Japanese United Front, from 1937 to 1945, the war of resistance was carried out on two fronts: the first, or "regular" front, engaged Chinese troops under Nationalist command in positional warfare; the second front engaged small units of the Communist Army in guerrilla and mobile warfare behind Japanese lines, complicated in some areas by fighting with KMT guerrillas. Aside from a notable victory at Taierhahuang, "Chinese forces on the regular front were passive, holding or avoiding advancing Japanese columns when they struck." The second front, on the other hand, "was constantly active, with even the smallest Chinese units constantly raiding Japanese garrisons nearest them."[50] Mao, over the war years, succeeded in seizing and holding the countryside over a wide area, while evading a frontal showdown with the Japanese in which his forces might have been destroyed. This was a major factor in the successful Communist expansion between 1937 and 1945.

The main "objective of the Chinese Communist Party throughout the war was the pursuit of power in China." All the propaganda which was designed to appeal to the various Chinese classes, the "limited character" of Communist military operations against Japanese forces, the consolidation of control in Communist-held areas, and the type of policies instituted in these areas, the hardening of "dictatorial discipline within the Communist Party itself" were all directed toward this "overriding tactical objective. And, by the close of World War II, [the Communists] had succeeded in laying the foundations for later achievement of this objective."[51]

In August 1945 the Japanese surrendered. Afterward, the Communists followed two parallel courses: negotiations at Chungking; and, in the entire country, a struggle for position in which Chu Teh, the commander of the 18th Army Group, rushed his troops into cities which were formerly under Japanese occupation and surrounded by Communist-held countryside, ignoring Chiang's command to await further orders. Although the Chinese civil war did not officially begin until July 1946, large-scale fighting between Communist and KMT forces had been going on in Manchuria since the Japanese surrender; and elsewhere sporadic clashes never ceased.

In the first stage of the civil war the Red Army avoided head-on encounters. The Kuomintang's objective was to regain control of Manchuria, the large cities, and the railroads. The Communist plan was to retreat and attack only at the opportune time. "The major objective is the annihilation of the enemy's fighting strength, and not the holding or taking of cities and places. The holding or taking of cities and places is the result of the annihilation of the enemy's fighting strength . . ."[52] Thus, in 1946, the Kuomintang offensives were quick and successful.

The area around Nanking and Shanghai was cleared quickly, and Chiang easily captured 165 towns in various provinces, including Kalgan, the only large city in North China that was in Communist hands.[53]

However, toward the end of the year, the peak of the Kuomintang successes had been reached and the KMT forces began to lose the initiative. An all-out Communist offensive started in May 1947, with an assault on the cities which were weakly defended. By the end of that year, almost all of Manchuria and Jehol had fallen into Communist hands. Red units—two, three, four, and sometimes five or six times the size of KMT units—dealt destructive blows to the Nationalist center and one or both flanks, and routed the remaining units. CCP forces were then swiftly transferred to other fronts. Communist strongholds were established only 50 miles from Nanking, the nation's capital. Chiang withdrew his troops to meet this threat and thereby greatly weakened his strength on other fronts.

During the autumn of 1948, the Communists opened an attack southward which annihilated the last strength of the Kuomintang. A decisive change took place in this phase of the Communist campaign: the Red forces shifted their emphasis from the countryside to the larger cities. The Nationalist forces had gained their positions without serious conflict in 1945 and 1946, and had never been thrown into major offensive operations. When it became necessary for the urban garrisons to support each other effectively, there was a clear lack of unity and delegated responsibility among the Nationalist leaders.[54] In April 1949 Mao ordered Chu Teh to cross the Yangtze, and shortly thereafter the Communists captured the rest of China.

The Communist north-to-south strategy which had paid off so well had been carefully planned. The first objective was to capture and consolidate Manchuria and north China. Manchuria is a vast area, rich in resources. During the Manchu period, and again during the Japanese occupation, it had been shown that possession of Manchuria was the key to the whole of China.[55] The Communists therefore had firmly established bases in north China during the war with Japan.

During the entire civil war period, Mao Tse-tung had never abandoned guerrilla warfare. In the initial stages of the revolution, that was the main pattern of fighting; and though the pattern evolved into mobile and positional warfare by 1948, guerrilla warfare was still considered important.[56]

During 1947, a program to consolidate the economy and government of the Communist-held territory was initiated. A land reform which liquidated the landlords was carried out; the Communists invoked Sun Yat-sen's principle that land should be given to the tillers. Farmers were encouraged and helped to raise production. Business enterprises which belonged to KMT bureaucrats were confiscated. New currency was introduced as a measure to establish a sound financial system. Political education and military training became equally important for both regular and irregular troops. The military effort was necessary, according to Mao, but it was always kept in mind that everything was for the enhancement of the Chinese Communist Party.[57] The civilian population was helped by the troops to overcome difficulties brought on by the war. The civilians in turn supported the front and provided troops with the necessary supplies.

Auxiliary forces or local militia were employed on a large scale. The employment of these forces was not new to the Communist Party, but they were more widely used and provided for in the 1945 party Constitution.[58] Greatly supplemented by local militia, the Red forces at the front generally represented only one-third of their total strength, whereas the Kuomintang generally employed over 90 percent. The militia facilitated Red attacks on any target without the need for large-scale movement of regular troops.

On the political front, the initial aim of the Communists was to establish a coalition with other parties and groups, forming a government which the CCP could later dominate. This became an extension of the classic strategy of the United Front; it gave Mao an opportunity to compete for support outside the government while infiltrating the government from within. In his 1945 report "On Coalition Government," Mao emphasized the need for free competition in politics.[59]

Methods for gaining support and countermeasures taken by government

The Chinese Communist Party employed all means of public information—radio, press, handbills, word-of-mouth, etc.—to win popular support for its political and military efforts. Propaganda themes changed as party policy changed. Considerable latitude was permitted in different localities. The items to be propagandized were loosely outlined by the higher authorities. More detailed outlines were provided for particular drives.

The CCP's propaganda techniques are summarized in the handbook, "Resolutions on Methods of Leadership," as follows:

1. First investigate and then propagandize.

2. Combine general slogans with actual local operations.

3. Always test the use of slogans in a small locality before applying them nationally.

4. Make the ideas of the party appear as if they come from the people.

5. Try to discover the activists or aggressive elements in every mass movement or propaganda campaign and use them to agitate or stimulate the moderate and "backward" elements in the locality.

6. Educate thoroughly the cadres before sending them out in a propaganda campaign.

7. Carry out only one propaganda campaign at a given time; support it with secondary or lesser drives.

8. Carefully review the successes and failures of the campaign after it is completed.[60]

In its reorientation program following the Kuomintang break with the Communists in 1927, the CCP emphasized the need to organize peasants "on as wide a scale as possible" in planned peasant insurrections. The only way to draw the masses into the struggle, the party concluded, was "to carry out the agrarian revolution in the villages." The "transfer of political power in the villages into the hands of the peasant associations" became the slogan. Land was confiscated and redistributed to the tillers. Rent exactions were eliminated. Usurious debts were wiped off the records.[61] These methods of gaining support were very successful. Again during the period of the Kiangsi Soviet, the need was felt for radical land reforms to amass peasant support. The "Land Law of the Soviet Republic" in 1931 launched a land reform which resulted in the confiscation and redistribution of land under soviet jurisdiction.

Following the Japanese invasion, the CCP waged a "patriotic" anti-Japanese campaign which also won many supporters. The "Long March," which was in reality a result of the extermination drives conducted by the Nationalist Government, was described by Mao as a drive against the Japanese forces. Many of the war lords who had supported Chiang before were won over by this appeal.

The CCP's work among the broad masses of the Chinese population was one of the main reasons for its success. Guerrilla warfare could not be carried out without peasant support. During the war against Japan, the CCP continued to win the support of the masses by "encouraging their active participation in local administrative work through popular elections, to improve their material welfare by reductions of rent, interest, and taxes, and to educate, indoctrinate, and stimulate them through popular organizations, cultural movements, and social reforms." The CCP also found active support among the intellectuals, especially the students, by "absorbing" them with propaganda, training, and work programs. The task was to transform the intellectuals into well-disciplined cadres for the party. The universities were often the victims of Nationalist oppression as part of Chiang's program to counter Communist successes, and because of this, "the Communist university at Yenan and other Red training institutions received a steady stream of refugee students: young, active, and professionally trained men from all over China."[62] The CCP was able to enlist the support of merchants, Kuomintang officers, and officials. By making themselves the champions of resistance during the final phases of the war, the Communists were "able to attract men of competence and talent and to set up fairly efficient administrations."[63]

Throughout the period of Nationalist rule, the Kuomintang attempted to counter Communist propaganda by regulating the flow of public information through a system of registration of all printed matter. The KMT controlled many of the more influential publications and subsidized others. A press law which governed Chinese publications was promulgated in 1930.[64]

The Nationalist Government also conducted propaganda drives and school indoctrination programs in its attempt to gain support. Propaganda themes, however, were confined to the subject of "nationalism more often concerned with criticism of Western nations than of the Japanese," and failed to enlist the support of the masses and the intellectuals.[65] Chiang wrote two books in an attempt to draw support from the professional and intellectual classes. One of them, *China's Destiny*, was made a "basic political textbook in military academies, universities, and schools. All administrative officials and candidates for government scholarships to study abroad were put through intensive courses in its principles."[66] Chiang avowed his belief in Sun Yat-sen's "progressive nationalism" and Western democracy; but his writings failed to convince the intellectuals.

Some positive steps were taken by the Nationalist Government toward gaining support by instituting reforms in 1947 and 1948. In April 1947, for example, the government reorganized the Executive Yuan and the State Council. Chiang hailed the reorganization as another step to end the period of "political tutelage" and begin the period of "constitutional government." At this time Chiang also expressed his willingness to settle his differences with the Communists if they would agree to put down their arms and renounce their policy of overthrowing his government by force. However, the struggle for power between the factions within the KMT continued to hamper real efforts toward improving the administration of the government.

There are indications that in 1947 Chiang began to realize the importance of instituting reforms to stem Communist advances and reverse Communist propaganda victories. In June the State Council passed a resolution that called for drastic reforms aimed at improving the lot of the people and at creating a better system for the production and distribution of industrial and agricultural goods. In July 1947 Chiang announced over the radio: "Unless drastic reforms are introduced, China may not be able to exist in the family of nations. Therefore, political, educational, economic, and social reforms, which should be made, shall not be delayed until the conclusion of the suppression campaign, but will be initiated right away . . ."[67]

However, the situation deteriorated further and adequate reforms could not be or were not carried out. In 1948, the government promulgated a series of drastic reform measures which produced a temporary boost in public morale. These reforms were tested in Shanghai. The government attempted to freeze the economic situation by imposing police measures on strong vested interests, but failed to hit the heart of the crisis. These government attempts at reform led to further deterioration of the economy.[68]

MANNER IN WHICH CONTROL OF GOVERNMENT WAS TRANSFERRED TO REVOLUTIONARIES

Chiang Kai-shek stepped down from the Presidency in January 1949 after the main Nationalist armies were destroyed at Hsuchow. Li Tsung-jen replaced Chiang, but fled from Nanking and eventually to the United States after refusing to sign a coalition agreement with the Communists. Chiang retreated to the island of Formosa and established his Nationalist Government there. "On February 3, 1949, the Communists made their triumphal entry into Peking, China's historical capital, and swiftly proceeded to impose on the whole country a totalitarian social structure, a centralized political regime, and a new ideology."[69]

THE EFFECTS OF THE REVOLUTION

CHANGES IN THE PERSONNEL AND INSTITUTIONS OF GOVERNMENT

From August to November 1948, representatives of various parties and groups convened and agreed on the general basis for a People's Political Consultative Conference. In September 1949 the Conference took these actions: it passed the Organic Law of the Central People's Government (CPG) of the People's Republic of China; it elected Mao Tse-tung Chairman of the Central People's Government and chose the Vice Chairman and the members of the CPG; it established the capital at Peking and selected a national flag; and it elected 180 members to the National Committee of the Chinese People's Political Consultative Conference prior to the convocation of the National People's Congress, which was to be elected through universal suffrage. In effect, the Organic Law established the CPG as the administrative arm of the party. No provisions were made for tenure of office, and the Chairman was given wide powers. The important governmental posts were assigned to Communist Party members.

The "Organic Law" which set up the central government was not a constitution and was not intended to be permanent. Instead, the new instrument of government was designed as a temporary guide for the transitional "new democratic" period that was to lead into socialism. Before the final victory of the Communists, they had tried to give the impression that their goal was not a dictatorship similar to the Russian model, but a "new democracy" which would be a coalition of all the "democratic" parties.[70]

Their real goals were not announced until 1949. At that time, Mao defined the new state as a "democracy for the people and dictatorship for the reactionaries."[71] The documents which served in lieu of a constitution remained in effect until 1954, when a new constitution was drafted to serve as a more permanent instrument of government "on the road to socialism."

MAJOR POLICY CHANGES

Communist China did not come into being suddenly when the Nationalist government was defeated. For that reason, there was no sudden transition of power, since the Communist regime already had implemented many of their policies in the areas under their control. However, the end of the civil war gave them a chance to remodel Chinese society on the whole mainland into a pattern designed to lead to a socialist state. One of the major instruments to achieve that goal was the Agrarian Reform Law of 1950. Land reform had been carried out before in Communist areas, but the Agrarian Reform Law applied uniformly to all of China and differed from previously instituted reforms. In the implementation a distinction was made between landlords, rich peasants, middle peasants, and poor peasants. The agrarian reform discriminated primarily against the landlords, even though they too were allowed to receive part of the land that was to be redistributed. The "rich" and "middle" peasants were not treated so harshly as to antagonize them at a time when they were still needed for political and economic reasons. The poor peasants benefited most. They gained land, and their debts to landlords were canceled.[72]

In other areas of the economy, too, the period immediately after the revolution was marked by a continuation of some form of capitalism. Until 1955 the transition of the "means of production" from private to public ownership was gradual. The gap between state-owned enterprises and privately-owned enterprises was bridged by a joint ownership plan. Private owners surrendered a share of their property to the state, which eventually took complete control. Mao had admitted in 1947 that capitalism would exist for a long time in China, "even after national victory." However, in 1955, the tempo of the transition was considerably increased.[73] In January 1956 the Communists announced that the "liquidation of private business" was, "in the main," complete.[74]

The Chinese Communist Party also extended its instruments of control to the social and cultural aspects of the Chinese society. In order to gain and hold popular support, the party instituted the "Communist massline policy" which was especially designed to "organize and direct" every key social group and movement in every area of activity.[75] Initially, the party established several major organizations to give the people a sense of participation in state activities, as well as to perform a number of welfare functions. Lesser organizations followed the major organizations so that within several years after the revolution there were virtually no aspects of Chinese life that were not directed and controlled by the party. Foreign policies, too, were altered radically. Communist China immediately established close links with the Soviet Union and the Soviet bloc. It relied on the Soviet Union for military and economic aid and had the support of the Soviet Union in its efforts to gain international recognition and a seat in the United Nations. Communist China's anti-Western orientation also found expression in the sympathetic attitude, if not outright aid, that it extended to anti-Western indigenous movements, particularly Southeast Asia. During the Korean War it felt strong enough to challenge the West on the battlefield in a limited war.

LONG RANGE SOCIAL AND ECONOMIC EFFECTS

Although communism envisages a classless society, the establishment of the Chinese People's Republic did not eliminate the class structure; it merely reshuffled it. In the order of primacy the structure was broken down in this manner: industrial workers, peasants, petty bourgeoisie (intellectuals, small holders, shopkeepers, traders, artisans), national capitalists (owners of enterprises), and the expendable (landowners, usurers).

Using the experience of the Soviet Union as a guide, China embarked upon an industrialization program which in theory was aimed at making China one of the major industrial powers of the world. Lacking materials and technical skills, it has had to rely heavily on the Soviet Union for imports of both. The two major features of economic reconstruction have been: the concentration on capital production, hand-in-hand with collectivization of agriculture; and the establishment of technical schools and institutes to create its own technical elite. The "Great Leap Forward," announced by the Communist Party in May 1958, was China's greatest attempt to speed its economic reconstruction.

The new regime concentrated on the nationwide problem of illiteracy. In order to promote a Communist-type educational system, shortly after the Communist takeover, the party launched an education campaign that established full-time, part-time, and night schools. This campaign was intensified in 1955, when a program was instituted that was designed to educate 70 percent of the illiterate young people in the villages, and an even higher proportion of the population in the cities by 1962. The chief motive behind these literacy campaigns was to educate and indoctrinate a new generation of Chinese into the new order. Communist statistics have shown a steady advance in education. Thus education has become a powerful weapon in the regime's propaganda machine.

OTHER EFFECTS

Communist China, though allied with the Soviet bloc, maintains a Trotskyist position on the Communist scale, which means that it holds to a strict anti-West and revolutionary attitude in foreign affairs. It strives to dominate the politics of the Asian world, and has recently extended its influence into the African and Latin American continents as well. This position poses a threat to the Western world. It also poses a threat to the "monolithic" structure of the Communist world. However, political expediency will probably continue to dictate close cooperation with the Soviet Union.[h]

---

[h] For further discussion on Sino-Soviet relations, see General Discussion of Far East, p. 539.

NOTES

1. Helmut G. Callis, *China, Confucian and Communist* (New York: Henry Holt and Co., Inc., 1959), p. 49.

2. Ibid., p. 56.

3. Ibid., pp. 3–8.

4. *A General Handbook of China* (New Haven: HRAF Subcontractor's Monograph, 1956), I, p. 211.

5. Callis, *China*, pp. 8–10.

6. Ibid., p. 11.

7. *A General Handbook of China*, II, p. 1489.

8. Ibid., p. 1821.

9. Ibid., pp. 1220–1224.

10. Ibid., pp. 1234–1235.

11. *A General Handbook of China*, I, p. 679.

12. Ibid., pp. 681–682.

13. Callis, *China*, p. 255.

14. *A General Handbook of China*, II, p. 1239.

15. Peter S. Tang, *Communist China Today: Domestic and Foreign Policies* (New York: Frederick A. Praeger, 1957), p. 31.

16. *A General Handbook of China*, II, p. 1052.

17. Robert C. North, *Moscow and Chinese Communists* (Stanford: Stanford University Press, 1953), p. 237.

18. Howard L. Boorman et al, *Moscow-Peking Axis, Strength and Strains* (New York: Harper and Brothers, 1957), p. 4.

19. *A General Handbook of China*, II, pp. 1271–1273.

20. Tang, *Communist China*, pp. 53–54.

21. *A General Handbook of China*, II, p. 1265.

22. Callis, *China*, p. 290.

23. Ibid., p. 284.

24. Ibid., p. 294.

25. C. P. Fitzgerald, *Revolution in China* (London: The Cresset Press, 1952), p. 73.

26. W. W. Rostow et al, *The Prospects for a Chinese Communist Society* (Cambridge: Center for International Studies, Massachusetts Institute of Technology, 1954), I, p. 54.

27. Harold C. Hinton, *Leaders of Communist China* (Santa Monica, Calif.: Rand Corporation, Astia Document Number AD 123522, 1956), pp. 148–153.

28. Ibid., pp. 231–232.

29. Callis, *China*, p. 272; see also Michael Lindsay, "Mao Tse-tung," *The Atlantic*, December 1959, 56–59.

30. Frank A. Kierman, Jr., *The Chinese Intelligentsia and the Communists* (Cambridge: Center for International Studies, Massachusetts Institute of Technology, 1954), p. 11.

31. Conrad Brandt, Benjamin Schwartz, and John K. Fairbank, *A Documentary History of Chinese Communism* (London: George Allen and Unwin, Ltd., 1952), p. 217.

32. Ibid.

33. Edgar Snow, *Red Star Over China* (New York: Random House, 1938), pp. 211–220.

34. See Brandt, *Documentary History*, pp. 419–421, for a commentary on the 1945 Constitution. The text of the Constitution follows on pp. 422–442; see also Tang, *Communist China*, pp. 113–124.

35. Tang, *Communist China*, p. 64.

36. Rostow, *Prospects*, pp. 32–33.

37. Tang, *Communist China*, p. 64.

38. Rostow, *Prospects*, p. 45.

39. A. Doak Barnett, *Communist China and Asia* (New York: Vintage Books, 1960), p. 339.

40. Ibid., pp. 343–344.

41. Ibid., p. 173.

42. Tang, *Communist China*, p. 50.

43. Brandt, *Documentary History*, p. 50.

44. Ibid., pp. 221–222.

45. Francis F. Fuller, "Mao Tse-tung: Military Thinker," *Canadian Army Journal*, XIII (1959), 62.

46. Mao Tse-tung, *Selected Works* (London: Lawrence and Wishart, 1954), I, pp. 175–253; see also Fuller, "Mao Tse-tung," pp. 64–65.

47. Snow, *Red Star*, pp. 191–192.

48. Fuller, "Mao Tse-tung," pp. 65–66.

49. Mao, *Works*, II, p. 135; see also Fuller, "Mao Tse-tung," pp. 66–68.

50. Israel Epstein, *The Unfinished Revolution in China* (Boston: Little Brown & Company, 1947), p. 86.

51. Rostow, *Prospects*, p. 35.

52. James Emmett Garvey, *Marxist-Leninist China: Military and Social Doctrine* (New York: Exposition Press, 1960), p. 241.

53. Jack Belden, *China Shakes the World*, (New York: Vintage Books, 1960), p. 351.

54. Rostow, *Prospects*, p. 45.

55. Fitzgerald, *Revolution*, p. 100.

56. Mao, *Works*, II, pp. 150–153.

57. Garvey, *Marxist-Leninist China*, p. 243.

58. Brandt, *Documentary History*, pp. 423–425.

59. Ibid., pp. 295–318.

60. Hsiao Hsia (ed.), *China, Its People, Its Society, Its Culture* (New Haven: HRAF Press, 1960), pp. 239–240.

61. Brandt, *Documentary History*, p. 50.

62. Callis, *China*, p. 290.

63. Ibid., p. 281.

64. Hsiao, *China, Its People*, p. 238.

65. Fitzgerald, *Revolution*, p. 71.

66. Epstein, *Unfinished Revolution*, p. 216.

67. *United States Relations with China*, Department of State Publication 3573, Far Eastern Series 30 (Washington: United States Department of State Office of Public Affairs, Division of Publications, 1949), p. 250.

68. Ibid., p. 278.

69. Callis, *China*, p. 307.

70. Hsiao, *China, Its People*, p. 208.

71. Ibid.

72. Peter S. H. Tang, *Communist China as a Developing Model for Underdeveloped Countries* (Washington: The Research Institute on the Sino-Soviet Bloc, 1960), pp. 29–31.

73. Ibid., p. 33.

74. Ibid., p. 34.

75. Ibid., p. 35.

RECOMMENDED READING

BOOKS:

Barnett, A. Doak. *Communist China and Asia*. New York: Vintage Books, 1960. Barnett describes the Communist regime and its background, and broadens his inquiry to include the recent problems Communist China presents to the Western world.

Belden, Jack. *China Shakes the World*. New York: Harper and Brothers, 1949. Jack Belden was in China during the period of Civil War. His personal observations, episodes, and interviews add an intimate ingredient to his analysis.

Brandt, Conrad, Benjamin Schwartz, and John K. Fairbank. *A Documentary History of Chinese Communism*. London: George Allen and

Unwin, Ltd., 1952. The authors have put together a score of Chinese Communist documents dating from 1921 to 1949 and have included excellent commentaries.

Chassin, L. M. *La Conquète de la Chine par Mao Tsé-toung (1945–1949).* Paris: Payot, 1952. The author breaks down the war from 1945 to 1949 into five phases of military and political activity. It presents an excellent account of military tactics and evaluation, supported by 12 maps.

Hsiao Hsia (ed.). *China, Its People, Its Society, Its Culture.* New Haven: HRAF Press, 1960. This volume contains a comprehensive study of China. It deals with all aspects of Chinese political, social, and economic life.

Fitzgerald, C. P. *Revolution in China.* London: The Cresset Press, 1952. Mr. Fitzgerald presents a very close analysis of the Chinese revolution, and includes very pertinent material on the revolutionary setting.

Garvey, James Emmett. *Marxist-Leninist China: Military and Social Doctrine.* New York: Exposition Press, 1960. Based on studies of CCP military doctrines and practice, the military writings of Mao, and the Marxist-Leninist theory and practice in China, the book analyzes CCP military doctrine as it has developed from practice.

Mao Tse-tung. *Selected Works.* London: Lawrence and Wishart, 1954 and 1956, I–IV. These selected works are basic. The papers were written between 1926 and 1945, and contain the political and military strategies and tactics of the Communist forces in China.

Schwartz, Benjamin I. *Chinese Communism and the Rise of Mao.* Cambridge: Harvard University Press, 1952. Mr. Schwartz deals primarily with the formative years of the Chinese Communist Party and the features of Maoist strategy.

Snow, Edgar. *Red Star Over China.* New York: Random House, 1938. Sympathetic with the Communist cause in China, the book has become controversial, but is still basic in the study of the Chinese Communist Revolution. The best source covering the period of the "Long March."

Tang, Peter S. H. *Communist China Today: Domestic and Foreign Policies.* New York: Frederick A. Praeger, 1957. Tang presents a very comprehensive study of the organizational structure of both the Communist Party and the government apparatus in China. A revised edition will be available in the near future.

United States Relations With China. Department of State Publication 3573, Far Eastern Series 30, Washington: United States Department of State Office of Public Affairs, Division of Publications,

1949. A compiled record of U.S. relations with China, with special emphasis on the years 1944–49.

Vinacke, Harold M. *A History of the Far East in Modern Times*. New York: Appleton-Century Crofts, Inc., 1959 (Sixth Edition). This is an excellent book for a general study of China. It offers a good deal of background.

PERIODICALS:

Fuller, Francis F., "Mao Tse-tung: Military Thinker," *Canadian Army Journal*, XIII, (1959), 62–71. A brief discussion of Mao's military doctrines.

Lindsay, Michael, "Mao Tse-tung," *The Atlantic*, December 1959, 56–59. A brief biography of the Chinese Communist leader. Mao is described as a doctrinaire Marxist.

Lindsay, Michael, "Military Strength in China," *Far Eastern Survey*, XVI, (April 9, 1947), 80–82.

Sellin, Thorsten (ed.), "Report on China," *The Annals of the American Academy of Political and Social Science*, 1951. The periodical reviews the Chinese revolution 2 years after its success. It includes 21 articles.

Tang, Peter S. H., "Stalin's Role in the Communist Victory in China," *American Slavic and East European Review*, XIII, 3 (October 1954), 375–388. The author explains the principal role of the peasant in the Chinese revolution.

OTHER:

Hinton, Harold C., *Leaders of Communist China*. Santa Monica, Calif.: Rand Corporation, 1956, Astia Document Number AD 123522. A study of 238 individuals, Communist and non-Communist, within the Chinese Government, Chinese Army, and Chinese Communist Party.

Rostow, W. W. et al, *The Prospects for a Chinese Communist Society*, I and II, Cambridge: Center for International Studies, Massachusetts Institute of Technology, 1954. The two-volume study is a detailed analysis of the results of the revolution.

SECTION VII

····················

EUROPE

General Discussion of Area and Revolutionary Developments
The German Revolution of 1933
The Spanish Revolution of 1936
The Hungarian Revolution of 1956
The Czechoslovakian Coup of 1948

TABLE OF CONTENTS

GENERAL DISCUSSION OF AREA AND REVOLUTIONARY DEVELOPMENTS

THE GERMAN REVOLUTION OF 1933

THE SPANISH REVOLUTION OF 1936

TABLE OF CONTENTS

619

THE HUNGARIAN REVOLUTION OF 1956

THE CZECHOSLOVAKIAN COUP OF 1948

TABLE OF CONTENTS

EUROPE

GENERAL DISCUSSION OF AREA AND REVOLUTIONARY DEVELOPMENTS

GENERAL DESCRIPTION

GEOGRAPHY

The subcontinent of Europe is located between the Mediterranean and Black Seas in the south, the Atlantic and Arctic Oceans in the north, the Atlantic Ocean in the west, and the Ural Mountains and Ural River in the east. "Europe" is an arbitrary geographic term referring to a part of the land mass of Europe and Asia, sometimes called Eurasia. It has a population of about 573 million living in an area a little larger than that of the United States.[a] The density of the population varies greatly from the thickly populated regions of Western and Central Europe to the sparsely settled steppes of European Russia. Climatic conditions range from the cold continental climate of Russia to the subtropical climate of southern Italy. All of Western Europe is affected by the warming influence of the Gulf Stream.

POLITICAL STRUCTURE

Europe is divided into 27 states, ranging in size from tiny Liechtenstein in the Alps to the vast expanse of the Soviet Union. Politically, it is divided into the Soviet bloc countries of Poland, Czechoslovakia, Hungary, Bulgaria, Romania, Albania, and East Germany, the members of the North Atlantic Treaty Organization (NATO), comprising France, West Germany, Great Britain, Portugal, Denmark, Norway, the Netherlands, Belgium, Luxembourg, Italy, Iceland, and Greece,[b] and the neutrals, Switzerland, Sweden, Ireland, Spain,[c] Austria, and several others. Though the last are nonaligned militarily, their ideological identification is clearly with the West. Yugoslavia and Finland occupy a special category: both are "neutral," but the former has a Communist

---

[a] Europe comprises 3,850,000 square miles compared to 3,022,387 for the U.S.

[b] Turkey is also a member of NATO, and factionally European, although most of its territory lies in Asia.

[c] Spain is not a member of NATO. Its defense agreement with the United States, however, modifies its neutrality.

government that frequently follows a course independent of Moscow,[d] while the latter has a democratic government which—due to its physical proximity to the Soviet Union and its alliance with Germany during World War II—has had to make considerable concessions to its powerful neighbor.

The governments of Europe represent many political systems, but there are three major types into which they can be grouped: the one-party totalitarian dictatorship, the parliamentary system, and the authoritarian government. The presidential system, in which the chief executive is elected directly by the people for a fixed term, has not been adopted by any European country, although some European governments reflect a compromise between the parliamentary and presidential types.

The Soviet-bloc nations have patterned their political structure after the one-party dictatorship of the Soviet Union. The most characteristic feature of the Communist system and other totalitarian governments is the absolute control of the executive over all other branches of government and the close interaction between the party and the state. All important government officials invariably hold high positions in the party hierarchy also. In the totalitarian systems of communism today and of national socialism before World War II, there is no sphere of human activity which cannot be made subject to the control of party and state. At the same time, there is an attempt to impose universal adherence to the political ideology and philosophical assumptions of the party. The degree of popular support varies, but one can only guess as to the number of people who oppose the regimes, since no free expression of political opinion is tolerated. Only the Soviet Union arrived at a Communist state through revolution. All other European Communist countries owe their government to the political disruption occasioned by World War II and the presence afterward of Soviet troops to back up indigenous Communist elements. In Yugoslavia and Albania the ground was prepared by Communist partisan movements which succeeded in eliminating their political rivals even before the end of the war.

Whether or not many of these regimes could survive without the backing of the Armed Forces of the Soviet Union is an open question. Even where they are not actually present, the Soviet Armed Forces are known to be ready to step in should any of the satellites be threatened by revolution. Still there has been evidence of unrest in Poland, East Germany, and Hungary, in 1953 and again in 1956. In 1953 Soviet

---

[d] Yugoslavia, however, has voted with the Soviet Union in the United Nations on most of the important issues.

troops quickly dealt with the situation, but in 1956 a violent revolution in Hungary overthrew the Communist regime and installed a more democratic one. The Soviet Union intervened, however, and crushed the revolt. These events proved that the East European Communist governments were not invulnerable, and the expression of popular discontent has resulted in some mild reforms.

The parliamentary system to which most of the Western-oriented nations subscribe shows many variations. There is an important difference between the two-party British political structure and the multiparty system of the now-defunct Fourth French Republic. In the former, the leader of the majority party is also the chief executive (Prime Minister) and wields considerable influence over Parliament. The two-party system assures the Prime Minister the support of a large percentage of the Parliament.[c] When this support is withdrawn, he has to resign and new elections are held. Another important stabilizing factor in Great Britain is the institution of the monarchy. Although the monarch's functions are largely ceremonial, the Crown serves as a unifying symbol which remains above party politics. At the other extreme were the Third and Fourth French Republics—the former ending during World War II and the latter beginning after World War II—and the Weimar Republic in Germany. In both countries the executive power was almost entirely dependent on shifting parliamentary coalitions, since no party could muster a legislative majority. As a result disproportionate power was held by smaller parties, on which the larger ones had to rely in trying to obtain a working majority. In both Germany and France the multiparty system gave way to a government in which the executive dispensed with normal constitutional parliamentary procedures and, instead, relied on emergency powers. In Germany this process led to a totalitarian dictatorship and in France to the Fifth Republic, which greatly strengthened the executive. President De Gaulle, who enjoys great personal prestige, uses his emergency powers to solve the problems confronting the country and to overcome the paralyzing effects of a disunited parliament.

Italy, another example of a multiparty parliamentary system, has experienced numerous government crises, caused at least partly by the existence of a large element in parliament which, as in Germany before 1933, in Spain before 1936, and in France during the Fourth Republic, was opposed to the idea of constitutional parliamentary government. In Italy, however, the ruling Christian Democrats have been able each time to find allies among one or more of the smaller politi-

---

[c] The "two-party" system does not mean that other parties cannot exist, but it does mean that there are only two parties contending for control over the government.

cal groups and thus to present an effective front against the Communist opposition and maintain orderly parliamentary procedures.

Not all multiparty systems show the same instability as the examples mentioned above. The governments of the Scandinavian countries and of the Netherlands, for instance, have been quite stable. In all of them the dependence of the executive on parliament is not so great as it has been in France or Germany and it is more than a coincidence that all of them still adhere to a monarchy. More important than the details of the political system in these countries is the fact that there is greater unity among the people and that nondemocratic ideologies have not been embraced by a large element of the population.

Spain and Portugal provide examples of authoritarian rule. Spain has at times been referred to as a totalitarian state, but it lacks an all-embracing ideology and the will to impose a rigid totalitarian system on the people. Spain had experienced a measure of democracy under the Republic (1931–1939) but the genuinely democratic element was destroyed by the competing factions of the right and left. Franco has been able to rule Spain in coalition with the Catholic Church, which supported his bid for power, but has not attempted to supplant the church as Hitler and the Russian Communist systems did. Lately the Spanish clergy have criticized the government's labor unrest and its failure to alleviate the condition of the poor: Portugal, one of the lesser developed European countries despite its glamorous history, has never developed democratic patterns and its present ruler can be called a benevolent autocrat.

HISTORICAL BACKGROUND

Some of the more important political patterns characteristic of European states have been briefly sketched above. The political history of Europe is inseparable from the major revolutions which have greatly affected the development of every European country. European revolutions, with few exceptions, have not been of the type, frequently found in the Middle East or in Latin America, which change the composition of the ruling group without affecting political or social patterns. Most European revolutions have been manifestations of serious imbalances in the social, economic, and political structure, and most of them can be related either to the religious revolution of the Reformation, to the French Revolution, or to the Russian Revolution. The relative importance and effects of these major upheavals will be discussed separately below.

THE REFORMATION

The Protestant conviction that each individual has the right and duty to read and interpret the Bible led people to question fundamental religious beliefs and the authority of the church. From here it was but a short step to questioning the legitimacy and justice of the political authorities, and revolting against unjust rule.

The Reformation also paved the way for the acceptance of the idea of the social contract between ruler and ruled as the basis for government and the ideas of natural law, which led to the definition of the natural rights of man. It assumed a political character when Protestant peoples tried to free themselves from the domination of Spain. The revolt of the Netherlands against Spanish rule was both nationalist and Protestant, and the Thirty Years War (1618–1648) destroyed the hope that a universal religious-political system could be established. Thus the ground was prepared for the idea of the nation state that was to mature after the French Revolution.

Only in the Netherlands and in England did the liberal ideas have an immediate effect on the role of government. The Netherlands and Northern Germany became independent of Spain. In England the Puritan revolution established the constitutional monarchy, which with the development of political theory during the 17th century developed gradually into the present parliamentary democracy. In France, the Huguenots fought to establish constitutional government in order to obtain religious tolerance, but were defeated, persecuted, and exiled. The absolutism of Louis XIV set the pattern of government for most of Europe. The power of the Catholic Church was reduced, but monarchs ruled their nation by "divine right."

Nevertheless, the liberal political ideas, generated primarily in the Netherlands and England, continued to exert their influence. At the same time fundamental changes took place in the social order as the middle class began to develop through the expansion of trade and the rise of great commercial centers. The aristocracy lost much of its importance, as the state was able to obtain revenue from other sources. The society and economy of nations were changing, but the political structure of the state remained rigid.

THE FRENCH REVOLUTION OF 1789

In France especially, conflicts of interest and aspiration between the absolute monarchy, the middle class, the aristocracy, and the peasantry resulted in governmental paralysis. The revolution of 1789 did more than change the political order. It brought to the fore liberal ideas on the rights of men and equality before law, and the concept that these should henceforth be guaranteed by constitutional government. It also introduced the principle that men ought to have the right to determine to which state they chose to owe allegiance. The revolutionists did not maintain that the rights to which they aspired were the prerogative of Frenchmen alone; they believed that liberty and equality were the birthright of all mankind.

From the ideas of the French Revolution subsequent revolutionary movements have drawn much of their doctrine. The spirit of the French revolution was spread all over Europe through the conquests of Napoleon, whose armies, composed of citizens as well as professional soldiers, annihilated the standing armies of the absolute rulers of Austria, Prussia, and Spain. Napoleon was finally defeated and the monarchy was restored, though without absolute power, but the revolution remained alive in all of Europe.

Immediately after Napoleon's defeat it appeared as if the conservative monarchies had won a major victory. The four major powers, Austria, Prussia, Russia, and England, agreed to stand together to supervise the affairs of Europe and to control the revolutionary ferment that remained just below the surface, not only in France, but everywhere. The ruling monarchs, through the so-called "Concert of Europe," established the principle that a revolution in any country was the affair of all the others and that intervention was justified to bring the erring nations again under control of those who believed in the divine rights of kings. The principle of intervention was used frequently to curb the liberal nationalism of the Italians, who wanted to free themselves of the control of the Pope and foreign powers and create a unified state.

The combination of nationalism and liberalism spread by the French Revolution was also promoted by the Industrial Revolution. After the fall of Napoleon the middle class, in France and elsewhere, continued to gain in importance through the rapid expansion of trade, industry, and finance, and the increase in the number and influence of intellectuals such as lawyers, doctors, journalists, and teachers. These men aspired to positions in society which traditionally had been held by the aristocracy and to political power which remained largely in the hands of monarchs. They agitated for an extension of suffrage rights and for written constitutions, modeled after the French revolutionary Constitution, and they championed freedom of speech, religion, and the press. It was the middle class that worked hardest for the unification of Italy and Germany, then divided into a number of petty kingdoms, principalities, and duchies, partly out of idealism, partly because of the economic and political advantages that a unified state was expected to bring about.

The first major challenge to the Concert of Europe came in 1830, when revolution broke out in France and toppled the Bourbon monarchy of Charles X, who had attempted to restore absolute rule. France did not revert to a republic, but Charles's successor, the Duke of Orleans who took the royal name of Louis Philippe, undertook to respect the constitutional limitations placed on the Crown. In Belgium, the political union with the Netherlands imposed by the Congress of Vienna was severed and Belgium became independent, chose a King, and established a constitutional monarchy. Revolutions in parts of Italy were suppressed through the intervention of Austrian and French troops. Unrest in some of the small German principalities led to some political reforms, but because of the dominant role of Austria, the spirit of liberalism was effectively stifled in Germany.

A more serious threat to Austria's conservative domination developed in 1848, when revolution flared all over Europe, as patriots imbued with the spirit of the French Revolution rose against the old social and political order. Disappointed by the autocratic rule of Louis Philippe, the French turned against him. The working classes, much more numerous now, were restive because of poor wages and working conditions. The King was forced to flee the country and a Republic with universal suffrage was proclaimed. Mass demonstrations in Austria forced the Chancellor, Prince Metternich, the main champion of the old order, to resign, and compelled the Emperor to grant his people a constitution. Revolutionary activity in Hungary also caused him to elevate that country to a semiautonomous constitutional monarchy, but the Austrian Emperor remained King of Hungary.

The success of the liberal forces was incomplete, however. In Italy the revolt was initially very successful—a Roman Republic was proclaimed and the Pope was driven from the Papal States—but unification was frustrated by Austrian troops, who maintained a foothold in the country. In Prussia the people of Berlin set up barricades and rioted, demanding a constitution. The King, wanting to avoid bloodshed, agreed, but the Constitution was heavily weighted in favor of the upper classes and landlords. In the other German states the revolutionary forces achieved considerable initial success. In 1848 a National Assembly was elected by universal male suffrage to meet in Frankfurt and draw up a constitution to unite the country. This attempt failed, however. Austria opposed the unification of Germany under a liberal constitution. Also, when the King of Prussia—the largest state and the natural leader of a united Germany—was offered the crown of a federated Germany, he refused to recognize the National Assembly, because he disapproved of its liberal spirit.

Most of the revolutions of 1848 had limited objectives. The revolutionary element was fighting for political reforms, not for complete demolition and reconstruction of the political system. Everywhere except in France the monarchy remained and the concessions that were granted could be either withdrawn or manipulated in such a manner that control remained with the old ruling groups. As a result, the old political order managed to survive and much that had been gained by the forces of liberalism was lost again. Alliances with other conservative forces reestablished the power of the monarchy in Prussia, Austria, and the smaller German states. Austria reestablished her control over Hungary and consolidated her hold on the rest of her empire. In France, the Republic failed to endure. The revolution of 1848 had brought a coalition between the workers and middle-class elements, but the middle class feared the potential power of the masses. A Socialist worker's rebellion was suppressed and, to restore order and unity, a nephew of Napoleon was elected President.

Yet the idea of constitutionalism could not be destroyed. Between 1852 and 1870 Italy and Germany were unified. In Italy the liberal elements won out and, through an alliance with France, Austria was driven out of the country and a constitutional monarchy was established. The unification of Germany, on the other hand, was accomplished through the efforts of the conservative Bismarck and the King of Prussia, and by the wars against Denmark (1864) and Austria (1866). In 1870 a German federation under Prussian leadership defeated France in war, thus overcoming a major obstacle to Prussia's hegemony over Germany. In 1848 the King of Prussia had refused to accept the crown from the hands of the German people; in 1871 he accepted it at the hands of the German princes. German nationalism had become divorced from liberalism and the spirit of the French Revolution, a fact which was to become of paramount importance after World War I.[f]

In Eastern Europe the forces of democracy, liberalism, and nationalism had made less progress than in Central and Western Europe. The Russian and Ottoman Empires remained thoroughly autocratic. Nationalism did develop in the Balkan territories under Turkish rule. Greece, with the help of England and France, revolted against Turkey and became an independent monarchy under a German prince. Other regions succeeded in achieving semiautonomous positions

---

[f] See summary of German Revolution of 1933, p. 639.

within the Ottoman Empire, but the spirit of liberalism had not really penetrated into the Balkans.

The progress of liberalism and democracy continued to dominate political developments in most European countries between 1870 and 1914. Germany, Austro-Hungary, Russia, and the Turkish Empire remained basically autocratic,[g] but even here opposition to absolutism and autocracy could not be stifled. Political parties were organized and the masses developed political consciousness. On the whole, the period between 1870 and 1914 was characterized by internal and external stability. The new political and economic theory of Marxism gained ground and supplanted the spirit of the French Revolution as the strongest revolutionary force in Europe and the rest of the world. Partly as a reaction to socialism, coalitions between bourgeois liberal factions and conservative groups were formed to stem the growth of the new threat to the social order.

THE RISE OF MARXISM

The French Revolution and the political developments in Europe that followed it had primarily benefited the middle class. The industrial revolution of the 19th century placed wealth and power in the hands of the bourgeoisie, but the new working class did not benefit. Workers in the new factories labored under miserable conditions. A tremendous rise in population had created abundant supplies of labor and working conditions were not yet regulated by the state. The conditions existing in factories would be unbelievable today.

Marx developed an economic interpretation of history and a historical determinism which led to the conclusion that the destruction of the capitalist system and the creation of a Socialist state were inevitable. In the Socialist state the means of production would be owned by the workers. Eventually the state would wither away, since the Socialist revolution would remove class conflict by eliminating all but the working class and thus would remove the need for authority. However, before this goal could be reached there would be a "dictatorship of the proletariat" until all remnants of capitalism were destroyed. Marx believed that under the capitalist system labor was treated as a commodity and that the trend in the direction of greater concentration of wealth in fewer hands would continue, while workers would be reduced closer and closer to the starvation level. This condition

---

[g] There was, however, a big difference between Germany and Austro-Hungary on the one hand, and Russia and the Ottoman Empire on the other.

would be relieved, he believed, by the revolution of the proletariat, which would shake off the chains imposed on it by the capitalist class.

Actual events contradicted Marx's theory, for the rich were not necessarily getting richer and the poor were not getting poorer. As a result of this contradiction a split occurred in the Socialist camp between those who adhered rigidly to Marx's idea of world revolution and those who believed that socialism could be arrived at through evolutionary processes. Moreover, as Marxist ideas gained more followers, labor and trade unions were organized and governments in Central and Western Europe as well as England began to regulate conditions under which labor could be employed.

World War I created the conditions under which socialism, both the evolutionary and the revolutionary varieties, could really prosper. The first successful Socialist revolution took place in Russia, a country in which under Marxist theory it would have been least expected because Russia had not yet become an industrial nation. Like the French Revolution of 1789, the Russian Revolution succeeded not so much because of the efforts of the revolutionaries as because of the inner weaknesses and impotence of the old regime, which the war magnified.

The Bolsheviks came to power in November 1917 and made peace with Germany. Marxists everywhere jubilantly hailed the birth of the first Socialist state, and turned to it for guidance and leadership, which the Bolsheviks were happy to provide. Moscow became the center of the international Communist movement, but it soon became evident that the national aims of the Soviet Union took precedence over the interests of Communist parties in other countries.

THE REVOLUTIONS OF THE RIGHT

World War I destroyed the equilibrium under which Europe had prospered before 1914. The real aims of the war continue to be debated even today, but the Allied nations at least professed to have fought, in President Wilson's words "to make the world safe for democracy." As a result of the war, the autocratic power of the Kaiser in Germany and the Emperor of Austro-Hungary was broken. Czechoslovakia and Yugoslavia emerged from the wreckage of the Austro-Hungarian Empire; Poland, Finland, and the three Baltic countries received their independence from Russia; and a number of new Arab states appeared in the territories long ruled by Turkey. The League of Nations was established to secure peace for the world.

But democracy did not prosper. In reality, World War I had the opposite effect, and in large areas of Europe reversed the gradual movement toward democratic reform which had been gaining impetus. The revolution in Russia had created a dictatorship and a totalitarian state even before the end of the war. In 1922 Mussolini established a Fascist state in Italy; in 1933 Hitler became dictator of Germany; and by 1939 General Franco destroyed Republican Spain. All of these events were accomplished through revolution; thus the major revolutions of postwar Europe were not Marxist or Communist, but nationalist and rightist in nature.[h]

In Italy, Socialist-inspired industrial and agrarian disorders were rampant in 1919 and 1920. Mussolini, despite the fact that he himself had been a Socialist, took advantage of the antirevolutionary sentiment to strengthen his Fascist Party and, in 1922, utilized the confusion created by a Socialist general strike to execute a bloodless revolution. Thirty thousand armed Blackshirts marched on Rome, and the King, rather than risk civil war, commissioned Mussolini to form a government.

While the rightist revolutions succeeded, the Communist revolutions failed. (The fall of the monarchy in Germany benefited not the Communists, who could not gather sufficient strength to bring their revolutionary efforts to a successful conclusion, but the evolutionary Social Democrats.) Yet the revolutionary Communist parties, through their antidemocratic spirit and destructive tactics did much to bring about the victories of the dictators. Moreover, the Fascists in Italy and the Nazis in Germany adopted many of the ideas and techniques originally developed by the Communists.[i] Though the one-party totalitarian regimes of the right were violently anti-Communist in their internal struggle for power, they adopted the Communist Party's basic organizational structure of tightly controlled cells and regional organizations; they also emulated Communist techniques of propaganda and terror. The regime established by Hitler was very similar to that of Stalin, though fundamental differences existed between the ideological basis of National Socialism and that of communism. If the totalitarianism of the right was short-lived, while Russian communism remains, it was due not to the internal weaknesses of the system, but to Hitler's and Mussolini's overestimation of the power of the states they controlled, coupled with their underestimation of the rest of the world.

Though World War II destroyed the revolutionary regimes of Hitler and Mussolini, it strengthened the position of the Soviet Union. The rightist regimes in Germany, Italy, and Spain were never strong

---

[h] The interwar years also witnessed the establishment of rightwing dictatorships in Hungary (1920), Poland (1926), Yugoslavia (1929), and Greece (1936).

[i] See the summaries of the Spanish Revolution of 1936, p. 677, and the German Revolution of 1933, p. 639.

revolutionary forces outside their countries. It is true that the success of Mussolini inspired Hitler and that both Hitler and Mussolini helped France to come to power, but each country conducted a nationalistic foreign policy and each emphasized its respective ideology as being right, not for the world in general, but only for that particular country.[j] Thus there was a major difference in contagious effect between the French and Russian Revolutions and those in Italy and Germany. When Germany and Italy were defeated, little remained to inspire other nations to follow a similar path. The Communist totalitarian ideology, on the other hand, grew as a result of the collapse of its nationalist rivals.[k]

RECENT TRENDS AND OUTLOOK FOR THE FUTURE

World War II brought a new development that was closely related to revolution: the emergence of national resistance movements against the German occupation. Whether or not a resistance movement is revolutionary is a theoretical question that need not be answered here. Of greater interest in the context of European revolutionary development is the role these resistance movements played after their initial objective of weakening the enemy had disappeared. Pre-World War II Communist parties outside Russia had to rely mostly on their own resources, as the Soviet Union was too much concerned with establishing socialism in its own country and defending it against counterrevolutionary forces, both internal and external, to be able to give substantial support. World War II, however, created an opportunity to organize forces ostensibly to fight Germany, but actually to be in a position to take over political control after Germany was defeated. The Communist resistance movements in Yugoslavia, Albania, Greece, France, and Italy were the most effectively organized and were potentially able to do the greatest damage to Germany. For this reason they were supported by the Western Allies. Thus their position, already relatively strong, was made even stronger vis-a-vis other groups that had organized for patriotic reasons.

At the end of the war, the Albanian partisan movement had little difficulty in taking over political control. In Greece, the Communist

---

[j] It must be noted here that the Thousand Year Reich, in the sense of a united right-wing Europe, appealed to certain European conservative circles.

[k] This is not meant to suggest that the Soviet Union or other Communist powers are not guided by purely national consideration, but only that the ideological orientation remains an international revolutionary force. In the post-World War II period it gained support primarily in the less-developed countries of Asia, Africa, and Latin America, rather than in the industrial powers of the West.

bid for power was blocked by the presence of Allied troops, but the wartime resistance movement became the nucleus for a postwar uprising that came close to success. France and Italy both were confronted with strongly organized Communist forces that constituted a serious threat to the democratic governments installed as the countries were freed of German troops. It remains a matter of speculation whether the Communists in Italy and France could have taken over the governments if British and American troops had not been present during the process of liberation.

In the eastern European countries other than Albania, Communist rule was established not through revolution but through the power of the Soviet Army. Communist regimes were installed in Poland, Romania, Bulgaria, Hungary, and Yugoslavia without much regard to the wishes of the population or the relative strength of the Communist movement. Only Czechoslovakia was allowed to set up a democratic form of government, probably because the political leaders of the country were known to be friendly toward the Soviet Union, and because the Communist Party was the strongest single party. However, after the cold war began, when Czechoslovakia indicated its desire to participate in the Marshall Plan, the Communists used their control of key ministries to take over the government and establish a Communist state. Thus democracy was destroyed by a revolution from within the government.[l]

At present, revolutions in the Communist countries do not seem likely. As long as the Soviet Union maintains her position, the example of Hungary is likely to deter anyone from making a similar attempt.[m] However, it is equally likely that revolutionary forces could organize quickly in such countries as Poland and Hungary if external conditions should neutralize the power of the Soviet Union, since there is little evidence that the majority of the population have adopted the Communist ideology. The fear of nuclear war will probably continue to deter the Western powers from actively supporting indigenous revolutionary movements, should these succeed in organizing.

Successful revolutions are contagious, as the example of the French and Russian Revolutions shows. By the same token, unsuccessful revolutions tend to inhibit other revolutionary movements. This has been the result of the Hungarian Revolution and also, to some extent, of the unsuccessful Communist revolution in Greece. Both events have contributed to the maintenance of the *status quo* in the East as well as the West.

---

[l] See summary of Czechoslovakian Revolution of 1948, p. 731.
[m] See summary of Hungarian Revolution of 1956, p. 711.

The least secure regimes in Europe appear to be the authoritarian governments of Portugal and Spain. In both, opposition has been evident and the fear of outside intervention is not likely to act as a deterrent (as is the case with Communist movements in the West and anti-Communist movements in the satellite countries). A more accurate method of predicting revolutions would need to be developed, however, before one would be safe in designating any country as either immediately threatened by revolution or immune to it.

The four revolutions summarized below are the national socialist Revolution in Germany, the Spanish Revolution, 1936–1939, the Czechoslovakian Revolution of 1948, and the Hungarian Revolution of 1956. Each offers a rich field for a study of revolutionary techniques, and of the interaction between techniques and the spirit which animated the revolution. While the ideology of the German Revolution is not likely to reappear, the Nazi story suggests that even modern "civilized" nations can be corrupted to an incredible degree. The Czechoslovakian Revolution is an example of a cold-blooded, calculated political maneuver that has little in common with the Marxian ideas of a proletarian revolution. One might suggest that it was more the result of power politics than of the revolutionary spirit within a nation. The Spanish Civil War illustrates numerous interesting aspects of revolution. On the one hand there was a basically conservative revolutionary force, on the other a democratic republic. But as the fighting dragged on, the issues became blurred by outside intervention—Italian and German on the side of the revolutionaries, Russian on the side of the Republic—and by the increasing penetration of the loyalist cause by Communists, native and imported. Thus what was initially a civil war took on the aspects of a struggle between rival revolutionary movements—Fascist and Communist—on Spanish soil.

THE GERMAN REVOLUTION OF 1933

SYNOPSIS

On August 19, 1934, the German people confirmed by plebiscite the fusion of the Reich Chancellorship and the Reich Presidency—in reality a fusion of civilian and military power in the person of Adolf Hitler. Reaching the Chancellorship in January 1933, through a mixture of legal political processes and backstage maneuvering, Hitler was able to consolidate his rule through the Enabling Act of March 24, 1933. Through coercive measures sanctioned by the Enabling Act, Hitler eliminated all vocal opposition. Upon President Hindenburg's death on August 2, 1934, Hitler declared himself President, and a plebiscite reaffirmed the declaration.

BRIEF HISTORY OF EVENTS LEADING UP TO AND CULMINATING IN REVOLUTION

The history of the Weimar Republic can be divided into three phases: the birth and crisis period from 1919 to 1923, the period of stability from 1924 to 1929, and the disintegration era from 1930 to 1934.

Promulgated on August 11, 1919, the Weimar Constitution provided for a quasi-federal republic with a parliamentary type of government. Initially the new government was dependent upon center and left-of-center political groups. The first problem it faced was carrying out the terms of the Treaty of Versailles. This treaty deprived Germany of at least 11 percent of its productive capacity, demanded the payment of reparations without setting the exact figure, stipulated that Germany disarm, and, finally, placed on it the burden of guilt for the war. The provisions greatly angered the German people in general, and nationalists and the army in particular. The fact that it was the responsibility of the Weimar Government to fulfill Germany's treaty obligations automatically focused on it the distrust of many Germans. Moreover, the fact that the victorious Allied armies never occupied Germany lent credence to the nationalists' assertions that the German people had been "stabbed in the back" by Weimar officials.

Nationalist suspicion and hatred were further intensified by Communist activity. Because the Weimar Government was based on a left-of-center complex, many felt that it condoned communism. Ex-servicemen's groups known as the *Freikorps* violently attacked Communist activities. This rightist violence against the Communists easily became translated into action against the government. The Kapp

Putsch of 1920 was an example of this. Although the rightist forces took over Berlin for 5 days, the *Putsch* failed because the army refused to join, considering it premature, and the working class countered with a general strike. There was a difference, however, between the treatment accorded to radical groups. On the one hand, the leftists received harsh sentences and little protection from the police. On the other hand, rightist violence, though directed against the regime, was not severely punished.

There were other events which caused trouble for the government. When reparations payments were finally set, the figure was so high that the government resigned. A new left-of-center and center coalition government agreed to the terms, which further incensed nationalist feeling. When in 1923 the French occupied the Ruhr in retaliation for nondelivery of certain reparations items, the government reacted with passive resistance. The threat of civil war arose as the French encouraged separatist movements. A tremendous inflation set in which wiped out the value of the mark and weakened and discouraged the lower-middle class.

The threats of civil war and socio-economic disintegration were arrested by the government of Gustav Stresemann. Another attempted uprising, Adolf Hitler's "Munich beerhall" *Putsch* of 1923, was crushed when the army obeyed Stresemann's orders. Stresemann introduced a new currency, and through this and other reforms stabilized the economy.

An era of prosperity and stability followed (1924–29). During this period, in which Stresemann was either Chancellor (1923–24) or Foreign Minister, internal as well as international conditions became more tranquil. Internally, the Stresemann period saw the growth of industry and an increase in general prosperity. Social welfare measures were greatly increased as the power of the moderate left grew.

In the field of foreign affairs, the period 1924–29 was marked by partial rapproachement with the Western powers. Up until 1924 Germany had been treated harshly by the victorious powers, especially France. The occupation of the Ruhr and the insistence on reparations payments had embittered many Germans. It was at least partly in reaction to this pressure that Germany concluded the Rapallo Treaty with Russia in 1922. This upset the Western powers because it signified that Germany was ready to follow an independent foreign policy and was able to play off the Soviet Union against the Western democracies. Though the treaty was ostensibly economic, the Germans were, at the same time, secretly discussing with the Soviet representatives the possibility of Russia's aid in helping the Germans build up their military forces in the fields which the Treaty of Versailles prohibited. The use of Russian territory for the training and development of German forces continued at least until 1930.

Under Stresemann Germany set about improving relations with the West. The Dawes Plan placed reparations payments on a more realistic basis and led to the evacuation of the Ruhr. The Locarno Pact established mutual guarantees of Germany's western borders and a quasirecognition of her eastern border. With admission to the League of Nations in 1926 and a permanent seat on its Council, Germany seemed to be on the way to becoming a satisfied power.

However, throughout the entire Stresemann period there were serious undercurrents countering the favorable trends. Hypernationalist groups like the Nazis and Alfred Hugenberg's Nationalist Party opposed every attempt to carry out the Treaty of Versailles. On the left the Communists attacked the increased power of business. The dissatisfaction of the army with a "defensive" policy and the alienation of the lower-middle classes by the 1923 inflation were other factors of unrest. Lastly, the economic boom rested on a precarious basis, international short-term credits.

With the death of Stresemann in 1929, the last phase of the Weimar Republic began. The battle over the Young Plan, which was to be the final and definitive reparations settlement and which provided for payment to be spread over 59 years, helped Hitler's party, in alliance with Hugenberg's Nationalists, to move into national prominence. Actually the Young Plan carried with it a highly advantageous *quid pro quo* for Germany—Allied withdrawal from the Rhineland, 4 years ahead of the final date set by the Versailles Treaty. In March 1930, nationalist unrest, intensified by the 1929 depression and growing unemployment, forced the resignation of the Hermann Mueller Cabinet. President Hindenburg appointed Heinrich Bruening, who, after trying vainly to get the *Reichstag* to vote his budget, was forced to call for elections. In these elections the power of the radicals increased; the Nazis captured 18 percent of the *Reichstag* seats while the Communists won 13 percent. With the newly won support of the Social Democrats, however, Bruening was able to form a government.

Bruening remained in office until the end of May with the support of the Social Democrats and the Center in the *Reichstag* and the backing of President Hindenburg. The failure of his attempt to effect an Austro-German Customs Union, which was blocked by French and Italian opposition, greatly weakened his hand and strengthened the parties of the right.

Meanwhile, the power of the Nazis was growing. When Hindenburg's term expired in 1932, Hitler audaciously announced his candidacy for the Presidency. Bruening persuaded the aged and increasingly senile Hindenburg to seek another term. He was elected, but Hitler demonstrated his strength by polling 40 percent of the vote.

For various reasons, Hindenburg's confidence in Bruening had been dwindling, and in June 1932 a scheming army officer, Gen. Kurt von Schleicher, who had the support of the army and the confidence of Hindenburg, persuaded the President to remove Bruening and replace him with Franz von Papen. During Papen's regime, which lasted 6 months, there were two *Reichstag* elections, one in July, the other in November. In the July elections, the Nazi Party obtained 230 seats out of 608. However, the November elections saw a significant decline in the Nazi popular vote, which was nearly 2 million less than in July, and in the party's *Reichstag* representation, which dropped to 196 seats. The Nazis were still much the largest party, however.

Papen was unable to obtain a parliamentary majority or put together a workable coalition, and Schleicher, tiring of his protegee, persuaded Hindenburg to drop Papen and name Schleicher Chancellor. He assumed office on December 2, 1932, but was equally unsuccessful in lining up a majority in the *Reichstag*. Hitler decided that his time had come. He convinced the aged President that he could succeed where Papen and Schleicher had failed, and on January 30, 1933, Hindenburg appointed him Chancellor. He had come to office legally, through the pressure exercised by his parliamentary plurality and the inability of the moderate parties to cooperate in support of a moderate cabinet.

From 1933 until August 1934 Hitler expanded his power into dictatorship. He had been given the Chancellorship on the promise that he would be able to form a coalition. However, the Center Party refused to cooperate, and Hitler called for an election, to be held in March of 1933. Hitler, as head of government, employed both official and unofficial instruments in coercing the German public. Although during the campaign many normal electioneering rights were denied the opposition parties, the actual voting was free. The most spectacular event during the campaign was the burning of the *Reichstag*. It was blamed on the Communists, but many historians feel that the Nazis were responsible for the fire—an effort to provide justification for future terrorism against all political opposition and minority groups.

The election, however, did not give the Nazi Party an absolute majority. Hitler was dependent on a coalition with the Nationalists. This did not satisfy him; neither was he willing to accept the restrictions of cabinet government. However, he still wanted to maintain the facade of legality. He found the answer in the Enabling Act proposed in March 1933. This was a constitutional amendment empowering the Chancellor to rule by decree without consulting the *Reichstag* or the President for 4 years. To get the bill passed, Hitler had to win a two-thirds majority in the *Reichstag*. His first step was to eliminate the 81 Communist deputies. Then, by conciliating the Nationalists and the Center he succeeded in getting the two-thirds majority, with only the Social Democrats in opposition, and his rule in civil matters was established without legal restraints.

However, Hitler still had to deal with the decentralizing forces of the Federal States, the trade unions, and the political parties, as well as the power of the army, which under the Weimar Constitution owed its allegiance to the President. Hitler abolished the state legislatures and concentrated all power in the *Reich* Government. He suppressed the unions and incorporated them in a Nazi "Labor Front." He dissolved all political parties except his own. There remained the control of the army. Hitler placated the army by eliminating the radical element from the command of the S.A. in the purges of June 1934. Upon the death of Hindenburg in August 1934, he declared himself President. On August 19, 1934, the German people, by an enormous majority (89.93 percent), affirmed the union of the Chancellorship and the Presidency in the person of Hitler. The revolution had been successful; Hitler was now Head of State, Head of Government, and Supreme Commander in Chief of the Army.

THE ENVIRONMENT OF THE REVOLUTION

DESCRIPTION OF COUNTRY

Physical characteristics

Germany is situated in the center of Europe with France, Luxembourg, Belgium, and the Netherlands on her western frontier; Denmark and the Baltic Sea on her northern boundary; Switzerland, Austria, and Czechoslovakia on her southern boundary; and Poland on her east. From 1919 to 1939 East Prussia was separated from the rest of Germany by the Polish Corridor, and touched the Baltic state, Lithuania, on the northeast. A consequence of Germany's geographical position is that the extent of German territory usually depends

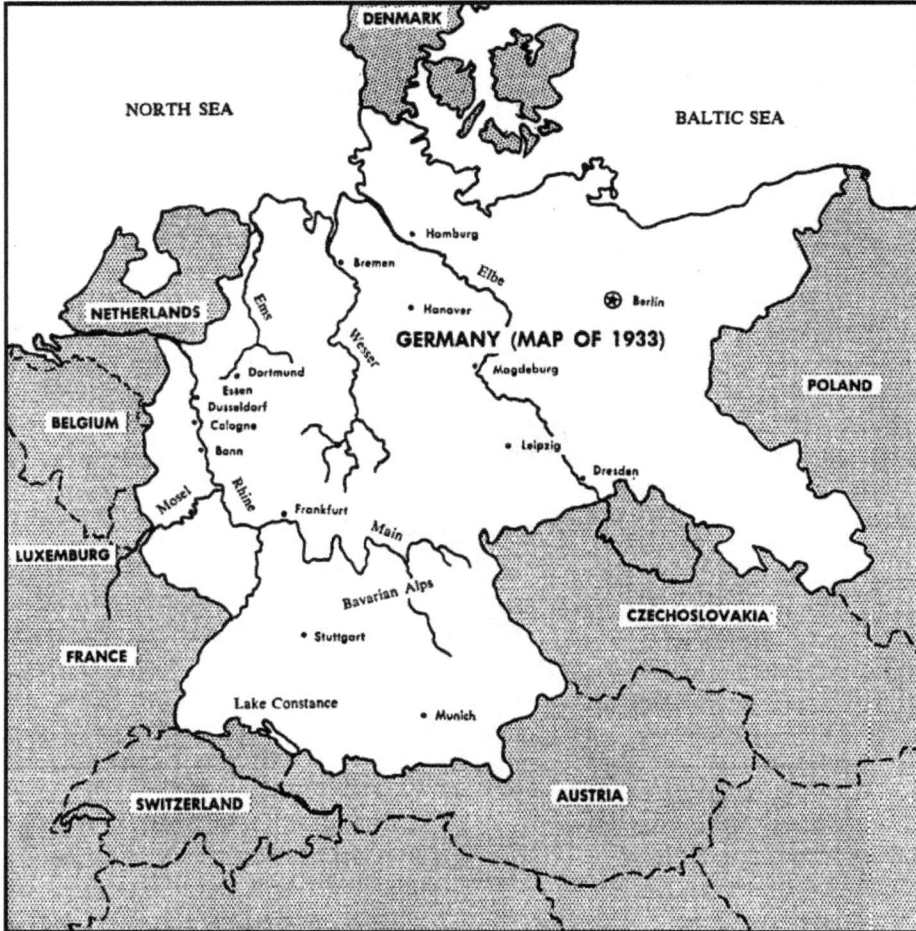

upon political considerations.[a] The Germany of the Weimar Republic had been deprived of 13 percent of its European territory and all of its colonies by the Treaty of Versailles.[1] It had an area of 180,934 square miles.

Germany can be divided into four regions: (1) the northern plains; (2) the mid-German hills; (3) southern Germany with the Main, Neckar, Swabian, Franconian Jura, and Danube basins, and the Bavarian Alps; and (4) the Rhine valley.

The people

Aside from the Jewish minority, the ethnic groups in Germany tend to follow geographical lines. In northwestern Germany the physical characteristics of the people are predominantly Nordic. "East of the

---

[a] The Versailles Treaty had, in fact, left Germany without clearly definable boundaries to the east and west, and without a clearly definable conception of what really constituted Germany.

Elbe Nordic features become less common and less pronounced. . . ." In the south the population is essentially Eurasiatic (Alpine). There is also much mixture of the Alpine and Nordic types. In the eastern areas there is some mixture of Slav types.

Aside from Yiddish and certain archaic forms like Frisian, Pomeranian, and Wendish, the German language is common to the entire country.

The population of Germany in 1933 was 66,027,000, with a density of 370 per square mile. Over 43 percent of the population lived in towns of over 20,000.[2] Some of the chief cities were Berlin, Breslau, Dresden, Duesseldorf, Essen, Frankfurt-am-Main, Cologne, Leipzig, Munich, Nuremberg, and Stuttgart.

Communications

The communication system was generally good. By the end of 1937 Germany had 33,878 miles of railroad lines in operation, more than any other European country except the Soviet Union.[3] The inland waterways were among the most highly developed in Europe. The Rhine, the Elbe, and the Oder all follow courses favorable to industry. Berlin, Hamburg, Stettin, and the cities of southwest Germany were served by waterways.[4] Although railways, rivers, and canals formed the principal means of transportation, the roads were beginning to become important, although it was not until the Nazi period that an extensive road development program was undertaken.[5] Germany's substantial international shipping was dependent upon Baltic and North Sea ports.[6] Lastly, the Weimar Republic saw the beginning of air travel.

Natural resources

Germany has only moderately fertile soil, which means that constant labor is required to maintain output. In 1927 there were over 28 million acres of forest land.[7] Fisheries are predominantly confined to the Baltic Sea, although there is some inland freshwater fishing.[8] Germany has large resources of coal, lignite, iron, and mineral salts.[9] Of somewhat lesser significance are lead ore, zinc ore, bauxite, pyrites, and magnesite.[10]

SOCIO-ECONOMIC STRUCTURE

Economic system

The two major economic developments under the Weimar Republic were the growth in governmental control and the increase in the size of both labor and business organizations.

The economic system became increasingly controlled by governmental action. Banks were set up by the *Reich* Government as well as by *Land*[b] (State) governments. The railroads, which previously had been operated by the State governments, were now centrally controlled by the Federal Government. There was also increased governmental spending in the field of social welfare.[11]

Other characteristics of the economy were increased monopolization of business and unionization of labor. The reorganization of industries after World War I resulted in the growth of powerful trusts while trade unions were active in almost all industries.[12]

The economy was primarily industrial. The most important industries were steel and iron, coal, textiles, electrical supplies, chemicals and dyes, and precision and optical instruments.[13] Germany imported foodstuffs, raw materials, and semimanufactured goods like wool, cotton, rubber, copper, and dairy products; it exported manufactured articles such as machinery, textiles, and chemical wares.[14] In terms of the balance of trade, the years from 1930 to 1933 showed a sizable surplus of exports for the first time since the war.[15]

In terms of monetary structure, the German economy's ability to expand and even to maintain itself rested on its ability to acquire credit. Most of the loans secured were essentially international short-term credits over which Germany had little control. The whole structure, therefore, was precarious.[16]

Class structure

The Weimar Republic saw a definite change in the ruling elite. While the army still maintained most of its influence, the Prussian aristocracy lost its official status and was replaced by the wealthy businessman and the bureaucrat. The lower class also began to have representatives in official circles, though this trend was more apparent in the State governments than in the Federal Government.

In spite of the increased power of the lower classes, there was little change in the class structure of society. No far-reaching land reform was undertaken, though there was some agitation for it in the 1930's.

---

[b] Singular form of *Laender.*

However, the distribution of wealth greatly affected the class structure during this period. The inflation of 1921–23 had pauperized many of the middle class, especially those on fixed incomes. The inflation did not affect—to any large extent—the position of the aristocratic landowners, big business, and organized labor, but it did alienate the middle class.

The increased educational opportunities provided a basis for increased social mobility. However, this factor was countered by the inflation of 1923, from which the middle class never recovered sufficiently to resume its traditional role of investor and entrepreneur. This fact, coupled with the growing class consciousness of the workers and capitalist-militarist bureaucratic groups, meant that the social mobility was not as great as could be expected. The middle class was no longer an independent source of social power, and the increased antagonism evident in the relationship between workers on the one hand and the capitalist-militarist-bureaucratic grouping on the other hand decreased the mobility of classes even more.

Literacy and education

Germany has always emphasized education. The Weimar Constitution affirmed this policy by stipulating that all children should go to school until they were 18 years old. The complex system of schooling provided vocational as well as academic training. Given this emphasis on education, literacy was high.

Major religions and religious institutions

By 1933 the Roman Catholics, found mostly in Bavaria, Rhenish Prussia, and Westphalia, numbered over 21 million, while the Evangelical Protestant Church, located primarily in the north, had over 40 million.[17] During the Weimar period both religious groups emphasized the social functions of religions. The Catholics were particularly influential in politics through their support of the Center Party, an important moderating force in Weimar politics.

The right of the state to control religion still existed in Weimar Germany. The State governments maintained their ultimate control over the Protestant churches.

GOVERNMENT AND THE RULING ELITE

Description of form of government

The Weimar Constitution, adopted August 11, 1919, established a complex system resting upon a federal structure, popular sovereignty, and a delicate balance between the executive and legislative branches.

The federal provisions stipulated that the decisive jurisdictions were given to the central government (*Reich*). Except for "police and internal administration, justice, and education," the *Reich* maintained complete legislative authority.[18] Moreover, the Constitution prescribed the internal structure of the governments of the *Laender* (i.e., states). It required that all *Laender* governments be republican, with identical election systems.[19] Not even "territorial sovereignty was left to the *Laender*, since their state boundaries could be changed by constitutional amendments, even against their will (Art. 18)."[20]

The role of the *Laender* was further set forth in provisions pertaining to the enforcement of federal supremacy, arbitration between the *Reich* and the *Laender*, and finally, the Federal Council or *Reichsrat*.

Federal authority over the states was enforced in two ways. As executive agents of the *Reich*, *Land* authorities were subject to far-reaching supervision by the Reich.[21] "If the *Land* failed to comply with a justified request of the Reich, or if the issue was controversial, the matter was referred to a decision of the Constitutional Tribunal (Art. 19, Par. 1)."[22] If the State "stubbornly refused to comply with the decision of the Constitutional Tribunal" or "failed in the fulfillment of its constitutional duties of maintaining orderly processes of government," the *Reich* Government, "after having exhausted all means of peaceful persuasion, could resort to the *ultima ratio* of applying coercion against the recalcitrant *Land* by the process of sanctions."[23]

The other two mechanisms determining the relationship between the federal unit and the central government, the Constitutional Tribunal and the Federal Council, attempted to establish some degree of equity and representation. The Constitutional Tribunal had jurisdiction over disputes within the States, between States and between the Federal Government and the member States, as well as the impeachment of the *Reich* President, Chancellor, and Ministers. The Federal Council or *Reichsrat* was originally designed to represent *Land* interests in the *Reich* Government. However, because the States did not have serious objections to government action, and because they had little power to check the *Reichsrat*, the organ soon became administrative.[24]

Aside from the fact that the Weimar Constitution established a definite relationship between the *Land* and the *Reich*, the Constitution created, for the first time in Germany, a government based on popular sovereignty. The degree of popular sovereignty was extensive.

The force of popular sovereignty was represented in three institutions. First, the Constitution provided for legislation by initiative and referendum. It was hoped that this would promote democratic legislation, but instead it introduced a degree of irresponsibility in the political process.[25] Secondly, the concept of popular sovereignty dominated the structure of the *Reichstag* or national parliament. A system of proportional representation was instituted whereby "each 60,000 voters elected one candidate from lists submitted by the parties."[26] This system, based on the democratic assumption that each vote should have meaning, led to a party structure which, as will be seen, undermined responsible and moderate party leadership. The last institution in which the concept of popular sovereignty was operative was the Presidency. By providing for his election by universal suffrage, the Constitution sought to give the President a separate source of power that would enable him to follow the will of the people.

The last important feature of the Weimar Constitution was the delicate balance between the executive and legislative powers. The balance was based on a complex relationship primarily because the executive function was divided between the President and the Chancellor. Given the traditional strand of monarchism in Germany, it was only natural that a strong executive should emerge. While the President had control during normal times over the military, all material decisions regarding civil matters were to be left up to the Cabinet. The Cabinet, which derived its power from the *Reichstag*, was to control the President in all of his actions. However, this control was limited by two factors. First of all, because the President was elected by popular suffrage he had great prestige and hence could influence the *Reichstag* by appealing over the heads of the Cabinet or could bypass the *Reichstag* entirely and appeal to the people. Secondly, the President, not the Chancellor, could dissolve the government and the parliament.[c] He could do this regardless of whether or not the parliament supported the Cabinet by claiming that the parliament did not represent the will of the people.[27]

All of these factors made the relationship between the legislature and the executive ambiguous. It became particularly delicate during times of crisis. Article 48 of the Constitution gave the President emergency powers, with the stipulation that the measures taken should be temporary and should not infringe on the Constitution. However, the power granted by Article 48 was quite extensive, sanctioning the abro-

---

[c] Dissolution of parliament, however, required the countersignature of the Chancellor.

gation of many civil rights. The only legal requirement on the right of the President was the countersignature of the Chancellor.[28] This gave the President an unusual amount of power during a crisis period. As will be seen, this represented a serious danger to the Constitution.

In summary, the Weimar Constitution clearly established a unified Germany with a quasi-satisfactory federal arrangement. However, the extreme degree of popular support called for by the Constitution made the government dependent upon stable political conditions. This dependence was increased by the ambiguous relationship between the executive and the legislative branches of the government, since a high degree of cooperation between the parties and between the Cabinet and the President was needed to make the government work.

Description of political process

At the outset of the Republic, 331 out of the 423 seats in the assembly were middle-of-the-road parties supporting the Republic.[29] There were, however, groups on the left as well as the right which challenged the existence of the Republic. As the history of the Republic unfolded, it was evident that these opponents would grow in strength. This was due primarily to the fact that the parties tended to solidify in their views and in their leadership. Because of the unusual system of elections, the party became more important than the individual politician. The result was an increase in the strength of party organizations. This helped the radical parties, since the most moderate parties did not possess the desire or the will to create a militant party organization. Hence, in the following analysis of the party politics of the Weimar Republic, it must be kept in mind that the tendency toward strong party organization, inherent in election laws, favored radical groups.

The political parties and power groups supporting the government represented the moderates of all classes. One of the most influential and stable parties was the Catholic Party. "It cut through all layers of the population, embracing Westphalian and Silesian aristocrats as well as the bulk of the peasants in Southern Germany, the Rhineland, and the Catholic parts of Prussia, in addition to the small Catholic bourgeoisie and the Catholic trade unions."[30] The Social Democratic Party, moderately Socialistic, also supported the Republic. While not particularly powerful on the national scene, this party had considerable power in many *Laender*, especially in the largest, Prussia. The German People's Party, which consisted of the conservative wing of the Liberals, supported the government, especially during the rule of Dr. Stresemann, their chief representative. It was a class party without mass basis, consisting mainly of big businessmen, industrial leaders, members of the propertied middle class, and the Protestant clergy.[31]

Lastly, there was the German Democratic Party, which represented the remnants of the old Liberal Party. Once the party of the middle class, this party increasingly lacked support in the Weimar Republic due to the disaffection of the middle class. In spite of this, a large number of the German intellectuals supported this party and it was represented in every cabinet from 1918 to 1931.[32] While these parties represented the majority of the German population at the beginning of the Republic, they did not increase their power, primarily because they neglected mass techniques and were bankrupt of effective, militant leadership.[33]

Hostile to the Republic and seeking to undermine it were parties of the radical right and left. The German National People's Party was made up of Prussian conservatives. "They drew their voting strength from the old ruling classes around the army and the civil service," as well as from those who feared the inroads of socialism.[34] Moreover, this party had a natural source of strength in the general German dissatisfaction over the Treaty of Versailles. At the opposite political pole were the Communists. The potential threat which they represented to the Republic served the National Socialist cause by spreading fear throughout the propertied classes. Lastly, the Bavarian Catholic Party, monarchist in outlook, opposed the Republic up to 1924. After that, however, it became a branch of the pro-Republic Catholic Center.

The National Socialist Party was in a sense an amalgam of many radical groups. The hypernationalism of Ernst Roehm was combined with the social reformism of the Strassers. The Nazis courted the industrial elite, the middle classes, and the masses. Given the difficulty of maintaining the allegiance of these diverse groupings, Hitler's techniques became all-important, for they covered up the inconsistencies in the party program. However, because his party did appeal to radicals of almost any sort, it became the center of the forces arrayed against the Weimar Republic. In fact, this desire to replace the Republic seemed to be the one cohesive factor in the National Socialist movement.

In comparing the parties supporting the Republic to the parties opposing it, one finds that the most outstanding differences were in organization and technique. The supporters of the government had few strong leaders, no myths, no militant creed with which to attract mass support. The opposition found latent support in the masses. Their radicalism was exploited on the one hand by the Communists, on the other by Hitler, a man who could vocalize the resentment against the existing order and give it militancy.

Legal procedure for amending constitution or changing government institutions

The Weimar Constitution was unusual in that it provided that governmental institutions could be changed without constitutional amendment. The mechanism for amendment was somewhat unwieldy. Qualified majorities were required in both the *Reichstag* and the Federal Council: the passage of an amendment in the *Reichstag* necessitated the presence of two-thirds of the total membership, and required that two-thirds of those present voted in favor of the proposed amendment; in the Federal Council, two-thirds of the members had to vote in favor. In special cases an amendment could be initiated by one-tenth of the electorate; it then required a majority vote of the *Reichstag* or of the registered voters.[35]

However, there was a simpler procedure. This involved passage of a statute which effectively circumvented the Constitution without, however, amending it. Such a "statute amending the Constitution" was passed "by way of the constitutional amendment procedure with the required majorities" and was not subject to judicial review.[36]

Aside from this, Article 48 was invoked to enable the President to replace a parliamentary cabinet with a presidential cabinet, i.e., a cabinet based on presidential rather than parliamentary consent. While some argued that this was illegal,[37] it seemed to be sanctioned under the power of the President to appoint the Cabinet and the authority given him through Article 48, and therefore represented some degree of constitutionality.[38]

Relationship to foreign powers

By 1930 Germany had reestablished itself in the family of great powers. In the Locarno Pact Germany accepted the western frontiers established by the Treaty of Versailles. In September 1926 Germany became a member of the League, with a permanent membership on the Council. In the east the situation was somewhat different. At Locarno Germany had made an effort to avoid an iron-clad guarantee of the eastern boundaries drawn at Versailles. Relations with Russia were formally cordial in terms of the Rapallo Treaty of 1922. Underneath this cordiality, there were ambiguous forces. On the one hand the army had been receiving Russian help in rebuilding its strength—not without the sanction of the republican government. On the other hand the Russians were directing the Communist Party in Germany in its efforts to destroy the Republic. In general, 1930 found Germany in relative peace with her neighbors.

The role of military and police powers

The German military occupied an ambiguous place in the life of the Weimar Republic. There were officers like Hindenburg, Gen. Kurt von Schleicher, and Gen. Wilhelm Groener, who supported the Republic mainly because they were in official positions. These men held great influence over the course of politics because they could speak for the army. Moreover, Hindenburg as President was the official as well as the symbolic head of the army.[39] There were other officers, however, who secretly supported movements for the overthrow of the Republic on the grounds that it had betrayed German honor. Roehm, before becoming head of the S.A., was one of these. General Ludendorff had actually found himself the leader of a revolutionary *Putsch* in 1923.[40]

The German Army was probably the most powerful force in politics. However, as long as Hindenburg remained President there was assurance that the Armed Forces would obey the Republic although many of the army men attempted to influence politics.

A word should be said here concerning the growth of paramilitary organizations like the S.S., the S.A., and the Communist "defense organization." Hitler built up his irregular forces by enlisting young men and "dissatisfied" veterans. This situation reached the point where these groups, especially the S.A., actually threatened the army. When Hitler came to power he eliminated the aggressive elements in the S.A. in order to maintain the allegiance of the army to his government. These paramilitary organizations did much to coerce the people of Germany into supporting Hitler.

The activity of the police left something to be desired. Because of the federal structure, the Republic did not have complete control over the police organizations of the states. Moreover, the strength of the paramilitary organizations and the apathy of the police led to something like collapse of law and order in the later days of the Republic.

WEAKNESSES OF THE SOCIO-ECONOMIC-POLITICAL STRUCTURE OF THE PREREVOLUTIONARY REGIME

History of revolutions or governmental instabilities

From the unification of Germany in 1871 until the end of World War I, although the imperial government had enjoyed ostensibly a substantial degree of stability, actually many radical and antigovernment groups had been evolving. "The lines of cleavage which were perceptible in the Bismarck period became much more clearly defined in the reign of William II." There were extreme leftist groups, like the

Revolutionary Socialists under Rosa Luxemburg, and extreme rightist groups, like the Pan-German League.[41]

After World War I the Weimar Republic was subject from its beginning to periods of extreme instability. Historians usually divide the years of the Republic into three periods: (1) 1919–23; (2) 1924–29; and (3) 1930–33.

The Republic was born when the fear of a leftwing takeover was prevalent. The extreme Socialists revolted during the formative period, but were crushed. However, after the Republic had been established, Nationalist groups represented the most serious threat to the stability of the regime. The Kapp *Putsch* was engineered in 1920 with the passive acquiescence of the German Army and with the active support of a naval brigade headed by Captain Hermann Erhardt. The government succeeded in quelling the revolt with the support of the workers who through the efforts of the Social Democrats, staged a general strike.[42] Shortly afterward, a Communist-led uprising in the Ruhr was quelled by the army. The episode forced the resignation of the Cabinet, and in subsequent elections the nationalist vote increased. The last uprising of the period was the Munich *Putsch* of November 1923, instigated by Hitler. His intention was to capture the Bavarian state government with the forced support of the Prime Minister Gustav von Kahr and Gen. von Lossow and then move on to Berlin. However, the Bavarian Government refused to cooperate, even though it had been in almost open revolt against the *Reich* Government. The insurrection was crushed and Hitler was sentenced to jail for 5 years, but his sentence was commuted to less than a year. This leniency toward the leader of a treasonous conspiracy was representative of the manner in which rightist revolutionaries were treated by the German Government.[43]

The period from 1924 to 1929 was relatively calm. Chancellor Stresemann, later Foreign Minister, maintained political stability while satisfactory economic conditions modified public discontent. This was the period, however, during which Hitler and other rightist leaders were strengthening their organizations and perfecting their techniques.

With the death of Stresemann in 1929, Germany entered a period of continual instability. Primarily because the election of 1930 had reduced the power of the center, the process of forming cabinets was extremely difficult. This difficulty tempted Hindenburg and Bruening to govern more or less without the *Reichstag*. Once Hindenburg became the center of power, a new element of instability quite different from parliamentary instability was introduced. The determining factor now was who could influence the aging President and toward what ends. Hence, this resulted in the complicated intrigues by which

Schleicher prevailed upon Hindenburg to replace Bruening with Papen, then maneuvered Papen out of the Chancellorship and himself into it, and in turn was supplanted by Hitler in a deal engineered by Papen.[44]

Therefore, the Weimar Republic, save for a period of relative stability from 1924 to 1929, was subject to the constant fear of revolution and continually plagued by instability.

Economic weaknesses

From 1929 to 1933 the German economy underwent severe crises in almost every sphere.

In the field of agriculture the depression practically eliminated exports.[45] This alienated the Prussian landowners, who were the chief commercial agriculturists. In spite of the animosity of these landowners to the Weimar Republic, they still received enormous subsidies from the government.[46]

In the field of capital accumulation, which was essential to economic expansion, the German economy lost much of its foreign credits around 1930.[47] This was due partly to the world depression and partly to the uncertainty of political life with the advent of the Nazis. In spite of the help extended by foreign countries in the form of debt moratoriums, German bank reserves dwindled.

Industrial exports contracted greatly as a result of the depression. The result was increased unemployment, which reached a peak in 1933 of over 6,000,000 (average unemployment for the years 1925–29 had been 1,200,000).[48]

The Bruening government attempted to meet the crisis by a deflationary policy; that is, by cutting salaries, by reducing imports, and by contracting credit.[49] This policy increased the animosity toward the government.

Thus, a Germany which had tremendous industrial capabilities suffered from monetary and fiscal insecurity as well as mass unemployment. The blame was placed in various quarters—on the Jews, on reparations payments, on the incompetence of the government. In actuality, much of the weakness of the German economy resulted from the instability of the international economic system.

Social tensions

A key to the degree of social tensions in Germany was the intensity of social hatred. Not only was the Jewish minority attacked but so were Communists and supporters of the Republic. Moreover, there was an increasing polarization into the conservatives (capitalist-militarist-bureaucrat) and the working class. Underlying all of this was the decline of the middle class. This was partly due to the concentration of industrial power in large trusts and the solidification of the workers into a class-conscious movement. Moreover, as has been pointed out, the middle class was dealt a sharp blow by the inflation of 1923. Therefore it was unable to serve as a moderating force.

The decline of the middle class left two giant social antagonists, the workers and the conservatives. Oddly enough it was the more diverse group, the conservatives, who included military men, civil servants, industrialists, and many of the middle class, which was the more cohesive and militant. The working class, despite its common economic interests and social status, never developed the degree of militancy and cohesiveness that the conservatives achieved under National Socialism. A prime reason for this was the divisive influence of the conflict between Communists and Social Democrats. The Social Democrats spent as much time fighting the Communists as they did opposing the German conservatives. Nevertheless, Germany in the 1930's was plagued with social tension arising out of the confrontation of two massive social groupings, the workers and the conservative classes. The Communists increased the tension by aggravating the conflict within the working classes.

In sum, Germany suffered from increased social tension as a result of the loss of moderating influences in the various classes. The middle class did not modify the nationalism, racism, and emotionalism of the groups which were supporting the Nazi movement. The working class became more radical as time went on, facilitating an increase in Communist strength. There was a heightening of radicalism on both the right and the left; the right, however, had an inherent advantage because it could play on German nationalism.

Government recognition of and reaction to weaknesses

Given the intensity of social tension, economic difficulty, and governmental weaknesses in the years immediately prior to Hitler's ascendancy, the governments of Bruening and Papen were well aware of the dangers but their policies were not always well chosen. Papen and Schleicher realized the power of the Nazis and sought to bring Hitler into the government in the hope that they could control him, since his party, though much the largest, did not have a majority in the *Reichstag*.[50] Papen persuaded Hindenburg to appoint Hitler Chancellor on the grounds that Hitler would give the government the popular support it so desperately needed. Papen was to have the Vice-Chancellorship and the power of Hindenburg behind him. Moreover, the Nazis were to have only three of eight Cabinet posts and none of the key Ministries. Hence, the government tried to make a last-second compromise to save the Republic by bringing Hitler into the Cabinet. From this point on, the real revolution began, for it was then that Hitler wrested power from those who had it and established his dictatorship.

FORM AND CHARACTERISTICS OF REVOLUTION

ACTORS IN THE REVOLUTION

The revolutionary leadership

Official and unofficial social and political positions

The leadership was made up primarily of persons from the middle or lower-middle classes and from the military, civil service, professional, and intellectual groups. Rudolf Hess, Ernst Roehm, Hermann Goering, and Gregor Strasser had all been army officers. Gottfried Feder, Alfred Rosenberg, Willy Ley, and Max Amann had been trained for professions like law and architecture. Dietrich Eckart, Hess, and Joseph Paul Goebbels can be classified as intellectuals. Heinrich Himmler and Julius Streicher had been school teachers; Wilhelm Frick had been a civil servant. Most of them had either never had a steady occupation, like Hess, Hitler, and Eckart, or had rejected their initial vocations, like Rosenberg, Streicher, Feder, Strasser, Himmler, Ley, and Roehm.[51]

By far the most important leader was Adolf Hitler. It was he who developed the organization and kindled the spirit of the Nazi movement, and it was he who knew how to win the support of the masses. Hitler was actually not a German, but an Austrian who became a German citizen only during the final stages of his movement. Born in 1889 at Braunau-on-the-Inn, a town near the German border, where his father was a customs official, Hitler by 1908 found himself poor, friendless, untrained, and unemployed in Vienna. He went to Munich in 1913 and joined the Bavarian Army a year later. He rose to the rank of corporal, was twice wounded, and received the Iron Cross. In his disappointment over the defeat of the Central Powers, Hitler developed a fanatical nationalism and a resolve to avenge the indignities of the peace treaty.

After the war Hitler joined a group of six men called the Committee of the German Workers' Party in Munich. Though there were other nationalist-socialist-type parties in various states, Hitler kept his group autonomous. By July 1921 Hitler was given unlimited power over the party and was made its President. Between 1921 and 1924 he established many of the associations that formed the core of his movement. Roehm, Hess, Goering, Feder, Eckart, Rosenberg, Heinrich Hoffmann, and Streicher all joined Hitler at this time.[52]

The Munich *Putsch* has been discussed earlier. Hitler was jailed for his role in it, but he was released within a year. He utilized his imprisonment time to put down the ideology and the program he had been developing over the years in his *Mein Kampf,* which was published in 1925. The failure of the *Putsch,* which in normal times would have meant the demise of Hitler's party if not of Hitler himself, served only as a temporary setback. Between 1929 and 1930 the party reached national prominence by taking the lead in opposing the Young Plan. Hitler succeeded in maintaining his leadership when the movement became nationwide, and in January 1933 Hitler became Chancellor of the *Reich.*

Political and ideological orientation

The leaders of the National Socialist movement were agreed on two points. First, all of them were nationalistic; they all wished to see a resurgent Germany. Secondly, most of them were radicals who wanted to make over the existing social and political order. Hitler himself was the symbol of the political orientation of the entire movement. His radicalism covered both ends of the political spectrum. His hypernationalism led him to enlist the support of the rightist camp of Hugenberg while his social reformism, particularly in his earlier years, placed him very near the Socialists. Like Hitler, Roehm was a militant nationalist who opposed the Weimar Republic. Hjalmar Schacht, although a militant nationalist, was not an early supporter of Hitler. His background in banking did not allow him to embrace the radical social reformism of the Nazi doctrine. There were, however, members of the leadership who were Socialists as well as nationalists. Goebbels and the Strasser brothers were constantly advocating the nationalization of big business. Although their nationalism did not allow them to become Communists, these three men held ideological positions that were very close to the Communist camp.[53]

Goals and aspirations

The goals and aspirations of the various members of the leadership were not completely homogeneous. All wanted to see the Nazis take over the German state. However, some, like the Strassers and Roehm, wanted to see certain other objectives realized. Roehm wished to have his beloved S.A. replace the German Army. Gregor Strasser wanted to see the Nazis institute a system of socialization. There were other motives operative on the Nazi leadership. Rosenberg wished to be pictured as the "philosopher" of the movement, while others, like Eckart, Himmler, Amann, Hess, and Ley seemed to be going along for adventure and material reward.[54]

The revolutionary following

The revolutionary following can be divided into two groups: those who participated actively in the movement and those who lent financial and political support. Men from the younger generation of the middle and lower-middle classes, eager for adventure, joined the S.A. and the more selective S.S.[d] This group, which grew to 400,000 before 1933, aided Hitler in his campaigns. The Germans who gave monetary and political support were from various groups. There were members of the lower and middle classes in both city and country who supported and voted for Hitler, members of the industrial class who aided him financially, and the army officers and civil servants who supported Hitler behind the scenes by exerting political pressure and acquiescing in his illegal techniques. Lastly, certain prominent political figures such as Schleicher and von Papen unwittingly aided Hitler in his rise to power by weakening the Weimar Government for their own political purposes.[55]

ORGANIZATION OF REVOLUTIONARY EFFORT

Internal organization

Though the Nazis called themselves a "party," one should understand that it was not a party in a democratic sense. It was really a congeries of groups, both legal and illegal, organized on the basis of age, occupation, or sex, which wanted to overthrow the government.[56] Together, the groups which worked for this purpose constituted the Nazi "Party"; hence, they were within an organizational framework. However, they can be studied separately if it is kept in mind that they were, at least in theory, under the control of Hitler and the "party."

These organizations can be divided into three groups. The first was the legitimate political party, which was represented in the *Reichstag* and which gave Hitler political power through legal means. The second comprised the paramilitary organizations: the *Sturm Abteilunger* (S.A.), or Storm Troopers, with its Nazi Motor Corps, and the *Schutz Stafflen* (S.S.), or Blackshirts. The third consisted of certain groups organized to contribute to the power of the party, such as the Hitler Youth, the Nazi Schoolchildren's League, the Student's League, the

---

d The S.A., unlike the S.S., included a sizable number of thugs and social misfits.

Order of German Women, the Nazi Teachers' Association, the Union of Nazi Lawyers, and the Union of Nazi Physicians.[57]

The Nazi Party which included all of these organizations was subject to the will of Hitler. However, five other men who had a great deal of power were Roehm, Strasser, Goering, Goebbels, and Frick. To oversee the entire party and to prevent minor squabbles from becoming a source of division, Hitler devised the USCHLA in 1926. This was a committee for investigation and settlement and was advantageously used by Hitler to strengthen his control over the party.[58]

The National Socialist Party divided the country into the district or *Gau*, which were in turn subdivided into circles (*Kreise*). Within each *Kreis* were local groups (*Ortsgruppe*), which in turn were divided into street cells and blocks. By 1929 there were 178,000 party members. A *Gauleiter* was appointed to head each *Gau*. He had control over all subordinate party members in his *Gau*.[59] Moreover, the *Gauleiter* was personally appointed by Hitler and owed allegiance directly to him.

The paramilitary organizations, the S.A. and the S.S., were placed under the control of Roehm in 1931. The geographical units were the same for the S.A. as for the party itself. The S.A. organization was similar to that of the German Army, with a general staff and its own headquarters. In fact, Roehm had so much power that he challenged Hitler's will on many occasions. Hitler's inability to control the S.A. led him to eliminate the top S.A. leadership in June 1934. Although the S.S., under Himmler, was subject to the will of Roehm in the Nazi organizational scheme, it served, in reality, as a counter to the S.A. The S.S. was a smaller and a much better disciplined group than the S.A.[60] It was the S.S. which carried out the liquidation of the Roehm leadership. Though on April 10, 1932, the S.A. and the S.S. were technically dissolved by the Bruening government with Hitler's*[e] acquiescence the organization itself remained intact.[61] There were close to 400,000 men in these two organizations by 1934.

The other Nazi organizations occupied an ambiguous place between the paramilitary organizations and the political parties. For example, the Hitler Youth, which was really a counter to the successive youth organizations of the *Reich* and which in 1932 numbered over 100,000, was designed to promote Nazi propaganda and enlist support among young people.[62] The other organizations were designed to organize and "educate" each group in German society. In addition, there were many separate departments, e.g., the Factory Cell Organization, the Economic Policy Department, the Propaganda Directorate under Goebbels, and the *Reich* Press Officer.

---

[e] Hitler acquiesced in the decision in order to pursue his policy of legality.

There were two important facts to note about the total Nazi organization. The first is that Hitler was the supreme commander, although men like Gregor Strasser held some grassroot power independent of Hitler. Secondly, the types of organization covered every field of German life. This made the transition to the Nazi state less difficult. Moreover, it gave important positions to men who felt they had no place in the society under the Republic.

External organization

There was some suspicion of outside help, perhaps French, when Hitler was a petty politician in Bavaria in 1923; however the amount was negligible. Yet the Nazi Party did organize groups which operated outside of German territory—seven external *Gau* areas, in Austria, Danzig, the Saar, and the Sudetenland in Czechoslovakia.[63] These organizations assumed more importance after Hitler came to power, and became valuable tools of Nazi foreign policy.[64]

GOALS OF THE REVOLUTION

Concrete political aims of revolutionary leaders

The Nazi movement aimed primarily at capturing the German state. Hitler hoped to replace the parliamentary system with strong leadership. This was what he called the "leadership principle": he asserted that the German "race" could reach its fulfillment only through strong leadership. Another aim was the addition of more territory to Germany and simultaneously, a rectification of the "injustices" of the Treaty of Versailles. Germany must be united with Austria and must procure more "living space" for the German "race."

There was some disagreement in the Nazi leadership over the proper political technique to be employed in reaching power. Hitler, having learned his lesson from the abortive Munich *Putsch*, decided that the only road to power was through legal methods. Roehm, who wanted an armed revolution, opposed him.[65]

Social and economic goals of leadership and following

The Nazi movement had various social and economic goals. In the early period Hitler's conception of the new social order was close to socialism. However, by 1930 Hitler was too dependent on the business elite to sponsor equalitarian social reform, and his conception of the new social order changed, in effect, to the corporate state. There must also be a social elite who directed the workings of society. Further, Hitler saw no reason to wipe out the existing industrial elite so long as they cooperated. Other elites had to be added to insure the growth of the German people.[66] The plans of the Nazis were more concerned with organization than with goals in the years preceding the takeover.

The positive program of the National Socialists was general to the point of ambiguity. However, the negative sections of the platform were forceful and specific. The Nazis' intense hatred of the Jews, Communists, and "republicans" was developed to a high degree. Hitler felt that these groups were responsible for the decline of Germany,[67] and a great portion of the Nazi platform was concerned with them.

The entire Nazi leadership was agreed on the negative part of the social program. However, there was a strong element led by Gregor Strasser, who had labor ties, which advocated an equalitarian social program and the nationalization of industry. This leftist group was forcefully silenced by Hitler.

REVOLUTIONARY TECHNIQUES AND GOVERNMENT COUNTERMEASURES

After 1924 Hitler's revolutionary techniques can be classified in two stages. The first stage was from 1924 until 1933 when he became Chancellor. The overall strategy was to gain control of Germany through constitutional means. The main instrument in this strategy was the Nazi Party, and Hitler concentrated on building up its electoral strength. The second stage, from January 1933 to August 1934, showed the role of force and coercion elevated from a mere propaganda tool to the prime instrument of strategy. After 1933 Hitler eliminated all opposition through terrorist methods. Because of the different roles of force in the two stages of Hitler's rise, they will be dealt with separately.

In the first stage the prime objective was to gain mass support for the Nazi Party. The tactics employed were to play on the insecurity of the masses, an insecurity which Hitler himself had helped to create. His strategy was to increase that insecurity through terrorism, riots, brawling, and disruption of the processes of law and order, while, at the same time, convincing the German people that the Nazi movement would offer the security and stability they longed for.

To create this feeling of insecurity was not a difficult task for Hitler's cohorts. Ever since the inception of the Republic, civil violence had frequently plagued German life; Hitler merely increased the civil disorder. Moreover, because the Communists were themselves practicing terrorist tactics, Hitler could easily blame the violence on them. The role of the S.A. was essential to this phase of Hitler's strategy. Hitler was able to recruit a force such as the S.A. because of the many unemployed men and "disposed" soldiers. He justified the existence of this paramilitary organization by claiming that it afforded protection to the National Socialist Party. This justification was partly valid; other German parties had the same type of organization.[68]

However, in "protecting" the Nazis, the S.A. and later the S.S. created civil unrest. The S.A. would march through the city, four abreast, singing "When Jewish blood gushes from our knives, things will be better," or songs like *Horst Wessel*, which glorified a dead S.A. man as a martyr.[69] Five hundred marching Storm Troopers, who held no direct allegiance to the *Reich*, instilled fear in the minds of the onlookers. These marchers were not always peaceful. In the 1930 election, for instance, the S.A. marched into the working class section of Berlin, whereupon Communists opened fire. A pitched battle ensued and spread all over the city.[70] In fact, wherever the Communists met, S.A. men could be found initiating a brawl.

Another S.A. technique for creating unrest was breaking up rival meetings. Sitting in the audience they harassed the speaker, or often beat up the speaker and dispersed the crowd.[71]

Lastly, Hitler deliberately increased the insecurity of the masses by promoting the breakdown of law and order. The acts of violence of his paramilitary organizations were only one phase of this understanding of the legal process. Men sympathetic to the Nazis were placed in strategic positions of both the police force and judiciary, especially in the state government.[72]

Hitler was able to carry on his tactics of violence and to gain the confidence of much of the public because he continually blamed the breakdown of law and order on the Communists. While the Nazis themselves were attacking newspapers of the opposition and breaking up political meetings, Hitler was proclaiming that the German state must be protected from the Communist threat and that—as in this July 1932 election—he should be given a free hand to do this.

While Hitler was successfully creating social and political chaos, he was also promoting hope, enthusiasm, and support for the party through Nazi organizations, propaganda, and mass rallies. Prime instruments in this were groups like the Nazi Lawyers, the Nazi Physicians, and Hitler Youth who gave a new loyalty to Germans alienated from their society. Moreover, these groups, which covered almost every element in German society, served as an important propaganda medium when they were mobilized during elections.

Hitler used propaganda both to create insecurity and to promote confidence in his party. Even during the period of economic and political stability, he told the masses that a breakdown was near.

When the depression came, Hitler seemed to be a prophet, offering the people hope. They wanted to hear that the evil around them was not permanent, that Germans could still attain a good life. By the clever use of myths and lies, Hitler gave them this assurance. He preached the concept that the Germans were a chosen race. The evils that had befallen Germany he blamed on devils in the body politic, certain groups which were depriving the Germans of their just due. He found two scapegoats—the Jews and the Communists. The Germans were ready to believe his accusations against the Communists because they had many times called for the overthrow of the Weimar regime. The charge that the Jews were conspiring for the "defeat" of the "German race" found credence because anti-Semitism was already deeply rooted in German minds. It was easy to identify the Jews with the Bolshevists.[73]

Hitler circulated his propaganda through various means: Like the other political parties, the Nazis had their own publications, their principal organ being the *Voelkischer Beobachter* (*Racist Observer*). They often distributed publications without charge. A constant stream of propaganda reached the various Nazi groups, which they in turn distributed. Election campaigns were also important transmitting devices. In the first presidential election in 1932, Hitler, Goebbels, and Strasser raced from town to town speaking to tremendous groups. Nazis "plastered the walls of the cities and towns with a million screeching colored posters, distributed 8 million pamphlets, and 12 million extra copies of their party newspapers, staged 3,000 meetings a day and, for the first time in a German election, made good use of films and gramaphone records. . . ."[74]

Hitler's special technique in gaining votes was the mass rally. Through these rallies he not only presented his views to thousands of people but also gave the masses an emotional release. The principal elements in these rallies were Hitler's showmanship—he timed his speeches and actions perfectly; the use of physical effects—the S.A. and S.S. in full dress and the blare of trumpets when the Fuehrer walked in; and finally, the coordination of actions in mass—shouting, singing, saluting, and standing. While it is difficult to assess the effect of these performances in terms of actual votes, it is safe to guess that the emotionalism of the masses impressed the government and the opposition parties.[75]

Once Hitler became Chancellor, his techniques shifted. Force became his primary reliance. He employed force openly, both in election campaigns and in changing the government of Germany. In the March 5, 1933, election he used the State-run radio stations to propagandize his campaign, suspended the operation of opposition presses, and broke up political meetings. He forced the police to cooperate with the S.A. and S.S., advising them to shoot if there were any "disturbances." After the Enabling Act was passed, Hitler's use of force became total. All opposition was crushed through the threat or the actual use of coercion.[76]

In both stages of the revolution, Hitler was able to hide behind the cloak of legality. Moreover, both stages were made possible by the lack of unity among his potential opponents and by the aid, both direct and indirect, of industry and the army. Aid from industry came after 1930.

Hitler did not court the business elite for financial help to the extent that he courted the masses for electoral help. He either addressed them individually or in small meetings arranged through the help of Goering, Himmler, Funk, and Dietrich. Hitler received sufficient financial aid from the industrialists—only a portion responded—because he promised them protection of their interests, economic prosperity, and, most important of all, the elimination of the Communist threat.

It is difficult to assess the amount of money given the Nazis by big business. In 1932 over 2,880,000 marks a day were needed to support the tremendous party organization. Although some money came in through the party itself—from the sale of newspapers, membership dues, and charges at mass rallies—it is safe to assume that the industrialists met a good part of the heavy party expenditures.

Given the important place of the army in German politics, it was necessary for Hitler to win its support. The technique he employed was a combination of an appeal to interest, emotionalism, and fear. He offered the army a vigorous foreign policy, a policy which would need military buildup. Emotionally, he appealed to the army's nationalism Also, he attempted to enlist the support of men whom he knew the army respected, such as Hindenburg and Blomberg. Lastly, perhaps unwittingly, he posed the threat of replacing the army with the S.A. Upon taking office, he frightened many of the officers by appealing directly to the rank-and-file soldier. However, he quickly quieted this fear by assuring the officers that "the army and navy would thenceforth be free to work entirely unhindered on training the development for the defense of the Reich."[77] Hitler assumed complete official control over the army in August 1934 when he became President, although he still felt it necessary to maintain its allegiance by the same tactics.

Hitler's overall strategy of maintaining a veneer of legality limited the counteractions which the government could take. Because he could claim that his party was a legitimate political party, he made direct governmental intervention difficult. Still, both Federal and State governments did try to limit Nazi activity. In the first 6 months of 1930 the authorities attempted a number of prohibitions. Outdoor meetings and parades were forbidden in Prussia, a law for the "protection of the Republic" was passed by the *Reichstag* in March, and in June the Prussian authorities forbade the S.A. to wear uniforms (brown shirts). The S.A. responded by wearing white shirts.[78]

In April of 1932 the Bruening government issued an order disbanding the S.A. and S.S. which was carried out at least in form. However, by June 1932 the Papen regime was forced to lift the ban because Papen needed the political support of Hitler to maintain his Cabinet.[79]

Hence, the few attempts to limit Nazi activities and check the growth of the party met with failure. Besides the party's and Hitler's skill in maintaining an appearance of legality, the ambiguous attitude of the German Army toward the Weimar Republic, and the size of the S.A. and S.S. troops made any attempt to crush the Nazi movement almost impossible.

MANNER IN WHICH CONTROL OF GOVERNMENT WAS TRANSFERRED TO REVOLUTIONARIES

It took Hitler from January 30, 1933, when he became Chancellor, until August 19, 1934, when the plebescite affirmed his assumption of the Presidency, to gain complete control of Germany. The three important events in the process were his appointment as Chancellor, the passage of the Enabling Act, and his capture of the Presidency.

Hindenburg appointed Hitler Chancellor on the condition that he could form a coalition with a parliamentary majority. Many pressures were brought on Hindenburg to do this. Von Papen, acting out of his own interests, convinced Hindenburg that Hitler could be kept under control. Hitler himself reminded Hindenburg of his popular support and of the strength of the S.A. and S.S. Moreover, through a mixture of bribes and blackmail, Hitler convinced Hindenburg's son Oskar that it was time for the Nazis to gain control. Therefore, on January 30, 1933, Hitler replaced Schleicher as Chancellor, with Papen becoming Vice-Chancellor. Only two other Nazis were in the Cabinet and they held inferior positions.[80]

After an unsuccessful attempt to form a majority coalition, Hitler dissolved the *Reichstag* and called for elections, which were set for March 5, 1933. The campaign saw the increased use of terror by the Nazis, capped by the burning of the *Reichstag*. However, Hitler failed to win an absolute majority; he depended on the votes of Hugenberg's Nationalist Party. Now Hitler took his second great step toward complete control of Germany, the Enabling Act. After banning the Communist and cajoling other parties, Hitler had only the Social Democrats in opposition, and they were not strong enough to deprive him of the two-thirds majority he needed. This act provided Hitler with complete control over civil affairs with no legal restraints whatsoever.[81]

The last phase of his takeover was the capture of the Presidency, for the President still officially held the allegiance of the army. Because Hindenburg was so popular with the army, Hitler had to wait until the old General died. Upon his death on August 2, 1934, Hitler immediately proclaimed himself President; Vice-Chancellor von Papen and five high ranking German officials signed the announcement. On August 19 Hitler called a plebiscite to affirm his new position. He was able to get Hindenburg's son to tell the German people that the dead General would have approved. With 95.7 percent of the 45½ million eligible Germans voting, Hitler received affirmation from 89.93 percent. Four and one-fourth million voted "No" and 800,000 spoiled their ballots.[82] The revolution completed, Hitler was now Head of State, Head of Government, and Supreme Commander in Chief of the Army.

THE EFFECTS OF THE REVOLUTION

CHANGES IN THE PERSONNEL AND INSTITUTIONS OF GOVERNMENT

Hitler consolidated the revolution by replacing the hostile personnel of certain institutions or by changing the institutions themselves. The *Reich* Government, which had ceased to be parliamentary government even before Hitler came to power, was now converted into a single-party government. Even the parties which were in the coalition were deprived of any voice in the decisionmaking.

The federal structure of the Weimar Republic was also abolished. Prussia was already under Goering's control, and between March 5 and 16 the other Federal States passed under Nazi control. The Nazi Minister of the Interior sent men to replace the existing state officials. In early April Hitler appointed State governors, who were subordinate to the Minister of the Interior. As Frick said, "The state governments from now on are merely administrative bodies of the *Reich*."[83]

Between April and July 1934 Hitler abolished all other political parties. The Communists had been suppressed previously. The Social Democrats made an attempt to appease Hitler, but he abolished the party on June 22. The middle-class parties—the Catholic Bavarian People's Party—dissolved themselves. The German National Party, which had enabled Hitler to gain power, was taken over on June 21, and officially "dissolved itself" on June 29. In a decree on June 14 the Nazi Party declared itself the sole party in Germany.[84]

The free trade unions, which had always opposed any hypernationalism, were eliminated. After a great mass demonstration on May 1, the Nazis occupied union headquarters on May 2, confiscated union funds, and arrested union leaders. The right of collective bargaining was also abolished.[85]

Lastly, Hitler replaced men in certain key positions with Nazis or Nazi sympathizers. Dr. Schacht, an extreme Nationalist who had helped finance the Nazis, became once more president of the *Reichsbank*, replacing the conservative Dr. Hans Luther.[86] Later Schacht became Minister of Economics.

Hitler's "aides" were given semiautonomous control over the areas for which they were responsible except when it came to major policy lines.[87] There were about 40 "national leaders," or *Reichsleitung*, by the end of the Second World War, who held positions which corresponded to heads of Ministries and other state positions in the Weimar Republic. These positions were usually created on an *ad hoc* basis rather than on the basis of an overall plan. Most of the bureaucratic personnel, provided they were not Jewish or overt opponents, were retained in office.[88]

MAJOR POLICY CHANGES

There were major policy changes in all spheres of German life. The government began an intense buildup of internal improvements which resulted in a rapid decrease in unemployment. Though private enterprise was subjected, initially, to only minor new controls, after 1937 Hitler increased control over business and over exports and imports. Socially, anti-Semitic and anti-Communist legislation was passed to eliminate the so-called "conspiracy" which Hitler said had plagued German life.

The greatest policy changes came in the field of international affairs. Some of the main German objections to the Treaty of Versailles were either a thing of the past, like reparations, or beyond Hitler's control at the moment, like frontier changes. He therefore directed his attack on the military clause of the treaty. On March 7, 1936, he repudiated parts of both the Locarno and the Versailles treaties by remilitarizing the Rhineland area. By 1936 Hitler had started the development of German military might on the land, on the sea, and in the air.[89] The policy of the treaty fulfillment which had been a political liability to all of the previous regimes was transformed by Hitler into the policy of revision.

LONG-RANGE SOCIAL AND ECONOMIC EFFECTS

When Hitler took power, Germany had more than 6,000,000 unemployed and was suffering from continual balance-of-payment difficulties. Hitler's economic objectives were the buildup of German industry, internal improvements, and autarchy (self-sufficiency). Hitler built up a powerful industrial machine, and eliminated unemployment, but Germany was still dependent on the outside world for many needed raw materials.[90]

Hitler wrought many changes in the social nature of Germany. Any deviation from the norms set by the Nazi regime was severely prosecuted. Not only did Hitler try to eliminate the "Jewish problem," he also brought force to bear upon academic circles, Protestant and Catholic churches, and political and literary figures who did not back his regime. Moreover, German life was regulated from an early age through education and through such groups as the Hitler Youth. The purpose of the educational system and the various Nazi organizations was to indoctrinate Germans so that they could better serve the state. The normal processes of law and order disintegrated under the Nazi terror, in the end making the individual more dependent upon the state. Lastly, there was a partial class upheaval with the influx of a new elite. It was partial because many members of the old elite, especially those in industry, were allowed to retain their wealth if not their position.

OTHER EFFECTS

Hitler's course of action in the international field led to war. Before the tide of battle had turned against him in 1943, Hitler had greatly increased the amount of land under German domination. Austria, Czechoslovakia, Upper Silesia, Posen, the Polish Corridor, Danzig, what corresponded to the old "Congress Poland," the Benelux countries, two-thirds of France, all of Norway, Denmark, Rumania, Hungary, Slovakia, and a 600-mile-deep strip of Russia were at one time under German control, in the form of military rule, integration with Germany, or under agreements which gave Germany virtual control.[91] However, Hitler was soon faced with a coalition of England, the U.S.S.R. and the United States which overpowered Germany and her allies. Hitler's phenomenal initial success was canceled and post-World War II Germany found herself partitioned into Eastern and Western zones, and under military control. Furthermore, she had lost territory in the east that had belonged to her prior to Hitler's expansionist drive.

In spite of Hitler's totalitarian tactics and his spectacular successes there were, from 1938 to 1944, movements among the German elite to remove him from power, plus some anti-Hitler activity among the military. During the war, plans to dispose of Hitler were discussed in various groups, among varied types of people—ex-diplomats, ex-Ministers, lawyers, Catholic and Protestant clergy, and a number of high-ranking army officers. The famous attempt on Hitler's life, led by Colonel Count von Stauffenberg, took place in 1944 at Wolfsschanze in East Prussia. Though badly shaken, Hitler survived an explosion set off not more than 6 feet away. All of these movements proved abortive, primarily due to the terroristic methods of the Gestapo which forced the opposition groups to operate in the strictest secrecy. This meant that there could be little chance for an organized, large-scale effort at counterrevolution.

Germany still feels the effects of Hitler's revolution, both politically, in terms of the partition, and psychologically, in terms of the attempt to live down the German past.

NOTES

1. James W. Angell, *The Recovery of Germany* (New Haven: Yale University Press, 1929), p. 15.
2. *Germany, Economic Geography,* Geographical Handbook Series, B.R. 529, HMSO, Vol. III, p. 19.
3. *Germany, Ports and Communications,* Geographical Handbook Series, B.R. 529C, HMSO, Vol. IV, 1945, p. 192.
4. Ibid., p. 505.
5. Ibid., p. 435.
6. *Germany, Economic Geography,* Vol. III, p. 8.
7. Ibid., p. 181.
8. Ibid., p. 224.
9. Ibid., p. 242.

10. Ibid., p. 243.

11. Ibid., p. 2.

12. Koppel S. Pinson, *Modem Germany* (New York: The Macmillan Co., 1954), p. 451.

13. Ibid., p. 450.

14. Robert E. Dickinson, Germany: *A General and Regional Geography* (New York: E. P. Dutton & Co., Inc., 1953), p. 289.

15. Paul Alpert, *Twentieth Century Economic History of Europe* (New York: Henry Schuman, 1951), p. 92.

16. Pinson, *Modern Germany*, p. 452.

17. Edmond Vermeil, *Germany in the Twentieth Century* (New York: Frederick A. Praeger, 1956), p. 14.

18. Gwendolen M. Carter, John H. Herz, and John C. Ranney, *Major Foreign Powers* (New York: Harcourt Brace and Company, 1957), p. 409.

19. Ibid., p. 409.

20. James T. Shotwell, (ed.), *Governments of Continental Europe* (New York: The Macmillan Co., 1957), p. 404.

21. Ibid., p. 405.

22. Ibid., p. 405.

23. Ibid., p. 405.

24. Ibid., p. 406.

25. Ibid., p. 410.

26. Carter, *Major Foreign*, p. 411.

27. Ibid., p. 413.

28. Shotwell, *Governments of Continental*, pp. 415–416.

29. Ibid., p. 397.

30. Ibid., p. 423.

31. Ibid., p. 422.

32. Ibid., pp. 422–424.

33. Carter, *Major Foreign*, p. 417.

34. Shotwell, *Governments of Continental*, p. 421.

35. Ibid., p. 423.

36. Ibid., p. 433.

37. Ibid., p. 429.

38. Lindsay Rogers, et al., "Aspects of German Political Institutions," Political Science Quarterly, XLVII, 3 (September, 1932), 321.

39. Alan Bullock, *Hitler: A Study in Tyranny* (New York: Bantam Books, Inc., 1961), p. 147.

40. Marshall Dill, Jr., *Germany: A Modern History* (Ann Arbor: The University of Michigan Press, 1961), p. 205.

41. Ibid., p. 204–205.

42. Ibid., p. 281.

43. Bullock, *Hitler*, p. 83.

44. Dill, *Germany*, p. 330–337.

45. Gustav Stolper, *The German Economy: 1870–1940* (New York: Reynal & Hitchcock, 1940), p. 182.

46. *Germany, Economic Geography*, vol. III, p. 10.

47. Ibid., p. 11.

48. Ibid., p. 9.

49. Ibid., p. 10.

50. Bullock, *Hitler*, p. 215.

51. Ibid.

52. Ibid., pp. 42–56.

53. William L. Shirer, *The Rise and Fall of the Third Reich* (New York: Simon and Schuster, 1960), p. 124.

54. Ibid.

55. Bullock, *Hitler*, p. 146.

56. Ibid., p. 142.

57. Ibid., p. 109.

58. Ibid., p. 107.

59. Shirer, *The Third Reich*, p. 120.

60. Bullock, *Hitler*, p. 136.

61. Ibid., p. 168.

62. Shirer, *The Third Reich*, p. 252.

63. Ibid., p. 120.

64. Bullock, *Hitler*, p. 281.

65. T. L. Jarman, *The Rise and Fall of Nazi Germany* (New York: Signet Books, 1961), p. 146.

66. Adolf Hitler, *Mein Kampf* (Boston: Houghton Mifflin Co., 1943), pp. 407–433.

67. Jarman, *The Rise*, p. 107.

68. Konrad Heiden, *Der Fuehrer: Hitler's Rise to Power* (Boston: Houghton Mifflin Company, 1944), p. 452.

69. S. William Halperin, *Germany Tried Democracy* (New York: Thomas Y. Crowell Company, 1946), p. 445.

70. Konrad Heiden, *A History of National Socialism* (New York: Alfred A. Knopf, 1954), p. 199.

71. Bullock, *Hitler*, p. 133.

72. Heiden, *A History of*, p. 42.

73. Shirer, *The Third Reich*, p. 158.

74. Jarman, *The Rise*, p. 126–127.

75. Shirer, *The Third Reich*, p. 194–195.
76. John W. Wheeler-Bennet, *The Nemesis of Power* (London: Macmillan and Co. Ltd., 1954), p. 290.
77. Bullock, *Hitler*, p. 134.
78. Ibid., p. 208.
79. Bullock, *Hitler*, p. 216.
80. Ibid., p. 229.
81. Ibid., pp. 265–267.
82. Ibid., p. 134.
83. Shirer, *The Third Reich*, pp. 198–201.
84. Ibid., p. 202.
85. Ibid., p. 204.
86. Bullock, *Hitler*, p. 268.
87. Dill, *Germany*, p. 353.
88. Ibid., pp. 354–357.
89. Ibid., p. 359.
90. Ibid., pp. 402–404.
91. Jarman, *The Rise*, p. 287.

RECOMMENDED READING

BOOKS:

Alpert, Paul. *Twentieth Century Economic History of Europe*. New York: Henry Schuman, 1951. Chapter 8 provides a good discussion of the German inflation of 1923.

Angell, James W. *The Recovery of Germany*. New Haven: Yale University Press, 1929. This book describes the recovery of Germany under the Weimar Republic from the defeat in World War I to 1928.

Brady, Robert A. *The Spirit and Structure of German Fascism*. New York: The Viking Press, 1937. Chapter I presents the background to the rise of national socialism. The rest of the book presents an acute analysis of the structure of the Fascist German state.

Bruck, W. F. *Social and Economic History of Germany from William II to Hitler: 1888–1938*. London: Oxford University Press, 1938. A technical survey of the economic and social conditions of Germany. Chapter II deals with the Weimar Republic.

Bullock, Alan. *Hitler: A Study of Tyranny*. New York: Bantam Books, Inc., 1961. This is one of the best accounts of the rise of nazism. It is especially worthwhile on the internal developments of the Nazi movement.

Butler, Rohan. *The Roots of National Socialism.* New York: E. P. Dutton and Co., 1942. A discussion of the intellectual background of the Nazi movement.

Carter, Gwendolen M., John H. Herz, and John C. Ranney. *Major Foreign Powers.* New York: Harcourt, Brace and Company, 1957. Part III discusses, in an elementary fashion, the nature of the Weimar Government.

Dickinson, Robert E. *Germany: A General and Regional Geography.* New York: E. P. Dutton and Company, Inc., 1953. An excellent survey of the people, territory, and industries of Germany.

Dill, Marshall Jr. *Germany: A Modern History.* Ann Arbor: The University of Michigan Press, 1961. Chapters 20–31 cover the history of the Weimar Republic and the rise of Hitler. Especially useful is his discussion of how Hitler consolidated the revolution.

Geographical Handbook Series: Germany. London: Her Majesty's Stationery Office, 1944–46, Volumes II, III, IV.

Goerlitz, Walter. *History of the German General Staff.* New York: Praeger, 1953. A survey of the role of German military in the history of Germany.

Halperin, S. William. *Germany Tried Democracy.* New York: Thomas Y. Crowell Company, 1946. A good account of the weaknesses of the Weimar Republic.

Heiden, Konrad. *A History of National Socialism.* New York: Alfred A. Knopf, growth of the movement. Also, the book provides some useful analyses of the propaganda of national socialism.

Heiden, Konrad. *Der Fuehrer: Hitler's Rise to Power.* Boston: Houghton Mifflin Company, 1944. A more concise account of Hitler's rise to power than his earlier book.

Hitler, Adolf. (trans. by Ralph Manheim). *Mein Kampf.* Boston: Houghton Mifflin Company, 1943. Essential to the understanding of Hitler.

Hitler, Adolf. *The Speeches of Adolf Hitler.* New York: Oxford University Press, 1942. Helpful in examining Hitler's use of the spoken word. The speeches are arranged topically.

Jarman, T. L. *The Rise and Fall of Nazi Germany.* New York: Signet Books, 1961. A clear, straightforward account of Hitler's career.

Jetsinger, Franz. *Hitler's Youth.* London: Hutchinson, 1958. An excellent study of the influence of Hitler's early experiences on his later career.

Lewis, W. Arthur. *Economic Survey, 1919–1939.* London: George Allen and Unwin Ltd., 1949. Chapter 6 examines the economic structure of Germany about 1930, with emphasis on national policies.

Pinson, Koppel S. *Modern Germany*. New York: The Macmillan Company, 1954. Chapter 22 is an excellent account of the results of Hitler's rise to the Chancellorship.

Peterson, Edward Norman. *Hjalmer Schacht: For and Against Hitler.* Boston: The Christopher Publishing House, 1954. A useful study of a man who aided Hitler because he believed Hitler would help Germany, but later turned against him.

THE SPANISH REVOLUTION OF 1936

SYNOPSIS

On July 17, 1936, the Army of Africa, led by Gen. Francisco Franco, revolted in Spanish Morocco against the Republican Government of Prime Minister Manuel Azana. On July 18, numerous elements of the National Army on the mainland, led by Generals Goded, Mola, and Sanjurjo, supported by the rightwing political parties of the National Front, revolted, in complicity with the Army of Africa, and opened hostilities against the central government. By the end of July, the revolutionaries had succeeded in consolidating their forces, and had seized the north and northwestern third of Spain. Thirty-two and a half months later, after bitter and bloody fighting, in which the rebels had the help of Italy, Germany, and Portugal, and the government received aid from the Soviet Union and volunteers from many countries, the revolutionaries finally gained control over the entire country.

BRIEF HISTORY OF EVENTS LEADING UP TO AND CULMINATING IN REVOLUTION

Spain has been the stage for a continual struggle between the broad mass of the people and a relatively small traditional ruling elite, composed of the institution of the monarchy, the high-ranking military officers, and the Roman Catholic Church. Up to the early part of the 20th century, the alliance of these three institutions had maintained an authoritarian regime.

As the ideas of democracy, socialism, anarchism, and communism began to take hold around the turn of the century, the situation began to change. The masses resorted to general strikes and other types of civil disturbance in an effort to secure the political rights enjoyed by their neighbors to the north. These strikes generally ended with the destruction of numerous churches.

The first Spanish Republic, which came into existence in 1873 and represented the first attempt at liberal democracy, was quickly suppressed by the army because of the resulting chaos. The monarchy was restored for lack of a better system. The period which followed was one of sham democracy. Demands for agrarian reform and for the curtailment of the growing power and wealth of the church were ignored. The beginning of industrialization, meanwhile, had led to the development of a vociferous proletariat, which pressed the government for reforms. Added taxation, to meet the cost of the war with the United States in 1898 and that in Morocco in 1909, had equally alienated the bourgeoisie, which now tended to support the masses against the government. The strength of the masses was threatening to overthrow the traditional ruling elite. In 1923 Gen. Miguel Primo de Rivera, with the knowledge of King Alfonso XIII, staged a coup d'etat and established a military dictatorship.

The regime of Primo de Rivera proved unpopular. By 1930 he had earned the opposition of all the classes. The upper class resented its loss of political power. The peasants were angered when the promised agrarian reforms were only partly implemented. The middle class resented the increased taxation imposed to finance the expansion of commerce and industry. Leftwing and intellectual agitation finally forced Primo de Rivera to resign in 1930, and continued agitation led to the King's abdication in 1931. In April 1931 the second Spanish Republic came into existence.

The right and the left developed into two antagonistic and equally powerful groups, while the center, composed of intellectuals and liberals, decreased rapidly in importance. The right factions included the Catholic, Monarchist, and later on, the Falangist Parties, and the army generals; the left included two moderate Republican parties, the Socialists, the Anarchists, and the Communists. The issues that divided them related to agrarian reform, social reform, the church question, and the problem of regionalism versus centralism.

The elections of 1931 brought the left to power. The Constitution which was drawn up during the year attempted to deal with the above issues. Church and state were separated, and Catalonia was granted local autonomy. The anticlerical attitude of the left earned it the enmity of the church, and the acceptance of regionalism irked the generals. The inability of the Republicans to implement fully the agrarian reforms estranged the masses. The leftist regime was crippled by disunity within its ranks. Internal rivalries between Anarchists and Socialists led to strikes and counterstrikes and these alienated the middle class. In the November 1933 elections the right was returned to power.

The new regime proved equally unpopular. It repealed the agrarian reforms and suppressed regional autonomy in Catalonia. But it enjoyed the support of the army and was able to maintain itself until internal conflicts and frequent cabinet changes led the President of the Republic to dissolve the Congress of Deputies (*Cortes*) on January 4, 1936.

The February 1936 elections witnessed the total polarization of the political parties into a Popular Front incorporating the parties of the left, and a National Front of the parties of the right. Once again, the left was returned to power. The moderate Republican parties of the Popular Front were called upon to form the government, in an effort to allay the fears of the middle class. However, these Republican parties, the Republican Left and the Republican Union, had no mass following, and relied heavily on the Socialists and Communists for a legislative majority. Socialist agitation and strikes, and the resulting chaos, convinced the National Front that a Socialist takeover was not far off, a prospect which they were not willing to accept. Agreement was finally concluded between the National Front and a number of generals to overthrow the government by force. After careful and deliberate planning, which lasted for more than a month, the date for the uprising was set for July 17, 1936. It was to begin in Spanish Morocco, and spread to the mainland.

THE ENVIRONMENT OF THE REVOLUTION

DESCRIPTION OF COUNTRY

Physical characteristics

Spain is situated on the Iberian Peninsula, which it shares with Portugal on the west, and is separated from France, in the north, by the Pyrenees Mountains. Gibraltar lies at its southernmost tip. It has an area of 194,945 square miles, including the Balearic and Canary Islands. Its Atlantic coastline is 422 miles long, and its Mediterranean coastline, 1,033 miles.

Spain can be divided into five topographical regions: the northern coastal belt, the central plateau, Andalusia in the south-southwest, Levante in the southeast, and Catalonia and the Ebro Valley in the northeast.

Spain has six major mountain ranges: the Pyrenees, the Carpato-Vetonica, the Oretana, the Marianica, the Penibetica, and the Ibérica; and five main rivers: the Ebro, the Duero, the Tagus, the Guadiana, and the Guadalquivir. The climate varies from region to region. It is cool and humid in the northern coastal belt, cold in winter and hot in summer in the central plateau, and temperate in Andalusia and Levante.[1]

The people

Basques, Cataláns, Galacians, and Andalusians comprised the major cultural groups. With the exception of the Basques, they did not constitute distinct ethnic groups, although the Cataláns enjoyed large

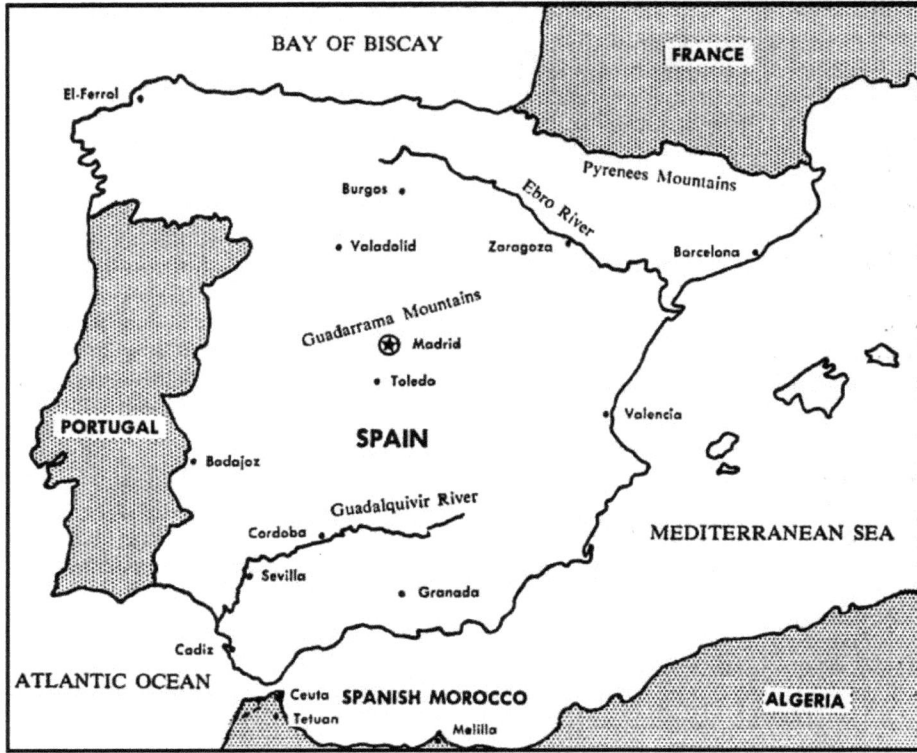

measures of autonomy. Spanish is the national language of Spain, Castilian the official dialect, and Galician the main dialect. Catalán represents a separate language.

The population of Spain was 24.5 million in 1936, with a density of about 139 per square mile. More than half the population lived in the major cities of Madrid (the Capital), Barcelona, Valencia, Sevilla, Malaga, Zaragoza, Bilbao, and Murcia.[2]

Communications

The overland communication system of Spain was generally poor. There were 74,994 miles of highways and roads, of which only 45,592 were macadamized, and 11,050 miles of railways. With the exception of the lower reaches of the Guadalquivir and sections of the Ebro, the rivers were not navigable. In 1936 the Spanish merchant marine consisted of about 1,000 seagoing vessels, operating from the major ports of Barcelona, Bilbao, Sevilla, and Valencia.

Natural resources

Spain has vast mineral deposits, fertile soil suitable for agriculture in some areas, and timber forests. The most abundant minerals are ore, coal, lead, potash salts, and mercury. Of somewhat lesser importance are iron pyrites, lignite, copper, zinc, sulphur, manganese, and tungsten.

SOCIO-ECONOMIC STRUCTURE

Economic system

The economy of Spain was largely based on a free enterprise system, even though the state played a limited role. Most public utilities were state-owned, and the state reserved the right to participate in all economic activities. However, most of the industrial complex was privately owned by Spanish and foreign nationals, and foreign capital investments were encouraged.

Agriculture and mining constituted the backbone of the economy. The principal industries were the processing of foodstuffs and ores; they did not, therefore, meet the requirements of the domestic markets. Spain's main agricultural products were cereals—especially wheat—oranges, rice, onions, olives, grapes, and filberts. Her major mining products were coal, iron ore, lead, and mercury.

Spain's principal exports were foodstuffs, minerals, tobacco, chemicals, metals, and timber. Her major imports were machinery, vehicles, chemicals, metals, petroleum, rubber, foodstuffs, cotton, and certain minerals.

Class structure

Socially, Spain was divided into two main groups. Roughly one-fifth of the population composed the upper and middle classes, which managed the affairs of the nation. Four-fifths of the population was composed of peasants and workers, with a high rate of illiteracy, and with little or no say in the affairs of the nation. The gap between these two groups was filled by a very small lower-middle class, composed of small shopkeepers and artisans. Spain was actually ruled by a trinity composed of the throne, the institution of the political generals, and the institution of the political church, supported by the nobility and the privileged classes—in fact, an even smaller elite within the upper and middle classes.[4]

The distribution of wealth reflected this disparity. The land and the business were in the hands of the upper and middle classes, while the rest of the population worked with their hands, either as tenant farmers or as miners and industrial workers.

The social structure tended to be rigid due to: (1) the traditional solidarity of the working class, which had given the workers a separate identity; (2) the "backwardness and inertness" of the economy, which had not provided the workers with new avenues of improvement; and (3) the lack of education.[5]

Literacy and education

Some 40 percent of the population were still illiterate in 1936. Most of the illiteracy was found among women, and in rural-agricultural areas. Primary education was made compulsory and free in an effort to reduce this high rate.[6]

Education was undertaken jointly by the church and the state. Schools were either state; or church-sponsored. There were nine state universities and two church-sponsored universities in 1936.

Major religions and religious institutions

Roman Catholicism was the religion of over 99 percent of the population. The few Protestant communities had no legal status, and were not allowed to proselytize.

It has been estimated that in the 1930's two-thirds of the population were not practicing Catholics—they did not confess or attend Mass. This figure varied from region to region and between the sexes. Church attendance was higher in rural areas and among women.[7]

The Constitution of 1931 provided for the separation of church and state. Prior to that date, Roman Catholicism had been the official religion. However, the church continued to play an important role in the affairs of the state because of its alliance with rightwing elements and the military.

GOVERNMENT AND THE RULING ELITE

Description of form of government

The 1931 Constitution provided for a parliamentary form of government. A unicameral system was adopted to bring the Parliament "closer to the people and avoid the pitfalls of a Senate sharing less liberal sentiments."[8] It was hoped that this parliamentary form of government would prevent the usurpation of executive powers and the establishment of a dictatorship.

The power of the executive was vested in the Prime Minister and his Cabinet, both appointed by the President. A vote of no confidence in the *Cortes* could force the prime Minister to resign.[9] The President of the Republic was elected jointly by the *Cortes* and a special electoral body for a term of 6 years. The functions of the President were mainly ceremonial, but he could dissolve the *Cortes*, on the advice of the Prime Minister, twice during his term of office, in cases of impasse.

Members of the parliament were elected by universal suffrage. The *Cortes* was elected every 2 years, and met twice a year. However, during the parliamentary recess, "a permanent representation of the *Cortes*, based upon the numerical strength of the various parties in the Chamber,"[10] met with the Cabinet regularly.

A Supreme Court, the Tribunal of Constitutional Guarantees, was empowered to rule on the constitutionality of the laws passed by the parliament. It could also arbitrate between autonomous regions, review the election of the President, and try the President of the Republic, the Prime Minister and members of his Cabinet, and the judges of the Supreme Court and the Attorney General.[11]

As noted above, the President, acting on the advice of the Prime Minister, could dismiss the *Cortes* in case of an impasse. However, if he exercised that power twice and the third *Cortes* judged the previous dismissal unjustified, it could impeach him. The Cabinet could be dismissed only if it failed to receive a majority vote in the *Cortes*, or upon the dissolution of the *Cortes*.[12]

Description of political process

In the elections of February 1936 the Popular Front, an electoral union of the political parties of the left, pledged to a platform of socialization, nationalization, and drastic agrarian and anticlerical reforms, emerged victorious with 4,716,156 votes and 256 seats in the *Cortes* out of a total of 475. The principal parties of the Popular Front were: the Socialists, headed by Francisco Largo Caballero and Miguel Indalecio Prieto; the Republican Left Party of Manuel Azana; the Republican Union under Diego Martínez Barrio; Luis Companys' Esquerra (Catalan Separatists), and the Communists. The Popular Front was supported by the General Union of Workers (UGT), a Socialist trade union of about 1,500,000[13] members, the semi-Trotskyist Workers and Peasants' Alliance, and the Republican Military Union (UMR), a small group of Republican officers. The Anarchists of the National Confederation of Labor (CNT) and the Iberian Anarchist Federation (FAI), a syndicalist trade union and a secret society respectively, with an estimated joint membership of more than 1,000,000,[14] did not join the Popular Front, but supported it in certain electoral districts.

Three weeks after the elections, Manuel Azana was officially designated by Niceto Alcala Zamora, President of the Republic, to form the new cabinet. This cabinet was to be composed entirely of Republican Left and Republican Union politicians. On May 10, 1936, after the impeachment of President Zamora by the *Cortes* in April 1936, Azana was elected President, and Santiago Casares Quiroga, another Republican Left politician, was called upon to form the government. His government also was composed of only Republicans. The Republican governments of Azana and Casares Quiroga were unable to function because of ideological differences with the other parties of the Popular Front. The Republican parties were moderate liberal parties. They supported only the agrarian and clerical reform program of the Popular Front, and had nothing in common with the Marxist-Socialist ideologies of the Anarchists, Socialists, and Communists. In fact, they had joined the Popular Front only to prevent a National Front victory.

The National Front received 3,793,601 votes, but won only 143 seats in the *Cortes*. It was composed of the Spanish Confederation of Autonomous Rightists (CEDA), a Catholic party, headed by Gil Robles; the Monarchist Party of Calvo Sotelo; the Carlist (traditionalist) Party led by the Count of Rodenzo and Fal Conde; the Agrarian (landowners) Party; and the Independent Party. The National Front was supported by the Spanish Military Union (UME), a group of rightwing officers, Catholic organizations such as the Defense of Catholic Interests (DIC), and José Antonio Primo de Rivera's Spanish Falange, a Fascist movement. As a whole, the National Front represented the middle and upper classes, the Spanish Catholic Church, and rightwing army generals. These groupings represented different vested interests, and lacked a common political platform. They were, nonetheless, united in their opposition to the Socialist-Marxist alliance which supported the government. However, the Carlist Party and the Spanish Falange were actually the two most important elements of the National Front, and special attention should be given to them because of the roles which they were to play in promoting and supporting the revolution.

The Carlist Party derived its name from the movement which supported the candidacy of Don Carlos, the brother of King Ferdinand VII, to the throne of Spain in 1834, and opposed the accession of Queen Isabella II, the only child of King Ferdinand. In 1936 it advocated the restoration of the monarchy, but supported the candidacy of Don Alfonso Carlos, a descendant of Ferdinand, while the Monarchist Party supported a descendant of King Alfonso. A religious party, closely associated with the Catholic Church, and opposed to modern trends of liberalism, the Carlists advocated the use of force in achieving their end. The Carlist Party was well established in the Navarre region of northern Spain, and its members represented generally the poorer aristocracy. In 1934 the Carlist Party organized the *Requete* (levies), a paramilitary body, which in 1936 numbered close to 14,000[15] men under arms, and trained openly.

The Spanish Falange, or Falanx, derived its name from the Macedonian battle unit that enabled Alexander the Great to conquer Greece in the 4th century B.C. It was founded in 1933 by José Antonio Primo de Rivera as a movement which would find "a national way of restraining the incoherence of liberalism." It advocated the National Socialist theory of the corporate state. But unlike the National Socialist movements in Germany and Italy, it accorded the church an eminent role in the affairs of the state. In 1934 it amalgamated with a number of other National Socialist movements, and organized a private army, the Blue Shirts, which at the outbreak of the revolution was to number about 50,000.[16] In February 1936 the Falange numbered about 25,000[17] members, half of whom were university students. Only one-fifth of the rest belonged to the working class. The membership of the Falange, between 1933 and early 1936, did not increase because the right, as a whole, tended to support the CEDA, which represented the right's only chance of coming into power by peaceful means. However, after the defeat of the National Front in the February 1936 elections, when the rightists "abandoned all ideas of peaceful and legal solutions and put their hopes in violent ones,"[18] the ranks of the Falange were swelled by great numbers of rightists, including the CEDA Youth Movement, who deserted the CEDA and joined the Falange.[19]

The Center parties, which included the Center Party; Lliga (Catalán businessmen); Radicals; Progressives; and Basques, decreased rapidly in importance, and split. The Lliga and the Basque Parties sided with the Popular Front and the rest with the National Front.

The political parties of the National Front attacked the government in their numerous newspapers, the principal of which were the *ABC* (Monarchist) and *Debate* (CEDA), and continually exhorted the people to revolt. They were, however, heavily censored, and suspended during most of the Republican era. Billboards were frequently used by both fronts during and after the elections. Radio broadcasting was not available to the National Front in 1936, because it was under government control. The government came equally under heavy attack from Catholic pulpits. In pastoral messages from the bishops, and in regular sermons, Catholics were warned against supporting the government, and threats of excommunication were leveled at leading politicians of the Popular Front.

Legal procedure for amending constitution or changing government institutions

Amendments to the Constitution of 1931 could be initiated by the government or by one-fourth of the members of the *Cortes*. The amendment proposal had to receive "the affirmative vote of two-thirds of the Deputies then holding office during the first four years of the constitutional regime, and an absolute majority thereafter."[20]

After voting in favor of an amendment proposal, the *Cortes* was to be dissolved automatically, and elections were to be held within 60 days. The new *Cortes* was to act as a constituent assembly to decide on the proposed amendment, and then resume its ordinary activities.[21]

Relationship to foreign powers

On the eve of the civil war, Spain was a member of the League of Nations. No major alliances had been concluded with other states. Officially, Spain's foreign policy orientation was one of collaboration with other state-members of the League. However, the Popular Front as a whole, especially the Socialists, was ideologically linked with the Popular Front in France, and the Socialist Party of Léon Blum, Prime Minister.[a]

The role of military and police powers

Because of the alliance of the institution of the political generals with the throne and the institution of the political church, the role of the army became one of maintaining the government in power. Army officers, who mostly belonged to the nobility or the upper class, took it upon themselves to "alter a regime when it no longer harmonized with the army's desires."[22]

In 1936 the army numbered 115,000 men and 15,000 officers. With the exception of the Army of Africa, which was composed of the Foreign Legion and native Moorish troops, most of the soldiers were conscripts doing their national service.

The Civil Guard (*Guardia Civil*) was established in 1844 to keep order in rural areas. Its organization was similar to that of the army, and it was led by officers with military rank. A large number of its rank and file were recruited from the army. In 1933 the *Guardia Civil* numbered 30,000 men and officers. In theory it was supposed to be loyal to the central government. However, because the *Guardia Civil* was closely associated with the army, and was traditionally loyal to the monarchy and to rightwing governments, the Republican Government created, in 1931, the Assault Guards (*Guardia de Asolto*), composed of men of known Republican and Socialist sympathies.[23]

---

[a] It should be noted here that some of the political parties of the National Front had established contacts with Italy, Germany, and Portugal to obtain financial and military assistance.

WEAKNESSES OF THE SOCIO-ECONOMIC-POLITICAL STRUC-
TURE OF THE PREREVOLUTIONARY REGIME

History of revolutions or governmental instabilities

The Republican Government of Manuel Azana came to power, in 1936, in a Spain divided against itself as a result of more than 150 years of bitter quarreling and civil unrest. The year 1808 marked the beginning of the disintegration of the institution of the monarchy; civil war broke out in 1834 over the issue of a liberal constitution; in 1868, the army expelled the monarch, and precipitated a religious and regional war; class hatred emerged in 1909, culminating in the antireligious riots of Barcelona; in 1917 a revolutionary strike was crushed by an insurrectionary army; in 1923 General Primo de Rivera established a military dictatorship; liberal protests and agitations led to his resignation in 1930, which was followed by the abdication of King Alfonso XIII in 1931; in 1932 the rightwing elements attempted to topple the first Republican Government of Spain in the 20th century; in 1934 leftist elements staged a revolt in Asturias in an effort to overthrow the rightwing government; and in 1936 the right and the left had polarized into fronts of equal strength, unwilling to cooperate, and ready to use force to gain their ends.

The governments of Azana and Casares Quiroga were doomed to failure because they did not represent either left or right and were, therefore, not accepted by either.[24] They were, in fact, minority governments which depended for their majority in the *Cortes* on the vote of the other political parties of the Popular Front, since the National Front had, from the very beginning, declared its determination not to support or collaborate with any Republican Government. Thus, these Republican Governments were, in reality, at the mercy of the Socialist and Communist Parties. Moreover, the Republican Governments not only were inherently weak, they were also the victims of the economic and social problems which had plagued Spain for more than a century.

Economic weaknesses

The inadequacy of agrarian reforms had been the major economic problem. In 1931, cadastre figures showed that about half the arable land was owned by 99 percent of the landowners, while the other half was owned by the remaining one percent. The solution to this problem had not been found by 1936. Furthermore, the economic crisis of 1929 had had serious repercussions on the agricultural economy. Agricultural prices had fallen, much land had gone uncultivated, and unemployment had reached phenomenal figures. Industry, still in its initial stage of development, could absorb only an insignificant number of the unemployed. These repercussions were still being felt in 1936, and added to the overall economic problem.[25]

Social tensions

Socially, Spain was sharply divided by quarrels between the church and the liberals, between those who advocated regionalism and those who advocated centralism and nationalism, and between the bourgeoisie and the working class.[26]

The church, the largest landowner and one of the wealthiest institutions, had, for more than a century, used its wealth, position, and prestige to maintain a *status quo* which would preserve its interests. It had thereby become identified with the traditional ruling elite, and was considered by the liberals to be reactionary, and, therefore, a major obstacle to political evolution. The efforts of the liberals to curtail the wealth of the church and its political activities culminated in bitter dissension.

The struggle between regionalists and centralists has its roots in the actual formation of the Spanish state. After the expulsion of the Moors around 1250, Spain broke up into a number of petty independent principalities owing nominal allegiance to the Spanish King. This symbolic unity at the top was not accompanied by a corresponding unity at the local level. The ultimate dissolution of the monarchy, and the establishment of a Republic, removed the vestiges of unity and resulted in demands for local rights. The Basques and the Cataláns were the elements agitating most vigorously for local autonomy. In 1931 the Republican Government granted the Cataláns a large measure of autonomy. This was resented by the army and the parties of the right, which interpreted it as an attempt to weaken the nation. These measures of autonomy were repealed by the rightwing government in 1934, thereby precipitating new waves of agitation.

The working class, as a whole, subscribed to the revolutionary ideology of the Socialist, Anarchist, and Communist Parties. As such, it was always ready to support revolutionary strikes and uprisings. This alienated the bourgeoisie, which lived in constant fear of a Russian-type revolution.

Government recognition of and reaction to weaknesses

Azana moved swiftly to stabilize the internal situation. Fearing an army coup d'etat, he removed General Franco, then Chief of Staff, and General Goded from the War Ministry in Madrid, and assigned them command posts in the Canaries and the Balearics respectively.

He appointed new civil governors of Republican sympathies throughout Spain to consolidate his position. Local rights were restored to the Cataláns, numerous members of the Falange were imprisoned, and the Falange Party was suspended. To satisfy the Popular Front, he released all leftist political prisoners, and pressed for agrarian reform. Land was distributed to more than 50,000 peasants. In an effort to pacify the bourgeoisie, he refrained from implementing the nationalization platform of the Popular Front.

These, however, proved to be half measures which failed to satisfy either the Popular Front or the bourgeoisie. Subsequently the popularity of Manuel Azana with the working class decreased, while that of the Socialists, now the revolutionary standard-bearers, increased. He accepted the Presidency of the Republic to "ensure that the socialists should never be allowed to form a government alone."[27]

The Quiroga government which succeeded Azana's found itself immobilized by the differences which had split the Popular Front on all matters relating to policy and tactics. The leftwing section of the Socialist Party, under the leadership of Largo Caballero, was fomenting a series of lightning strikes. Supported by the Communists, Caballero was openly talking of supplanting the Republican Government with a dictatorship of the proletariat, and had clashed with the Anarchists.[28] The National Front, meanwhile, had closed ranks, and the final plans for a revolt were being laid. The ever-increasing wave of unrest, terror, and political assassinations was to provide them with the immediate reasons. The murder of Calvo Sotelo, a staunch monarchist and one of the church's leading spokesmen, on July 12, 1936, by Assault Guards proved to be the signal. On July 17, 1936, the Army of Africa revolted, and was followed by similar garrison revolts in Spain. Henceforth, Spain was to be divided between the Nationalists, representing the revolutionaries, and the Republicans, representing the Popular Front.

FORM AND CHARACTERISTICS OF REVOLUTION

ACTORS IN THE REVOLUTION

The revolutionary leadership

The revolution of July 17, 1936, was a joint civilian and military undertaking. The military leadership included large numbers of high-ranking officers and generals, and principal instigators being Generals Sanjurjo, Goded, Mola, and Franco. However, the capture of General Goded by the Republicans on July 19, and the accidental death of

General Sanjurjo on July 20, left Generals Mola and Franco as the outstanding figures. These two were not among the so-called "political generals." The civilian leadership was composed of Fal Conde, Ramón Serrano Suner, and José Antonio Primo de Rivera.

General Mola was Director-General of Security under King Alfonso XIII, and earned the nickname of "Shoot-Mola" and the enmity of the Republican intellectuals for his role in suppressing the 1930–31 riots. As a result, he was retired by the first Republican Government of Manuel Azana, in 1932. In 1934 he was recalled by the rightwing and Catholic government, and appointed Commanding General of the Army of Africa. In 1934 he was transferred by Azana and appointed Military Governor of Pamplona. Prior to 1936 he had not been involved in plots against the Republicans.

General Franco, the scion of a good Spanish family, achieved his rank at the age of 35. In 1917 he participated in crushing a general strike. He served in Morocco, and commanded the Foreign Legion from 1923–27. There he earned a reputation for military brilliance for the role he played in bringing the Moroccan War to a successful end. In 1934 he was appointed Joint Chief of Staff and was given the task of suppressing the miners' rebellion in Asturias. He accomplished this with the help of the Foreign Legion. Strict, patient, cautious, and a brilliant organizer, Franco was completely dedicated to his career. Prior to 1936 he had refused to meddle in politics, but he was accused in 1937 by Portela Valladares, Premier of the caretaker government during the February 1936 elections, of plotting to prevent the Popular Front from entering office following the elections. In 1936, just before departing for the Canaries, General Franco paid Premier Azana a visit in which he warned him of the threat of communism. The nonchalant attitude of the Prime Minister seemed to have prompted Franco to join the conspiracy.

Fal Conde, a young Andalusian lawyer, and one of the leaders of the Carlist Party, was involved in numerous plots against the Republican regime in 1932. He was jailed by the Republicans, and was released in 1934 when the right entered office.

Ramón Serrano Suner, the brother-in-law of General Franco, was the CEDA Youth leader. He is credited with having promoted the merger of the CEDA Youth with the Falange.

José Antonio Primo de Rivera, the son of General Primo de Rivera, began his career as a monarchist. A practicing Roman Catholic, he founded the Falange in 1933. He was jailed by the Republicans in 1932 for his political activities, and was released in 1934 by the rightist government. He lost his electoral seat in the 1936 elections, and was again jailed by Azana in March 1936.

Politically, these men had one thing in common. They were all opposed to parliamentary government, and advocated an authoritarian government of one type or another. In ideology, however, they differed radically. The military were interested only in restoring their rule over Spain. Fal Conde was a Carlist, interested in restoring the monarchy. Ramón Serrano Suner was interested in restoring a Catholic rightwing government; and José Antonio Primo de Rivera advocated a Fascist corporate state.

An agreement was reached, prior to the revolution, to establish a joint military and civilian dictatorship under General Sanjurjo, which would include representatives of all the parties joining in the revolution, and which would satisfy their basic demands and aspirations.[29]

The revolutionary following

The military following of the Nationalists, on July 18, 1936, included the 15,500 officers and men of the Army of Africa (Moorish troops and the Foreign Legion), and the 34,000 officers and men of the Civil Guard. Most of the 7,228 officers of the National Army sided with the revolution, but the rank and file could not be counted on because of their sympathies for the Popular Front. In many instances, the rank and file of the National Army turned against their officers and sided with the Republicans. The naval officers, as a whole, supported the Nationalists. The attempts of these naval officers to sail their ships into Nationalist-held ports were, in most cases, thwarted by mutinous crews of distinct Republican sympathies, who either imprisoned their officers or shot them. However, the Nationalists acquired one battleship, several cruisers, a destroyer, and a number of gunboats. The air force sided with the Republicans.[30]

The middle and upper classes tended to side with the Nationalists. The lower-middle class was divided in its sympathies. The revolutionary following included the members of the political parties of the National Front and their paramilitary organizations, the hierarchy of the Spanish Catholic Church, industrialists, big businessmen, and financiers. Separatist-minded businessmen from Catalonia, financiers, and industrialists were sympathetic to the Republican cause.

ORGANIZATION OF REVOLUTIONARY EFFORT

Internal organization

Early opposition to the Republican Government came from two different groups: the political parties of the National Front, and the army generals. Prior to the agreement concluded between them during the last 2 months preceding the revolution, these groups prepared for a revolution on separate bases. The army generals began earnest preparations, after the elections of February 1936, by secretly contacting other officers, in command positions and in the War Ministry, and enlisting their support. They were thus able to compile a list of those officers and garrisons who would support them, and of those who would remain loyal to the government.

Some of the political parties began their revolutionary preparations in 1934. The Falange and the Carlists created adjunct paramilitary bodies which trained and drilled openly.

Secret contact between the political parties and the army generals was first established in March 1936. Fal Conde got in touch with General Sanjurjo, while Serrano Suner acted as liaison between the CEDA and the Falange, on the one hand, and General Franco on the other. Bickering and bargaining between the generals and the political parties hampered the conclusion of an agreement; and an army uprising, set for the end of April 1936, was postponed. The Carlists demanded that all political parties in Spain be dissolved after the revolution, and that a government of three men, under General Sanjurjo, be established. Further demands concerning the restoration of the monarchy, and the use of the monarchist flag as the flag of the revolution, complicated matters. The Falange, on the other hand, was initially unenthusiastic in its support of an army uprising, and tended to consider such an uprising as reactionary. However, tentative agreement was finally achieved between these two groups in May–June 1936, and it was decided that the leadership of the revolution would go to the generals, and that a government would be established under General Sanjurjo. Plans were made to create military and civil branches of the revolution in all the provinces of Spain. These "provincial branches were instructed to work out detailed plans for seizing public buildings in their areas, particularly lines of communications, and prepare a declaration announcing a state of war."[31] The political parties were to enroll their paramilitary bodies in the Nationalist Army; and on the day following the revolution, the rank and file of these parties were to place themselves at the disposal of the provincial branches in areas where the takeover had been successful. In areas that had remained loyal to the Republican Government, they were to lie low, to emerge later as a fifth column.[b]

External organization

Prior to July 17, 1936, the support of foreign countries to the revolutionary cause was limited. The solicitation of foreign aid was not pursued by the Nationalist Front as a matter of policy. Contacts with foreign powers were made separately by the generals and by some of the political parties of the National Front. In 1934 the Carlists secured financial and military aid from Italy with which to train their *Requetes*. Portugal offered the generals the safety of its territory from which to plot the overthrow of the Republican Government. In February 1936, General Sanjurjo was received in Berlin by Admiral Canaris, then head of the German *Abwehr* (army counterintelligence), and "assured himself that German military aid; if it - should be necessary to secure the success of the rising, would be contemplated by Canaris at least."[33] As soon as the revolution broke out, Germany, Italy, and Portugal abandoned their cautious policies of limited and indirect involvement in Spanish affairs, and supported the Nationalists actively.

After some initial hesitation, Germany and Italy lent their complete support to the Nationalist cause. At the end of the civil war, German aid amounted to $199,520,000. The number of German nationals who participated and fought in the war totaled 16,000 men, and included the *Condor Legion* (German air force unit) and 30 antitank companies. Italian aid amounted to $372,200,000; and Italian forces in Spain numbered 50,000 men. Portugal never hesitated. On July 24, 1936, President Salazar offered General Mola Portugal's unconditional support. Henceforth, German aid to the Nationalists was to be channeled through Portugal, and Portuguese soil was to provide the Nationalists with vital staging areas. Ultimately, 20,000 Portuguese were to fight on the side of the Nationalists.[34]

On the international scene, the diplomatic maneuvering of Germany and Italy proved effective in preventing France and England from supporting the Republican Government, thus partly isolating the Republic. By joining the Non-Intervention Committee, and actively supporting it, Germany and Italy were able to cloak their aid to Spain, while seeing to it that Republican Spain would not be able to buy war material from the other members of the commission, which

---

[b] The term "fifth column," which has become so familiar and which we now use without thinking, originated here. It was first used by General Mola in a radio address in which he warned residents of Madrid that a "fifth column" of rebel sympathizers was waiting in the city to help the rebels take it over.[32]

included France, England, and the Soviet Union. However, the Soviet Union also used the committee cloak its aid to the Republicans.

GOALS OF THE REVOLUTION

Concrete political aims of revolutionary leaders

The Nationalists claimed, at the outbreak of the revolution, that it was undertaken "in order to forestall a projected 'Red' revolution which was to have taken place at the end of July or in August with Russian support."[35] However, the major aim of the revolution seems to have been to eliminate any possibility of a Socialist-Communist takeover. The National Front had reached the conclusion that the Republican Government had lost control of the working class to the other members of the Popular Front, and that the only safety against a Socialist-Communist coup d'etat lay in a revolution of their own.[36]

Social and economic goals of leadership and following

There was no social or economic goal. The Nationalists represented a political union of groups with different ideologies and different social and economic platforms. They appear to have been agreed only in their determination to prevent the socialization of the economy which a Socialist-Communist regime might have initiated.

REVOLUTIONARY TECHNIQUES AND GOVERNMENT COUNTERMEASURES

Methods for weakening existing authority and countermeasures by government

Prerevolutionary activities

In the few months before the revolution, the National Front lost no opportunity to discredit and render impotent the new Republican Government, and create conditions which would justify its taking power, by revolution if necessary. The National Front was aided in bringing about these conditions by the disunity within the Popular Front. The Socialist Party had split into two factions, and fighting had erupted between the Socialist UGT and the Anarchist CNT and FAI; this conflict culminated in a series of strikes, riots, and assassinations.

To weaken the government, the National Front resorted to spreading rumors and adverse propaganda. It alleged that the notorious Hungarian Communists Bela Kun and Erno Gero, and a number of Spanish Communists who had fled after the Asturias revolt, had arrived

694

in Spain to instigate a revolution. The Republican Government was likened to the Kerensky government of Russia, which had preceded the Bolshevik revolution, and the violent incidents occurring daily were enumerated in the *Cortes* to show that the government had lost control of the situation.[37] At the same time, the Falange contributed to the atmosphere of unrest by carrying out, independently of the National Front, a policy of assassination and other forms of violence.

The government did not expect to gain the support of the National Front, but nevertheless took certain conciliatory measures. It gave the opposition parties representation in the councils, named a moderate, Martínez Barrio, to the Presidency of the *Cortes*, and refused to take punitive action against Generals Franco, Goded, and Mola, despite evidence that they were planning a revolution.[38] The allegations of the right relating to the presence of foreign Communist instigators were scrupulously investigated and proved false. But the government could do little to stop the wave of terror, and the assassination of independent leader, José Calvo Sotelo, caused the final break.

The outbreak of the revolution

In the afternoon of July 17, 1936, the garrison of the city of Melilla, in Spanish Morocco, revolted and seized the town. Word was then passed to the garrison commanders in Tetuán and Ceuta, the two other major cities in Spanish Morocco, and these towns were seized. General Franco, at Las Palmas in the Canaries, was informed of the uprising and asked to take command of the Army of Africa, in accordance with prearranged plans. Simultaneously, the signal was flashed to the army garrisons on the mainland.

The techniques for carrying out the revolution were relatively simple. Early in June, the Nationalist generals were assigned major cities which were to be secured when the signal for the revolution had been given. Sevilla was assigned to General Llano, Valadolid to General Saliquet, Burgos and Pamplona to General Mola, Madrid to General Villegas, Zaragoza to General Cabanellas, Valencia to General Goded, Barcelona to General Carrasco, and the Army of Africa to General Franco. Smaller cities were assigned to other officers. These officers were to assume command of the garrisons, seize the cities, occupy all government buildings and strategic locations, arrest all leftist leaders, deal summarily with all who resisted, and declare war on the Republican Government. When this first objective was secured, they were then to attempt to link up and come to the aid of revolutionary forces in cities in which the uprising had been only partly successful. In cities judged to be predominantly leftist in sympathy, the Nationalist officers were to remain with their garrisons and await relief.[39]

When news of the Morrocco uprisings reached the government, it attempted to suppress and isolate these outbreaks by ordering navy and air force units to bombard the insurgent cities, and prevent the Army of Africa from crossing over to Spain. These measures proved ineffective. The aerial bombardment inflicted little damage, and the naval units were paralyzed by mutinies. When the Civil Guard and a majority of the elements of the National Army revolted the government found itself deprived of the only effective means by which it could have put down the revolution. The UGT, CNT, and FAI trade unions represented the only force at the disposal of the government capable of resisting the Nationalists.[40] "Yet for the Government to use this force would mean that it accepted the inevitability of a leftwing revolution."[41] Casares Quiroga, a moderate liberal, refused to take this step, and chose to adopt, instead, a policy of accommodation and compromise, in an attempt to avoid civil war and the overthrow of his regime. Thus, on the morning of the 18th, Casares Quiroga dispatched a loyal general to negotiate a compromise with General Cabanellas, in Zaragoza, with the promise that "a forthcoming change of Ministry will satisfy all the General's demands and obviate the necessity for a rising."[42] When this attempt failed, Casares Quiroga resigned rather than order the distribution of arms to the members of the UGT, CNT, and FAI. Martínez Barrio, his immediate successor, also attempted to compromise with the rebels by offering General Mola a post in the Cabinet. This and a similar offer to General Cabanellas were rejected.

Martinez Barrio then resigned to make way for a new government which would authorize the arming of leftist trade unions. In the night of July 18–19, Dr. José Giral became Prime Minister. His Cabinet, like that of his predecessor, was composed solely of Republicans. On the morning of the 19th arms were distributed to the people of Madrid, and in other cities where no uprising had yet taken place.

On July 18–19, and in the following few days, Nationalist uprisings were successful in Córdoba, Granada, Cadiz, Corunna, El-Ferrol, and León. The uprisings in Madrid and Barcelona started on the 19th and were suppressed by the now armed members of the Socialist and Anarchist trade unions. By the end of 1936, the Nationalists were in possession of the northwestern third of Spain. The dividing line ran halfway up the Portuguese-Spanish frontier in a northwesterly direction, bypassing Madrid, taking a southeasterly course at the Guadarrama Mountains to Teruel, and then north to the Pyrenees, halfway across the Franco-Spanish frontier. A long strip of the coastline on the Bay of Biscay, which included Asturias, Santander, and the two Basque provinces, remained loyal to the Republican government. Elsewhere on the mainland, Nationalists held the isolated cities of

Seville, Granada, Córdoba, Cadiz, and Algeciras. Spanish Morocco, the Canary Islands, and the Balearics, with the exception of Minorca, were also held by the Nationalists.

In Nationalist-held territory, martial law was imposed. Those suspected of having opposed the revolution were executed, while all known leftists were jailed. In Republican territory, the government lost much of its powers to the leftist trade unions. These had organized military committees which executed and imprisoned rightists, rightist-sympathizers, and Nationalist army officers who had fallen into their hands. A number of churches were burned. The military committees expropriated some property, nationalized the commercial establishments, and banned public worship in churches. The large landowners, in particular, suffered as their land was confiscated by the government.

Methods for gaining support and countermeasures taken by government

Nationalists

The Nationalist revolt was not a popular uprising. To give it some form of mass following, the Nationalists were obliged to expand the political sphere of the revolution by establishing some semblance of government, and by initiating a social program. On October 1, 1936, General Franco was appointed head of the Spanish state by Nationalist officers and Falange leaders. A technical council of state was created to serve as cabinet. The social program was based mostly on the Falangist platform. Universal suffrage, trade unions, strikes, and regional autonomies were abolished. The separation of church and state was maintained, but the rights of the Jesuits, which had been suspended by the Republicans, were restored. In the spring of 1937, the Falange, the Carlists, and the other rightwing political parties were amalgamated, by decree, into a new Falange Party. An intense recruitment campaign for the new party was initiated. Army noncommissioned officers, officers, and high Nationalist officials were given automatic membership, and the population was encouraged to join now or never.[43] The Nationalists intensified their propaganda in an attempt to identify their cause with that of Spanish nationalism. All anti-Republican activities were referred to as anti-Red, and great pains were taken to show that the Republican Government was Communist-dominated. Furthermore, the monarchist flag, and such slogans as "One State," "One Country," "One Chief," "One," "Great," and "Free" were adopted, in an effort to generate nationalistic feelings within their own ranks. The Roman Catholic Clergy, throughout the world, was to aid the Nationalist cause, by portraying it as a crusade against communism and atheism.

Republicans

The Republican Government, to give its cause a greater popular following, to attract the middle class, and to allay the fears of the peasantry, adopted social and economic policies not at all in accordance with the ideology of the Socialist-Marxist wing of the Popular Front. Despite some of the initial steps taken by the trade unions, the government continued to show respect for the property of the peasant. There was to be no interference in the affairs of the small businessmen, and no general socialization of industry. "At the same time they took up a national and patriotic attitude of defense of their country against the foreign invader."[44] The Nationalists were pictured as Fascist foreigners, composed of Moors, Italians, and Germans. Cabinet reshuffles, which brought in the Socialists, Communists, and Anarchists, were made to give the Republican Government greater representation and unity, while members of the UGT, CNT, and FAI were enrolled in the Armed Forces of the Republic. Because of the appeal of the Republican intellectuals to the intellectuals of the world, the cause of Republican Spain was championed and romanticized by writers and thinkers throughout the world, as a struggle between democracy and legality on the one hand, and Fascist totalitarianism on the other. This factor, more than any other, caused nationals of other countries to enlist in the International Brigades to help defend the Republic.

Foreign intervention

Immediately after the outbreak of the revolution, the Nationalists turned to Italy, Germany, and Portugal for military, economic, and financial assistance. These countries responded for different reasons. Italy's intervention was motivated by three main factors: (1) alarm over the possible establishment of a Communist-dominated government in Spain, and its repercussions on the Fascist regime of Mussolini; (2) a desire for military prestige which would enhance the prestige Mussolini had gained by his successful campaign in Ethiopia; and (3) the wish to obtain a more favorable strategic position for the Italian Navy over France and England, her rivals in the Mediterranean.[45]

Germany's intervention was motivated by similar considerations, and by others. Like Italy, Germany feared a possible Communist takeover, and was interested in acquiring strategic facilities in Spain. But because of Spain's remoteness, those considerations tended to be of less importance. Germany, a highly industrialized country, was also interested "in securing the economic command over Spanish raw materials and Spanish markets."[46] German military commanders, moreover, saw the civil war as an ideal opportunity to train their men, and to apply their new concepts of warfare.[c]

Portugal was motivated by ideological considerations. The regime of Dr. Salazar was akin, in many ways, to the totalitarian form of government which the Falange advocated. An ideological affinity, therefore, existed. A leftist victory in Spain, moreover, might have encouraged Dr. Salazar's political foes, hoping for aid from the Spanish left, to attempt his overthrow. Such aid might be granted—"The more so because it was one of the dreams of the Iberian Left to incorporate the whole of the peninsula into a federation of national states, remapped on a linguistic basis of which Portugal-cum-Galicia would have been one."[48]

Figures published in *The New York Times*, on February 28, 1941, show that Italy sent Nationalist Spain 50,000 men of her regular forces, "763 aircraft, 141 aircraft motors, 1,672 tons of bombs, 9,250,000 rounds of ammunition, 1,930 cannon, 10,135 automatic guns, 240,747 small arms, 7,514,537 rounds of artillery ammunition, and 7,663 motor vehicles." The Nationalists' debt to Italy was officially placed at 7½ billion lire, or $372,200,000.[49]

German military assistance to the Nationalists amounted to 500 million reichsmarks, or $199,520,000. However, no documents have been produced to show the exact number and character of that material, although it has now been established that the Condor Legion and some 30 antitank companies saw action in the civil war. With the exception of the members of the Condor Legion, most of the 16,000 German nationals sent to Spain were experts, technicians, and instructors.[50] In the economic and financial field, two holding companies, ROWAK and HISMA, were created to channel German exports to Spain and handle Spanish repayments in raw material. These two companies helped stabilize the Nationalist currency[d] by providing Nationalist Spain with the necessary currency backing and credit.

Portugal, a relatively poor country, could not give the Nationalists any material aid, but some 20,000 Portuguese saw action as part of the *Legion de Virago*. Portugal did, however, support the Nationalists by giving them refuge and staging areas. It was also through Portugal

---

[c] Herman Goering, among many other things also chief of the *Luftwaffe*, stated at the Nuremberg Trials that he had urged Hitler to aid General Franco in order to prevent the spread of communism, and also to afford the Luftwaffe a chance to test its new technical developments.[47]

[d] The Republican Government began the civil war in possession of all the currency gold backing, thus depriving the Nationalists of a currency backing of their own, and the financial means of obtaining credit abroad.[51]

that most of the German aid was channeled in the early phases of the revolution.[52]

On July 19, 1936, the Giral government of Republican Spain turned to France and other governments for material assistance. France did supply 200 planes, but later, in agreement with England, adopted a policy of nonintervention.[c] The Soviet Union chose to support the Republicans in a limited way. Mexico, motivated by ideological considerations, was the only country in the American continent to give its full support to the Republican cause.[57]

The Soviet Union did not send large troop contingents. Most of the 1,000 Soviet nationals who were despatched to Spain were technicians, military instructors, airmen, artillery officers, and staff officers. Soviet material aid that reached Spanish ports, according to German intelligence sources, included 242 aircraft, 703 cannon, 27 antiaircraft guns, 731 tanks, 1,386 motor vehicles, 29,125 tons of ammunition, and 69,200 tons of miscellaneous war material totaling $408,920,000. According to Nationalist estimates made in October 1938, Comintern (Communist International) aid amounted to 198 cannon, 200 tanks, 3,247 machineguns, 4,000 trucks, 47 artillery units, 4,565 tons of ammunition, 9,579 vehicles of different types, and 14,889 tons of fuel. Mexico gave the Republicans an estimated $2,000,000 worth of war materiel, which included a cargo of German arms destined for the Mexican Army.[58]

A unique feature of the civil war was the International Brigades which fought on the side of the Republicans. These brigades were

---

[c] The advent of the Blum government gave rise to a wave of unrest among the working class which alarmed the French right. The specter of a "Red Revolution" came to replace the specter of German militarism in the minds of the right, and the resulting fear led it to oppose any attempt by the government to support Republican Spain. Blum, on the other hand, had hoped "that a Government of the Left in France might be able to achieve that reconciliation with Germany which previous Governments of the Right had been unable or unwilling to bring about."[53] Any attempt, therefore, to aid the Republicans might have reduced such hopes. Furthermore, the Blum government feared that a decision to take unilateral action on behalf of Republican Spain would have prompted the British Conservative government of Stanley Baldwin to shy away from a leftist France and ally itself with Germany.[54]

England, at the time, was purusing its sincere quest for peace. The British Government believed that the reoccupation of the Rhineland by Germany in February 1936, and the conquest of Ethiopia by Italy, had satisfied these countries, and would induce them to "help create a new European order."[55] A nonintervention policy was therefore adopted by the British Government in the hopes that it would localize the Spanish issue, and prevent it from embroiling the rest of Europe in war. The British government also believed that the outcome of the Spanish revolution would not in any way affect the strategic interests of England.

The decision of the Soviet Union to give the Republic only limited aid was probably based on Stalin's decision that "he would not permit the Republic to lose, even though he would not help it win. The mere continuance of the war would keep him free to act in any way. It might even make possible a world war in which France, Britain, Germany and Italy would destroy themselves, with Russia, the arbitrator, staying outside."[56]

"organized by unsponsored voluntary efforts,"[59] but were actively supported and led by Comintern agents. They consisted of volunteers, recruited throughout the world, and reached a maximum number of about 20,000 men.

On August 3, 1936, France proposed a nonintervention pact which would prohibit the export of war materiel to Spain. By mid-August, 26 nations had joined, including Great Britain, France, Italy, Germany, and the Soviet Union. Portugal joined on September 28, 1936. At the first meeting, in London on September 9, a Non-Intervention Committee was created to supervise the embargo of arms to Spain. Acts of intervention by Germany, Italy, Portugal, and the Soviet Union, and the resulting accusations and counteraccusations, led to interminable delays in the establishment of an effective supervisory system. In February 1937, after the arrival of German, Italian, and Soviet Union military personnel on Spanish soil, in the guise of volunteers, the member nations undertook to prohibit the recruiting of volunteers for Spain; and in April 1937, international observers were sent to the Franco-Spanish and Spanish-Portuguese borders, and a naval cordon was established along the Spanish coast. In May 1937, after the *Deutschland* incident[f], Germany and Italy withdrew their ships from the cordon. In June 1937, Portugal refused to allow observers on its frontiers. In September 1937, a new naval patrol was set up by a nine-power conference, including the main European powers, to deal with the sinking of ships outside Spanish waters by unidentified submarines. The patrol was successful in putting an end to this type of warfare, but the Non-Intervention Committee remained ineffectual, due to repeated infractions by Germany, Italy, Portugal, and the Soviet Union. In April 1939, 1 month after the end of the revolution, the committee was disbanded. It had been successfully used by Germany, Italy, Portugal, and the Soviet Union to hide their illegitimate activities, and its existence had prevented the Republican Government from securing from other sources the war materiel to which it was legally entitled.[61]

Military and political developments, 1936–39

By the end of July 1936, it had become apparent that the Nationalists had failed to achieve a total and complete revolution. Two-thirds of the country remained loyal to the Republic. In the succeeding 32 ½ months, characterized by bloody and bitter fighting, the Nationalists were finally able to achieve their end.

---

[f] Germany claimed that Republican aircraft had bombed the German battleship *Deutschland*, in the evening of May 29, 1937, killing 22 and wounding 83. The German fleet took its revenge by bombing the Republican city of Almeria.[60]

On July 19, 1936, the Army of Africa, to be known later on as the Army of the South, landed at Cadiz, under the command of General Franco. Subsequently, it conquered the southern part of Spain, along the Spanish-Portuguese border, and relieved the beleaguered cities of Sevilla, Granada, Córdoba, and Huelva. On August 14, the Army of the South captured Badajoz, and linked up with the Nationalist Army in the north, to be known as the Army of the North, under General Mola. The Army of the South then marched on Madrid, and the Republican Government fled to Valencia. But the Army of the South was repulsed by the International Brigade and Republican militiamen and Madrid remained in Republican hands. In September 1936 the Army of the North captured Irun and San Sebastian, thereby cutting the lines of communication of the Basque Republic with France, and on June 19, 1937, the Basque Provinces were subdued.[g] By July 1937, the Nationalists "held 35 of Spain's 50 provincial capitals and 119,690 square miles of territory."[63] On January 9, 1938, the Republicans captured Teruel, but lost it on February 22. In the spring of 1938, a Nationalist offensive, which drove through to the Mediterranean, split Republican Spain into two parts, forcing the Republican Government to move from Valencia to Barcelona. On December 23, 1938, the Nationalists launched another offensive in Catalonia. Barcelona fell on January 26, 1939, and the Republican government moved to Figueras. When Gerona fell on February 4, the Republicans lost Catalonia, and the Republican Government moved to the French border.

Political developments in Republican Spain, during the civil war period, were marked by dissension, power struggles, and open strife. The split of the Socialist Party into an extreme left, pro-Communist wing, headed by Largo Caballero, and a moderate leftwing, headed by Indalecio Prieto, had enabled the Communist Party, small but efficient, to subvert the Socialist Party, and completely dominate its youth movement. Ensuing machinations by Largo Caballero and the Communists led to the fall of the Giral government, on September 4, 1936, and the formation of a new cabinet, under Largo Caballero, which included the Communists but excluded the Anarchists. The advent to power of the Communists was marked by Communist attempts to eliminate the Anarchists and the Trotskyists through terrorism. The resulting power struggle forced Largo Caballero to reconstitute his government; and in November 1936 four members of the CNT were included in the Cabinet. In May 1937 a brief 7-day civil war broke out in Catalonia between the Anarchists and the Trotskyists, on the one hand, and the Socialists and the Communists on the other. This

---

[g] General Mola was killed in this campaign on June 3, 1937.[62]

resulted in the exclusion of the Anarchists from the autonomous Catalán government. This change was reflected in the new government of Dr. Juan Negrin, which succeeded that of Largo Caballero on May 15, 1937. The Anarchists were excluded, while the Communists gained in power and influence. In April 1938, the seriously strained relations between the Anarchists and the Socialists were somewhat eased by the inclusion of the Anarchists in the reshuffle of the Negrin government. Peace conditions were submitted by Dr. Negrin to the Nationalists, but were rejected by General Franco.

MANNER IN WHICH CONTROL OF GOVERNMENT WAS TRANSFERRED TO REVOLUTIONARIES

On February 1, 1939, the Negrin government, meeting at Figueras in northern Catalonia, submitted new peace proposals to the Nationalists. When these, too, were rejected, Dr. Negrin returned to Madrid to continue the fight. On February 28, 1939, however, President Azana decided to resign rather than fight on, after France and England had accorded General Franco official recognition. At this time, everybody except Dr. Negrin believed that peace with the Nationalists could be concluded if the Communists were expelled from the Republican Government. A Defense Council was secretly formed, and on March 5, 1939, it deposed the government of Dr. Negrin. Fighting broke out, 2 days later, between the Communists, supporting Dr. Negrin, and the supporters of the Defense Council. This short civil war within a civil war dominated the last few days of Republican Spain. Surrender negotiations initiated by the Defense Council failed, and the Nationalist advance on Madrid continued practically unopposed. Republican leaders of all ranks and colors fled into exile. On March 28, 1939, the Nationalist Army entered Madrid, and in the following few days, all of Spain was under Nationalist control.

THE EFFECTS OF THE REVOLUTION

CHANGES IN THE PERSONNEL AND INSTITUTIONS OF GOVERNMENT

With the cessation of hostilities, Spain entered into a period of transition which resulted in drastic changes in government personnel and institutions. The government personnel and political leaders who emerged after the civil war were new men, civilian and military, who belonged almost entirely to the Falange Party. Generals Sanjurjo and

Mola, and numerous politicians such as Primo de Rivera and Calvo Sotelo, had perished during the civil war. Catholic politicians, such as Gil Robles, had been discarded;[64] and Republican leaders and government personnel had either fled or had been jailed. The military and the Falange were to assume prominent roles in this transitional period.

The parliamentary form of government was replaced with a dictatorial form of government. During the first 3 years of this transitional period, Spain was ruled by General Franco. His dictatorship was established by virtue of the compromise, concluded during the civil war, which united the Falangist, Catholic, and Monarchist Parties. The *Cortes* was not convened until July 17, 1942, when it was recreated as a body subservient to the will of the chief of state.

MAJOR POLICY CHANGES

In September 1938, at the height of the Czechoslovakian crisis, General Franco was urged by France and England to declare his neutrality. It was intimated then that France would, in case of war, attack in Morocco and across the Pyrenees. General Franco acquiesced. At the outbreak of World War II, in September 1939, General Franco, "partly from poverty but more so from policy,"[65] kept Spain officially neutral, although his sympathies lay with the Axis Powers. Shortly after the entry of the United States into the war, General Franco declared Spain a nonbelligerent state, although the Spanish volunteer Blue Division was sent to fight alongside the Germans on the Russian front.

Internally, General Franco set about to consolidate his revolution and his dictatorship by establishing a corporate state, as advocated by the Falange Party. Political parties and political activities, save those of the Falange, were abolished. Regional autonomy and provincialism were suppressed. The whole of Spain came under the domination of the Falange, which in turn was controlled by the government. The church alone escaped Falange domination. It retained some of its past influence and prestige, but found itself, nonetheless, subservient to General Franco.

However, in 1947, the new Constitution proclaimed Spain a kingdom and Catholic; and a limited form of representative government was introduced.

LONG RANGE SOCIAL AND ECONOMIC EFFECTS

Economically and financially, the picture in Spain, in 1939, was bleak. Agriculture, transportation, and public works had suffered great damage. Industry, which had survived the civil war practically intact, was able to resume production on only a limited scale because the majority of the skilled workers either had fled or were in jail. The reconstruction plans announced by the government could be implemented only on a very limited scale, because the outbreak of World War II prevented the inflow of foreign capital investment and materials. The shortage of foodstuffs was made worse during the next 2 years by successive droughts, and a majority of the Spanish people suffered privation during 1941–42. The shipment of a greater part of the gold reserves to the Soviet Union by the Republicans, in payment for war materiel, badly weakened the finances and contributed in part to the inflationary situation.[66]

The class composition of society presaged future tensions. Spain, in 1939, had become a two-class society A relatively small number of army officers and Falange leaders had replaced the upper and middle classes as the ruling elite. The mass of the population, peasants, and workers, overwhelmed by military forces and leaderless, had lost none of its outlook and solidarity. The bourgeoisie, hardest hit by the revolution and virtually eliminated, could no longer bridge the gap between the working class and the ruling elite. Socially, Spain in 1939 was more strikingly divided than it had been in 1936.

Some of the policies adopted by the Franco regime tended to increase, rather than decrease, the danger of further labor troubles. The agrarian reforms initiated by the Republicans were repealed in 1940. Trade unions were replaced by a Falange-dominated system which represented both employers and employees, but which in reality favored the employers; and wages were kept low, while the prices of essentials rose.[67]

OTHER EFFECTS

Shortly after fleeing from Spain, the Republican leaders set up a government-in-exile in Paris. At the outbreak of World War II it moved to Mexico, but returned to Paris at the end of the war. Later on, it moved to Toulouse, in southern France, to be closed to Spain. Contacts with Republican elements in Spain were reestablished, and it is believed that the first general strike which took place in the Basque region, in May 1947, was instigated by the government-in-exile. On the international scene, the government-in-exile sought to obtain international condemnation of the Franco regime. In 1946, Dr. Giral presented the case of the Republican government-in-exile before the

United Nations. It is, however, very difficult to estimate the efficacy of this group, and the role which it may play in the future of Spain.

At the end of World War II, in 1945, Spain was ostracized by a majority of nations. The leftist governments which came to power in most of the European countries attempted to force the resignation of the Franco regime by imposing international sanctions. In December 1946, the United Nations passed a resolution, by 34 votes to 6, demanding the replacement of the Franco regime "by one deriving its authority from the governed, coupled with a recommendation that all member nations should recall their ambassadors or ministers from Madrid."[68] As a consequence of this resolution, Spain was denied Marshall aid. In November 1950, however, this resolution was rescinded by the United Nations; and on December 14, 1955, Spain was admitted to the United Nations. Meanwhile, on September 26, 1953, a bilateral agreement was concluded between the United States and Spain. It provided for the establishment and joint use of U.S. bases in Spain, and the granting of economic aid to Spain.[69]

Today, some 22 years later, Franco still rules Spain. His regime has brought the country unprecedented stability, and prosperity has ameliorated the economic situation. Prosperity, however, partially a byproduct of U.S. aid and American military construction and expenditures, has tended to be very unevenly distributed.

NOTES

1. "Spain," *The Worldmark Encyclopedia of the Nations* (New York: Harper & Brothers, 1960), 892.

2. Ibid.

3. Gerald Brenan, *The Spanish Labyrinth* (Cambridge: Cambridge University Press, 1950), p. 81.

4. Lawrence Fernsworth, *Spain's Struggle for Freedom* (Boston: Beacon Press, 1957), p. 3.

5. Brenan, *Labyrinth*, p. 88.

6. Maximillian Olay, "Spain's Swing to the Left," in *Recovery Through Revolution*, ed. Samuel D. Schmalhauser (New York: Avici-Friede, 1933), p. 111.

7. Hugh Thomas, *The Spanish Civil War* (New York: Harper and Brothers, 1961), p. 31.

8. Fernsworth, *Struggle*, p. 134.

9. Frank E. Manuel, *The Politics of Modern Spain* (New York: McGraw-Hill, Inc., 1938), p. 71.

10. Ibid., p. 72.

11. Ibid.

12. Ibid.

13. Thomas, *Civil War*, p. 99, footnote.

14. Brenan, *Labyrinth*, p. 145.

15. Katherine Duff, "Course of the War in Spain," in Arnold J. Toynbee's *Survey of International Affairs* 1937, vol. II, *The International Repercussions of the War in Spain (1936–7)*, (Oxford: Oxford University Press, 1938), p. 44.

16. Thomas, *Civil War*, p. 205.

17. Ibid., p. 99.

18. Brenan, *Labyrinth*, p. 309.

19. Ibid.

20. Arnold R. Verduin, *Manuscript of Spanish Constitutions 1808–1931* (Ypsilanti, Michigan: Michigan State University Lithoprinters, 1941), p. 99.

21. Ibid.

22. Manuel, *Politics*, p. 42.

23. Fernsworth, *Struggle*, p. 18.

24. Thomas, *Civil War*, p. 111.

25. Manuel, *Politics*, p. 98.

26. Thomas, *Civil War*, passim.

27. Brenan, *Labyrinth*, p. 302.

28. Thomas, *Civil War*, pp. 107–108.

29. Ibid., pp. 101–102.

30. Duff, "Course," pp. 42–43.

31. Thomas, *Civil War*, p. 102.

32. *Webster's New Collegiate Dictionary* (Springfield, Massachusetts: G. & C. Merriam Company, 1956), p. 309.

33. Thomas, *Civil War*, p. 101.

34. Ibid., p. 634.

35. Arnold J. Toynbee, *Survey of International Affairs 1937*, vol. II, *The International Repercussions of the War in Spain (1936–7)* (Oxford: Oxford University Press, 1938), p. 22.

36. Ibid.

37. Thomas, *Civil War*, pp. 104–105.

38. Fernsworth, *Struggle*, p. 14.

39. Thomas, *Civil War*, pp. 102–109, 126.

40. Ibid., p. 141.

41. Ibid.

42. Ibid., p. 140.

43. Francis Noel-Baker, *Spanish Summary* (London: Hutchinson & Company, Ltd., 1948), pp. 51–52.
44. Brenan, *Labyrinth*, p. 316.
45. Toynbee, *Repercussions*, pp. 177–185.
46. Ibid., p. 186.
47. Thomas, *Civil War*, p. 228.
48. Toynbee, *Repercussions*, p. 205.
49. Thomas, *Civil War*, p. 634.
50. Ibid.
51. Ibid., p. 273.
52. Ibid., p. 231.
53. Ibid., p. 141, footnote.
54. Ibid., p. 219.
55. Ibid., p. 220.
56. Ibid., p. 216.
57. Toynbee, *Repercussions*, p. 212.
58. Thomas, *Civil War*, pp. 635–643.
59. Noel-Baker, *Summary*, p. 61.
60. Thomas, *Civil War*, p. 440.
61. Noel-Baker, *Summary*, p. 61.
62. "Spain," *Encyclopaedia Britannica* (1958), XXI, 138.
63. Thomas, *Civil War*, p. 441.
64. Ibid., p. 618.
65. Ibid.
66. Noel-Baker, *Summary*, p. 66.
67. Ibid., pp. 66–67.
68. *Encyclopaedia Britannica*, p. 140.
69. Ibid., pp. 140–141.

RECOMMENDED READING

BOOKS:

Brenan, Gerald. *The Spanish Labyrinth*. Cambridge: Cambridge University Press, 1950. A general historical survey of events leading to the civil war. Contains a good discussion of Spanish politics, and the political parties of the prerevolutionary era.

Cattell, David T. *Communism and the Spanish Civil War*. Berkeley: University of California Press, 1955. Gives a detailed description of the roles played by the Spanish Communist Party, the Comintern, and the Soviet Union in the civil war.

Fernsworth, Lawrence. *Spain's Struggle for Freedom*. Boston: Beacon Press, 1957. A journalist's account of the civil war period, with a good section on the role of the traditional elite.

Manuel, Frank E. *The Politics of Modern Spain*. New York: McGraw-Hill Inc., 1938. Contains a good discussion of the political, social, and economic problems of Spain prior to the civil war.

Thomas, Hugh. *The Spanish Civil War*. New York: Harper and Brothers, 1961. The first thorough and most objective study of the civil war. Discusses in detail the politics of Spain, the development of the civil war, and the intervention of the European powers.

Toynbee, Arnold J., *Survey of International Affairs, 1937*, vol. II, *The International Repercussions of the War in Spain (1936–7)*. Oxford: Oxford University Press, 1938. Contains an excellent discussion of the motivations for intervention and nonintervention by the European powers, and the role and purpose of the Non-Intervention Committee.

THE HUNGARIAN REVOLUTION OF 1956

SYNOPSIS

A series of peaceful demonstrations beginning in the summer of 1956 were followed by a sudden uprising in October 1956 which led to the overthrow of a Communist regime that was subordinate to the policies of the Soviet Union. Led by students, workers, soldiers, and intellectuals, the uprising apparently was not planned beforehand by an organized group of revolutionary leaders. Soviet forces intervened and terminated the existence of the revolutionary government a few days after it was established.

BRIEF HISTORY OF EVENTS LEADING UP TO AND CULMINATING IN REVOLUTION

Hungary was the last of the Axis countries to conclude an armistice with the major allies in 1945. An Allied Control Commission to supervise reconstruction was established, headed by Marshal Klementy Voroshilov of the Red Army. The agreement in effect put into the hands of the Soviet Union two instruments with which to control the domestic and foreign affairs of Hungary: the Allied Control Commission itself and the Soviet High Command. Under Soviet auspices a coalition of Hungarian parties, including the Communist Party, under the leadership of Matyas Rakosi and Enro Gero, was formed to set up a new government. Although the Communist Party did not at first control this government, it was able to place key men in several ministries and extend its influence to all levels of the administration. Between 1946 and 1949 the Communist Party eliminated all opposition by persecuting other political parties and expelling parliamentary leaders, thereby establishing a *de jure* one-party totalitarian regime.

During the years 1949–53, rapid industrialization and forced collectivization depressed the living standard of the workers and peasants; the Hungarian economy was subordinated to the interests of the U.S.S.R.; rigid controls frustrated and limited the potentialities of the intellectuals; and Soviet-style purges created instability within the Communist Party elite. In 1953 Imre Nagy, a critic of collectivization and forced industrialization, was named Premier. He introduced the "new course," which relaxed political, social, and economic tensions and gave the people a period of social democracy.

In 1955 Moscow severely criticized the "soft line" introduced by Nagy and instructed the Hungarians to "correct" their policies. Rakosi replaced Nagy and the new course gave way to a harsher policy. The oppressive and repressive measures of the 1949–53 period were reimposed. Hostility toward the small Communist ruling group mounted in all occupational categories even among those who enjoyed higher incomes and a comfortable standard of living.[1] Overt signs of discontent, expressed in open defiance of the Communist Party leadership, came first from within the party apparatus itself and gradually increased. Popular bitterness, on the other hand, expressed itself in sudden and total insubordination.[2]

In October 1956 a peaceful demonstration followed by a spontaneous and unplanned national uprising of students, workers, soldiers, and intellectuals overthrew the ruling elite and established a network of Revolutionary and Workers' Councils throughout the country. A multiparty government was established; one of its first acts was to proclaim Hungary's neutrality. Its downfall followed quickly. The Soviet Union, unwilling to tolerate so close to its frontier a government not subservient to its will, sent in tank units on November 4, and stamped out the revolution.

THE ENVIRONMENT OF THE REVOLUTION

DESCRIPTION OF COUNTRY

Physical characteristics

Approximately the size of Indiana, Hungary is in the eastern central part of Europe. It is landlocked and borders on Austria, Czechoslovakia, the Soviet Union, Rumania, and Yugoslavia.

Hungary is divided into three geographical sections, each having its own peculiarities: the Trans-Danubian, which has a diversity of land forms, hills alternating with valleys; the great plain, which is a flat expanse of grassland and fertile soil; and the northern upland, which is a well-forested mountain area. The flat plain covers almost two-thirds of the central and eastern parts of the country. Mt. Kekes, the highest elevation, is 3,330 feet in altitude.

The country is a merging point for three different climatic zones: the Oceanic, the Mediterranean, and the Asiatic or Continental.[3]

The people

The Hungarian people for the most part are Magyars, whose ancestors came from the Eurasian steppes in the ninth century, but mixed racially with the Slavs and Rumanians living in the Danube val-

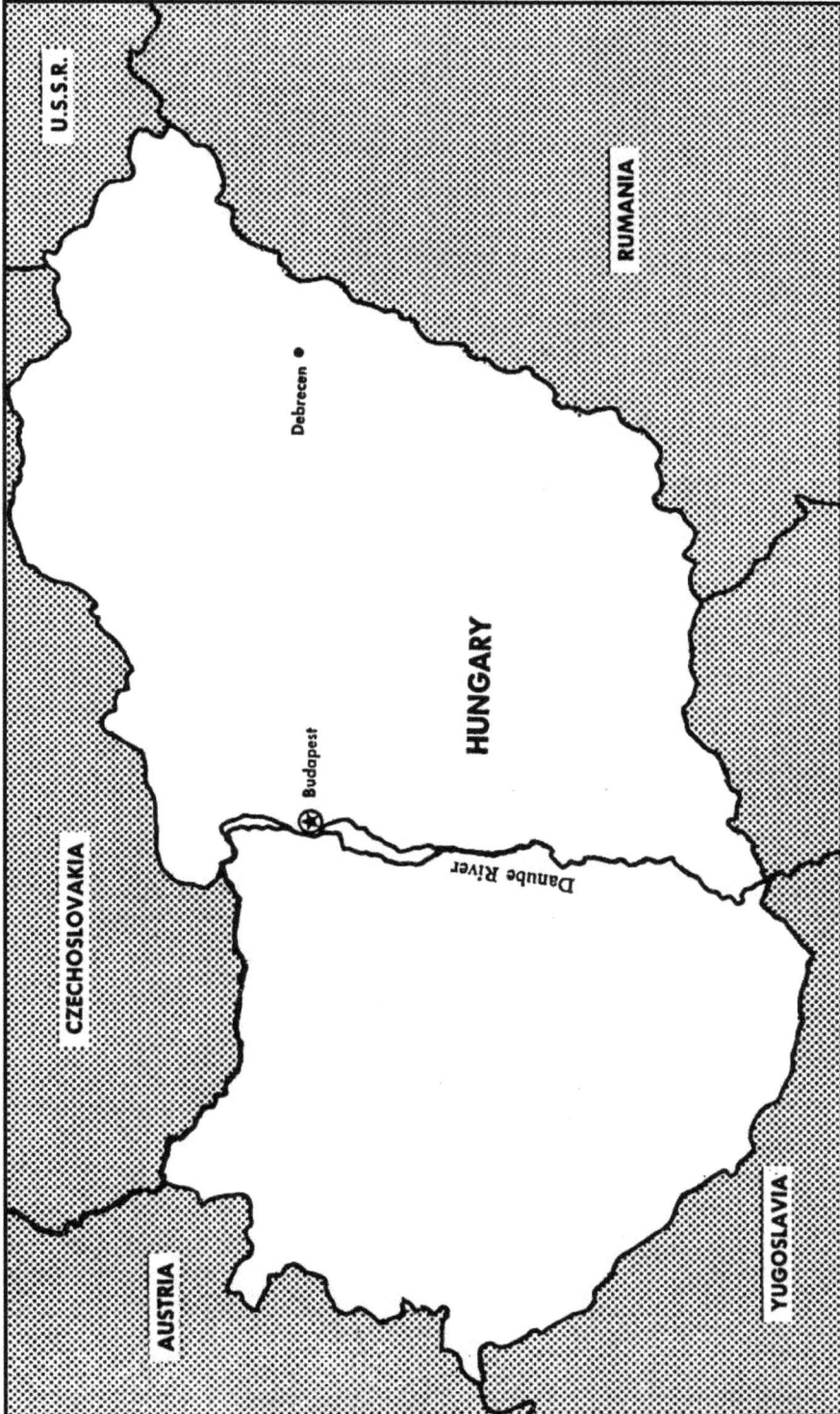

ley and the Carpathians. Only one-half of 1 percent of the population is non-Hungarian. The language, which has been preserved through the years, belongs to the Finno-Ugrian family.

The most densely populated country in East Central Europe, Hungary had slightly under 10 million people in 1955,[4] an average of 271.6 per square mile. The river valleys, particularly those in the eastern parts, were more densely populated. Budapest, the capital as well as the cultural and economic center, had approximately 1,725,000 inhabitants. Other cities of economic importance were Miskolc, Gyor, Pecs, Szeged, and Debrecen.

Communications

The most important roads in the early 1950's were either asphalt or concrete. Though highly developed and adequate for most purposes, they were not entirely suitable for heavy motor traffic. The railroad system, on the other hand, was the best in East Central Europe. It consisted of a series of main lines radiating from Budapest, and had recently been expanded to improve connections with the Soviet Union. Hungary also had a series of navigable waterways. The Danube, which forms a natural boundary with Czechoslovakia and takes a 90° turn to divide the country in half, and the Tisza, cover a distance of 601 miles. Budapest was the largest river port, handling sea vessels from the Black Sea and the Mediterranean. Most large cities had airports; there was air service between Budapest and the important provincial cities.[5]

Natural resources

Hungary is well endowed with natural resources. The oil fields near the Yugoslav border, for example, more than adequately supplied domestic needs and provided a surplus for export. Hungary had 9 percent of the world's reserves of bauxite in the early 1950's.[6] Uranium ore, which was discovered on the Slovak frontier in the early 1950's, was completely taken over by the Soviet Union. Hungary also has deposits of coal, natural gas, iron ore, manganese, copper, lead, and zinc.

SOCIO-ECONOMIC STRUCTURE

Economic system

Since the end of World War II, the economic has been sub-ordinated to the economic interests of the Soviet Union. Under Soviet guidance, Hungarian production was geared to meet Soviet requirements. Prior to 1956, Hungary embarked upon two major plans. The first, a 3-year plan designed to raise production to prewar levels, was instituted in 1947 and was fulfilled ahead of schedule by the end of 1949. The second, aimed at transforming Hungary from an "agrarian-industrial" country to an "industrial-agrarian" one, was instituted in 1950, but it became a source of controversy between groups of Communist planners and was recast several times.[7] The burden of economic expansion was largely borne by the workers and the peasants.

The Communist regime attempted to socialize all sectors of the economy. Nationalization of industrial enterprises employing 10 or more workers was completed in 1949.[8] In agriculture, the regime confiscated large estates without compensation and redistributed the land to the peasantry. The peasants' holdings were later collectivized, however, into cooperatives and state farms. In neither the industrial nor the agricultural sectors was the regime able to coerce the workers and the peasants into putting forth their maximum productive effort.

Hungary's industrial enterprises produced large quantities of rolling stock, such as locomotives, turbines and generators, trains, and coaches. In agriculture, cereal grain was traditionally the main crop; but the regime's price policies discouraged the production of that crop and encouraged the production of industrial crops such as textile plants and tobacco. The livestock industry was of secondary importance.

Class structure

In Hungary as elsewhere, communism did not create a classless society as Karl Marx had envisaged; it merely reshuffled the class structure. The middle class lost its dominant economic and social position. As most of its members had derived their income from owning property or operating industrial enterprises, they were forced to become wage earners when their properties were confiscated. The workers—supposedly the favorite sons of the Communist system—while not subject to arbitrary arrest, as were members of the middle class, were not among the privileged. They were forced to meet high productive norms, and their level of living dropped considerably because wages were kept low to limit purchasing power and fewer consumer goods were made available to them. The peasants were uprooted and forced to join the collective farms. Resistance exposed them to economic pressure. Large landholders became "enemies of the people" and most of them were arbitrarily imprisoned.

The privileged were the Communist Party leaders, functionaries, cadres, and anyone else who served the party, regardless of class origin. This political bureaucracy, the "new class," acquired all the characteristics of the older privileged class. The new class followed in the footsteps of the "aristocracy" and "capitalists" and occupied the homes of the "exploiters" in the stylish quarters of large cities.

Literacy and education

The Communist regime placed great emphasis on education. To meet the demands of industrial expansion, the regime extended the school system and directed students toward technical specialization rather than liberal arts. The Communist methods were highly successful and by 1955 illiteracy had been reduced to slightly above 4 percent. In the 1953–54 school year, 165,000 students were enrolled in secondary education and 53,000 attended higher institutions. Ninety percent of the medical students were financially supported by the regime; after graduation they were obligated to serve the state for a specified period. Although the quality of education was open to question, the Communists believed that their attempt to create a "toiling intelligentsia" had been successful, until the 1956 uprising demonstrated that their indoctrination in Marxism-Leninism had largely failed.[9]

Major religions and religious institutions

Roman Catholicism was the predominating religion, incorporating approximately 65 percent of the population in the early 1950's. The Roman Catholic Church had a number of educational institutions; these were suppressed by the regime in 1948, but some were restored in 1951. Protestants made up about 27 percent of the population. There was a Protestant university and seminary. Other religious minorities included those of the Jewish and Orthodox faiths. Many of the social functions performed by the churches prior to Communist ascendancy were taken over by the state. Churches performed only religious functions.[10]

GOVERNMENT AND THE RULING ELITE

Description of form of government

The Hungarian Constitution of 1949 declares that the power of the state originates in the workers of cities and villages and that they in turn are led by the Vanguard (the party), supported by the democratic union of the people. The Hungarian Government is a soviet-type government similar to that of the Soviet Union. Councils at the village or town level appoint delegates to a higher level council, and so on up the council hierarchy to the Supreme Soviet, the council at the national level (Parliament). Executive Committees are appointed by these delegates to perform the necessary administrative functions. Paralleling this structure is the party hierarchy. The Hungarian Workers' (Communist) Party—the result of a merger between the Social Democrats and the Communists in 1948—was the only legal party in Hungary in the early fifties, and completely controlled the government at all levels by placing party functionaries in responsible positions such as the Presidium (presidential council). An overlap between party and government is common and members of the party may also assume power in the Presidium. The Prime Minister and the Cabinet members are generally party leaders and many times in the past the First Secretary of the party assumed the Premiership. The state controls all the means of coercion and persuasion and thus governs the political, economic, and social life of every person.

The ruling elite in the early fifties were the members of the Central Committee of the Hungarian Workers' Party. This type of collective leadership was called "democratic centralization." The party was responsible to no legislature or judiciary. The Supreme Soviet rubberstamped the decrees of the Central Committee without altering them; institutions controlled by the party enforced them; and judges appointed by the party applied the law according to party wishes.

There was a procedure for electing delegates to the different councils at all levels, but these councils had no actual policy-making functions.

Description of political process

The Hungarian Communist Party was split in two factions when the provisional national government, after the German withdrawal, was established in Debrecen in December 1944. One group, the "leftists," were members of an "indigenous" underground anti-Fascist movement who had remained in Hungary during the war and who were led by Laszlo Rajk. The other group, the "moderates," were Moscow-trained émigrés who returned to Hungary in 1944 under Soviet auspices and who were led by Matyas Rakosi. Since major policy was dictated by Stalin, and since Rakosi adulated Stalin, Rakosi became the leader of the party. The party received 3 portfolios out of 12 in this first postwar cabinet.

The Communist Party gradually assumed more power over the period of several years but did not begin to eliminate opposition (see below) until after the election of November 1945. The results of this election were disappointing to the Communists: The Smallholders' Party whose views varied from liberal conservatism to semi-Socialist radicalism, had received a clear majority of the votes and several other parties had won seats in the Parliament. The Smallholders' Party had the support of the peasants, who at one time formed over one-half of the population. Most of the leaders were middle-class intellectuals. The Social Democratic Party had a much smaller mass base; it favored cooperation with the Communist Party. The National Peasant Party was a populist party subservient to the Communist Party. After the defeats suffered by the Communists in the 1945 elections, only 4 out of 18 cabinet posts were assigned to Communists. However, the important Ministry of Interior, which controlled the police system, was in Communist hands.

The Smallholders' Party was the greatest obstacle to the Communist Party in its bid for power. Thus the Communists proceeded to eliminate the Smallholders by taking two steps: It formed a coalition with the Social Democratic Party and the National Peasant Party in a "leftist bloc" against the Smallholders; and then it proceeded to attack the individual members of the Smallholders' Party with false charges of "reaction" and "conspiracy." Many were tried, convicted, and imprisoned. By May 1947, the Prime Minister, who was a Smallholder, was in exile, the Secretary General of the Party was under house arrest, and many prominent members of the party had to flee the country. When the Social Democratic Party merged with the Communists in 1948 to form the Hungarian Workers' Party, the new party had *de facto* control of the government, and after the elections in 1949, it gained *de jure* control.[11]

After Rajk had helped Rakosi purge the non-Communist elements by 1948, his radicalism was challenged by Rakosi, and he himself became the victim of a party purge. By October 1949 Rajk had been expelled from the party, arrested, tried, and executed on trumped-up charges. This resulted in a wave of terror in which an estimated 200,000 Rajk followers were purged, some executed, others imprisoned.[12]

Following the purge of the "left," Rakosi's policies became extremely radical. A "right" opposition trend developed under the leadership of Imre Nagy. Between 1947 and 1951, Nagy's career was in eclipse: He was removed from the Cabinet and dropped from the Politburo and the Central Committee. He was reinstated in 1951 and continued to criticize Rakosi's industrialization and collectivization programs.

After Stalin died in 1953, the new government in Moscow, then headed by Malenkov and Beria, forced Rakosi and his faction to abandon their policies and relinquish their governmental posts. Rakosi remained First Secretary of the party, but Nagy became Premier. This situation lasted until 1955, when the Hungarian Workers' Party was instructed by Moscow to change its course again. Rakosi returned and Nagy was again stripped of his governmental and party posts.

Another wave of relaxation was signaled by the 20th Party Congress in February 1956. Opposition to the Communist leadership became more public and more defiant. The "right" faction now demanded the rehabilitation of Nagy and a reorganization of the government under his leadership. Very shortly this faction was joined by the purge victims of the 1941–51 era and the situation became more intense. In October 1956 Nagy was reinstated.

To ensure the wide participation of the Hungarian people in its programs, the party and the government incorporated within a Patriotic People's Front all religious, social, and scientific organizations and institutions. The National Council of Trade Unions, the Federation of Working Youth (DISZ), the Hungarian Federation of Democratic Women (MNDSZ), plus scores of other organized groups became the transmission belts of the party and the major vehicles of Communist agitation and propaganda.

Opposition to the party was, of course, illegal. Nevertheless, shortly after the collapse of Stalinism, and particularly after the 20th Party Congress in the Soviet Union, criticism increased dramatically. The nonpartisan Hungarian Writers' Union, an organization whose activities were not entirely limited to the literary field, became one of the government's most persistent critics. It must be kept in mind, however, that the Hungarian Writers' Union was not an anti-Communist organization. On the contrary, many of the intellectuals who belonged to it were party members, but considered themselves as constituting a progressive Leninist wing of the party. The increasing popularity of the group was reflected in the wide circulation of its organ, the *Literary Gazette*.

Somewhat similar to the Writers' Union was the Petofi Circle. Established during the Nagy period of 1953–55, it drew its members mostly from among "like-minded Communist intellectuals" who had also been critical of the ruling group under the leadership of Matyas Rakosi from 1948 to 1953. Providing a forum for literary discussions, the Petofi Circle soon took on a political character and "attracted adherents from all walks of life." Debates and reunions held from March to June 1956 brought together Communist youth from a variety of organizations. A debate on June 27 attracted 6,000; 3 months earlier a similar debate had drawn only a dozen. Summaries of the debates were widely circulated in the form of leaflets and pamphlets. Later members of the Circle secured control of the official party organ, *Szabad Nep*.[13]

The criticism was an expression of the fear and anxiety instilled by the party's ruling group under the leadership of Rakosi as First Secretary. Frequent arbitrary arrests by the secret police and purges of party members executed by the ruling group intensified press criticism and provoked public demonstrations.

Legal procedure for amending constitution or changing government institutions

Constitutional changes required the vote of two-thirds of the members of the national council, according to the provisions of the 1949 Constitution. Amendments were promulgated by the president of the Presidential Council and became effective upon publication in the official gazette.

Relationship to foreign powers

The Soviet Union firmly established its influence immediately following World War II; first within the Allied Control Commission, and second within the Communist Party, which was dominated by Soviet-trained Hungarians. Although it professed to favor a democratic coalition of all political parties in 1945, the Soviet Union took active part in the political plots which removed "noncooperative" leaders and enhanced the position of the Communist Party. Soviet occupation troops were effective in crushing or discouraging opposition.

Gradually, as in most other bloc nations, strong political and economic ties were fashioned between Hungary and the Soviet Union. This relationship produced strange patterns of behavior. Whenever a political shakeup took place within the party leadership in the Soviet Union, the repercussions were transmitted to the party leadership in Hungary, where a similar reorganization would take place. The two most striking events in the Soviet Union which brought about drastic changes in Hungary were the death of Stalin and the 20th Party Congress. Both events encouraged a measure of democratization in Hungary and intensified the power struggle within the Communist Party leadership.

A treaty of friendship and mutual assistance between Hungary and the Soviet Union was signed in Moscow on February 18, 1948. Similar treaties were concluded with Yugoslavia, Rumania, Poland, and Bulgaria. In 1955 Hungary became a member of the Warsaw Pact, a Soviet counterpart of NATO.

The role of military and police powers

The Hungarian Army was gradually infiltrated by the Communists between 1945 and 1948, through propaganda tactics and purges, until they had established the control needed to remold the army as a "pillar of the party."

The police were controlled by the party in fact. Including a Frontier Guard equipped with tanks and heavy artillery, the police (AVH) numbered approximately 100,000 and were modeled after the Soviet MVD. The AVH had a network of agents throughout Hungary; a large concentration was needed in Budapest, and was used largely to round up the intellectual opposition after 1948. Some elements of the middle class also were arrested. Approximately one percent of the population of Budapest consisted of AVH personnel. During the revolution, this organization proved faithful to the ruling group.[14]

WEAKNESSES OF THE SOCIO-ECONOMIC-POLITICAL STRUCTURE OF THE PREREVOLUTIONARY REGIME

History of revolutions or governmental instabilities

The Hungarian Government's greatest liability was its one-party characteristic. This may not have been a weakness in itself, but the one party that did control the government apparatus was constantly racked by factional struggles. These were both ideological and personal. Before Stalin's death, party members could adulate him and remain in favor. After his death, they had to choose sides—Stalinists or anti-Stalinists—and the position of those who chose wrong became untenable. Rakosi, for example, was a Stalinist and lost his position in 1953 when Stalin died, and again in 1956 when Stalin was denounced by Khruschev at the 20th Party Congress. In any case the struggle between the Soviet leadership and the factions led by Rakosi, the Stalinist, and Nagy, on the right, undermined the party's stability. After October 6, 1956, the party showed its "inability to rule."[15]

Economic weaknesses

The economy of Hungary was weakened to begin with by the destruction of industries by the German troops during World War II. The Russians later marched in and stripped Hungary of industrial machinery, taking it as war booty and reparation payments. Thus the starting point for postwar economic reconstruction was very low.

The program of rapid industrialization instituted in the late forties was highly unjustified: Hungary lacked the necessary raw materials. Nevertheless, the rapid creation of capital goods was embarked upon. The intensified industry began to make demands upon the workers for increases in norms, but failed to meet their demands for consumer goods. Wages were controlled by the piece rate system, the premium wage system, and the constant readjustment of norms. This, along with the demands for higher production, resulted in a lower level of living for the workers. The economic plenty promised by the Communist propaganda machine failed to materialize.

The peasants were subjected to all sorts of discrimination if they defied the forced collectivization program. Somewhat like the workers, they were faced with increasing norms and compulsory deliveries of agricultural products and their level of living also deteriorated.

Social tensions

Following the period of nationalization, socialization, industrialization, and collectivization, according to Communist doctrine, the destruction of capitalist ownership was to result in a classless society. In Hungary, however, as in all other Communist countries, this did not take place: one class was eliminated but was replaced by another that imposed privations on every citizen and even purged its own elite. The new class was rooted in the Communist Party and was made up of those who had special privileges and economic preference because of the administrative monopoly they held. The members of this new class were hated by those they replaced and feared by those they ruled.

Government recognition of and reaction to weaknesses

Imre Nagy introduced the "new course," which slowed the pace of industrialization and halted collectivization. He ushered in a period of relaxation, and in general alleviated the harshness of the preceding years. By 1955, however, Rakosi had managed to persuade the members of the Central Committee that Hungary was in great need for industrialization and the control mechanisms that Nagy had removed.

Rakosi reintroduced many of his harsh economic policies, but did not reinstitute political terror to the degree he had practiced before. Some private enterprises established during the Nagy regime were able to remain in business, and the secret police organization, weakened under Nagy, did not regain its power. Those who were not considered politically reliable were removed from the government administration, university faculties, and some industries. However, compulsory teaching of Russian was abolished, and welfare, health protection, pension, and vacation plans for workers were introduced. The Rakosi regime was unable to revoke the freedoms that Nagy had allowed, and discontent grew with the opportunity to articulate it.

FORM AND CHARACTERISTICS OF REVOLUTION

ACTORS IN THE REVOLUTION

The revolutionary leadership

It is difficult to label the revolutionary leadership in Hungary as instigators and organizers. The outstanding characteristic of this revolution is that it was totally unplanned and unorganized, and that its leadership developed during the course of the revolution rather than before. The quiet demonstrations which took place in Budapest in October 1956 perhaps generated a psychological or crowd force which was finally released when the AVH fired into the gathering at one of the demonstrations. From there on the extemporaneous revolution took its course and developed its *ad hoc* organizations.

Imre Nagy was the popular figure in the revolution. He had been a Communist Party member since 1918. He lived in Moscow for 15 years, and returned to Hungary in 1944 with the Soviet Army. He held several posts in the Cabinet and became Premier in 1953. He improved the life of the workers and peasants and became known as a "just" man. He was expelled from the party in 1955 because his European orientation prevented him from being sufficiently militant. After his dismissal he held many intellectual exchanges at his villa. He was reinstated in October 1956. The mass demonstrations which drew him into the revolutionary leadership found him unprepared. He had no program or plan of action.

The right opposition represented by Nagy allied itself with the anti-Communist Hungarians to provide some sort of leadership. This strange combination consisted mostly of intellectuals from the Hungarian Writers' Union and the Petofi Circle. These intellectuals were not fighters; and those who were Communists were very poor "Bolsheviks." The political center which this group formed challenged the Central Committee of the Hungarian Workers' Party.

The revolutionary following

The revolutionary following was young, averaging perhaps 35 years or younger. Its members were discontented Communist intellectuals, disillusioned party functionaries, dissatisfied university students, and disgruntled workers and peasants. The demonstrations which began October 23 in the public squares of Budapest were led by students from three universities. Soon large crowds began to gather. Workers left their factories to join the students, and members of the Armed Forces also joined them. In less than 24 hours the masses were in full rebellion. The intellectuals attempted to quiet the crowds, but little attention was paid to them. Motley bands of workers, soldiers, teenagers, and riffraff coalesced, under self-appointed leaders. The workers showed the greatest courage in the fighting, while the peasantry performed the auxiliary services of supplying food and giving comfort. Members of the former middle class were much less vigorous in revolutionary activity.

ORGANIZATION OF REVOLUTIONARY EFFORT

As the revolutionary pace increased, Revolutionary and Workers' Councils took over the responsibilities of the government all over Hungary. They assumed control of the town councils, borough councils, district councils, etc., which were constitutionally established organs of the party. Workers' councils also attempted to take control of factories, mines, and other industrial enterprises. Both types of councils sprang up without coordination and without central control. Both were still in the formative stage at the time of the second Soviet intervention on November 4. These councils were direct expressions of popular dissatisfaction and represented the first practical steps to restore order and organize the Hungarian economy without party control or terror. Their members were chosen at meetings, and in most cases were freedom fighters, workers, and politicians.[16]

GOALS OF THE REVOLUTION

Concrete political aims of revolutionary leaders

Prior to the outbreak of the revolution, the different civic organizations drew up their demands; these varied little in their general character. One of the chief political aims was to restore national sovereignty. Since World War II Hungary had remained under the control of the Soviet Union; now those who headed the revolution desired to return their country's political and economic destiny to Hungarian control. Restoration of democracy, human dignity, and the rights of man was another demand.

As the pace of the revolution heightened, the political demands became bolder and more specific: reintroduction of the multiparty system, free elections, and withdrawal from the Warsaw Pact. Some groups demanded freedom, liberty, and solidarity with Poland; others wanted the restoration of Hungarian symbols and celebrations which had been suppressed by the Communist regime; most of the leaders wanted independence within a Marxist framework.

The ideological orientation of the revolutionary leadership was Marxist. The Communists wanted to retain one-party rule, but they also wanted to dislodge the entrenched party leadership. They wanted more freedom and the restoration of democratic practices which would assure both Communists and non-Communists of legal protection against persecution. They wanted more equitable economic policies that would take account of the immediate desires of the people. They proposed to revive the Patriotic People's Front as a symbol of mass participation in political affairs. They wanted the enjoyment of actual liberties. They wanted to govern for the people.

Social and economic goals of leadership and following

The suppression of creative arts had greatly frustrated certain classes of the intellectuals. Writers particularly wanted greater artistic and creative freedom. They wanted more "truth" and "honesty." They wanted to set up an "idea model" state adapted to human needs generally and to Hungarian national requirements specifically. They wanted more autonomy within the universities and a free exchange of information.

The workers and peasants demanded radical economic changes. Workers wanted reforms within the factory management; they wanted trade unions through which they could voice their opinions and influence policy. They wanted the norm system abolished and working conditions improved. They demanded more information concerning Hungary's foreign trade, economic difficulties, and its uranium deposits which had been under Soviet control. The peasants wanted to abolish collectivization along with the co-ops and compulsory deliveries. Above all they wanted a truly national economy and a more equitable socialist system.

REVOLUTIONARY TECHNIQUES AND GOVERNMENT COUNTERMEASURES

Methods for weakening existing authority and countermeasures by government

The political instability which made the Hungarian revolution possible was the result of two contributing forces. The first developed gradually from political rivalry within the Communist Party apparatus and manifested itself in party purges and open defiance toward party leadership. The second was a mass or popular force developing from general discontent, incubated over a long period, but released during the demonstrations. The revolution was neither organized nor directly inspired by the intraparty rivalry. It began with a sudden clash between the police and the masses attracted by student demonstrations.[17]

On October 23 a group of university students decided to hold a quiet demonstration. The Minister of Interior, Laszlo Piros, issued a proclamation prohibiting demonstrations. However, the students, defying this prohibition, went on with their preparations. Piros reversed his decision when two demonstrations were well on the way: one in Buda[a] before the statue of General Bem, and the other in Pest under the Petofi statue. Placards affirming Hungarian nationalism were raised. At the Parliament building the crowd of students and workers had grown by 6 p.m. to proportions estimated variously at 200,000 to 300,000. Some of the demonstrators went to work on the Stalin statue and by 9:30 p.m. it fell from its pedestal. Others marched to the radio building to have their demands broadcast. The building was guarded by the AVH, however, who threw gas bombs from upper windows to disperse the crowds, and then fired upon the demonstrators reportedly inflicting casualties.[b] In several instances the AVH used white ambulances with Red Cross plates and white uniformed AVH agents to transport arms and supplies. This infuriated the crowd. Hungarian tanks sent in by the regime refused to fire upon the people. Late that evening the insurgents finally succeeded in occupying the building of the party organ *Szabad Nep*.

Many police and army units were aiding the insurgents and supplying them with arms. The arrival of revolutionary forces strengthened the armed workers from the Csepel and Ujpest areas. Whether the Soviet tanks were officially asked to intervene is a matter of conjecture. Fraternization with Russian troops was reported. At 2 a.m. on October 24 the first Soviet tank patrol was seen. Fierce fighting developed later that day where these units were employed.

The fighting continued and on October 25 the activities were intensified in front of the Parliament building. Soviet tanks fired upon unarmed crowds. Molotov bombs were used effectively by the freedom fighters, and the tanks had difficulty in maneuvering in the narrow streets. On October 27 Nagy was able to assume control of the government apparatus and the next day he ordered a cease-fire. A few skirmishes continued, but by October 30 the new Cabinet had taken office and all was quiet. Soviet armed forces began to withdraw from Budapest. The revolution was over.

---

[a] Budapest is physically divided by the Danube and the two halves of the city are respectively called Buda and Pest.

[b] There was intermittent fighting around the radio building for the next few days after which it was finally seized by the revolutionaries.

Methods for gaining support and countermeasures taken by government

Once the people realized that a revolution was in progress the only elements not supporting the movement were those party members holding power and the AVH. It was, therefore, unnecessary for the revolutionaries to campaign for support. However, printed matter was kept circulating, particularly after the party organ was seized. Most radio stations also supported the movement.

Between October 23 and 25, the regime struggled to keep control. Party leaders announced that Imre Nagy had been appointed Premier, that he had invited Soviet units to come in and that he had declared martial law. Evidence shows that this was a tactic employed to quiet the crowds: The AVH abducted Nagy and reportedly forced him to order the people to go home. This measure did not work, however, and the fighting in Budapest continued. There was little fighting in the provinces. In most cases, the Revolutionary and Workers' Councils, without much effort, were able to assume control of the local administrations.

MANNER IN WHICH CONTROL OF GOVERNMENT WAS TRANSFERRED TO REVOLUTIONARIES

The transfer of power from the Communist bureaucracy to the Revolutionary and Workers' Councils took place in the early days of the revolution and in most cases without opposition. There was a swift and complete collapse of the Communist Party and its machinery. The party lost all its support, and the only institution which did not stop functioning was the secret police. However, popular resentment against the AVH was so overwhelming that Nagy was obliged to abolish it on October 29.

Nagy moved into the Parliament building and on October 30 announced that the new Cabinet, composed of Communists and non-Communists, had abolished the one-party system. Janos Kadar, who replaced Erno Gero as First Secretary of the Communist Party, agreed with this step in order to avoid further bloodshed. Zoltan Tildy, former leader of the Smallholders' Party, announced that elections would be held soon throughout the country.

THE EFFECTS OF THE REVOLUTION

CHANGES IN THE PERSONNEL AND INSTITUTIONS OF GOVERNMENT

On November 3 the government was again reorganized. Several Communists were dismissed, and three Cabinet portfolios were given to the independent Smallholders' Party, and three to the Social Democratic Party. The Petofi Party was given two ministries. Kadar had reorganized the Communist Party into the Hungarian Socialist Workers' Party on November 1, and his party received three portfolios.

MAJOR POLICY CHANGES

The most drastic policy changes were in foreign affairs. On November 1 the Council of Ministers proclaimed Hungarian neutrality. This in effect cancelled Hungary's commitments under the Warsaw Pact. At the same time the new Defense Minister, Pal Maleter, was negotiating for the withdrawal of Soviet troops.

There were no long range social and economic effects. On November 4, Russian troops returned, and within a week defeated the revolutionary forces. A new government was organized by Janos Kadar, and he and his new Cabinet were sworn in on November 7. A one-party system was reestablished, all organizations providing a forum for exchange of ideas were disbanded, and the state information office assumed control of the press. Although the new regime may not have been as harsh as the Rakosi or Gero regimes, it still deprived the Hungarian people of "the exercise of their fundamental political right to participate in the function of government through elected representatives of their own choice."[18]

NOTES

1. Paul E. Zinner, "Revolution in Hungary: Reflections on the Vicissitudes of a Totalitarian System," *The Journal of Politics*, XXI (February 1959), 13.

2. Paul Kecskemeti, *The Unexpected Revolution: Social Forces in the Hungarian Uprising* (Stanford, California: Stanford University Press, 1961), p. 2.

3. Julius Rezler, "The Land," *Hungary*, Ernst C. Helmreich (New York: Frederick A. Praeger, 1957), pp. 34–40.

4. Frederick Pisky, "The People," *Hungary*, Ernst C. Helmreich (New York: Frederick A. Praeger, 1957), p. 45.

5. Gabrie Racz, "Transportation and Communications," *Hungary*, Ernst C. Helmreich (New York: Frederick A. Praeger, 1957), pp. 316–333. ·

6. Jan Wszelaki, "Mining," *Hungary*, Ernst C. Helmreich (New York: Frederick A. Praeger, 1957), p. 289.

7. Jan Wszelaki, "Industry," *Hungary*, Ernst C. Helmreich (New York: Frederick A. Praeger, 1957), p. 294.

8. Ibid., pp. 293–294.

9. William Juhasz, "Education," *Hungary*, Ernst C. Helmreich (New York: Frederick A. Praeger, 1957), pp. 190–211.

10. Pisky, "The People," pp. 65–66.

11. Melvin J. Lasky, *The Hungarian Revolution, A White Book* (New York: Frederick A. Praeger, 1957), pp. 15–18.

12. Kecskemeti, *Revolution*, p. 18.

13. Zinner, "Revolution," p. 22.

14. Randolph Braham, "State Security," *Hungary*, Ernst C. Helmreich (New York: Frederick A. Praeger, 1957), pp. 132–150.

15. Zinner, "Revolution," p. 24.

16. Hannah Arendt, *The Origins of Totalitarianism* (New York: Meridian Books, Inc., 1958), pp. 497–499.

17. Kecskemeti, *Revolution*, p. 79.

18. Neal V. Buhlar, "The Hungarian Revolution," *Hungary*, Ernst C. Helmreich (New York: Frederick A. Praeger, 1957), p. 371.

RECOMMENDED READING

BOOKS:

Arendt, Hannah. *The Origins of Totalitarianism*. New York: Meridian Books, Inc., 1958. Arendt covers some of the organizational phases of the revolution with a few excellent observations on the Revolutionary and Workers' Councils.

Fetjo, Francois. *Behind the Rape of Hungary*. New York: David McKay Co., Inc., 1957. The origins and implications of the Revolution are well covered.

Helmreich, Ernst C. *Hungary*. New York: Frederick A. Praeger, 1957. A comprehensive breakdown of Hungary by categories, which includes sketches of its history, people, government, economy, the revolution, etc.

Kecskemeti, Paul. *The Unexpected Revolution: Social Forces in the Hungarian Uprising.* Stanford, California: Stanford University Press, 1961. The author sees the revolution as a result of the political instability caused by two "processes:" an elite process, or political rivalry within the Communist Party; and a mass process, or sudden and complete popular insubordination. The book is brief and the material well presented.

Lasky, Melvin J. *The Hungarian Revolution, A White Book.* New York: Frederick A. Praeger, 1957. Based on Hungarian sources, leaflets, broadcasts, as well as dispatches of foreign correspondents and eyewitness accounts, this collection documents the sequence of events.

Vali, Ferenc A. *Rift and Revolt in Hungary.* Cambridge, Massachusetts: Harvard University Press, 1961. A solid, informed, and scholarly analysis of events in Hungary since 1945. The core of the book is the uprising of 1956 and its background, but the author brings the melancholy narrative of the aftermath up to 1961.

PERIODICALS:

Zinner, Paul E. "Revolution in Hungary: Reflections of the Vicissitudes of a Totalitarian System," *The Journal of Politics,* XXI (February 1959), 3–37. A very concise and detailed preliminary study resulting from the examination of 250 interviewees.

OTHERS:

United Nations. *Report of the Special Committee on Hungary,* General Assembly Official Records: 11th Sess., Supplement No. 18, (A/3592), New York, 1957. This report is a most complete general study of the revolution presented in a manner best suited for the study of revolutions and has been relied on heavily in this report.

THE CZECHOSLOVAKIAN COUP OF 1948

SYNOPSIS

The independent Third Czechoslovak Republic, inaugurated in 1945 after the defeat of Nazi Germany, came to an end in February 1948. The Communists, using a coalition government in which they were strongly represented, gained high governmental office and proceeded to infiltrate the police and security forces, suppress political opposition, and generally subvert the government through both constitutional and unconstitutional means. They thus won control of the government, and in July 1948 proclaimed a Soviet-type constitution.

BRIEF HISTORY OF EVENTS LEADING UP AND TO CULMINATING IN REVOLUTION

Czechoslovakia was established as an independent state in 1918 out of fragments of the old Austro-Hungarian Empire. A Constitution ratified in 1920 set the basis for a centralized government with legislative power residing in the parliament. This political system survived until 1938, when Czechoslovakia was forced, under the Munich agreement between Germany, Italy, Britain, and France, to cede to Germany the western third of its territory, where the population was predominantly German-speaking. Slovakia was given an independent status but "chose" to become a protectorate of Germany 6 months later. By March 1939 the Second Republic, formed after Munich, had ceased to exist.

At the start of World War II, former President Eduard Beneš established a Czech Government-in-exile in London. Disillusioned by France's failure to fulfill its alliance obligations at the time of Munich, Beneš turned to Russia. He met with Stalin in Moscow in 1943 and formed a Soviet-Czech alliance for future protection of the Czechoslovak Republic.

As soon as the Red Army had freed Slovakia, a meeting of the Czech exiles took place in Košice in Slovakia. These leaders, who had spent the war years in London and Moscow, formulated a program for the restoration of constitutional authority and reconstruction of the country. The Košice program laid down the guidelines to be followed until a government could be formed. It called for a coalition government of the legally recognized political parties, each of which was pledged to the maintenance of the National Front. It provided for the election of a Provisional National Assembly by national committees, as representatives of the people. The Assembly was to remain in existence until the election of a Constituent National Assembly in May 1946. The plan further provided for the expulsion of all Germans and confiscation of their property, incorporation of all partisan units into the "democratized" army, and close association with Soviet Russia for the purposes of foreign trade and protection. It affirmed the principle of popular participation in government, and stated that Czechs and Slovaks would have autonomy in local affairs but would be subordinate to the central government. Collaborators and traitors were to be excluded from the government, and political parties with an anti-Soviet background or collaborationist leanings suppressed. The plan pledged the government to speedy reconstruction of industry and agriculture and to the support of individual initiative through loans. Protection of civil liberties and advancement of social welfare and education were also promised.[1] This program received the support of all the recognized political parties.

The Provisional National Assembly was elected in October 1945. It reinstated by decree the Constitution of 1920 as the law of the land, and confirmed the social and economic provisions that had been set forth in the Košice program. Some changes were made in the Constitution, however, such as reduction of the voting age from 21 to 18, and provision for a single-chamber parliament of 300 members.

The 1946 elections were remarkably free of internal or external interference. The returns showed an increase in Communist strength; the party received 38 percent of the vote cast, won 114 seats in the Assembly, and was able to claim the premiership and the key Ministries of Interior, Agriculture, Information, Finance, Labor and Welfare, and National Defense.

Czech affairs were free of outside interference until July 1947. At that time, Stalin spoke out against Czech participation in the Marshall Plan negotiations in Paris and against a proposed Franco-Czech alliance. The Communist Party was able to block both projects. Communist activity increased and the Czechs did not know "a calm day up to the coup d'etat in February 1948."[2]

The activities of the Communist-controlled police and security forces provided one of the main rallying points for the non-Communist parties. Disclosures of police brutality, illegal possession of arms and munitions, and the use of former Nazis as witnesses against opponents of the Communists were presented in the Assembly and the Cabinet. The Cabinet debate over the replacement by Communists of non-Communist division chiefs of police resulted in the adoption of a resolution, carried over Communist opposition, demanding that the replacements be rescinded. But the Communist Minister of Interior disregarded the resolution.

In protest, 12 non-Communist Ministers resigned. They had obtained the promise of the President, however, that he would not accept the resignations. This was intended to leave the Cabinet inoperative and to focus responsibility and popular resentment on the Communists. Under the agreement establishing the National Front, all legal parties had to be represented in the Cabinet. So long as the President declined to accept the resignations, their parties could not replace the 12 Ministers without disavowing them. The non-Communists expected that public pressure would shortly force the Communists to capitulate on the police issue; if they did not, the opposition would have a powerful argument in the coming elections. It did not work, however. The Communists organized mass meetings and letterwriting campaigns denouncing the resigning Ministers and demanding that the President accept their resignations. Communist-controlled security forces and *gendarmes* moved into the capital. Finally, on February 25, 1948, the President yielded, accepted the resignations, and approved a new Cabinet, with the 12 vacant portfolios going to renegade members of the non-Communist parties to maintain the pretense that the National Front was still in effect. The May elections were a farce, with only one slate submitted to the voters. In June a new Soviet-type Constitution, declaring Czechoslovakia a "people's democracy," was adopted. President Beneš refused to sign the new Constitution and subsequently resigned.

THE ENVIRONMENT OF THE REVOLUTION

DESCRIPTION OF COUNTRY

Physical characteristics

Czechoslovakia, approximately the size of the state of New York, is located in Central Europe, and is surrounded by Germany, Poland, the U.S.S.R., Hungary, and Austria. The western areas are essentially urban and industrialized; there is some highly developed agriculture on the Moravian plains in the south central part. The eastern area is primarily agricultural, although strides have been taken toward industrialization. The western section is mostly highlands and plains, while the eastern section, located in the sub-Carpathian range, is primarily mountainous, with a small fringe of plains in the south.

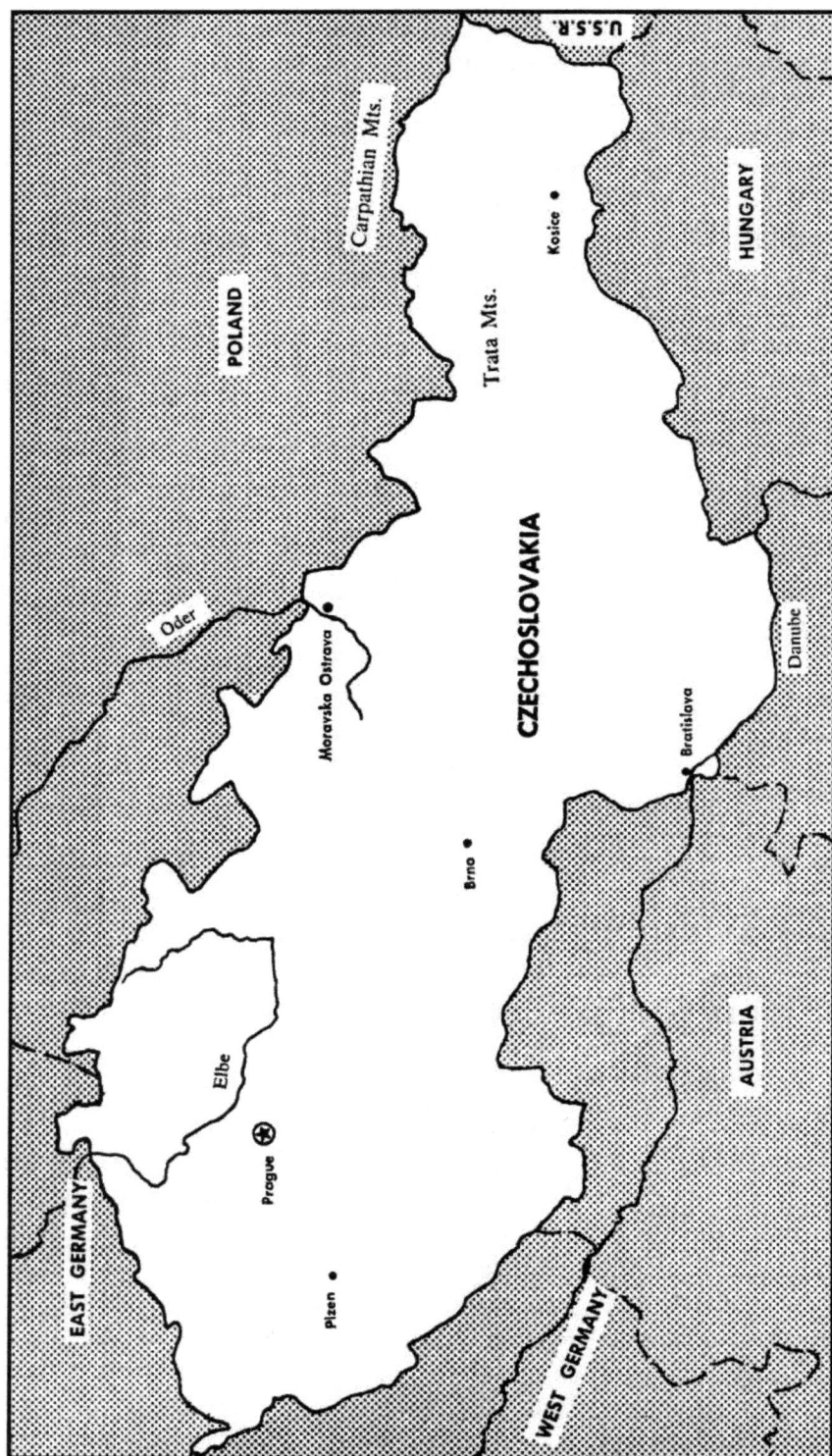

Czechoslovakia's location makes it subject to both the oceanic and the continental climate systems, which account for the wide local differences in temperature and rainfall.[3]

The people

The Czechs and the Slovaks, the predominant ethnic groups in Czechoslovakia, were originally Slavs who migrated to the area in the fifth century. There were German and Magyar invasions in the 10th and 11th centuries; the Slovaks fell under Magyar domination and were considered a subject people. They bore the brunt of the Avar and the Turkish invasions and of the determined Magyarization programs. The Czechs emerged as Bohemians under their own King, with closer ties to western Europe than to the Slovaks. Although their languages were similar, these peoples developed a strong awareness of their historical and political differences. Antagonisms between the two have been present throughout their histories and have been a disruptive influence within the united Czechoslovak Republic. The official language today is Czech, although concessions are made to other languages in local areas.

The census of 1947 listed the population at 12,164,095 persons, of whom approximately 50 percent were urban dwellers. Because of the broken terrain the distribution is unequal. Approximately two-thirds of the population live in the western half of the country. Prague (Praha), the capital city, has a population of slightly less than one million, and Brno, the second largest city, slightly more than 250,000. Both of these cities and Ostrava, the third largest, are in the western section.[4]

Communications

Pre-World War II Czechoslovakia had the most dense rail and roadway network of all the states between the U.S.S.R. and Germany from the Baltic to the Adriatic. Much of this transportation system was destroyed during the war, but it was fast being rebuilt at the time of the Revolution. The major portions of the rail, road, and river networks are in the western industrialized areas. In the east—due to the rugged terrain and neglect—the transportation system is skeletal.

The Vltava, the Otava-Blanice, the Elbe (Labe), and the Oder river systems, all located in the west, are heavily used for barge traffic and timber rafting. These systems give Czechoslovakia access to the plains of Europe and to the North and Baltic Seas.[5] The Danube in the southeast also is used as a waterway. At the end of the war most of the barges and river boats had been destroyed or confiscated.

Natural resources

Czechoslovakia's wealth in natural resources lies mainly in coal, mined at Plzeň and Brno, and its prosperity has been based on the quantity and quality of this resource.[6] Lumber is another important natural resource, and there are also deposits of iron, silver, gold, copper, and graphite.

SOCIO-ECONOMIC STRUCTURE

Economic system

Nationalization of heavy industry, war industry, and potential war industry, stressed by the Košice program, began with the expropriation of former German-owned industry and business. To raise the standard of living and improve internal conditions, material and developmental priorities were established by an Economic Council set up by the Provisional National Assembly. This gave the economy a measure of central planning. Although the Košice program and the later 2-Year Plan stressed government planning, private enterprise was encouraged by means of loans, grants, and credits.

The three major categories of Czech exports in 1946 were light manufacturing products, machinery, and foodstuffs. Light manufacturing products included glass and glassware, textiles, steel bars, sawmill products, and footwear; these products made up approximately 23 percent of the export trade. Light machinery made up 5 percent of the exports, and foodstuffs (potatoes, hops, and refined sugar) approximately 23 percent. In 1948 exports of foodstuffs declined to 3.8 percent, while those of manufactured goods rose to 27 percent, due to increased exports of footwear. After the war, destinations of the exports shifted from the Western countries to Soviet Russia and her allies. Imports consisted mainly of raw materials and semifinished goods.

Class structure

There was a large peasant and working class, and a middle class consisting of civil servants, small and middle-sized business entrepreneurs, landowners with middle-sized holdings, and the professional groups—teachers, doctors, and engineers. The ruling elite consisted of the political leaders, most of whom had come from the middle and the upper portions of the lower class. Mobility from one class to another was possible and an accepted aim in life.

Literacy and education

Like most highly industrialized countries, Czechoslovakia enjoys a high literacy rate. The census of 1938 claimed 96 percent of the population to be literate; the 1947 census estimate was 97 percent.[7]

The Košice program pledged the government to support education and cultural advancement.[8] It called for deletion of anti-Bolshevik remarks from the textbooks and a removal of German and Hungarian influences in the educational system.

Education today is compulsory between the ages of 6 and 14, is state-controlled and free.

Major religions and religious institutions

Approximately 75 percent of the Czechoslovakians are Roman Catholic. Catholicism is strongest in Slovakia. The remaining 25 percent of the population are Protestant, Jewish, agnostic, or adhere to other minority sects.

GOVERNMENT AND THE RULING ELITE

Description of form of government

The form of government after the war was essentially that in operation under the Constitution of 1920. Instead of a bicameral parliament, however, elected for 6 and 8 years, the government instituted in 1945 consisted of a single-chamber parliament, a cabinet, and a president elected on a temporary basis. The Provisional National Assembly was to serve until the Constituent National Assembly was elected; the latter was to serve until the drafting of a new constitution, to be completed not later than 1948. The Cabinet consisted of the leaders of the National Front parties. The President was elected by the Assembly.

Legislative power was vested in the Assembly; bills became law when signed by the President. The real legislative work was done within the individual party councils and each party maintained strict control of its delegates. Executive power was held by the Cabinet and the President. A vote of no confidence by the Chamber of Deputies could compel the resignation of the government. The single-chamber parliament assumed this power in 1945.[9] The judiciary was independent and headed by the Minister of Justice. Judges were appointed for life.

Description of political process

In prewar Czechoslovakia there were as many as 14 political parties; postwar Czechoslovakia recognized only six. Under the Košice program collaborationist parties were outlawed and anti-Soviet parties were suppressed. The six legal parties were: (1) the Communist Party in Czechoslovakia, (2) the Communist Party in Slovakia, (3) the People's Party, (4) the National Socialists, (5) the Social Democrats, and (6) the Slovak Democrats. The parties were generally left of center in ideology. The two Communist Parties followed the Marxist doctrine of complete socialization of the state; the Social Democrats held views close to those of the Communists and, in fact, generally voted with them; the National Socialists believed in the gradual or evolutionary development of socialism but recognized free private enterprise; the People's Party was generally non-Socialist and believed in a greater degree of free enterprise; the Slovak Democrats attracted anti-Soviet persons who were concerned with Slovak autonomy and who, therefore, could not organize openly.

The Provisional National Assembly was elected indirectly. Local committees—the mechanism by which the people were to participate in government—selected representatives to provincial congresses. These in turn elected national committees, and the final selection of the national legislature was made by these committees. The political parties operated at all three levels, and the provincial and national committees reflected party strength in the local committees. However, Communist infiltration in the provincial congresses and national committees allowed them to secure 98 of 300 seats in the Provisional National Assembly and the important Ministries of Interior, Information, Agriculture, National Defense, and Education. Zdeněk Fierlinger, the Social Democratic leader, was made Premier.

The elections of 1946 for the Constituent National Assembly were free and popular. Voting was compulsory under penalty of fine. The Communist platform of 1946 had mass appeal. It promised protection and support of the property rights of farmers and small and middle-sized entrepreneurs, and guaranteed the security of existence for the middle class in the cities.[10] The Communists concealed their plans for excessive expropriation and nationalization. The democratic parties offered basically the same program.

These elections brought only slight changes in the makeup of the Assembly. Because of the existence of two Communist Parties, the Communists emerged with 114 seats as the strongest delegation. The willingness of the Social Democrats to vote with the Communists gave the leftwing of the government a slim majority of 153 votes, "adequate to secure ratification of ordinary measures although not of constitutional changes."[11] This was a definite shift in power to the left.

Reflecting individual party strength in the Assembly, "the new ministry contained nine Communists and seventeen non-Communists. . . . The National Socialists, the People's Party, and the Slovak Democrats now had four seats each; the Social Democrats remained with three."[12] In addition, two nonparty experts, Jan Masaryk and Gen. Ludvík Svoboda, continued in office. (Svoboda, though supposedly nonpartisan, actually had Communist leanings.)

Since all the parties pledged support of the government and the Košice program, there was no opposition to the government and its policies.

Legal procedure for amending constitution or changing government institutions

Constitutional amendments could be initiated by the individual parties through the Cabinet Ministers and required a two-thirds vote of the Assembly for ratification. On policy regarding Slovakia, a majority of the Slovak delegation was needed in addition to a majority of the Assembly.[13]

Relationship to foreign powers

The Košice program and subsequent National Assemblies reaffirmed the Soviet-Czech alliance concluded in 1943. This alliance, to run for 20 years, was one of friendship and mutual assistance. Czech policy called for closer ties with Russia. This was due in part to the fear of a resurrected Germany on the Western frontier; in part to the feeling that the Western powers were neither strong enough, interested enough, nor consistent enough for the maintenance of Czechoslovakian independence; and in part to the recognition of the proximity of Russia and Russian physical force. Russia, in the eyes of the Czechs, was to take the place of the former European allies and was not to be alienated. The Soviet-Czech alliance of 1943 included a passage calling for mutual noninterference in the internal affairs of the other state.[14] This pact was essentially an anti-German, nonaggression treaty, with the notable addition that it pledged mutual aid in case of involvement in hostilities.

Russia seems to have respected the internal noninterference clause until July 1947, when Stalin decided that Czech participation in the Marshall Plan negotiations in Paris were incompatible with the Soviet-Czech alliance.[15] Again, the degenerating relations between France and the U.S.S.R., culminating in the expulsion of the Communists from the new French Government in 1947, brought remarks from Stalin against a pending Franco-Czech alliance; the Czech Communists forced rejection of the treaty.

The role of military and police powers

The police of Czechoslovakia were concerned with enforcement of internal civil and criminal law. The Corps of National Security (SNB) was a unit similar to the state police forces in the United States. A third police organization, the *Gendarmerie*, was a mobile police force somewhat similar to the Nazi SS. This force had been created in 1945 and 1946 to maintain order in the frontier regions against Germans still in the area,[16] but was used as an adjunct of the Communist Party. These troops were to occupy strategic positions in Prague during the last days of the crisis.

All of these forces were under the control of the Minister of Interior, who had the power to replace officials on his own authority.

The army's prime mission was protection of the country from outside attack, but could be called upon to aid the police in an internal crisis. Control of the army rested with the Minister of National Defense.

WEAKNESSES OF THE SOCIO-ECONOMIC-POLITICAL STRUCTURE OF THE PREREVOLUTIONARY REGIME

History of revolutions or governmental instabilities

The First Czechoslovak Republic, organized in 1920, was an example of a viable democracy. It suffered, however, from differences between the western and eastern sections in education and industrial and agricultural development, and from the presence of minority groups within the country. It was a heterogeneous state and all minorities were not represented in the National Assembly that formed the Constitution of 1920. Although the Constitution guaranteed the protection of minority rights, local administration did not always implement those guarantees. The unusually large German minority sought German affiliations rather than Czech, and the Slovaks, who felt discriminated against, often worked against the central government in their desire for greater autonomy, and failed to implement its regulations.

The western section of the country, formerly under Austrian rule, had a relatively high educational and industrial level of development. The eastern section, formerly under Hungarian rule, was made up predominantly of backward peasants, with a tiny educated class. The Slovaks in the eastern area were restive under Czech leadership and guidance, and aspired to a greater degree of self-government.

Since the Constitution guaranteed political recognition, most minority groups formed political parties. However, these were not effective and the five main parties controlled the government.

Economic weaknesses

The economic disparity between the two sections was a major weakness. The new nation took over the former Austrian manufacturing plants, coal production, and transportation facilities. There was no similar level of development in the former Hungarian lands. The lack of private investment capital forced recourse to insurance companies and banks for funds. In the late 1930's the German minority, supported by Nazi Germany, was able to obtain control of major sections of the economy merely by securing key positions in the lending banks and insurance companies. This led to virtual foreign control of Czech industrial development.

Land, under the former Austro-Hungarian denomination, had been concentrated into semifeudal holdings. Although a land reform program was initiated, land ownership was not effectively distributed.

With the advent of World War II, Germany took control of all facets of the economy. Industry was forced to conform to German war needs. Wartime destruction, in addition to the wartime disruption of trade, aggravated the situation of the Czech economy, as did the poor harvests in 1945.

Social tensions

Tensions were present between the Czechs and the Slovaks, and between the government and the minority groups. The Slovak desire for autonomy and resentment of Czech educational and economic superiority clashed with the Czech desire for national unity. The Czechs, in many instances, held themselves to be superior to the Slovaks, thus adding fuel to the fires of Slovak nationalism.

The small minority groups who felt unrepresented in the Assembly resisted government attempts to make them a part of a nation made up primarily of Czechs and Slovaks.

Government recognition of and reaction to weaknesses

The Košice program was an attempt to alleviate the socio-economic and political weaknesses. The German minority was expelled from the country and all German influence in business and education was erased.

Slovak nationalism was recognized and political control in Slovakia was vested in the Slovak National Council. Although the Council was subordinate to the National Assembly, the agreement of a majority of the Slovak delegates in the Assembly was required to effect any change in Slovakia. Other minorities were guaranteed protection and representation.

To preclude foreign control of the economy, all industries of more than 500 employees were nationalized. Banking, insurance, and credit facilities were also nationalized.

Reconstruction of industry, agriculture, and transportation was not only promised, but given first priority. The government negotiated with the United Nations Relief and Rehabilitation Agency for aid and received "747,000 tons of foodstuffs, 26,000 tons of agricultural supplies, 68,000 tons of transport equipment, 137,000 tons of petroleum products, and 111,000 tons of industrial raw materials and machinery." Proceeds from the sale of this equipment were used "in reconstruction and welfare work."[17] By mutual consent of the National Front strikes were declared to be detrimental to the reconstruction effort and were to be avoided. This was merely a declaration against strikes in principle, and no legal ban was placed on the actual use of strikes.

The 2-Year Plan, officially announced in January 1947, had the specific purposes of surpassing the prewar levels of industrial and agricultural production and increasing transportation efficiency. The industrialization of Slovakia, the underdeveloped, rural area of the country, was stressed.

To make up for the bad harvests in 1947, the government attempted to negotiate for Marshall Plan aid. When this was blocked by the Communists under pressure from Stalin, the government turned to the U.S.S.R. for wheat shipments.[a]

FORM AND CHARACTERISTICS OF REVOLUTION

ACTORS IN THE REVOLUTION

The revolutionary leadership

The leaders of the revolution were the Communist Cabinet Ministers, the leftwing Social Democrats, and the leaders of the Communist Party itself. Most of these had spent the war years in Moscow and were

---

[a] In all probability the Czech Government was motivated by something more than an effort to make up for bad harvest. The Czech industrial economy was oriented toward Western trade, and had made rapid strides in restoring pre-World War II contacts. Furthermore, as with most Western European countries, the Czechs were faced with the "dollar" crisis, aggravated by the overly ambitious 2-Year Plan. Under the circumstances, Marshall Plan aid appeared as enticing to the Czech Government as it had to the Western European countries.

completely in sympathy with Communist doctrines. The Premier, Klement Gottwald, was chairman of the Czech Communist Party. Rudolph Slánsky, the Secretary General of the Communist Party, was Moscow-trained, as was Václav Kopecky, the Minister of Information. Both had spent the war years in Moscow. Gen. Ludvík Svoboda, the Minister of National Defense, had commanded Czech troops in the U.S.S.R. during the war. Zdeněk Fierlinger, a Social Democrat, had been the London provisional Government's Ambassador to Moscow during the war and was Gottwald's predecessor as Premier from 1945 to 1946. He had taken over the leadership of the Social Democrats and had brought them into close alliance with the Communists. He was displaced from the leadership in 1947, but retained his strong influence with the leftwing of the party. Václav Nosek, the Communist Minister of Interior, had spent the war years in England.

The background of the leaders was varied. Gottwald had been a joiner by trade, had deserted from the Austrian Army before the end of World War I, and with Zapotocky, the trades union leader, had helped to found the Czechoslovak Communist Party. He later became an editor of *Pravda,* a member of the Central Committee of the Communist Party, and, finally, the Secretary General of that organization.

Nosek had been a coal miner in his youth, but upon graduation from Charles University in Prague had entered the foreign service. Kopecky had been a member of the Socialist Student Movement and chairman of the International Federation of Marxist Students. He worked in the Communist Party as a journalist. Svoboda had studied agriculture before enlisting in the army during World War I. He remained in the army after the war and was a professor at the military academy at Hranice.

Zapotocky, the chairman of the Unified Trades Union (URO), had a secondary school education and had been a stonemason. He had a stormy career that included imprisonment in 1920 for strike activities, active participation in the founding and maintenance of the Czechoslovak Communist Party, and imprisonment by the Nazis during World War II.

The revolutionary following

The members of the unit committees, or factory councils, the Committees of Action, the officials of the police forces, and the Communist Party cadres made up the revolutionary following. These groups were the only active participants during the crisis and the period immediately afterward. They were members of the Communist Party and the majority were from the working classes.

ORGANIZATION OF REVOLUTIONARY EFFORT

The Communists worked mainly through legal organizations. Both the Communist Party itself and the factory unit committees were legal. The unit committees were formed within each factory, business, and governmental service for the purpose of representing the workers and assisting the management in planning and production.[18] Their organization paralleled that of the trade unions. They were used as propaganda organs of the Communist Party.

The Committees of Action were an imitation of the Russian Workers' and Peasants' Soviets. These committees were ostensibly composed of representatives of all parties and national organizations, but were really made up of "approved" appointees controlled by the local Communist Party secretariats.[19] The Communists called them into existence during the crisis to give a show of broadly-based support for their demands.

The small unit committees "spoke" for the large numbers of employees they represented. The factory unit councils were elected by the workers from a list of candidates prepared by the factory unions, and were directed by the unions. Since the chairman of the Unified Trades Union (URO) was a Communist, this gave the Communist Party effective control over the factory councils. The Committees of Action were more militant and evidently had a more extensive role than the unit committees. The Action Committees were later to assume the functions of local administration and to occupy the governmental offices.

The connecting link between these groups and the leadership was the Communist Party cadre. The cadre and local leaders carried out the directives of the party headquarters.

GOALS OF THE REVOLUTION

The immediate avowed goals were the preservation of "peace and internal security." This shibboleth was used to give a façade of legality and constitutionality to the actions taken between the resignation of 12 Ministers and the final coup. The overall aim of the revolution was the complete communization of the Republic, installation of a "people's democracy," and its inclusion in the Soviet bloc of states. Destruction of political opposition was also a goal and the most important one if the major aims were to be secure.

The economic goals were: (1) complete nationalization of all enterprises employing more than 50 persons; (2) complete central control of financial matters, including banking, insurance, and credit; and (3) collectivization of the land, with the concomitant destruction of private ownership.

Nationalization and collectivization of land were intended to raise the standard of living of all the citizens, ruin the large or middle-sized landholders, and give security to the landless. The resulting society would be classless.

REVOLUTIONARY TECHNIQUES AND GOVERNMENT COUNTERMEASURES

Methods for weakening existing authority and countermeasures by government

The techniques used by the Communists in the takeover were, on the whole, legal because the Communist Party was a recognized member of the existing government. Some anti-Soviet parties had been suppressed immediately after the war by the simple expedient of accusing them of treason or collaboration. The liberation of Czechoslovakia by the Red Army troops and their initial presence aided the Communists in securing positions in the newly-formed National Committees on all levels. Possession of the important Ministries of Interior, Information, Finance, Labor and Welfare, National Defense, and Agriculture placed the Communists in a position where they could legally replace non-Communists with Communists in the police, the security forces, the labor unions, and financial institutions, and could deny newsprint to opposition newspapers or severely limit their supply. Working within the general framework of the National Front, the Communists organized factory councils, trade unions, and farmers unions.

The illegalities consisted mainly of coercion and the abuse of civil rights by the police force. The police used "confessions" of former Nazis, collaborationists, and persons of questionable character to remove any opposition. Terrorist tactics (intimidation, blackmail, and the actual use of force) were used against non-Communists who voiced opposition.

A congress of unit committees was called February 22, 1948, for the purpose of ". . . voting a program of radical socialism"—the nationalization of all enterprises employing more than 50 persons.[20] The Communists expected to attract the full support of the Social Democrats by this maneuver. Two days before the meeting, 12 non-Communist Ministers (representing the National Socialists, the People's Party, and the Slovak Democrats) resigned. The Communists denounced the resignation as a "reactionary" attempt to work outside the National Front by forming a new government that would exclude the Communist Party. They called on all the workers and peasants to stand behind the legal government (the Communists and the Social Democrats) and demand that the President accept the resignations. Letterwriting campaigns were organized and the President's office was flooded with letters and telegrams, most of them identical in wording. By pressure and coercion, the Communists were able to persuade some members of the opposition parties to defect and agree to serve in the government—without authorization from their party leaders, of course. Thus Gottwald was able to submit to the President a revised Cabinet with nominal representation of all the National Front parties and thus maintain the façade of legality. Beneš signed his acceptance of the list on February 25.

Methods for gaining support and countermeasures taken by government

The democratic elements within the government (the National Socialists, the People's Party, and the Slovak Democrats) recognized the extent of Communist infiltration and attempted to resist. However, the Communist control of vital functions of the government and the parliamentary majority they held in conjunction with the Social Democrats made this impossible.

The continued communization of the police force was the issue for which the other parties attempted to enlist the support of the Social Democrats. Their purpose in creating a Cabinet crisis was twofold: (1) to show the extent of Communist infiltration and protest the refusal of the Minister of the Interior to comply with the Cabinet resolution demanding that this infiltration cease, and (2) to weaken the Communist voter appeal in the May elections. Since the President had agreed not to accept the resignations or to accept a cabinet that excluded any member of the National Front, the Communists were to be confronted with the choice of yielding and acting upon the Cabinet resolution regarding the police, or exposing themselves to the public as working outside of the National Front program, and hence, against the state. The democrats felt that they were working within the legal and constitutional framework and fully expected the Communists to do the same on pain of defeat in the May elections. A Communist defeat would still allow the party to participate in the government, though on a much reduced basis. This would preserve the National Front.

When it was announced that the President had signed the new Cabinet list, thousands of students marched on the President's palace to protest, but were brutally dispersed by the police. A few were killed and hundreds imprisoned.

THE EFFECTS OF THE REVOLUTION

CHANGES IN THE PERSONNEL AND INSTITUTIONS OF GOVERNMENT

The coup resulted in the complete takeover of the government by the Communist Party. The Cabinet submitted to Beneš by Gottwald contained renegade members of the National Socialist Party, the People's Party, and the Slovak Democratic Party.

The coup was followed by a series of purges in the newspapers, the universities, and the Assembly. Committees of Action took control of the democratic newspapers and magazines and governmental offices. Students were expelled for "disloyal" tendencies, democratic leaders were arrested, and all social and athletic clubs were taken over by the Committees of Action.

The elections in May were a farce. The candidates were picked by the Communists and there were no opposition slates. Although maintaining the pretense of the National Front, the Communists considered a coalition government "a temporary arrangement, only a step in the struggle for exclusive power."[21]

A new Constitution was adopted on May 8. It proclaimed Czechoslovakia to be a "people's democracy." It called for "one national assembly of 368 members selected for 6 years; a presidium of 24 members . . ." which elected a president who served for 7 years. A decree of December 21 abolished the "old provinces and replaced them with 19 regions."[22] President Beneš refused to sign the Constitution and resigned his office on June 7. He died 3 months later. On June 9 the Constitution was ratified by the newly elected President, Klement Gottwald.

Under the Constitution of 1948, the President retains most of his wide powers. There was no attempt to imitate the 1936 Soviet Constitution. The President summons and adjourns the National Assembly, appoints the Premier, and signs into law acts of the Assembly. Executive power is vested in the Council of Ministers, which is appointed by the President.

MAJOR POLICY CHANGES

After the coup Czechoslovakia maintained its membership in the United Nations, but as a member of the Communist bloc. In 1955 Czechoslovakia signed the Warsaw Pact, which called for joint use of armed forces with the U.S.S.R. It is also a member of the Council of Mutual Economic Assistance (CEMA), which coordinates economic planning in the East European countries, and has extended credits for industrialization to other East European members of the Council.

LONG RANGE SOCIAL AND ECONOMIC EFFECTS

A 5-year economic recovery plan was instituted in 1949, the chief feature of which was ". . . enormous development of heavy industry to the complete neglect of the light industries and consumers' goods industries. . . ."[23] Industries employing more than 50 persons were nationalized, as were most small businesses. Those that were not nationalized were considered not necessary for the completion of the production process. However, they were subject to rigid controls and governmental pressures.

Under CEMA, Czech industry was forced to adapt to requirements decided upon in Moscow. In effect, it became an extension of Soviet Russian industry. By 1950 the U.S.S.R. supplied 29.4 percent of Czechoslovak imports and received 28 percent of her exports.

Land reform was instituted, but President Gottwald urged caution in its implementation. By December 1954, although collectivization was a declared aim, only 40 percent of the arable land had been collectivized.

Imposition of a "people's democracy" deprived the bourgeois ". . . with complete finality of all remnants of power and influence on further developments in Czechoslovakia."[24]

NOTES

1. William Diamond, *Czechoslovakia Between East and West* (London: Stevens and Sons, Ltd., 1947), pp. 2–5.

2. Hubert Ripka, *Czechoslovakia Enslaved* (London: Gollancz, Ltd., 1950), p. 94.

3. Harriet Wanklyn, *Czechoslovakia* (New York: Praeger, 1954), pp. 8, 64.

4. *Encyclopaedia Brittanica World Atlas* (London: William Benton, 1957), 114.

5. Wanklyn, *Czechoslovakia*, pp. 311, 343; and Diamond, *Czechoslovakia*, p. 62.

6. Wanklyn, *Czechoslovakia*, p. 265.

7. UNESCO, *World Literacy at Mid-Century* (Geneva: UNESCO, 1957), p. 43.

8. Diamond, *Czechoslovakia*, p. 5.

9. R. W. Seton-Watson, *A Short History of Czechoslovakia* (London: Hutchinson and Company, Ltd., 1943), ch. XVI; and Vratislav Busek and Nicholas Spulber (eds.), *Czechoslovakia* (New York: Praeger, 1957).

10. Josef Korbel, *The Communist Subversion of Czechoslovakia, 1938–1948* (Princeton: Princeton University Press, 1959), p. 152.

11. H. Gordon Skilling, "Revolution and Continuity in Czechoslovakia, 1945–1946," *Journal of Central European Affairs*, XX, 4 (January 1961), 374–375.

12. Ibid., p. 375.

13. Ripka, *Czechoslovakia*, p. 179.

14. Jan Taborsky, "Beneš and Stalin in Moscow—1943 and 1945," *Journal of Central European Affairs*, XIII, 3 (July 1953), 162.

15. Ripka, *Czechoslovakia*, p. 62.

16. Ibid., p. 259.

17. Diamond, *Czechoslovakia*, p. 68.

18. Korbel, *Communist Subversion*, p. 158.

19. Ripka, *Czechoslovakia*, p. 232.

20. Korbel, *Communist Subversion*, p. 206.

21. Ibid., p. 237.

22. Busek and Spulber, *Czechoslovakia*, pp. 48, 49, 87.

23. Hugh Seton-Watson, *The East European Revolution* (London: Methuen and Company Ltd., 1956), p. 249.

24. Korbel, *Communist Subversion*, p. 239.

RECOMMENDED READING

BOOKS:

Borkenau, Franz. *European Communist*. London: Faber and Faber, Ltd., 1953. The author traces the successes and failures of communism in Europe through a comparison of countries that have experienced Communist activities.

Busek, Vratislav, and Nicholas Spulber (eds.). *Czechoslovakia*. New York: Praeger, 1957. An excellent discussion of Czechoslovakian history, government, economy, and society.

Diamond, William. *Czechoslovakia Between East and West*. London: Stevens and Sons, Ltd., 1947. A study made by an UNRRA official, this volume covers the economic and historic aspects of Czechoslovakia from 1945 through the 2-Year Plan of 1947.

Korbel, Josef. *The Communist Subversion of Czechoslovakia, 1938–1948*. Princeton: Princeton University Press, 1959. Subtitled "The Failure of Coexistence," this volume is an excellent study of the fall of Czechoslovak democracy in the face of determined Communist infiltration.

Ripka, Hubert. *Czechoslovakia Enslaved*. London: Gollancz, Ltd., 1950. Ripka, a journalist and a member of the government until the February coup, gives an inside and emotional view of the events leading to the coup.

Seton-Watson, Hugh. *The East European Revolution*. London: Methuen and Company, Ltd., 1956. A scholar's approach to the "Sovietization" of East Europe. Factual, concise, and analytical.

Seton-Watson, R. W. *A Short History of Czechoslovakia*. London: Hutchinson and Company, Ltd., 1943. A useful account of the past history of Czechoslovakia, especially from the First Republic to the Munich Agreement of 1938–1939.

Stransky, Jan. *East Wind Over Prague*. London: Hollis and Carter, 1950. This volume is an emotional account, by an emigree, of the violence and subversion of his country.

Wanklin, Harriet. *Czechoslovakia*. New York: Praeger, 1954. The author, a geographer, covers the history, economics, and demography of Czechoslovakia quite concisely, without forgetting that she is a geographer.

PERIODICALS:

Skilling, H. Gordon. "Revolution and Continuity in Czechoslovakia, 1945–1946." *Journal of Central European Affairs*, XX, 4 (January 1961), 357–377. This article stresses the concern of Beneš for legal continuity of government from the war years through the election of the Constituent National Assembly.

Taborsky, Jan. "Beneš and Stalin in Moscow—1943 and 1945," *Journal of Central European Affairs*, XIII, 3 (July 1953). The author was Beneš secretary during the meetings with Stalin. He discusses the conditions under which Beneš agreed to full Communist representation in the postwar Cabinet, and the decision to cede Ruthenia to the U.S.S.R.

OTHER:

Kaplan, Morton. *The Communist Coup in Czechoslovakia.* Research Monograph No. 5. Center of International Studies, Princeton University, 1960. An attempt to find the conditions which allowed the Communist absorption of a democratic government.